Past, Present, and Future of Statistical Science

Past, Present, and Future of Statistical Science

Edited by

Xihong Lin, Christian Genest,
David L. Banks, Geert Molenberghs,
David W. Scott, and Jane-Ling Wang

CRC Press is an imprint of the
Taylor & Francis Group, an **informa** business
A CHAPMAN & HALL BOOK

COPSS
Committee of Presidents
of Statistical Societies

Cover images **Top row:** Andrew Gelman, Alice S. Whittemore, Tze Leung Lai, Marie Davidian, Peter Hall, Michael A. Newton, Jianqing Fan. **Second row:** Mary Gray, Bruce G. Lindsay, C. F. Jeff Wu, R. Dennis Cook, Nancy Reid, Noel A. Cressie, Donald A. S. Fraser. **Third row:** Ingram Olkin, Lynne Billard, Donald B. Rubin, Bradley Efron, Iain M. Johnstone, Juliet Popper Shaffer, Herman Chernoff. **Fourth row:** James O. Berger, Xiao-Li Meng, Elizabeth A. Thompson, Peter J. Bickel, Louise Ryan, Raymond J. Carroll, Xihong Lin. **Fifth row:** David B. Dunson, Kathryn Roeder, Ross L. Prentice, Jeffrey S. Rosenthal, Robert J. Tibshirani, Nancy Flournoy, Arthur P. Dempster. **Sixth row:** Terence P. Speed, Roderick J. Little, Norman E. Breslow, Grace Wahba, Mark J. van der Laan, Stephen E. Fienberg, Mary E. Thompson. **Seventh row:** Donna Brogan, Pascal Massart, Nilanjan Chatterjee, Larry A. Wasserman, Theodore W. Anderson, Nan M. Laird, Rafael A. Irizarry.

CRC Press
Taylor & Francis Group
6000 Broken Sound Parkway NW, Suite 300
Boca Raton, FL 33487-2742

Printed on acid-free paper
Version Date: 20140213

International Standard Book Number-13: 978-1-4822-0496-4 (Hardback)

Visit the Taylor & Francis Web site at
http://www.taylorandfrancis.com

and the CRC Press Web site at
http://www.crcpress.com

Contents

Preface

Statistics is the science of data collection, analysis, and interpretation. It plays a pivotal role in many disciplines, including environmental, health, economic, social, physical, and information sciences. Statistics not only helps advance scientific discovery in many fields but also influences the development of humanity and society. In an increasingly data-driven and information-rich world, statistics is ever more critical in formulating scientific problems in quantitative terms, accounting for and communicating uncertainty, analyzing and learning from data, transforming numbers into knowledge and policy, and navigating the challenges for making data-driven decisions. The emergence of data science is also presenting statisticians with extraordinary opportunities for increasing the impact of the field in the real world.

This volume was commissioned in 2013 by the *Committee of Presidents of Statistical Societies* (COPSS) to celebrate its 50th anniversary and the International Year of Statistics. COPSS consists of five charter member societies: the American Statistical Association (ASA), the Institute of Mathematical Statistics (IMS), the Statistical Society of Canada (SSC), and the Eastern and Western North American Regions of the International Biometric Society (ENAR and WNAR). COPSS is best known for sponsoring prestigious awards given each year at the Joint Statistical Meetings, the largest annual gathering of statisticians in North America. Through the contributions of a distinguished group of statisticians, this volume aims to showcase the breadth and vibrancy of statistics, to describe current challenges and new opportunities, to highlight the exciting future of statistical science, and to provide guidance for future generations of statisticians.

The 50 contributors to this volume are all past winners of at least one of the awards sponsored by COPSS: the R.A. Fisher Lectureship, the Presidents' Award, the George W. Snedecor Award, the Elizabeth L. Scott Award, and the F.N. David Award. Established in 1964, the Fisher Lectureship honors both the contributions of Sir Ronald A. Fisher and a present-day statistician for their advancement of statistical theory and applications. The COPSS Presidents' Award, like the Fields Medal in mathematics or the John Bates Clark Medal in economics, is an early career award. It was created in 1979 to honor a statistician for outstanding contributions to statistics. The G.W. Snedecor Award, founded in 1976 and bestowed biennially, recognizes instrumental theoretical work in biometry. The E.L. Scott Award and F.N. David Award are also given biennially to commend efforts in promoting the role of women in statistics and to female statisticians who are leading exemplary careers; these awards were set up in 1992 and 2001, respectively.

This volume is not only about statistics and science, but also about people and their passion for discovery. It contains expository articles by distinguished authors on a broad spectrum of topics of interest in statistical education, research, and applications. Many of these articles are accessible not only to professional statisticians and graduate students, but also to undergraduates interested in pursuing statistics as a career, and to all those who use statistics in solving real-world problems. Topics include reminiscences and personal reflections on statistical careers, perspectives on the field and profession, thoughts on the discipline and the future of statistical science, as well as advice for young statisticians. A consistent theme of all the articles is the passion for statistics enthusiastically shared by the authors. Their success stories inspire, give a sense of statistics as a discipline, and provide a taste of the exhilaration of discovery, success, and professional accomplishment.

This volume has five parts. In Part I, Ingram Olkin gives a brief overview of the 50-year history of COPSS. Part II consists of 11 articles by authors who reflect on their own careers (Ingram Olkin, Herman Chernoff, Peter Bickel), share the wisdom they gained (Dennis Cook, Kathryn Roeder) and the lessons they learned (David Brillinger), describe their journeys into statistics and biostatistics (Juliet Popper Schaffer, Donna Brogan), and trace their path to success (Bruce Lindsay, Jeff Rosenthal). Mary Gray also gives an account of her lifetime efforts to promote equity.

Part III comprises nine articles devoted to the impact of statistical science on society (Steve Fienberg), statistical education (Iain Johnstone), the role of statisticians in the interplay between statistics and science (Rafael Irizarry and Nilanjan Chatterjee), equity and diversity in statistics (Mary Thompson, Nancy Reid, and Louise Ryan), and the challenges of statistical science as we enter the era of big data (Peter Hall and Xihong Lin).

Part IV consists of 24 articles, in which authors provide insight on past developments, current challenges, and future opportunities in statistical science. A broad spectrum of issues is addressed, including the foundations and principles of statistical inference (Don Fraser, Jim Berger, Art Dempster), nonparametric statistics (David Dunson), model fitting (Andrew Gelman), time series analysis (Ted Anderson), non-asymptotic probability and statistics (Pascal Massart), symbolic data analysis (Lynne Billard), statistics in medicine and public health (Norman Breslow, Nancy Flournoy, Ross Prentice, Nan Laird, Alice Whittemore), environmental statistics (Noel Cressie), health care and finance (Tze Leung Lai), statistical genetics and genomics (Elizabeth Thompson, Michael Newton), survey sampling (Rod Little), targeted learning (Mark van der Laan), statistical techniques for big data analysis, machine learning, and statistical learning (Jianqing Fan, Rob Tibshirani, Grace Wahba, Larry Wasserman). This part concludes with "a trio of inference problems that could win you a Nobel Prize in statistics" offered by Xiao-Li Meng.

Part V comprises seven articles, in which six senior statisticians share their experience and provide career advice. Jeff Wu talks about inspiration, aspiration, and ambition as sources of motivation; Ray Carroll and Marie Davidian give tips for success in research and publishing related to the choice of research topics and collaborators, familiarity with the publication process, and effective communication; and Terry Speed speaks of the necessity to follow one's own path and to be enthusiastic. Don Rubin proposed two possible topics: learning from failure and learning from mentors. As they seemed equally attractive, we asked him to do both. The book closes with Brad Efron's "thirteen rules for giving a really bad talk."

We are grateful to COPSS and its five charter member societies for supporting this book project. Our gratitude extends to Bhramar Mukherjee, former secretary and treasurer of COPSS; and to Jane Pendergast and Maura Stokes, current chair and secretary/treasurer of COPSS for their efforts in support of this book. For their help in planning, we are also indebted to the members of COPSS' 50th anniversary celebration planning committee, Joel Greenhouse, John Kittelson, Christian Léger, Xihong Lin, Bob Rodriguez, and Jeff Wu.

Additional funding for this book was provided by the International Chinese Statistical Society, the International Indian Statistical Association, and the Korean International Statistical Society. We thank them for their sponsorship and further acknowledge the substantial in-kind support provided by the Institut des sciences mathématiques du Québec.

Last but not least, we would like to express our deep appreciation to Heidi Sestrich from Carnegie Mellon University for her technical assistance, dedication, and effort in compiling this volume. Thanks also to Taylor and Francis, and especially senior editor Rob Calver, for their help and support. With the publisher's authorization, this book's content is freely available at `www.copss.org` so that it can benefit as many people as possible.

We hope that this volume will inspire you and help you develop the same passion for statistics that we share with the authors. Happy reading!

The editors

Xihong Lin
Harvard University
Boston, MA

Christian Genest
McGill University
Montréal, QC

David L. Banks
Duke University
Durham, NC

Geert Molenberghs
Universiteit Hasselt and KU Leuven
Belgium

David W. Scott
Rice University
Houston, TX

Jane-Ling Wang
University of California
Davis, CA

Contributors

Theodore W. Anderson
Stanford University
Stanford, CA

James O. Berger
Duke University
Durham, NC

Peter J. Bickel
University of California
Berkeley, CA

Lynne Billard
University of Georgia
Athens, GA

Norman E. Breslow
University of Washington
Seattle, WA

David R. Brillinger
University of California
Berkeley, CA

Donna J. Brogan
Emory University
Atlanta, GA

Raymond J. Carroll
Texas A&M University
College Station, TX

Nilanjan Chatterjee
National Cancer Institute
Bethesda, MD

Herman Chernoff
Harvard University
Cambridge, MA

R. Dennis Cook
University of Minnesota
Minneapolis, MN

Noel Cressie
University of Wollongong
Wollongong, NSW, Australia

Marie Davidian
North Carolina State University
Raleigh, NC

Arthur P. Dempster
Harvard University
Cambridge, MA

David B. Dunson
Duke University
Durham, NC

Bradley Efron
Stanford University
Stanford, CA

Jianquing Fan
Princeton University
Princeton, NJ

Stephen E. Fienberg
Carnegie Mellon University
Pittsburgh, PA

Nancy Flournoy
University of Missouri
Columbia, MO

Donald A.S. Fraser
University of Toronto
Toronto, ON

Andrew Gelman
Columbia University
New York, NY

Mary W. Gray
American University
Washington, DC

Peter G. Hall
University of Melbourne, Australia
University of California, Davis, CA

Rafael A. Irizarry
Dana-Farber Cancer Institute and
Harvard School of Public Health
Boston, MA

Iain M. Johnstone
Stanford University
Stanford, CA

Tze Leung Lai
Stanford University
Stanford, CA

Nan M. Laird
Harvard School of Public Health
Boston, MA

Xihong Lin
Harvard School of Public Health
Boston, MA

Bruce G. Lindsay
Pennsylvania State University
University Park, PA

Roderick J. Little
University of Michigan
Ann Arbor, MI

Pascal Massart
Université de Paris-Sud
Orsay, France

Xiao-Li Meng
Harvard School of Public Health
Boston, MA

Michael A. Newton
University of Wisconsin
Madison, WI

Ingram Olkin
Stanford University
Stanford, CA

Ross Prentice
Fred Hutchinson
Cancer Research Center
Seattle, WA

Nancy M. Reid
University of Toronto
Toronto, ON

Kathryn Roeder
Carnegie Mellon University
Pittsburgh, PA

Jeffrey S. Rosenthal
University of Toronto
Toronto, ON

Donald B. Rubin
Harvard University
Cambridge, MA

Louise M. Ryan
University of Technology Sydney
Sydney, Australia

Juliet Popper Shaffer
University of California
Berkeley, CA

Terry Speed
University of California
Berkeley, CA
Walter and Eliza Hall Institute of Medical Research
Melbourne, Australia

Elizabeth A. Thompson
University of Washington
Seattle, WA

Mary E. Thompson
University of Waterloo
Waterloo, ON

Robert J. Tibshirani
Stanford University
Stanford, CA

Mark van der Laan
University of California
Berkeley, CA

Grace Wahba
University of Wisconsin
Madison, WI

Larry A. Wasserman
Carnegie Mellon University
Pittsburgh, PA

Alice S. Whittemore
Stanford University
Stanford, CA

C.F. Jeff Wu
Georgia Institute of Technology
Atlanta, GA

Part I

The history of COPSS

1

A brief history of the Committee of Presidents of Statistical Societies (COPSS)

Ingram Olkin
Department of Statistics
Stanford University, Stanford, CA

Shortly after becoming Chair of COPSS in 1992, I collated some of the organization's archival history. At that time there was already a 1972 document prepared by Walter T. Federer who was Chair starting in 1965. The following is a composite of Federer's history coupled with my update in 1994, together with a review of recent activities.

1.1 Introduction

In 1958–59, the American Statistical Association (ASA), the Biometric Society (ENAR), and the Institute of Mathematical Statistics (IMS) initiated discussions to study relationships among statistical societies. Each of the three organizations often appointed a committee to perform similar or even identical duties, and communication among these and other groups was not always what was desired. Thus, in order to eliminate duplication of work, to improve communication among statistical societies, and to strengthen the scientific voice of statistics, the ASA, under the leadership of Rensis Likert, Morris H. Hansen, and Donald C. Riley, appointed a Committee to Study Relationships Among Statistical Societies (CONTRASTS). This committee was chaired by Frederick Mosteller. A series of campus discussions was initiated by members of CONTRASTS in order to obtain a broad base of opinion for a possible organization of statistical societies.

A grant of $9,000 was obtained from the Rockefeller Foundation by ASA to finance a series of discussions on the organizational needs of North American statisticians. Subsequently, an inter-society meeting to discuss relationships among statistical societies was held from Friday evening of September 16 through Sunday morning of September 18, 1960, at the Sterling Forest

Onchiota Conference Center in Tuxedo, New York. An attempt was made at inclusiveness, and 23 cognate societies sent representatives in addition to Wallace O. Fenn of the American Institute of Biological Sciences, G. Baley Price of the Conference Board of Mathematical Sciences, and May Robinson of the Brookings Institute. See *The American Statistician*, 1960, vol. 14, no. 4, pp. 2–3, for a complete listing of the representatives.

Mr. Fenn described the origin of the American Institute of Biological Sciences (AIBS) and pointed out their accomplishments after organization. Mr. Price discussed the background and importance of the Conference Board of Mathematical Sciences. The main motion for future action that was passed was to the effect that a federation of societies concerned with statistics should be organized to consider some or all of the following items: (1) publicity; (2) publication; (3) bulletin; (4) newsletter; (5) subscription exchange; (6) abstracts; (7) translations; (8) directors; (9) national roster; (10) recruitment; (11) symposia; (12) visiting lecturer program and summer institutes; (13) joint studies; (14) films and TV; (15) Washington office; (16) nominations for national committees; (17) fellowships at national levels; (18) cooperative apprentice training program; (19) international cooperation. It is satisfying to note that many of these activities came to fruition. A committee was appointed to draft a proposal for a federation of statistical societies by January 1, 1961.

1.1.1 The birth of COPSS

It was not until December 9, 1961, that a meeting was held in New York to discuss the proposed federation and the roles of ASA, ENAR–WNAR, and IMS in such an organization. A more formal meeting of the presidents, secretaries, and other members of the ASA, IMS, and the Biometric Society (ENAR) was held at the Annual Statistics Meetings in December of 1961. At this meeting the Committee of Presidents of Statistical Societies (COPSS) was essentially born. It was agreed that the president, the secretary, and one society-designated officer of ASA and IMS, the president and secretary of ENAR, the president of WNAR, and one member-at-large would form the COPSS Committee (*Amstat News*, October 1962, p. 1). The Executive Committee of COPSS was to be composed of the presidents of ASA, ENAR, and IMS and the secretary of ASA, with each president to serve as Chairman of the Committee for four months of the year.

Philip Hauser, then President of the ASA, reported on the deliberation of COPSS at the Annual Meeting. Six joint committees were established with representatives from ASA, IMS, ENAR, and WNAR. The charges are described in his report as follows.

1. *The Joint Committee on Professional Standards* is charged with considering the problems relating to standards for professional statisticians and with recommending means to maintain such standards.

2. *The Joint Committee on the Career Brochure* will be charged generally with considerations relating to statistics as a career. A specific continuing task will be the preparation and revision of a Career Brochure.

3. *The Joint Committee on Educational Opportunities* will be charged with the preparation of a brochure designed to inform students and others interested in statistics as a career of the appropriate available training facilities, both graduate and undergraduate.

4. *The Joint Committee for Liaison with the American Association for the Advancement of Science* (AAAS) will represent the statistics profession in a new section within the AAAS.

5. *The Joint Committee on Organizational Changes* is charged with the study of constitutional changes among the societies and with recommending those which may facilitate more effective and efficient collaboration between them and between members of the profession generally.

6. *The Joint Committee on News and Notes* has the task of working out arrangements for *The American Statistician* to carry news and notes for the Institute of Mathematical Statistics and the Biometric Society as well as for ASA.

As the report continued, creation of these six joint committees was to be hailed as a major step forward toward more effective collaboration between statistical societies in pursuing common ends.

The joint publication of a directory of members was part of the initial thrust for avoiding duplication, and for a cooperative venture on the part of statistical societies. However, the needs of the member societies did not always coincide, and directories have been published jointly in 1973, 1978, 1987, and 1991, and separately by IMS in 1981. In 1996 it was generally agreed to plan for a joint directory to appear no less frequently than every three years. With the advent of computer technology, an up-to-date member directory became available at any time. However, the joint directory was concrete evidence of the benefits of collaboration between the statistical societies.

A meeting of the full COPSS Committee was held on Monday, September 10, 1962, to discuss a memorandum prepared by H.L. Lucas on (7) COPSS and its activities; (8) membership of COPSS Committees; (9) charges to COPSS Committees; and (10) Committee reports. The revised version of this report became the official document of COPSS. In addition to discussing the topics listed, COPSS established the following committees: (11) Standardization of Symbols and Notations; and (12) Memorial Session in Honor of Sir Ronald A. Fisher, who had died in Adelaide, Australia, on July 29.

The COPSS Committee on Organizational Changes and COPSS Executive Committee held a meeting on Friday, March 1, 1963, to discuss a number of items dealing with cooperative arrangements among societies in COPSS, including certain possible structural changes within the societies.

The items discussed were: (1) a national committee on statistics; (2) financial needs of COPSS; (3) liaison with related societies; (4) improvement of inter-society arrangements with respect to *The American Statistician*; (5) Mathematical Statistics Section of ASA; (6) implications of reorganization to IMS; (7) progress in coordination of ENAR and WNAR; (8) recommendations on Sections of ASA, other societies affiliated with ASA, and for improvement in structure and activities of COPSS; and (9) joint billings of ASA, ENAR, IMS, and WNAR.

Two meetings of COPSS on August 26, 1963, and September 6, 1963, were held to consider a national committee on statistics, action on a report on availability of new statisticians, review of reports from COPSS Committees, a committee on liaison with related societies, the problem of recruiting census enumerators, and a proposal for publishing statistical tables. The 13th COPSS Committee, Liaison with Statistical Societies, was appointed during the summer of 1963.

At their meeting on Thursday, January 16, 1964, the COPSS Executive Committee considered the distribution of the new edition of *Careers in Statistics*, a national committee on statistics, recommendations of the Committee on Standardization of Symbols and Notation, suggestions regarding the Liaison Committee, availability of statisticians, and other items.

At the meeting of COPSS held on Tuesday, December 29, 1964, the member-at-large, Walter T. Federer, was elected as Chairman and Executive Secretary of COPSS for a three-year term, 1965–67. Federer was subsequently reappointed for a second three-year term, 1968–70, a one-year term in 1971, and a second one-year term in 1972. After this change the Executive Committee no longer met, so the ASA and IMS both designated the President-Elect as the other officer on COPSS, and some committees were added and others disbanded.

1.2 COPSS activities in the early years

It was decided that the minutes of COPSS meetings were of sufficient general interest to be published in *The American Statistician*. Meeting minutes are found in the following issues: February 1964, p. 2 (meeting of January 16, 1964); April 1969, p. 37 (meeting of August 21, 1968); June 1970, p. 2 (meeting of August 21, 1969); April 1972, p. 38 (meeting of December 27, 1970); April 1972, p. 40 (meeting of August 23, 1971). In addition, the minutes of the Onchiota Conference Center meeting appear on p. 2 of the October 1960 issue (see also p. 41).

Membership lists of COPSS Committees were published in *Amstat News* as follows: February 1963, p. 31; April 1965, p. 61; April 1966, p. 44; April

1967, p. 48; April 1968, p. 40; February 1969, p. 39; April 1970, p. 48; April 1971, p. 47; April 1972, p. 55.

Other *Amstat News* citations relating to work on COPSS Committees are the "Brochure on Statistics as a Career" (April 1962, p. 4), the "AAAS Elects Statisticians as Vice Presidents" (April 1962, p. 4), "Statistics Section (U) of the AAAS, " by M.B. Ullman (February 1964, p. 9), and "Recommended Standards for Statistical Symbols and Notation," by M. Halperin, H.O. Hartley (Chairman), and P.G. Hoel (June 1965, p. 12).

The Academic Programs Committee was most beneficial, especially through the work of Franklin Graybill in several publications of the Committee on the Undergraduate Program in Mathematics (CUPM) relating to statistics, and of Paul Minton in preparing a list for *Amstat News* (October 1970, December 1971) of US and Canadian schools that offer degrees in statistics. The work of the Bernard Greenberg committee on preparing the "Careers in Statistics" brochure was particularly commendable as evidenced by the fact that several hundred thousand copies of the brochure were distributed. The work of other Committees as well helped COPSS to achieve its goals. To further their efforts, it was suggested that the committee members be placed on a rotating basis of three-year terms whenever possible, that certain members be considered ex-officio members of the committees, that the Committee for Liaison with Other Statistical Societies be studied to find more effective means of communication between societies, and that the Executive Committee of COPSS consider holding additional meetings to consider ways of strengthening statistics in the scientific community and nationally.

In 1965 the Conference Board of the Mathematical Sciences (CBMS) began conducting a survey of the state of undergraduate mathematical and statistical sciences in the nation. This survey is conducted every five years and ten full reports have been issued, the latest being in 2010. With its statistical expertise, COPSS participated in the 1990 and 1995 surveys.

COPSS was represented at meetings of the Conference Board of Mathematical Sciences, and this served to bring concerns of the statistical community to the attention of the cognate mathematics societies.

A chronic concern related to the annual meetings of the statistical sciences. For many years the IMS alternated meeting with the ASA in one year and with the American Mathematical Association and Mathematical Association of America in another year. Somehow schedules did not always mesh geographically. This led to a considerable amount of negotiation. At one point having Joint Statistical Meetings (JSM) with all societies included solved this problem.

In the early days of COPSS the position of Chair rotated among the Societies. Later a Chair and Secretary/Treasurer were chosen by the chairs of the member societies. These are listed in Table 1.1.

TABLE 1.1
List of COPSS Chairs and Secretary/Treasurers.

Period	Chair	Secretary/Treasurer
1994–96	Ingram Olkin	Lisa Weissfeld
1997–00	Marvin Zelen	Vickie Hertzberg
2001–03	Sallie Keller	Aparna Huzurbazar
2004–06	Linda Young	Karen Bandeen-Roche
2007–09	Jessica Utts	Madhuri Mulekar
2010–12	Xihong Lin	Bhramar Mukherjee
2013–15	Jane Pendergast	Maura Stokes

1.3 COPSS activities in recent times

From the beginning of the 1990s, COPSS has been an established organization with a host of activities. The main program is described below, along with a number of notable projects.

The joint "Careers" brochure was revised in 1991 and again in 1993, and was mailed to over 100,000 individuals, schools, and institutions. It has been subsequently revised by a committee consisting of Donald Bentley (Chair), Judith O'Fallon (ASA), Jeffrey Witner (IMS), Keith Soyer (ENAR), Kevin Cain (WNAR), and Cyntha Struthers representing the Statistical Society of Canada (SSC). The latter society joined COPSS as a member in 1981.

A task force was formed to recommend a revision of the 1991 Mathematics Subject Classification in Mathematical Reviews (MR). A committee (David Aldous, Wayne Fuller, Robb Muirhead, Ingram Olkin, Emanuel Parzen, Bruce Trumbo) met at the University of Michigan with Executive Editor Ronald Babbit and also with editors of the *Zentralblatt für Mathematik*.

In 2011 the US Centers for Medicare and Medicaid Services (CMS) asked COPSS to prepare a white paper on "Statistical Issues on Assessing Hospital Performance." Then Chair, Xihong Lin formed a committee consisting of Arlene Ash, Stephen Fienberg, Thomas Louis, Sharon-Lise Normand, Thérèse Stukel, and Jessica Utts to undertake this task. The report was to provide evaluation and guidance on statistical approaches used to estimate hospital performance metrics, with specific attention to statistical issues identified by the CMS and stakeholders (hospitals, consumers, and insurers). Issues related to modeling hospital quality based on outcomes using hierarchical generalized linear models. The committee prepared a very thoughtful and comprehensive report, which was submitted to CMS on November 30, 2011, and was posted on the CMS website.

1.3.1 The Visiting Lecturer Program in Statistics

The Visiting Lecturer Program (VLP) in Statistics was a major undertaking by COPSS. At the Annual Meeting of the IMS held at Stanford University in August 1960, discussion in the Council of the Institute, program presentations, and general comment forcibly re-emphasized the need to attract many more competent young people to professional careers in statistics. Governmental, educational, and industrial requirements for trained statisticians were not being met, and new positions were being created at a faster rate than the output of new statisticians could address. The difficulty was compounded by a paucity of information about careers in statistics, by the loss of instructional personnel to higher-paying, non-academic employment, and by competition from other sciences for students with mathematical skills. A proposal for a program of visiting scientists in statistics covering the years 1962–67 was drawn up and presented to the National Science Foundation. The program was funded in 1962 by means of a three-year NSF grant to set up the Visiting Lecturer Program in Statistics for 1963–64 and 1964–65 under the Chairmanship of Jack C. Kiefer. The VLP was administered by the IMS but became a COPSS Committee because of the nature of its activities.

The original rationale for the VLP was described as follows.

> "Statistics is a very broad and exciting field of work. The main purpose of this program is to convey this excitement to students and others who may be interested. Specifically, we hope that this program will:
>
> 1. Provide information on the nature of modern statistics.
> 2. Illustrate the importance of statistics in all fields of scientific endeavor, particularly those involving experimental research, and to encourage instruction in statistics to students in all academic areas and at all levels.
> 3. Create an awareness of the opportunities for careers in statistics for students with high quantitative and problem-solving abilities and to encourage them to seek advanced training in statistics.
> 4. Provide information and advice to university and college faculties and students on the present availability of advanced training in statistics.
> 5. Encourage the development of new courses and programs in statistics."

Over the years the objectives changed somewhat, and by 1995 the Program had five similar main objectives: (1) to provide education and information on the nature and scope of modern statistics and to correct misconceptions held in regard to the science; (2) to establish and emphasize the role that statistics plays in research and practice in all fields of scientific endeavor, particularly those involving experimental research, and to encourage instruction in statistical theory and application to students in all academic areas; (3) to create an awareness of the opportunities for careers in statistics among young men and women of high potential mathematical ability and to encourage them to seek advanced training in statistics; (4) to provide information and advice to students, student counselors, and university and college faculties on the present availability of advanced training in statistics; (5) to encourage and stimulate new programs in statistics both to supplement programs in other curricula and to develop the further training of statisticians.

The 1963–64 VLP was highly successful. There were about 100 requests for lectures and over 70 visits were made by 31 lecturers. Almost every school which was assigned a speaker received the first of its three choices. Lecturers were urged to give two talks, one at a technical level and one on statistics as a career. The program continued into the 1990s. A 1991–93 report by then-Chair Lynne Billard noted that in the two-year period there were over 40 visits. These were mainly to universities in which there was no statistics department. Examples are Allegheny College, Carson Newman College, Bucknell University, Furman University, Moorhead State University, Memphis State University, University of Puerto Rico, and Wabash University, to name but a few. In general the program was not designed for universities with an active statistics department. The chairs of the VLP are listed in Table 1.2.

TABLE 1.2
List of Chairs of the Visiting Lecturer Program (VLP).

Period	Chair
1962–63	Jack C. Kiefer
1964–65	W. Jackson Hall
1966–67	Shanti S. Gupta
1970–71	Don B. Owen
1974–75	Herbert T. David
1984–86	Jon Kettenring
1987–89	Fred Leysieffer
1990–94	Lynne Billard

For a list of the lecturers and number of their visits, see *The American Statistician*: 1964, no. 4, p. 6; 1965, no. 4, p. 5; 1966, no 4, p. 12; 1967, no. 4, p. 8; 1968, no. 4, p. 3; 1970, no. 5, p. 2; 1971, no. 4, p. 2.

1.4 Awards

COPSS initiated a number of awards which have become prestigious hallmarks of achievement. These are the Presidents' Award, the R.A. Fisher Lectureship, the George W. Snedecor Award, the Elizabeth L. Scott Award, and the Florence Nightingale David Award. The COPSS Awards Ceremony usually takes place on Wednesdays of the Joint Statistical Meetings. The Presidents' Award, the Snedecor or Scott Award, and the David Award are announced first, and are followed by the Fisher Lecture. All awards now include a plaque and a cash honorarium of $1,000.

The George W. Snedecor and Elizabeth L. Scott Awards receive support from an endowment fund. In the early days of these awards, financing was in a precarious state. This led to a solicitation for funds in August 1995 that raised $12,100. Subsequently a fiscal policy was established for the support of awards.

This history of COPSS provides us with an opportunity to thank the following donors for their assistance at that time: Abbott Laboratories ($500); Biopharmaceutical Research Consultants ($100); Bristol-Myers Squibb Pharmaceutical Research Institute ($500); Chapman & Hall ($1,000); COPSS ($4,250); Duxbury Press ($500); Institute of Mathematical Statistics ($500); Institute for Social Research Survey Research Center, University of Michigan ($500); Iowa State University ($750); Procter & Gamble ($500); Springer-Verlag ($500); Section on Statistical Graphics ($500); SYSTAT ($500); Trilogy Consulting Corporation ($500); John Wiley and Sons ($1,000).

1.4.1 Presidents' Award

COPSS sponsors the Presidents' Award and presents it to a young member of the statistical community in recognition of an outstanding contribution to the profession of statistics. The Presidents' Award was established in 1976 and is jointly sponsored by the American Statistical Association, the Institute of Mathematical Statistics, the Biometric Society ENAR, the Biometric Society WNAR, and the Statistical Society of Canada operating through COPSS. (In 1994 the Biometric Society became the International Biometric Society.) The first award was given in 1979, and it is presented annually.

According to the award description, "The recipient of the Presidents' Award shall be a member of at least one of the participating societies. The Presidents' Award is granted to an individual who has not yet reached his or her 40th birthday at the time of the award's presentation. The candidate may be chosen for a single contribution of extraordinary merit, or an outstanding aggregate of contributions, to the profession of statistics."

The Presidents' Award Committee consists of seven members, including one representative appointed by each of the five member societies plus the COPSS Chair, and a past awardee as an additional member. The Chair of the Award Committee is appointed by the Chair of COPSS.

Prior to 1988 the COPSS Presidents' Award was funded by the ASA. When the 1987 award depleted this account, COPSS voted to reallocate a fraction of membership dues to fund future awards. Recipients are listed in Table 1.3 with their affiliation at the time of the award.

TABLE 1.3
List of COPSS Presidents' Award winners.

Year	Winner	Affiliation (at the Time of Award)
1979	Peter J. Bickel	University of California, Berkeley
1981	Stephen E. Fienberg	Carnegie Mellon University
1983	Tze Leung Lai	Columbia University
1984	David V. Hinkley	University of Texas, Austin
1985	James O. Berger	Purdue University
1986	Ross L. Prentice	Fred Hutchinson Cancer Research Center
1987	Chien-Fu Jeff Wu	University of Wisconsin, Madison
1988	Raymond J. Carroll	Texas A&M University
1989	Peter Hall	Australian National University
1990	Peter McCullagh	University of Chicago
1991	Bernard W. Silverman	University of Bristol, UK
1992	Nancy M. Reid	University of Toronto, Canada
1993	Wing Hung Wong	University of Chicago
1994	David L. Donoho	Stanford University
1995	Iain M. Johnstone	Stanford University
1996	Robert J. Tibshirani	University of Toronto, Canada
1997	Kathryn Roeder	Carnegie Mellon University
1998	Pascal Massart	Université de Paris-Sud, France
1999	Larry A. Wasserman	Carnegie Mellon University
2000	Jianqing Fan	University of North Carolina, Chapel Hill
2001	Xiao-Li Meng	Harvard University
2002	Jun Liu	Harvard University
2003	Andrew Gelman	Columbia University
2004	Michael A. Newton	University of Wisconsin, Madison
2005	Mark J. van der Laan	University of California, Berkeley
2006	Xihong Lin	Harvard University
2007	Jeffrey S. Rosenthal	University of Toronto, Canada
2008	T. Tony Cai	University of Pennsylvania
2009	Rafael Irizarry	Johns Hopkins University
2010	David B. Dunson	Duke University
2011	Nilanjan Chatterjee	National Cancer Institute
2012	Samuel Kou	Harvard University
2013	Marc Suchard	University of California, Los Angeles

1.4.2 R.A. Fisher Lectureship

The R.A. Fisher Lectureship was established in 1963 to honor both the contributions of Sir Ronald Aylmer Fisher and a present-day statistician for their advancement of statistical theory and applications. The list of past Fisher Lectures well reflects the prestige that COPSS and its constituent societies place on this award. Awarded each year, the Fisher Lectureship represents meritorious achievement and scholarship in statistical science, and recognizes highly significant impacts of statistical methods on scientific investigations. The Lecturer is selected by the R.A. Fisher Lecture and Award Committee which is chosen to reflect the interests of the member societies. The Lecture has become an integral part of the COPSS program, and is given at the Joint Statistical Meeting.

In the early days of COPSS, the award of the Lectureship was governed by the following conditions: (1) The Fisher Lectureship is awarded annually to an eminent statistician for outstanding contributions to the theory and application of statistics; (2) the Fisher Lecture shall be presented at a designated Annual Meeting of the COPSS societies; (3) the Lecture shall be broadly based and emphasize those aspects of statistics and probability which bear close relationship to the scientific collection and interpretation of data, areas in which Fisher himself made outstanding contributions; (4) the Lecture shall be scheduled so as to have no conflict with any other session at the Annual Meeting; (5) the Chair of the Lecture shall be the Chair of the R.A. Fisher Lecture and Award Committee or the Chair's designee: the Chair shall present a short statement on the life and works of R.A. Fisher, not to exceed five minutes in duration, and an appropriate introduction for the Fisher Lecturer; (6) the Lecturer is expected to prepare a manuscript based on the Lecture and to submit it to an appropriate statistical journal. There is an additional honorarium of $1,000 upon publication of the Fisher Lecture.

The recipients of the R.A. Fisher Lectureship are listed in Table 1.4, together with the titles of their lectures and their affiliations at the time of the award.

TABLE 1.4
Recipients of the R.A. Fisher Lectureship and titles of their lectures.

Year	Winner and Affiliation (*Title of the Talk*)
1964	Maurice S. Bartlett, University of Chicago and University College London, UK *R.A. Fisher and the last fifty years of statistical methodology*
1965	Oscar Kempthorne, Iowa State University *Some aspects of experimental inference*
1966	(none)
1967	John W. Tukey, Princeton University and Bell Labs *Some perspectives in data analysis*

TABLE 1.4
Recipients of the R.A. Fisher Lectureship (cont'd).

Year	Winner and Affiliation (*Title of the Talk*)
1968	Leo A. Goodman, University of Chicago *The analysis of cross-classified data: Independence, quasi-independence, and interactions in contingency tables with or without missing entries*
1970	Leonard J. Savage, Princeton University *On rereading R.A. Fisher*
1971	Cuthbert Daniel, Private Consultant *One-at-a-time plans*
1972	William G. Cochran, Harvard University *Experiments for nonlinear functions*
1973	Jerome Cornfield, George Washington University *On making sense of data*
1974	George E.P. Box, University of Wisconsin, Madison *Science and statistics*
1975	Herman Chernoff, Massachusetts Institute of Technology *Identifying an unknown member of a large population*
1976	George A. Barnard, University of Waterloo, Canada *Robustness and the logic of pivotal inference*
1977	R.C. Bose, University of North Carolina *R.A. Fisher's contribution to multivariate analysis and design of experiments*
1978	William H. Kruskal, University of Chicago *Statistics in society: Problems unsolved and unformulated*
1979	C.R. Rao, The Pennsylvania State University *Fisher efficiency and estimation of several parameters*
1980	(none)
1981	(none)
1982	Frank J. Anscombe, Yale University *How much to look at the data*
1983	I. Richard Savage, University of Minnesota *Nonparametric statistics and a microcosm*
1984	(none)
1985	T.W. Anderson, Stanford University *R.A. Fisher and multivariate analysis*
1986	David H. Blackwell, University of California, Berkeley *Likelihood and sufficiency*
1987	Frederick Mosteller, Harvard University *Methods for studying coincidences* (with P. Diaconis)
1988	Erich L. Lehmann, University of California, Berkeley *Model specification: Fisher's views and some later strategies*
1989	Sir David R. Cox, Nuffield College, Oxford *Probability models: Their role in statistical analysis*
1990	Donald A.S. Fraser, York University, Canada *Statistical inference: Likelihood to significance*
1991	David R. Brillinger, University of California, Berkeley *Nerve cell spike train data analysis: A progression of technique*

TABLE 1.4
Recipients of the R.A. Fisher Lectureship (cont'd).

Year	Winner and Affiliation (*Title of the Talk*)
1992	Paul Meier, Columbia University *The scope of general estimation*
1993	Herbert E. Robbins, Columbia University *N and n: Sequential choice between two treatments*
1994	Elizabeth A. Thompson, University of Washington *Likelihood and linkage: From Fisher to the future*
1995	Norman E. Breslow, University of Washington *Statistics in epidemiology: The case-control study*
1996	Bradley Efron, Stanford University *R.A. Fisher in the 21st century*
1997	Colin L. Mallows, AT&T Bell Laboratories *The zeroth problem*
1998	Arthur P. Dempster, Harvard University *Logistic statistics: Modeling and inference*
1999	John D. Kalbfleisch, University of Waterloo, Canada *The estimating function bootstrap*
2000	Ingram Olkin, Stanford University *R.A. Fisher and the combining of evidence*
2001	James O. Berger, Duke University *Could Fisher, Jeffreys, and Neyman have agreed on testing?*
2002	Raymond J. Carroll, Texas A&M University *Variability is not always a nuisance parameter*
2003	Adrian F.M. Smith, University of London, UK *On rereading L.J. Savage rereading R.A. Fisher*
2004	Donald B. Rubin, Harvard University *Causal inference using potential outcomes:* *Design, modeling, decisions*
2005	R. Dennis Cook, University of Minnesota *Dimension reduction in regression*
2006	Terence P. Speed, University of California, Berkeley *Recombination and linkage*
2007	Marvin Zelen, Harvard School of Public Health *The early detection of disease: Statistical challenges*
2008	Ross L. Prentice, Fred Hutchinson Cancer Research Center *The population science research agenda:* *Multivariate failure time data analysis methods*
2009	Noel Cressie, The Ohio State University *Where, when, and then why*
2010	Bruce G. Lindsay, Pennsylvania State University *Likelihood: Efficiency and deficiency*
2011	C.F. Jeff Wu, Georgia Institute of Technology *Post-Fisherian experimentation: From physical to virtual*
2012	Roderick J. Little, University of Michigan *In praise of simplicity not mathematistry!* *Simple, powerful ideas for the statistical scientist*
2013	Peter J. Bickel, University of California, Berkeley *From Fisher to "Big Data": Continuities and discontinuities*

1.4.3 George W. Snedecor Award

Established in 1976, this award honors George W. Snedecor who was instrumental in the development of statistical theory in biometry. It recognizes a noteworthy publication in biometry appearing within three years of the date of the award. Since 1991 it has been given every other year in odd years.

George W. Snedecor was born on October 20, 1881, in Memphis, TN, and was educated at the Alabama Polytechnic Institute, the University of Alabama, and the University of Michigan. He joined the faculty of Iowa State College (University) in 1913 and taught there for 45 years. In 1924, he and his colleague Henry Wallace (who became Secretary of Agriculture, 1933–40, 33rd Vice President of the United States, 1941–45, and Secretary of Commerce, 1945–46) organized a seminar to study regression and data analysis. He formed the Iowa State Statistics Laboratory in 1933 and served as Director. His book *Statistical Methods* was published in 1937, and later, with William G. Cochran as co-author, went through seven editions. Iowa State's Department of Statistics separated from the Mathematics Department in 1939; it offered a Master's in statistics, the first of which was given to Gertrude Cox.

The F distribution, which is central to the analysis of variance, was obtained by Snedecor and called F after Fisher. Snedecor served as president of the American Statistical Association in 1948, was named an Honorary Fellow of the Royal Statistical Society in 1954, and received an honorary Doctorate of Science from North Carolina State University in 1956. Further details about Snedecor are contained in "Tales of Statisticians" and "Statisticians in History" (*Amstat News*, September 2009, pp. 10–11).

The recipients of the George W. Snedecor Award are listed in Table 1.5, along with references for the awarded publications.

TABLE 1.5
Recipients of the George W. Snedecor Award and publication(s).

1977	A. Philip Dawid Properties of diagnostic data distribution. *Biometrics*, 32:647–658.
1978	Bruce W. Turnbull and Toby J. Mitchell Exploratory analysis of disease prevalence data from survival/sacrifice experiments. *Biometrics*, 34:555–570.
1979	Ethel S. Gilbert The assessment of risks from occupational exposure to ionizing radiation. In *Energy and Health*, SIAM–SIMS Conference Series No. 6 (N. Breslow, Ed.), SIAM, Philadelphia, PA, pp. 209–225.
1981	Barry H. Margolin, Norman Kaplan, and Errol Zeiger Statistical analysis of the *Ames salmonella*/microsome test. *Proceedings of the National Academy of Science*, 78:3779–3783.

TABLE 1.5
Recipients of the George W. Snedecor Award (cont'd).

Year	Recipient(s) and Publication
1982	Byron J.T. Morgan Modeling polyspermy. *Biometrics*, 38:885–898.
1983	Cavell Brownie and Douglas S. Robson Estimation of time-specific survival rates from tag-resighting samples: A generalization of the Jolly–Seber model. *Biometrics*, 39:437–453; and
1983	R.A. Maller, E.S. DeBoer, L.M. Joll, D.A. Anderson, and J.P. Hinde Determination of the maximum foregut volume of Western Rock Lobsters (*Panulirus cygnus*) from field data. *Biometrics*, 39:543–551.
1984	Stuart H. Hurlbert Pseudoreplication and the design of ecological field experiments. *Ecological Monographs*, 54:187–211; and
1984	John A. Anderson Regression and ordered categorical variables. *Journal of the Royal Statistical Society, Series B*, 46:1–30.
1985	Mitchell H. Gail and Richard Simon Testing for qualitative interactions between treatment effects and patients subsets. *Biometrics*, 41:361–372.
1986	Kung-Yee Liang and Scott L. Zeger Longitudinal data analysis using generalized linear models. *Biometrika*, 73:13–22; and Longitudinal data analysis for discrete and continuous outcomes. *Biometrics*, 42:121–130.
1987	George E. Bonney Regressive logistic models for familial disease and other binary traits. *Biometrics*, 42:611–625; Logistic regression for dependent binary observations. *Biometrics*, 43:951–973.
1988	Karim F. Hirji, Cyrus R. Mehta, and Nitin R. Patel Exact inference for matched case-control studies. *Biometrics*, 44:803–814.
1989	Barry I. Graubard, Thomas R. Fears, and Mitchell H. Gail Effects of cluster sampling on epidemiologic analysis in population-based case-control studies. *Biometrics*, 45:1053–1071.
1990	Kenneth H. Pollack, James D. Nichols, Cavell Brownie, and James E. Hines Statistical inference for capture-recapture experiments. *Wildlife Monographs*, The Wildlife Society 107.
1993	Kenneth L. Lange and Michael L. Boehnke Bayesian methods and optimal experimental design for gene mapping by radiation hybrid. *Annals of Human Genetics*, 56:119–144.

TABLE 1.5
Recipients of the George W. Snedecor Award (cont'd).

1995	Norman E. Breslow and David Clayton Approximate inference in generalized linear models. *Journal of the American Statistical Association*, 88:9–25.
1997	Michael A. Newton Bootstrapping phylogenies: Large deviations and dispersion effects. *Biometrika*, 83:315–328; and
1997	Kathryn Roeder, Raymond J. Carroll, and Bruce G. Lindsay A semiparametric mixture approach to case-control studies with errors in covariables. *Journal of the American Statistical Association*, 91:722–732.
1999	Daniel Scharfstein, Anastasios Butch Tsiatis, and Jamie Robins Semiparametric efficiency and its implications on the design and analysis of group sequential studies. *Journal of the American Statistical Association*, 92:1342–1350.
2001	Patrick J. Heagerty Marginally specified logistic-normal models for longitudinal binary data. *Biometrics*, 5:688–698.
2003	Paul R. Rosenbaum Effects attributable to treatment: Inference in experiments and observational studies with a discrete pivot. *Biometrika*, 88:219–231; and Attributing effects to treatment in matched observational studies. *Journal of the American Statistical Association*, 97:183–192.
2005	Nicholas P. Jewell and Mark J. van der Laan Case-control current status data. *Biometrika*, 91:529–541.
2007	Donald B. Rubin The design versus the analysis of observational studies for causal effects: Parallels with the design of randomized trials. *Statistics in Medicine*, 26:20–36.
2009	Marie Davidian Improving efficiency of inferences in randomized clinical trials using auxiliary covariates. *Biometrics*, 64:707–715 (by M. Zhang, A.A. Tsiatis, and M. Davidian).
2011	Nilanjan Chatterjee Shrinkage estimators for robust and efficient inference in haplotype-based case-control studies. *Journal of the American Statistical Association*, 104:220–233 (by Y.H. Chen, N. Chatterjee, and R.J. Carroll).
2013	John D. Kalbfleisch Pointwise nonparametric maximum likelihood estimator of stochastically ordered survival functions. *Biometrika*, 99:327–343 (by Y. Park, J.M.G. Taylor, and J.D. Kalbfleisch).

1.4.4 Elizabeth L. Scott Award

In recognition of Elizabeth L. Scott's lifelong efforts to further the careers of women, this award is presented to an individual who has helped foster opportunities in statistics for women by developing programs to encourage women to seek careers in statistics; by consistently and successfully mentoring women students or new researchers; by working to identify gender-based inequities in employment; or by serving in a variety of capacities as a role model. First awarded in 1992, it is given every other year in even-numbered years.

Elizabeth Scott was born in Fort Sill, Oklahoma, on November 23, 1917. Her family moved to Berkeley, where she remained for the rest of her life. She was in the UC Berkeley astronomy program and published more than ten papers on comet positions. She received her PhD in 1949. Her dissertation was part astronomy and part statistics: "(a) Contribution to the problem of selective identifiability of spectroscopic binaries; (b) Note on consistent estimates of the linear structural relation between two variables." She collaborated with Jerzy Neyman on astronomical problems as well as weather modification.

In 1970 Elizabeth Scott co-chaired a university sub-committee which published a comprehensive study on the status of women in academia. Subsequently she led follow-up studies concerning gender-related issues such as salary discrepancies and tenure and promotion. She developed a toolkit for evaluating salaries that was distributed by the American Association of University Professors and used by many academic women to argue successfully for salary adjustments. She often told of her history in the Astronomy Department which provided a telescope to every male faculty, but not to her. She received many honors and awards, and served as president of the IMS, 1977–78, and of the Bernoulli Society, 1983–85. She was Chair of the Statistics Department from 1968 to 1973. She was a role model for many of the women who are our current leaders. She died on December 20, 1988.

TABLE 1.6
Recipients of the Elizabeth L. Scott Award.

Year	Winner	Affiliation (at the Time of the Award)
1992	Florence N. David	University of California, Riverside
1994	Donna Brogan	University of North Carolina, Chapel Hill
1996	Grace Wahba	University of Wisconsin, Madison
1998	Ingram Olkin	Stanford University
2000	Nancy Flournoy	University of Missouri, Columbia
2002	Janet Norwood	Bureau of Labor Statistics
2004	Gladys Reynolds	Centers for Disease Control and Prevention
2006	Louise Ryan	Harvard University
2008	Lynne Billard	University of Georgia
2010	Mary E. Thompson	University of Waterloo, Canada
2012	Mary W. Gray	American University

For more details of her life and accomplishments, the web site "Biographies of Women Mathematicians" (http://www.agnesscott.edu/lriddle/women) recommends: (1) "Elizabeth Scott: Scholar, Teacher, Administrator," *Statistical Science*, 6:206–216; (2) "Obituary: Elizabeth Scott, 1917–1988," *Journal of the Royal Statistical Society, Series A*, 153:100; (3) "In memory of Elizabeth Scott," *Newsletter of the Caucus for Women in Statistics*, 19:5–6. The recipients of the Elizabeth L. Scott Award are listed in Table 1.6.

1.4.5 Florence Nightingale David Award

This award recognizes a female statistician who exemplifies the contributions of Florence Nightingale David, an accomplished researcher in combinatorial probability theory, author or editor of numerous books including a classic on the history of probability theory, *Games, Gods, and Gambling*, and first recipient of the Elizabeth L. Scott Award. Sponsored jointly by COPSS and the Caucus for Women in Statistics, the award was established in 2001 and consists of a plaque, a citation, and a cash honorarium. It is presented every other year in odd-numbered years if, in the opinion of the Award Committee, an eligible and worthy nominee is found. The Award Committee has the option of not giving an award for any given year.

F.N. David was born in the village of Irvington in Herefordshire, England, on August 23, 1909. She graduated from Bedford College for Women in 1931 with a mathematics degree. She sought advice from Karl Pearson about obtaining an actuarial position, but instead was offered a research position at University College, London. David collaborated with Pearson and Sam Stouffer (a sociological statistician) on her first paper, which appeared in 1932. Neyman was a visitor at this time and urged her to complete her PhD, which she did in 1938. During the war, she served as a statistician in military agencies. She remained at University College until 1967 when she joined the University of California at Riverside, serving as Chair of Biostatistics which was later renamed the Department of Statistics. Her research output was varied and included both theory and applications. She published *Probability Theory for Statistical Methods* in 1949, and jointly with D.E. Barton, *Combinatorial Chance* in 1962. David died in 1993 at the age of 83. The recipients of the F.N. David Award are listed in Table 1.7.

TABLE 1.7
Recipients of the Florence Nightingale David Award.

Year	Recipient	Affiliation
2001	Nan M. Laird	Harvard University
2003	Juliet Popper Shaffer	University of California, Berkeley
2005	Alice S. Whittemore	Stanford University
2007	Nancy Flournoy	University of Missouri, Columbia
2009	Nancy M. Reid	University of Toronto, Canada
2011	Marie Davidian	North Carolina State University
2013	Lynne Billard	University of Georgia

Part II

Reminiscences and personal reflections on career paths

2

Reminiscences of the Columbia University Department of Mathematical Statistics in the late 1940s

Ingram Olkin
Department of Statistics
Stanford University, Stanford, CA

2.1 Introduction: Pre-Columbia

Every once in a while in a dinner conversation, I have recalled my student days at Columbia, and have met with the suggestion that I write up these recollections. Although present-day students may recognize some of the famous names such as Hotelling, Wald, and Wolfowitz, they won't meet many faculty who were their students. The following is the result, and I hope the reader finds these reminiscences interesting. Because recollections of 60 years ago are often inaccurate, I urge readers to add to my recollections.

I started City College (CCNY) in 1941 and in 1943 enlisted in the US Army Air Force meteorology program. After completion of the program, I served as a meteorologist at various airports until I was discharged in 1946. I returned to CCNY and graduated in 1947, at which time I enrolled at Columbia University. As an aside, the professor at CCNY was Selby Robinson. Although not a great teacher, he somehow inspired a number of students to continue their study of statistics. Kenneth Arrow, Herman Chernoff, Milton Sobel, and Herbert Solomon are several who continued their studies at Columbia after graduating from CCNY.

Harold Hotelling was a key figure in my career. After receiving a doctorate at Princeton, Hotelling was at Stanford from 1924 to 1931, at the Food Research Institute and the Mathematics Department. In 1927 he taught three courses at Stanford: mathematical statistics (among the very early faculty to teach a rigorous course in statistics), differential geometry, and topology (who would tackle this today?). In 1931 he moved to Columbia, where he wrote his most famous papers in economics and in statistics (principal components, canonical correlations, T^2, to mention but a few). His 1941 paper on the teaching of statistics had a phenomenal impact. Jerzy Neyman stated that it was one of the most influential papers in statistics. Faculty attempting to convince university administrators to form a Department of Statistics often used this paper as an argument why the teaching of statistics should be done by statisticians and not by faculty in substantive fields that use statistics. To read more about Hotelling, see Olkin and Sampson (2001a,b).

2.2 Columbia days

At Columbia, Hotelling had invited Abraham Wald to Columbia in 1938, and when Hotelling left in 1946 to be Head of the Statistics Department at Chapel Hill, Wald became Chair of the newly formed department at Columbia. The department was in the Faculty of Political Economy because the Mathematics Department objected to statistics being in the same division. The two economists F.E. Croxton and F.C. Mills taught statistics in the Economics Department and insisted that the new department be the Department of Mathematical Statistics to avoid any competition with their program. The other faculty were Ted Anderson, Jack Wolfowitz, later joined by Howard Levene and Henry Scheffé; Helen Walker was in the School of Education. (Helen was one of a few well-known, influential female statisticians. One source states that she was the first woman to teach statistics.) For a detailed history of the department, see Anderson (1955).

In the late 1940s Columbia, Chapel Hill, and Berkeley were statistical centers that attracted many visitors. There were other universities that had an impact in statistics such as Princeton, Iowa State, Iowa, Chicago, Stanford, and Michigan, but conferences were mostly held at the top three. The first two Berkeley Symposia were in 1946 and 1950, and these brought many visitors from around the world.

The Second Berkeley Symposium brought a galaxy of foreign statisticians to the US: Paul Lévy, Bruno de Finetti, Michel Loève, Harold Cramér, Aryeh Dvoretzky, and Robert Fortet. Domestic faculty were present as well, such as Richard Feynman, Kenneth Arrow, Jacob Marshak, Harold Kuhn, and Albert Tucker. Because some of the participants came from distant lands, they often visited other universities as part of the trip. During the 1946–48 academic years the visitors were Neyman, P.L Hsu, J.L. Doob, M.M. Loève, E.J.G. Pitman, R.C. Bose, each teaching special-topic courses. Later Bose and Hsu joined Hotelling at Chapel Hill.

With the GI Bill, I did not have to worry about tuition, and enrolled at Columbia in two classes in the summer of 1947. The classes were crowded with post-war returnees. One class was a first course in mathematical statistics that was taught by Wolfowitz. Some of the students at Columbia during the 1947–50 period were Raj Bahadur and Thelma Clark (later his wife), Bob

Bechhofer, Allan Birnbaum, Al Bowker, Herman Chernoff (he was officially at Brown University, but worked with Wald), Herb T. David, Cyrus Derman, Sylvan Ehrenfeld, Harry Eisenpress, Peter Frank, Leon Herbach, Stanley Isaacson, Seymour Jablon, Jack Kiefer, Bill Kruskal, Gerry Lieberman, Gottfried Noether, Rosedith Sitgreaves, Milton Sobel, Herbert Solomon, Charles Stein, Henry Teicher, Lionel Weiss, and many others. Columbia Statistics was an exciting place, and almost all of the students continued their career in statistics. There was a feeling that we were in on the ground floor of a new field, and in many respects we were. From 1950 to 1970 *The Annals of Mathematical Statistics* grew from 625 to 2200 pages, with many articles from the students of this era.

Some statistics classes were held at night starting at 5:40 and 7:30 so that students who worked during the day could get to class. However, math classes took place during the day. I took sequential analysis and analysis of variance from Wald, core probability from Wolfowitz, finite differences from B.O. Koopman, linear algebra from Howard Levi, differential equations from Ritt, a computer science course at the Columbia Watson Lab, and a course on analysis of variance from Helen Walker. Anderson taught multivariate analysis the year before I arrived. Govind Seth and Charles Stein took notes from this course, which later became Anderson's book on multivariate analysis.

Wald had a classic European lecture style. He started at the upper left corner of the blackboard and finished at the lower right. The lectures were smooth and the delivery was a uniform distribution. Though I had a lovely set of notes, Wald treated difficult and easy parts equally, so one did not recognize pitfalls when doing homework. The notion of an application in its current use did not exist. I don't recall the origin of the following quotation, but it is attributed to Wald: "Consider an application. Let $X_1, \ldots, X_n$ be i.i.d. random variables." In contrast to Wald's style, Wolfowitz's lectures were definitely not smooth, but he attempted to emphasize the essence of the topic. He struggled to try to explain what made the theorem "tick," a word he often used: "Let's see what makes this tick." However, as a novice in the field the gems of insight that he presented were not always appreciated. It was only years later as a researcher that they resurfaced, and were found to be illuminating.

Wolfowitz had a number of other pet phrases such as "It doesn't cut any ice," and "stripped of all baloney." It was a surprise to hear Columbia graduates years later using the same phrase. In a regression class with Wolfowitz we learned the Gauss–Seidel method. Wolfowitz was upset that the Doolittle method had a name attached to it, and he would exclaim, "Who is this Doolittle?" Many years later when Wolfowitz visited Stanford a name might arise in a conversation. If Wolfowitz did not recognize the name he would say "Jones, Jones, what theorem did he prove?"

In 1947–48 the only serious general textbooks were Cramér, Kendall, and Wilks' soft-covered notes. This was a time when drafts of books were being written. Feller's Volume I appeared in 1950, Doob's book on stochastic processes in 1953, Lehmann's notes on estimation and testing of hypotheses in 1950, Scheffé's book on analysis of variance in 1953. The graduate students at Columbia formed an organization that duplicated lecture notes, especially those of visitors. Two that I remember are Doob's lectures on stochastic processes and Loève's on probability.

The Master's degree program required a thesis and mine was written with Wolfowitz. The topic was on a sequential procedure that Leon Herbach (he was ahead of me) had worked on. Wolfowitz had very brief office hours, so there usually was a queue to see him. When I did see him in his office he asked me to explain my question at the blackboard. While talking at the blackboard Wolfowitz was multi-tasking (even in 1947) by reading his mail and talking on the telephone. I often think of this as an operatic trio in which each singer is on a different wavelength. This had the desired effect in that I never went back. However, I did manage to see him after class. He once said "Walk me to the subway while we are talking," so I did. We did not finish our discussion by the time we reached the subway (only a few blocks away) so I went into the subway where we continued our conversation. This was not my subway line so it cost me a nickel to talk to him. One of my students at Stanford 30 years later told me that I suggested that he walk with me while discussing a problem. There is a moral here for faculty.

Wald liked to take walks. Milton Sobel was one of Wald's students and he occasionally accompanied Wald on these walks. Later I learned that Milton took his students on walks. I wonder what is the 21st century current version of faculty-student interaction?

2.3 Courses

The Collyer brothers became famous for their compulsive collecting. I am not in their league, but I have saved my notes from some of the courses that I took. The following is an excerpt from the Columbia course catalog.

> Mathematical Statistics 111a — Probability. 3 points Winter Session. Professor WOLFOWITZ.
>
> Tu. Th. 5:40–6:30 and 7:30–8:20 p.m. 602 Hamilton.
>
> Fundamentals. Combinatorial problems. Distribution functions in one or more dimensions. The binomial, normal, and Poisson laws. Moments and characteristic functions. Stochastic convergence and the law of large numbers. Addition of chance variables and limit theorems.
>
> This course terminates on Nov. 18. A thorough knowledge of calculus is an essential prerequisite. Students are advised to study higher algebra simultaneously to obtain a knowledge of matrix algebra for use in more advanced mathematical statistics courses.

Milton Sobel was the teaching assistant for the 111a course; Robert Bechofer and Allan Birnbaum were also TAs. I remember that Milton Sobel sat in on Wald's class on analysis of variance. Because he was at least one year ahead of me I thought that he would have taken this course earlier. He said he did take it earlier but the course was totally different depending on who was teaching it. It was depressing to think that I would have to take every course twice! As the course progressed Wald ran out of subscripts and superscripts on the right-hand side, e.g., $x_{ij}^{k\ell}$, and subsequently added some subscripts on the left-hand side.

Wolfowitz recommended three books, and assigned homework from them:

(a) H. Cramér (1945): *Mathematical Methods of Statistics*

(b) J.V. Uspensky (1937): *Introduction to Mathematical Probability*

(c) S.S. Wilks (1943): *Mathematical Statistics*

He mentioned references to Kolmogorov's *Foundation of Probability* and the Lévy and Roth book *Elements of Probability.*

Wolfowitz used the term "chance variables" and commented that the Law of Small Numbers should have been called the Law of Small Probabilities. As I look through the notes it is funny to see the old-fashioned factorial symbols $\lfloor n$ instead of $n!$. As I reread my notes it seems to me that this course was a rather simplified first course in probability. Some of the topics touched upon the use of independence, Markov chains, joint distributions, conditional distributions, Chebychev's inequality, stochastic convergence (Slutsky's theorem), Law of Large Numbers, convolutions, characteristic functions, Central Limit Theorem (with discussion of Lyapunov and Lindeberg conditions). I have a comment in which Wolfowitz notes an error in Cramér (p. 343): (a) if $y_1, y_2, \ldots$ is a sequence with $\sum y_i = c_i$ for all c and $\sigma^2(y_i) \to 0$ as $i \to \infty$, then $p\lim_{i\to\infty}(y_i - c_i) = 0$; (b) the converse is not true in that it may be that $\sigma^2(y_i) \to \infty$ and yet $p\lim(y_i - c_i) = 0$.

The second basic course was 111b, taught by Wald. The topics included point estimation, consistency, unbiasedness, asymptotic variance, maximum likelihood, likelihood ratio tests, efficiency. This course was more mathematical than 111a in that there was more asymptotics. In terms of mathematical background I note that he used Lagrange multipliers to show that, for $w_1, \ldots, w_n \in [0, 1]$, $\sum_{i=1}^n w_i^2/(\sum_{i=1}^n w_i)^2$ is minimized when $w_i = 1/n$ for all $i \in \{1, \ldots, n\}$. Apparently, convexity was not discussed.

There is a derivation of the chi-square distribution that includes a discussion of orthogonal matrices. This is one of the standard proofs. Other topics include Schwarz inequality (but not used for the above minimization), and sufficiency. The second part of the course dealt with tests of hypotheses, with emphasis on the power function (Wald used the term "power curve"), acceptance sampling, and the OC curve.

My Columbia days are now over 65 years ago, but I still remember them as exciting and an incubator for many friendships and collaborations.

References

Anderson, T.W. Jr. (1955). The Department of Mathematical Statistics. In *A History of the Faculty of Political Science, Columbia*, R.G. Hoxie, Ed. Columbia University Press, pp. 250–255.

Olkin, I. and Sampson, A.R. (2001a). Hotelling, Harold (1895–1973). In *International Encyclopedia of the Social & Behavioral Sciences* (N.J. Smelser and P.B. Baltes, Eds.). Pergamon, Oxford, 6921–6925. URL `http://www.sciencedirect.com/science/article/pii/B0080430767002631`.

Olkin, I. and Sampson, A.R. (2001b). Harold Hotelling. In *Statisticians of the Centuries*. Springer, New York, pp. 454–458.

3

A career in statistics

Herman Chernoff
Department of Statistics
Harvard University, Cambridge, MA

3.1 Education

At the early age of 15, I graduated from Townsend Harris high school in New York and made the daring decision to study mathematics at the City College of New York (CCNY) during the depression, rather than some practical subject like accounting. The Mathematics faculty of CCNY was of mixed quality, but the mathematics majors were exceptionally good. Years later, one of the graduate students in statistics at Stanford found a copy of the 1939 yearbook with a picture of the Math Club. He posted it with a sign "Know your Faculty." At CCNY we had an excellent training in undergraduate mathematics, but since there was no graduate program, there was no opportunity to take courses in the advanced subjects of modern research. I was too immature to understand whether my innocent attempts to do original research were meaningful or not. This gave me an appetite for applied research where successfully confronting a real problem that was not trivial had to be useful.

While at CCNY, I took a statistics course in the Mathematics Department, which did not seem very interesting or exciting, until Professor Selby Robinson distributed some papers for us to read during the week that he had to be away. My paper was by Neyman and Pearson (1933). It struck me as mathematically trivial and exceptionally deep, requiring a reorganization of my brain cells to confront statistical issues. At that time I had not heard of R.A. Fisher, who had succeeded in converting statistics to a theoretical subject in which mathematicians could work. Of course, he had little use for mathematicians in statistics, on the grounds that they confronted the wrong problems and he was opposed to Neyman–Pearson theory (NP).

Once when asked how he could find the appropriate test statistic without recourse to NP, his reply was "I have no trouble." In short, NP made explicit the consideration of the alternative hypotheses necessary to construct good tests. This consideration was implicit for statisticians who understood their problem, but often unclear to outsiders and students. Fisher viewed it as an unnecessary mathematization, but the philosophical issue was important. Years later Neyman gave a talk in which he pointed out that the NP Lemma was highly touted but trivial. On the other hand it took him years of thinking to understand and state the issue.

Just before graduation I received a telegram offering me a position, which I accepted, as Junior Physicist at the Dahlgren Naval Proving Grounds in Virginia. After a year and a half I left Dahlgren to study applied mathematics at Brown University in Providence, Rhode Island. Dean Richardson had set up a program in applied mathematics where he could use many of the distinguished European mathematician émigrés to teach and get involved in research for the Defense Department, while many of the regular faculty were away working at Defense establishments. There was a good deal of coming and going of faculty, students and interesting visitors during this program.

During the following year I worked very hard as a Research Assistant for Professor Stefan Bergman and took many courses and audited a couple. I wrote a Master's thesis under Bergman on the growth of solutions of partial differential equations generated by his method, and received an ScM degree. One of the courses I audited was given by Professor Willy Feller, in which his lectures were a preliminary to the first volume of his two volume outstanding books on probability.

During the following summer, I took a reading course in statistics from Professor Henry Mann, a number theorist who had become interested in statistics because some number theory issues were predominant in some of the work going on in experimental design. In fact, he had coauthored a paper with Abraham Wald (Mann and Wald, 1943) on how the o and O notation could be extended to o_p and O_p. This paper also proved that if X_n has as its limiting distribution that of Y and g is a continuous function, then $g(X_n)$ has as its limiting distribution that of $g(Y)$.

Mann gave me a paper by Wald (1939) on a generalization of inference which handled that of estimation and testing simultaneously. Once more I found this paper revolutionary. This was apparently Wald's first paper on decision theory. Although it did not resemble the later version of a game against nature, it clearly indicated the importance of cost considerations in statistical philosophy. Later discussions with Allen Wallis indicated that Wald had been aware of von Neumann's ideas about game theory. My theory is that in this first paper, he had not recognized the relevance, but as his work in this field grew, the formulation gradually changed to make the relationship with game theory clearer. Certainly the role of mixed strategies in both fields made the relation apparent.

At the end of the summer, I received two letters. One offered me a predoctoral NSF fellowship and the other an invitation I could not decline, to join the US Army. It was 1945, the war had ended, and the draft boards did not see much advantage in deferring young men engaged in research on the war effort. I was ordered to appear at Fort Devens, Massachusetts, where I was given three-day basic training and assigned to work as a clerk in the separation center busy discharging veterans. My hours were from 5 PM to midnight, and I considered this my first vacation in many years. I could visit Brown on weekends and on one of these visits Professor Prager suggested that I might prefer to do technical work for the army. He arranged for me to be transferred to Camp Lee, Virginia, where I was designated to get real basic training and end up, based on my previous experience, as a clerk in the quartermaster corps in Germany. I decided that I would prefer to return to my studies and had the nerve to apply for discharge on the grounds that I was a scientist, a profession in good repute at the end of the war. Much to everyone's surprise my application was approved and I returned to Brown, where much had changed during my brief absence. All the old European professors were gone, Prager had been put in charge of the new Division of Applied Mathematics, and a new group of applied mathematicians had replaced the émigrés.

I spent some months reading in probability and statistics. In particular Wald's papers on sequential analysis, a topic classified secret during the war, was of special interest.

During the summer of 1946 there was a six-week meeting in Raleigh, North Carolina, to open up the Research Triangle. Courses were offered by R.A. Fisher, J. Wolfowitz, and W. Cochran. Many prominent statisticians attended the meeting, and I had a chance to meet some of them and young students interested in statistics, and to attend the courses. Wolfowitz taught sequential analysis, Cochran taught sampling, and R.A. Fisher taught something.

Hotelling had moved to North Carolina because Columbia University had refused to start a Statistics Department. Columbia realized that they had made a mistake, and started a department with Wald as Chair and funds to attract visiting professors and faculty. Wolfowitz, who had gone to North Carolina, returned to Columbia. I returned to Brown to prepare for my preliminary exams. Since Brown no longer had any statisticians, I asked Wald to permit me to attend Columbia to write my dissertation in absentia under his direction. He insisted that I take some courses in statistics. In January 1947, I attended Columbia and took courses from T.W. Anderson, Wolfowitz, J.L. Doob, R.C. Bose and Wald.

My contact with Anderson led to a connection with the Cowles Commission for Research in Economics at the University of Chicago, where I was charged with investigating the possible use of computing machines for the extensive calculations that had to be done with their techniques for characterizing the economy. Those calculations were being done on electric calculating machines by human "computers" who had to spend hours carrying 10 digits inverting matrices of order as much as 12. Herman Rubin, who had received his PhD working under T. Koopmans at Cowles and was spending a postdoctoral year at the Institute for Advanced Study at Princeton, often came up to New York to help me wrestle with the clumsy IBM machines at the Watson

Laboratories of IBM, then located near Columbia. At the time, the engineers at Watson were working on developing a modern computer.

3.2 Postdoc at University of Chicago

I completed my dissertation on an approach of Wald to an asymptotic approximation to the (nonexistent) solution of the Behrens–Fisher problem, and in May, 1948, I went to Chicago for a postdoc appointment at Cowles with my new bride, Judith Ullman. I was in charge of computing. Among my new colleagues were Kenneth Arrow, a former CCNY mathematics major, who was regarded as a brilliant economist, and Herman Rubin and several graduate students in economics, one of whom, Don Patinkin, went to Israel where he introduced the ideas of the Cowles Commission and later became President of the Hebrew University.

Arrow had not yet written a dissertation, but was invited to visit Rand Corporation that summer and returned with two outstanding accomplishments. One was the basis for his thesis and Nobel Prize, a proof that there was no sensible way a group could derive a preference ordering among alternatives from the preference orderings of the individual members of the group. The other was a proof of the optimum character of the sequential probability ratio test for deciding between two alternatives. The latter proof, with D. Blackwell and A. Girshick (Arrow et al., 1949), was derived after Wald presented a proof which had some measure theoretic problems. The backward induction proof of ABG was the basis for the development of a large literature on dynamic programming. The basic idea of Wald and Wolfowitz, which was essential to the proof, had been to use a clever Bayesian argument.

I had always been interested in the philosophical issues in statistics, and Jimmie Savage claimed to have resolved one. Wald had proposed the minimax criterion for deciding how to select one among the many "admissible" strategies. Some students at Columbia had wondered why Wald was so tentative in proposing this criterion. The criterion made a good deal of sense in dealing with two-person zero-sum games, but the rationalization seemed weak for games against nature. In fact, a naive use of this criterion would suggest suicide if there was a possibility of a horrible death otherwise. Savage pointed out that in all the examples Wald used, his loss was not an absolute loss, but a regret for not doing the best possible under the actual state of nature. He proposed that minimax regret would resolve the problem. At first I bought his claim, but later discovered a simple example where minimax regret had a similar problem to that of minimax expected loss. For another example the criterion led to selecting the strategy A, but if B was forbidden, it led to C and not A. This was one of the characteristics forbidden in Arrow's thesis.

Savage tried to defend his method, but soon gave in with the remark that perhaps we should examine the work of de Finetti on the Bayesian approach to inference. He later became a sort of high priest in the ensuing controversy between the Bayesians and the misnamed frequentists. I posed a list of properties that an objective scientist should require of a criterion for decision theory problems. There was no criterion satisfying that list in a problem with a finite number of states of nature, unless we canceled one of the requirements. In that case the only criterion was one of all states being equally likely. To me that meant that there could be no objective way of doing science. I held back publishing those results for a few years hoping that time would resolve the issue (Chernoff, 1954).

In the controversy, I remained a frequentist. My main objection to Bayesian philosophy and practice was based on the choice of the prior probability. In principle, it should come from the initial belief. Does that come from birth? If we use instead a non-informative prior, the choice of one may carry hidden assumptions in complicated problems. Besides, the necessary calculation was very forbidding at that time. The fact that randomized strategies are not needed for Bayes procedures is disconcerting, considering the important role of random sampling. On the other hand, frequentist criteria lead to the contradiction of the reasonable criteria of rationality demanded by the derivation of Bayesian theory, and thus statisticians have to be very careful about the use of frequentist methods.

In recent years, my reasoning has been that one does not understand a problem unless it can be stated in terms of a Bayesian decision problem. If one does not understand the problem, the attempts to solve it are like shooting in the dark. If one understands the problem, it is not necessary to attack it using Bayesian analysis. My thoughts on inference have not grown much since then in spite of my initial attraction to statistics that came from the philosophical impact of Neyman–Pearson and decision theory.

One slightly amusing correspondence with de Finetti came from a problem from the principal of a local school that had been teaching third graders Spanish. He brought me some data on a multiple choice exam given to the children to evaluate how successful the teaching had been. It was clear from the results that many of the children were guessing on some of the questions. A traditional way to compensate for guessing is to subtract a penalty for each wrong answer. But when the students are required to make a choice, this method simply applies a linear transformation to the score and does not provide any more information than the number of correct answers. I proposed a method (Chernoff, 1962) which turned out to be an early application of empirical Bayes. For each question, the proportion of correct answers in the class provides an estimate of how many guessed and what proportion of the correct answers were guesses. The appropriate reward for a correct answer should take this estimate into account. Students who hear of this approach are usually shocked because if they are smart, they will suffer if they are in a class with students who are not bright.

Bruno de Finetti heard of this method and he wrote to me suggesting that the student should be encouraged to state their probability for each of the possible choices. The appropriate score should be a simple function of the probability distribution and the correct answer. An appropriate function would encourage students to reply with their actual distribution rather than attempt to bluff. I responded that it would be difficult to get third graders to list probabilities. He answered that we should give the students five gold stars and let them distribute the stars among the possible answers.

3.3 University of Illinois and Stanford

In 1949, Arrow and Rubin went to Stanford, and I went to the Mathematics Department of the University of Illinois at Urbana. During my second year at Urbana, I received a call from Arrow suggesting that I visit the young Statistics Department at Stanford for the summer and the first quarter of 1951. That offer was attractive because I had spent the previous summer, supplementing my $4,500 annual salary with a stint at the Operations Research Office of Johns Hopkins located at Fort Lesley J. McNair in Washington, DC. I had enjoyed the visit there, and learned about the Liapounoff theorem about the (convex) range of a vector measure, a powerful theorem that I had occasion to make use of and generalize slightly (Chernoff, 1951). I needed some summer salary. The opportunity to visit Stanford with my child and pregnant wife was attractive.

The head of the department was A. Bowker, a protégé of the provost F. Terman. Terman was a radio engineer, returned from working on radar in Cambridge, MA during the war, where he had learned about the advantages of having contracts with US Government agencies and had planned to exploit such opportunities. Essentially, he was the father of Silicon Valley. The Statistics Department had an applied contract with the Office of Naval Research (ONR) and I discovered, shortly after arriving, that as part of the contract, the personnel of the department supported by that contract were expected to engage in research with relevance to the ONR and to address problems posed to them on annual visits by scientists from the NSA. We distributed the problems posed in mathematical form without much background. I was given the problem of how best to decide between two alternative distributions of a random variable X when the test statistic must be a sum of integers Y with $1 \leq Y \leq k$ for some specified value of k and Y must be some unspecified function of X. It was clear that the problem involves partitioning the space of X into k subsets and applying the likelihood ratio. The Liapounoff theorem was relevant and the Central Limit Theorem gave error probabilities to use to select the best procedure.

In working on an artificial example, I discovered that I was using the Central Limit Theorem for large deviations where it did not apply. This led me to derive the asymptotic upper and lower bounds that were needed for the tail probabilities. Rubin claimed he could get these bounds with much less work and I challenged him. He produced a rather simple argument, using the Markov inequality, for the upper bound. Since that seemed to be a minor lemma in the ensuing paper I published (Chernoff, 1952), I neglected to give him credit. I now consider it a serious error in judgment, especially because his result is stronger, for the upper bound, than the asymptotic result I had derived.

I should mention that Cramér (1938) had derived much more elegant and general results on large deviations. I discovered this after I derived my results. However, Cramér did require a condition that was not satisfied by the integer-valued random variables in my problem. Shannon had published a paper using the Central Limit Theorem as an approximation for large deviations and had been criticized for that. My paper permitted him to modify his results and led to a great deal of publicity in the computer science literature for the so-called Chernoff bound which was really Rubin's result.

A second vaguely stated problem was misinterpreted by Girshick and myself. I interpreted it as follows: There exists a class of experiments, the data from which depend on two parameters, one of which is to be estimated. Independent observations with repetitions may be made on some of these experiments. The Fisher information matrix is additive and we wish to minimize the asymptotic variance, or equivalently the upper left corner of the inverse of the sum of the informations. We may as well minimize the same element of the inverse of the average of the informations. But this average lies in the convex set generated by the individual informations of the available experiments. Since each information matrix has three distinct elements, we have the problem of minimizing a function on a convex set in three dimensions. It is immediately obvious that we need at most four of the original available experiments to match the information for any design. By monotonicity it is also obvious that the optimum corresponds to a point on the boundary, and we need at most three of the experiments, and a more complicated argument shows that a mixture of at most two of the experiments will provide an asymptotically optimal experiment. This result (Chernoff, 1953) easily generalizes to the case of estimating a function of r of the k parameters involved in the available experiments.

The lively environment at Stanford persuaded me to accept a position there and I returned to settle in during the next academic year. Up to then I had regarded myself as a "theoretical statistical gun for hire" with no long-term special field to explore. But both of the problems described above have optimal design implications. I also felt that the nature of scientific study was to use experiments to learn about issues so that better experiments could be performed until a final decision was to be made. This led me to have sequential design of experiments as a major background goal.

At Stanford, I worked on many research projects which involved optimization and asymptotic results. Many seemed to come easily with the use of Taylor's theorem, the Central Limit Theorem and the Mann–Wald results. A more difficult case was in the theorem of Chernoff and Savage (1958) where we established the Hodges–Lehmann conjecture about the efficiency of the nonparametric normal scores test. I knew very little about nonparametrics, but when Richard Savage and M. Dwass mentioned the conjecture, I thought that the variational argument would not be difficult, and it was easy. What surprised me was that the asymptotic normality, when the hypothesis of the equality of the two distributions is false, had not been established. Our argument approximating the relevant cumulative distribution function by a Gaussian process was tedious but successful. The result apparently opened up a side industry in nonparametric research which was a surprise to Jimmie Savage, the older brother of Richard.

One side issue is the relevance of optimality and asymptotic results. In real problems the asymptotic result may be a poor approximation to what is needed. But, especially in complicated cases, it provides a guide for tabulating finite-sample results in a reasonable way with a minimum of relevant variables. Also, for technical reasons optimality methods are not always available, but what is optimal can reveal how much is lost by using practical methods and when one should search for substantially better ones, and often how to do so.

Around 1958, I proved that for the case of a finite number of states of nature and a finite number of experiments, an asymptotically optimal sequential design consists of solving a game where the payoff for the statistician using the experiment e against nature using θ is $I(\hat{\theta}, \theta, e)$ and I is the Kullback–Leibler information, assuming the current estimate $\hat{\theta}$ is the true value of the unknown state (Chernoff, 1959). This result was generalized to infinitely many experiments and states by Bessler (1960) and Albert (1961) but Albert's result required that the states corresponding to different terminal decisions be separated.

This raised the simpler non-design problem of how to handle the test that the mean of a Normal distribution with known variance is positive or negative. Until then the closest approach to this had been to treat the case of three states of nature $a, 0, -a$ for the means and to minimize the expected sample size for 0 when the error probabilities for the other states were given. This appeared to me to be an incorrect statement of the relevant decision problem which I asked G. Schwarz to attack. There the cost was a loss for the wrong decision and a cost per observation (no loss when the mean is 0). Although the techniques in my paper would work, Schwarz (1962) did a beautiful job using a Bayesian approach. But the problem where the mean could vary over the entire real line was still not done.

I devoted much of the next three years to dealing with the non-design problem of sequentially testing whether the mean of a Normal distribution with known variance is positive or negative. On the assumption that the payoff for each decision is a smooth function of the mean μ, it seems natural to measure the loss as the difference which must be proportional to $|\mu|$ in the neighborhood of 0. To facilitate analysis, this problem was posed in terms of the drift of a Wiener process, and using Bayesian analysis, was reduced to a free boundary problem involving the heat equation. The two dimensions are Y, the current posterior estimate of the mean, and t, the precision of the estimate. Starting at (t_0, Y_0), determined by the prior distribution, Y moves like a standard Wiener process as sampling continues and the optimal sequential procedure is to stop when Y leaves the region determined by the symmetric boundary.

The research resulted in four papers; see Chernoff (1961), Breakwell and Chernoff (1964), Chernoff (1965a), and Chernoff (1965b). The first was preliminary with some minor results and bounds and conjectures about the boundary near $t = 0$ and large t. Before I went off on sabbatical in London and Rome, J.V. Breakwell, an engineer at Lockheed, agreed to collaborate on an approach to large t and I planned to concentrate on small t. In London I finally made a breakthrough and gave a presentation at Cambridge where I met J. Bather, a graduate student who had been working on the same problem. He had just developed a clever method for obtaining inner and outer bounds on the boundary.

Breakwell had used hypergeometric functions to get good asymptotic approximations for large t, but was unhappy because the calculations based on the discrete time problem seemed to indicate that his approximations were poor. His letter to that effect arrived just as I had derived the corrections relating the continuous time and discrete time problems, and these corrections indicated that the apparently poor approximations were in fact excellent.

Bather had impressed me so that I invited him to visit Stanford for a postdoc period. Let me digress briefly to mention that one of the most valuable functions of the department was to use the contracts to support excellent postdocs who could do research without teaching responsibilities and appreciate courses by Stein and Rubin that were often too difficult for many of our own students.

Breakwell introduced Bather and me to the midcourse correction problem for sending a rocket to the moon. The instruments measure the estimated miss distance continuously, and corrections early are cheap but depend on poor estimates, while corrections later involve good estimates but are expensive in the use of fuel. We found that our methods for the sequential problem work in this problem, yielding a region where nothing is done. But when the estimated miss distance leaves that region, fuel must be used to return. Shortly after we derived our results (Bather and Chernoff, 1967), a rocket was sent to the moon and about half way there, a correction was made and it went to the desired spot. The instrumentation was so excellent (and expensive) that our refined method was unnecessary. Bather declined to stay at Stanford as Assistant Professor and returned with his family to England to teach at Suffolk University. Later I summarized many of these results in a SIAM monograph on sequential analysis and optimal design (Chernoff, 1972).

A trip to a modern factory in Italy during my sabbatical gave me the impression that automation still had far to go, and the study of pattern recognition and cluster analysis could be useful. There are many methods available for clustering, but it seemed that an appropriate method should depend on the nature of the data. This raised the problem of how to observe multidimensional data. It occurred to me that presenting each n-dimensional data point by a cartoon of a face, where each of the components of the data point controlled a feature of the face, might be effective in some cases. A presentation of this idea with a couple of examples was received enthusiastically by the audience, many of whom went home and wrote their own version of what are popularly called "Chernoff faces." This took place at a time when the computer was just about ready to handle the technology, and I am reasonably sure that if I had not done it, someone else would soon have thought of the idea. Apparently I was lucky in having thought of using caricatures of faces, because faces are processed in the brain differently than other visual objects and caricatures have a larger impact than real faces; see Chernoff (1973).

3.4 MIT and Harvard

At the age of 50, I decided to leave Stanford and start a statistics program at MIT in the Applied Mathematics Section of the Mathematics Department. For several years, we had a vital but small group, but the School of Science was not a healthy place for recognizing and promoting excellent applied statisticians, and so I retired from MIT to accept a position at Harvard University, from which I retired in 1997, but where I have an office that I visit regularly even though they don't pay me.

I am currently involved in a collaboration with Professor Shaw-Hwa Lo at Columbia University, who was inspired by a seminar course I offered at Harvard on statistical issues in molecular biology. We have been working on variable selection methods for large data sets with applications to biology and medicine; see Chernoff (2009).

In review, I feel that I lacked some of the abilities that are important for an applied statistician who has to handle problems on a daily basis. I lacked the library of rough and ready techniques to produce usable results. However, I found that dealing with real applied problems, no matter how unimportant, without this library, required serious consideration of the issues and was often a source of theoretical insight and innovation.

References

Albert, A.E. (1961). The sequential design of experiments for infinitely many states of nature. *The Annals of Mathematics Statistics*, 32:774–799.

Arrow, K.J., Blackwell, D., and Girshick, M.A. (1949). Bayes and minimax solutions of sequential design problems. *Econometrica*, 17:213–244.

Bather, J.A. and Chernoff, H. (1967). Sequential decisions in the control of a space ship. *Proceedings of the Fifth Berkeley Symposium, University of California Press*, 3:181–207.

Bessler, S. (1960). *Theory and Application of the Sequential Design of Experiments, k-actions and Infinitely Many Experiments: Part I–Theory.* Technical Report 55, Department of Statistics, Stanford University, Stanford, CA.

Breakwell, J.V. and Chernoff, H. (1964). Sequential tests for the mean of a Normal distribution II (large t). *The Annals of Mathematical Statistics*, 35:162–173.

Chernoff, H. (1951). An extension of a result of Liapounoff on the range of a vector measure. *Proceedings of the American Mathematical Society*, 2:722–726.

Chernoff, H. (1952). A measure of asymptotic efficiency for tests of a hypothesis based on the sum of observations. *The Annals of Mathematical Statistics*, 23:493–507.

Chernoff, H. (1953). Locally optimal designs for estimating parameters. *The Annals of Mathematical Statistics*, 24:586–602.

Chernoff, H. (1954). Rational selection of decision functions. *Econometrica*, 22:422–443.

Chernoff, H. (1959). Sequential design of experiments. *The Annals of Mathematical Statistics*, 30:755–770.

Chernoff, H. (1961). Sequential tests for the mean of a Normal distribution. *Proceedings of Fourth Berkeley Symposium on Mathematical Statistics and Probability, University of California Press*, 1:79–95.

Chernoff, H. (1962). The scoring of multiple choice questionnaires. *The Annals of Mathematical Statistics*, 35:375–393.

Chernoff, H. (1965a). Sequential tests for the mean of a Normal distribution III (small t). *The Annals of Mathematical Statistics*, 36:28–54.

Chernoff, H. (1965b). Sequential tests for the mean of a Normal distribution IV (discrete case). *The Annals of Mathematical Statistics*, 36:55–68.

Chernoff, H. (1972). Sequential analysis and optimal design. *Eighth Regional Conference Series in Applied Mathematics*. Society for Industrial and Applied Mathematics, Philadelphia, PA.

Chernoff, H. (1973). The use of faces to represent points in k-dimensional space graphically. *Journal of the American Statistical Association*, 68:361–368.

Chernoff, H., Lo, S.H., and Zheng, T. (2009). Discovering influential variables: A method of partitions. *The Annals of Applied Statistics*, 3:1335–1369.

Chernoff, H. and Savage, I.R. (1958). Asymptotic normality and efficiency of certain non-parametric test statistics. *The Annals of Mathematical Statistics*, 29:972–994.

Cramér, H. (1938). Sur un nouveau théorème-limite de la théorie des probabilités. *Actualités scientifiques et industrielles*, F36, Paris, France.

Mann, H.B. and Wald, A. (1943). On stochastic limit and order relationships. *The Annals of Mathematical Statistics*, 14:217–226.

Neyman, J. and Pearson, E.S. (1933). On the problem of the most efficient tests of statistical hypotheses. *Philosophical Transactions of the Royal Society*, 231:289–337.

Schwarz, G. (1962). Asymptotic shapes of Bayes sequential testing regions. *The Annals of Mathematical Statistics*, 33:224–236.

Wald, A. (1939). Contributions to the theory of statistical estimation and testing of hypotheses. *The Annals of Mathematical Statistics*, 10:299–326.

4

"... how wonderful the field of statistics is..."

David R. Brillinger
Department of Statistics
University of California, Berkeley, CA

4.1 Introduction

There are two purposes for this chapter. The first is to remind/introduce readers to some of the important statistical contributions and attitudes of the great American scientist John W. Tukey. The second is to take note of the fact that statistics commencement speeches are important elements in the communication of statistical lore and advice and not many seem to end up in the statistical literature. One that did was Leo Breiman's 1994.[1] It was titled "What is the Statistics Department 25 years from now?" Another is Tukey's[2] presentation to his New Bedford high school. There has been at least one article on how to prepare such talks.[3]

Given the flexibility of this COPSS volume, in particular its encouragement of personal material, I provide a speech from last year. It is not claimed to be wonderful, rather to be one of a genre. The speech below was delivered June 16, 2012 for the Statistics Department Commencement at the University of California in Los Angeles (UCLA) upon the invitation of the Department Chair, Rick Schoenberg. The audience consisted of young people, their relatives and friends. They numbered perhaps 500. The event was outdoors on a beautiful sunny day.

The title and topic[4] were chosen with the goal of setting before young statisticians and others interested the fact that America had produced a great scientist who was a statistician, John W. Tukey. Amongst other things he created the field Exploratory Data Analysis (EDA). He gave the American statistical community prestige, and defined much of their work for years.

Further details on specific remarks are provided in a Notes section. The notes are indexed by superscripts at their locations. A brief bibliography is also provided.

4.2 The speech (edited some)

I thank Rick for inviting me, and I also thank the young lady who cheered when my name was announced. She has helped me get started. It is so very nice to see Rick and the other UCLA faculty that I have known through my academic years.

Part of my time I am a sports statistician, and today I take special note of the Kings[5] winning the Stanley Cup[6] five days ago. I congratulate you all for surely your enthusiasm energized your team. I remark that for many years I have had a life-size poster of Wayne Gretzky,[7] wearing a Kings uniform, in my Berkeley office,[8] Rick would have seen it numerous times. All of you can enjoy this victory. I can tell you that I am still enjoying my Leafs[9] victory although there has been a drought since.

Rick asked me to talk about "how wonderful the field of statistics is." No problem. I welcome the opportunity. I have forever loved my career as a statistical scientist, and in truth don't understand why every person doesn't wish to be a statistician,[10] but there is that look. I mean the look one receives when someone asks what you do, and you say "statistics." As an example I mention that a previous University of California President once told me at a reception, seemingly proudly, that statistics had been the one course he had failed in his years at the University of California. Hmmh.

My talk this afternoon will provide a number of quotations associated with a great American scientist, Rick's statistical grandfather,

John Wilder Tukey (1915–2000)

To begin, Rick, perhaps you know this already, but in case not, I mention that you owe John Tukey for your having an Erdős number of 4.[11]

Mr. Tukey had a number of aliases including: John Tukey, Mr. Tukey, Dr. Tukey, Professor Tukey, JWT, and my favorite — The Tuke. The Tuke was born June 16, 1915 in New Bedford, Massachusetts, and in some ways he never left. He was a proud New Englander, he ate apple pie for breakfast, and he bought his Princeton house putting cash on the barrelhead.

Dr. Tukey was a unique individual during his childhood, as a professor, as an advisor, as an executive, and as a consultant. He learned to read at a very young age and was home schooled through high school. His higher education included Bachelor's and Master's degrees in Chemistry from Brown University in 1936 and 1937, followed by a Master's and a Doctorate in Mathematics from Princeton in 1938 and 1939.

He went on to be Higgins Professor at Princeton and Associate Executive Director of Research Information Sciences at Bell Telephone Laboratories. As a graduate student at Princeton he drew attention by serving milk instead of the usual beer at his doctoral graduation party.[12]

John Tukey was quick, like Richard Feynman.[13] He could keep track of time while reciting poetry and seemingly do three different things simultaneously. I watched him continually, I guess, because I had never seen anyone quite like him before. He was called upon continually to provide advice to presidents and other decision makers. He created words and phrases like: bit, software, saphe cracking, the jackknife and his marvelous creation, EDA.[14] He delighted in vague concepts, things that could be made specific in several ways, but were often better left vague. He worked in many fields including: astronomy, cryptography, psephology, information retrieval, engineering, computing, education, psychology, chemistry, pollution control, and economics.

John Tukey was firmly associated with Princeton and Bell Labs.[15] Moreover, he had associations with UCLA. For example, I can mention his friendship and respect for Will Dixon. Will started your Biostat/Biomath group here in 1950 and had been John's colleague at the Fire Control Research Office (FCRO)[16] in World War II.

John had the respect of scientists and executives. The Princeton physicist John Wheeler[17] wrote:

> "I believe that the whole country — scientifically, industrially, financially — is better off because of him and bears evidence of his influence. [· · ·] John Tukey, like John von Neumann, was a bouncy and beefy extrovert, with interests and skills in physics as well as mathematics."

A former President of Bell Labs, W.O. Baker[18] said in response to a personal question:

> "John was indeed active in the analysis of the Enigma[19] system and then of course was part of our force in the fifties which did the really historic work on the Soviet codes as well. So he was very effective in that whole operation. [· · ·] John has had an incisive role in each major frontier of telecommunications science and technology: uses of transistors and solid state; digital and computers."

Dr. Tukey was involved in the construction of the von Neumann computer. In particular, A. Burks wrote:

> "John Tukey designed the electronic adding circuit we actually used in the Institute for Advanced Studies Computer. In

> this circuit, each binary adder fed its carry output directly into the next stage without delay."

John Tukey was renowned for pungent sayings.

> "The best thing about being a statistician," he once told a colleague, "is that you get to play in everyone's backyard."

> "The collective noun for a group of statisticians is a quarrel."

> "Perhaps because I began in a hard but usually non-deductive science — chemistry — and was prepared to learn 'facts' rather than 'proofs', I have found it easier than most to escape the mathematician's implicit claim that the only real sciences are the deductive ones."

> "Doing statistics is like doing crosswords except that one cannot know for sure whether one has found the solution."

> "A consultant is a man who thinks with other people's brains."

> "The stronger the qualitative understanding the *data analyst* can get of the subject matter field from which his data come, the better — just so long as he does not take it too seriously."

> "Most statisticians are used to winning arguments with subject-matter colleagues because they know *both* statistics and the subject matter."

> "The first task of the analyst of data is quantitative detective work."

> "Well, what I think you need is folk dancing."[20]

Tukey had a quick wit. For example the seismologist Bruce Bolt and I developed a method to estimate certain Earth parameters following a great earthquake. I half-boasted to John, that with the next one Bruce and I would be in the morning papers with estimates of the parameters and their uncertainties. John's response was,

> "What if it is in Berkeley?"

Indeed.

Tukey wrote many important books, and many papers. A selection of the latter may be found in his Collected Works.[21]

Some advice for the students

Learn the theory for the theory becomes the practice.

Learn the math because that is the hard part of the other sciences.

In consulting contexts ask, 'What is the question?' Ask it again, and again, and...

Answer a question with, 'It depends,' followed by saying what it depends upon.

Be lucky, remembering that you make your luck.

Don't forget that statisticians are the free-est of all scientists — they can work on anything. Take advantage.

Closing words

Congratulations graduates.

May your careers be wonderful and may they emulate John Tukey's in important ways.

Thank you for your attention.

4.3 Conclusion

In my academic lifetime, statistical time series work went from the real-valued discrete time stationary case, to the vector-valued case, to the nonstationary case, to the point process case, to the spatial case, to the spatial-temporal case, to the generalized function case, to the function-valued time parameter case. It proved important that robust/resistant variants[22] followed such cases.

In summary there has been a steady progression of generalization and abstraction in modeling and data analysis of random processes. Learning the mathematics and continuing this progression is the challenge for the future.

For more details on John Tukey's life, see Brillinger (2002a) and Brillinger (2002b). This work was partially supported by the NSF Grant DMS–100707157.

Notes

1. www.stat.berkeley.edu/~dpurdy/Breiman-1994-commencement.html
2. See p. 306 in Anscombe (2003).
3. See Rodriguez (2012).
4. These words come from an email of Rick's describing his wishes for the talk.
5. The Kings are the National Hockey League (NHL) team in Los Angeles. They won the Stanley Cup in 2012.
6. The Stanley Cup is the trophy awarded to the NHL championship team each year.
7. Wayne Gretzky is a renowned Canadian hockey player holding many NHL records.
8. Room 417 in Evans Hall on the UCB campus.
9. The NHL team based in Toronto, Canada, where I grew up.
10. In Sacks and Ylvisaker (2012) one reads, "But seriously why would one choose to be something other than a statistician?"
11. A mathematician's Erdős number provides the "collaborative distance" from that person to Paul Erdős.
12. The famous mathematician John von Neumann is reputed to have said, "There is this very bright graduate student, and the remarkable thing is that he does it all on milk."
13. Richard Feynman was an American physicist known for his work in the theoretical areas. With Julian Schwinger and Sin-Itiro Tomonaga, he received the Nobel Prize in Physics in 1965.
14. Exploratory data analysis (EDA): 1. It is an attitude; *and* 2. A flexibility; *and* 3. Some graph paper (or transparencies, or both). See Tukey (1965).
15. Bell Labs was an institution sponsored by AT&T. It was the birthplace for many scientific and development advances.
16. The Fire Control Research Office (FRCO) located in Princeton during the Second World War.
17. John Archibald Wheeler (1911–2008) was an American theoretical physicist, and colleague of Tukey at Princeton. He worked in general relativity.

18. Personal communication from W.O. Baker.

19. Enigma was an important code employed by the Germans in World War II.

20. Personal communication from Leo Goodman.

21. See Cleveland (1984–1994). There are eight volumes spread over the years 1984–1994.

22. Robust refers to quantities not strongly affected by non-normality, and resistant refers to those not strongly affected by outliers.

References

Anscombe, F.R. (2003). The civic career and times of John W. Tukey. *Statistical Science*, 18:287–360.

Brillinger, D.R. (2002a). John W. Tukey: His life and professional contributions. *The Annals of Statistics*, 30:1535–1575.

Brillinger, D.R. (2002b). John W. Tukey's work on time series and spectrum analysis. *The Annals of Statistics*, 30:1595–1618.

Cleveland, W.S., Ed. (1984–1994). *The Collected Works of John W. Tukey.* Eight volumes, Wadsworth Publishing, Belmont, CA.

Rodriguez, R.W. (2012). Graduation time: Is your commencement speech ready? *Amstat News*, No. 419 (May), pp. 3–4.

Sacks, J. and Ylvisaker, D. (2012). After 50+ years in statistics, an exchange. *Statistical Science*, 27:308–318.

Tukey, J.W. (1965). We need both exploratory and confirmatory. *The American Statistician*, 34:23–25.

5

An unorthodox journey to statistics: Equity issues, remarks on multiplicity

Juliet Popper Shaffer
Department of Statistics
University of California, Berkeley, CA

The progression to my statistics career was anomalous, and not to be recommended to anyone interested initially in statistics. A fuller account of my earlier studies, as well as information on my childhood, can be found in an interview I gave to Robinson (2005) as former Editor of the *Journal of Educational* (now *Educational and Behavioral*) *Statistics* (JEBS). It is available for download on JSTOR and probably through many academic libraries.

In this paper I will recount briefly some pre-statistical career choices, describe a rather unorthodox way of becoming a statistician, introduce my major area, multiplicity, and note briefly some of my work in it, and make some general remarks about issues in multiplicity. I'll discuss the more technical issues without assuming any background in the subject. Except for a few recent papers, references will be only to some basic literature on the issues and not to the recent, often voluminous literature.

5.1 Pre-statistical career choices

At about 14 years of age I read a remarkably inspiring book, "Microbe Hunters," by Paul de Kruif (1926), and decided immediately to be a scientist. Since then, several other scientists have noted a similar experience with that book.

In addition to an interest in science, mathematics was always attractive to me. However, I thought of it wrongly as something very remote from the real world, like doing crossword puzzles, and wanted to be more engaged with that world.

My high school courses included a year each of beginning biology, chemistry, and physics. I wanted to take the four years of mathematics available, but that turned out to be a problem. At that time in my Brooklyn public high school, and probably at the other public high schools in New York City, boys were automatically enrolled in mathematics in the first semester of 9th grade, and girls in a language of their choice. That left me with only 3 1/2 semesters, and course availability made it impossible to take more than one mathematics class at a time.

This was just one example of the stereotyping, not to speak of outright discrimination, against women in those years, especially in mathematics and science. In fact, Brooklyn Technical High School, specializing in the science–technology–engineering–mathematics (STEM) area, was restricted to boys at that time. (It became co-ed in 1972.) My interview in JEBS discusses several other such experiences.

I solved the problem by taking intermediate algebra as an individual reading course, meeting once a week with a teacher and having problems assigned, along with a geometry course. In that way I managed to take all four years offered.

I started college (Swarthmore College) as a Chemistry major, but began to consider other possibilities after the first year. Introductory psychology was chosen in my second year to satisfy a distribution requirement. In the introductory lecture, the professor presented psychology as a rigorous science of behavior, both animal and human. Although some of my fellow students found the lecture boring, I was fascinated. After a brief consideration of switching to a pre-med curriculum, I changed my major to psychology.

In graduate school at Stanford University I received a doctorate in psychology. I enjoyed my psychological statistics course, so took an outside concentration in statistics with several courses in the mathematics and statistics departments.

There was then much discrimination against women in the academic job world. During the last year at Stanford, I subscribed to the American Psychological Association's monthly *Employment Bulletin.* Approximately half the advertised jobs said "Men only." Of course, that was before overt sex discrimination was outlawed in the Civil Rights act of 1964. After an NSF Fellowship, used for a postdoctoral year working in mathematical psychology at Indiana University with one of the major contributors to that field (William Estes), I got a position in the Department of Psychology at the University of Kansas, thankfully one of the more enlightened departments.

5.2 Becoming a statistician

I taught part-time during several years while my three children were small. There were no special programs for this at the time, but fortunately my department allowed it. There was no sabbatical credit for part time, but finally, with enough full-time years to be eligible, I decided to use the sabbatical year (1973–74) to improve my statistics background. I chose the University of California (UC) Berkeley, because I had been using a book, "Testing Statistical Hypotheses," by E.L. Lehmann, as background for my statistics teaching in the Psychology Department.

As life sometimes evolves, during that year Erich Lehmann and I decided to marry. After a one-year return to Kansas, and a one-year visiting appointment at UC Davis in the Mathematics Department, I joined the Statistics Department at UC Berkeley as a lecturer, later becoming a senior lecturer. Although I had no degree in statistics, my extensive consulting in the Psychology Department at Kansas gave me greater applied statistical experience than most of the Statistics faculty at UC Berkeley, and I supervised a Berkeley Statistics Department consulting service for many years. We probably had about 2000 clients during that time, mostly graduate students, but some faculty, retired faculty, and even outside individuals of all kinds. One of the most interesting and amusing contacts was with a graduate student studying caterpillar behavior. The challenge for us was that when groups of caterpillars were being observed, it was not possible to identify the individuals, so counts of behaviors couldn't be allocated individually. He came to many of our meetings, and at the end invited all of us to a dinner he was giving in his large co-op, giving us a very fancy French menu. Can you guess what kind of a menu it was? Much to my regret, I didn't have the courage to go, and none of the consultants attended either.

During this early time in the department, I was also the editor of JEBS (see above) for four years, and taught two courses in the Graduate School of Education at Berkeley.

It's interesting to compare teaching statistics to psychologists and teaching it to statisticians. Of course the level of mathematical background was far greater among the statisticians. But the psychologists had one feature that statisticians, especially those going into applied work, would find valuable. Psychological research is difficult because the nature of the field makes it possible to have many alternative explanations, and methods often have defects that are not immediately obvious. As an example of the latter, I once read a study that purported to show that if shocks were given to a subject while a particular word was being read, the physiological reactions would generalize to other words with similar meanings. As a way of creating time between the original shock and the later tests on alternative words, both with similar and dissimilar meanings, subjects were told to sit back and think of other things. On thinking about this study, it occurred to me that subjects that had just been shocked on a particular word could well be thinking about other words with similar meanings in the interim, thus bringing thoughts of those words close to the time of the shock, and not separated from it as the experimenters assumed.

Thus psychologists, as part of their training, learn to think deeply about such alternative possibilities. One thing that psychologists know well is that individual differences are so important that it is essential to distinguish between multiple methods applied to the same groups of individuals and methods applied to different independent groups. I found that the statisticians were less sensitive to this than the psychologists. In the consulting classes, they sometimes forgot to even ask about that. Also, they sometimes seemed to think that with the addition of a few variables (e.g., gender, ethnicity, other distinguishing variables) they could take care of individual differences, and treat the data as conditionally independent observations, whereas psychologists would be quite wary of making that assumption. There must be other fields in which careful experimental thinking is necessary. This is not statistics in a narrow sense, but is certainly important for applied statisticians who may be involved in designing studies.

5.3 Introduction to and work in multiplicity

In addition to teaching many psychology courses during my time at Kansas, I also taught most of the statistics courses to the psychology undergraduate and graduate students. Analysis of variance (ANOVA) was perhaps the most widely-used procedure in experimental psychology. Consider, for example, a one-way treatment layout to be analyzed as an ANOVA. Given a significant F statistic, students would then compare every treatment with every other to see which were different using methods with a fixed, conventional Type I error rate α (usually .05) for each. I realized that the probability of some false conclusions among these comparisons would be well above this nominal Type I error level, growing with the number of such comparisons. This piqued my interest in multiplicity problems, which eventually became my major area of research.

The criterion most widely considered at that time was the family-wise error rate (FWER), the probability of one or more false rejections (i.e., rejections of true hypotheses) in a set of tests. If tests are carried out individually with specified maximum (Type I) error rates, the probability of one or more errors increases with the number of tests. Thus, the error rate for the whole set should be considered. The statistical papers I read all referred to an unpublished manuscript, "The Problem of Multiple Comparisons," by John W. Tukey (1953). In those days, before `Xerox`, it was impossible to get copies of that manuscript. It was frustrating to have to use secondary sources. Fortunately, with the advent of `Xerox`, that problem has disappeared, and now, in addition, the manuscript is included in Tukey's Collected Works (Braun, 1994).

Tukey's treatment was extremely insightful and organized the field for some time to follow. In his introduction, he notes that he should have published it as a book at that time but "One reason this did not happen was the only piece of bad advice I ever had from Walter Shewhart! He told me it was unwise to put a book out until one was sure that it contained the last word of one's thinking."

My earlier work in multiple comparisons is described in the interview I gave to Robinson (2005). Since the time of Tukey's manuscript, a suggested alternative measure of error to be controlled is the False Discovery Rate (FDR), introduced by Benjamini and Hochbert (1995). This is the expected proportion of false rejections, defined as zero if there are no false rejections. Controlling FDR implies that if there are a reasonable number of true rejections a small proportion of false ones is tolerable. John Tukey himself (personal remark) was enthusiastic about this new approach for some applications.

One of my early pieces of work that started a line of research by others arose from an interest in directional inference. When one tests a hypothesis of the form $\theta = \theta_0$, where θ_0 is a specific value, it is usually of interest, if the hypothesis is rejected, to decide either that $\theta > \theta_0$ or $\theta < \theta_0$, although strictly speaking the permitted alternative is $\theta \neq \theta_0$. In fact, most researchers automatically conclude that the direction of sample departure from θ_0 is the correct direction. Do these more detailed inferences lead to the probability of errors (often called Type III errors) beyond the acceptable bounds? In some methods, the sample outcome can be reformulated as a confidence interval about a sample value of θ, in which case, if θ_0 is not included in the interval, the usual directional inference is acceptable. However, in some more complex methods for testing such hypotheses, the results cannot be so reformulated. Although it seemed intuitively clear that the usual directional inference would still be valid, it turned out that I couldn't prove it, and in fact ended up proving it was not necessarily true. This began a whole line of studies. A paper by Finner (1999) summarized what was known up to that point. There is some extended work on directional inference since then (see, e.g., Guo et al. 2010), but there are still unsolved problems in this area.

Another early line of work that has had many applications is the consideration of logical as well as statistical restrictions, in a way that can strengthen procedures by eliminating some steps, thereby increasing power. An example of some of the more recent work in this area is Westfall and Tobias (2007). Both the above lines of research are discussed in my JEBS paper, so won't be further explained here.

Recently, I have revisited an earlier research interest, interpretability, with a new approach that is somewhat easier to apply than my earlier formulation. Outcomes are considered relatively more interpretable if the rejected hypotheses of equality result in separation of treatments into nonoverlapping clusters. A recent publication (Shaffer et al., 2013) describes the approach in detail. It deals with both FWER and FDR, as well as the per-family error rate (PFER), the expected number of false rejections, also defined in the Tukey manuscript.

In the context of pairwise comparisons of population means, for each control criterion (FWER or FDR), one can compare methods based on individual p-values, one for each hypothesis, with methods based on considering the range of sample treatment outcomes between the two means being compared. Our work shows that range-based methods lead to a higher proportion of separations than individual p-value methods. In connection with the FDR, it appears that an old test procedure, the Newman–Keuls, which fell out of favor because it did not control FWER, does control FDR. Extensive simulation results support this conclusion; proofs are incomplete. Interpretability is the main issue I'm working on at present.

5.4 General comments on multiplicity

Although championed by the very eminent John Tukey, multiplicity was a backwater of research and the issues were ignored by many researchers. This area has become much more prominent with the recent advent of "Big Data."

Technological advances in recent years, as we know, have made massive amounts of data available bringing the desire to test thousands if not millions of hypotheses; application areas, for example, are genomics, proteomics, neurology, astronomy. It becomes impossible to ignore the multiplicity issues in these cases, and the field has enjoyed a remarkable development within the last 20 years. Much of the development has been applied in the context of big data. The FDR as a criterion is often especially relevant in this context. Many variants of the FDR criterion have been proposed, a number of them in combination with the use of empirical Bayes methods to estimate the proportion of true hypotheses. Resampling methods are also widely used to take dependencies into account. Another recent approach involves consideration of a balance between Type I and Type II errors, often in the context of simultaneous treatment of FDR and some type of false nondiscovery rate.

Yet the problems of multiplicity are just as pressing in small data situations, although often not recognized by practitioners in those areas. According to Young (2009), many epidemiologists feel they don't have to take multiplicity into account. Young and others claim that the great majority of apparent results in these fields are Type I errors; see Ioannidis (2005). Many of the newer approaches can't be applied satisfactorily in small data problems.

The examples cited above — one-way ANOVA designs and the large data problems noted — are what might be called well-structured testing problems. In general, there is a single set of hypotheses to be treated uniformly in testing, although there are variations. Most methodological research applies in this context. However, there have always been data problems of a very different kind, which might be referred to as ill-structured. These are cases in which there are hypotheses of different types, and often different importance, and it isn't clear how to structure them into families, each of which would be treated with a nominal error control measure.

A simple example is the division into primary and secondary outcomes in clinical research. If the primary outcomes are of major importance, how should that be taken into account? Should error control at a nominal level, for example the usual .05 level, be set separately for each set of outcomes? Should there be a single α level for the whole set, but with different weights on the two different types of outcomes? Should the analysis be treated as hierarchical, with secondary outcomes tested only if one (or more) of the primary outcomes shows significant effects?

A more complex example is the analysis of a multifactor study by ANOVA. The standard analysis considers the main effect of each factor and the interactions of all factors. Should the whole study be evaluated at the single nominal α level? That seems unwise. Should each main effect be evaluated at that level? How should interactions be treated? Some researchers feel that if there is an interaction, main effects shouldn't be further analyzed. But suppose one high-order interaction is significant at the nominal α level. Does that mean the main-effect tests of the factors involved aren't meaningful?

Beyond these analyses, if an effect is assumed to be significant, how should the ensuing more detailed analysis (e.g., pairwise comparisons of treatments) be handled, considering the multiplicity issues? There is little literature on this subject, which is clearly very difficult. Westfall and Young (1993, Chapter 7) give examples of such studies and the problems they raise.

Finally, one of the most complex situations is encountered in a large survey, where there are multiple factors of different types, multiple subgroups, perhaps longitudinal comparisons. An example is the National Assessment of Educational Progress (NAEP), now carried out yearly, with many educational subjects, many subgroups (gender, race-ethnicity, geographical area, socio-economic status, etc.), and longitudinal comparisons in all these.

A crucial element in all such ill-structured problems, as noted, is the definitions of families for which error control is desired. In my two years as director of the psychometric and statistical analysis of NAEP at Educational Testing Service, we had more meetings on this subject, trying to decide on family definitions and handling of interactions, than any other. Two examples of difficult problems we faced:

(a) Long term trend analyses were carried out by using the same test at different time points. For example, nine-year-olds were tested in mathematics with an identical test given nine times from 1978 to 2004. At first it was planned to compare each time point with the previous one. In 1982, when the second test was given, there was only one comparison. In 2004, there were eight comparisons (time 2 with time 1, time 3 with time 2, etc.). Treating the whole set of comparisons at any one time as a family, the family size increased with the addition of each new testing time. Thus, to control the FWER, each pairwise test had to reach a stricter level of significance in subsequent analyses. But it would obviously be confusing to call a change significant at one time only to have it declared not significant at a later time point.

(b) Most states, as well as some large cities, take part in state surveys, giving educational information for those units separately. Suppose one wants to compare every unit with every other. With, say, 45 units, the number of comparisons in the family is $45 \times 44/2$, or 990. On the other hand, people in a single unit (e.g., state) are usually interested mainly in how it compares with other units; these comparisons result in a family size of 44. Results are likely to differ. How can one reconcile the different decisions, when results must be transmitted to a public that thinks that there should be a single decision for each comparison?

Extensive background material and results for NAEP are available at nces.ed.gov/nationsreportcard. In addition, there is a book describing the development of the survey (Jones and Olkin, 2004). For information specifically on handling of multiplicity issues, see `nces.ed.gov/nationsreportcard/tdw/analysis/2000\_2001/infer\_multiplecompare\_fdr.asp`.

In summary, the work on multiplicity has multiplied with the advent of big data, although the ill-structured situations described above have been around for a long time with little formal attention, and more guidance on handling of family size issues with examples would be a contribution that could result in wider use of multiple comparison methods.

References

Benjamini, Y. and Hochberg, Y. (1995). Controlling the false discovery rate: A practical and powerful approach to multiple testing. *Journal of the Royal Statistical Society, Series B*, 57:289–300.

Braun, H., Ed. (1994). *The Collected Works of John W. Tukey.* Vol. 8: *Multiple Comparisons, 1948–1983.* Chapman & Hall, London.

de Kruif, P. (1926). *Microbe Hunters.* Harcourt, Brace, Jovanovich, New York.

Finner, H. (1999). Stepwise multiple test procedures and control of directional errors. *The Annals of Statistics*, 27:274–289.

Guo, W., Sarkar, S.K., and Peddada, S.D. (2010). Controlling false discoveries in multidimensional directional decisions, with applications to gene expression data on ordered categories. *Biometrics*, 66:485–492.

Ioannidis, J.P.A. (2005). Why most published research findings are false. *PLoS Medicine*, 2(8):e124.

Jones, L.V. and Olkin, I., Eds. (2004). *The Nation's Report Card: Evolution and Perspectives.* Phi Delta Kappa Educational Foundation, Bloomington, IN.

Robinson, D. (2005). Profiles in research: Juliet Popper Shaffer. *Journal of Educational and Behavioral Statistics*, 30:93–103.

Shaffer, J.P., Kowalchuk, R.K., and Keselman, H.J. (2013). Error, power, and cluster-separation rates of pairwise multiple testing procedures. *Psychological Methods*, 18:352–367.

Tukey, J.W. (1953). The problem of multiple comparisons. In *The Collected Works of John W. Tukey*, Vol. 8: *Multiple Comparisons, 1948–1983* (H. Braun, Ed.). Chapman & Hall, London.

Westfall, P.H. and Tobias, R.D. (2007). Multiple testing of general contrasts: Truncated closure and the extended Shaffer–Royen method. *Journal of the American Statistical Association*, 102:487–494.

Westfall, P.H. and Young, S.S. (1993). *Resampling-based Multiple Testing: Examples and Methods for p-value Adjustment.* Wiley, New York.

Young, S.S. (2009). Health findings and false discoveries. Paper presented in the symposium "False Discoveries and Statistics: Implications for Health and the Environment," at the 2009 Annual Meeting at the American Association for the Advancement of Science.

6

Statistics before and after my COPSS Prize

Peter J. Bickel
Department of Statistics
University of California, Berkeley, CA

This is largely an account of my research career, the development of different interests as I moved along, and the influences, people and ideas that determined them. It also gives an idiosyncratic view of the development of the field and ends with some words of advice.

6.1 Introduction

I was fortunate enough to be young enough to receive the first COPSS prize in 1979. It was a fairly rushed affair. I flew back from France where I was giving some lectures, mumbled that I felt like the robber in a cops and robbers drama since I didn't feel I had done enough to deserve the prize, and then returned to France the next day.

This is partly the story of my life before and after the prize and my contributions, such as they were, but, more significantly, it describes my views on the changes in the main trends in the field that occurred during the last 30+ years. In addition, given my age of 72, I can't resist giving advice.

6.2 The foundation of mathematical statistics

During the period 1940 to 1979 an impressive theoretical edifice had been built on the foundations laid by Fisher, Pearson and Neyman up to the 1940s and then built up by Wald, Wolfowitz, LeCam, Stein, Chernoff, Hodges and Lehmann, and Kiefer, among others, on the frequentist side and by L.J. Savage on the Bayesian side, with Herbert Robbins flitting in between. There were, of course, other important ideas coming out of the work of people such as

I.J. Good, H.O. Hartley, J.W. Tukey and others, some of which will come up later. The focus of most of these writers was on classical inference applied to fairly standard situations, such as linear models, survey sampling and designed experiments as explored in the decision-theoretic framework of Wald, which gave a clear infrastructure for both frequentist and Bayesian analysis. This included the novel methodology and approach of sequential analysis, introduced by Wald in the 1950s, and the behavior of rank-based nonparametric tests and estimates based on them as developed by Hodges and Lehmann. Both of these developments were pushed by World War II work. Robustness considerations were brought to the fore through Tukey's influential 1958 paper "The Future of Data Analysis" and then the seminal work of Hampel and Huber.

6.3 My work before 1979

My thesis with Erich Lehmann at the University of California (UC) Berkeley was on a robust analogue of Hotelling's T^2 test and related estimates. I then embarked on a number of contributions to many of the topics mentioned, including more robustness theory, Bayesian sequential analysis, curve estimation, asymptotic analysis of multivariate goodness-of-fit tests, and the second-order behavior of rank test power and U-statistics. Several of the papers arose from questions posed by friends and colleagues. Thus, some general results on asymptotic theory for sequential procedures as the cost of observation tended to zero was prompted by Yossi Yahav. The second-order analysis of nonparametric tests grew out of a question asked by Hodges and Lehmann in their fundamental paper published in 1970 (Hodges and Lehmann, 1970). The question was picked up independently by van Zwet and myself and we then decided to make common cause. The resulting work led to the development of second-order theory for U-statistics by van Zwet, Götze and myself; see Bickel (1974) and subsequent papers. The work on curve estimation originated from a question posed by Murray Rosenblatt.

I even made an applied contribution in 1976 as a collaborator in the analysis of "Sex bias in graduate admission at Berkeley" which appeared in *Science* (Bickel et al., 1975). Thanks to my colleague David Freedman's brilliant textbook *Statistics*, it garnered more citations than all my other work. This was initiated by a question from Gene Hammel, Professor of Anthropology and then Associate Dean of the Graduate Division.

Two opportunities to work outside of UC Berkeley had a major impact on my research interests, the second more than the first, initially.

6.3.1 Imperial College

In 1965–66 I took a leave of absence from UC Berkeley and spent a year at Imperial College, London during the last year that George Barnard held the Statistics Chair. My job was mainly tutoring undergraduates in their classes in probability theory, but I became acquainted with John Copas in addition to Barnard at Imperial and with some of the other British statisticians and probabilists, including David Cox, Henry Daniels, Allan Stuart, Jim Durbin, the two very different Kendalls, David and Maurice, and others. The exposure to data analysis, after the theory-steeped Berkeley environment, was somewhat unsettling but I wonder if, at least subconsciously, it enabled me to deal sensibly with the sex bias in graduate admissions questions.

I returned to Imperial in 1975–76 during David Cox's tenure as chair. I interacted with an active group of young people including Tony Atkinson, Agnes Herzberg, Ann Mitchell, and Adrian Smith. Agnes and I initiated some work on robust experimental design but my knowledge of martingales was too minimal for me to pursue David Cox's suggestion to further explore the Cox survival analysis model. This again was an opportunity missed, but by then I was already involved in so many collaborations that I felt unready to take on an entirely new area.

6.3.2 Princeton

My second exposure to a very different environment came in 1970–71 when, on a Guggenheim Fellowship, I joined David Andrews, Peter Huber, Frank Hampel, Bill Rogers, and John Tukey during the Princeton Robustness Year. There I was exposed for the first time to a major computational simulation effort in which a large number of estimates of location were compared on a large number of possible distributions. I found to my pleased surprise that some of my asymptotic theory based ideas, in particular, one-step estimates, really worked. On the other hand, I listened, but didn't pay enough attention, to Tukey. If, for instance, I had followed up on a question of his on the behavior of an iteration of a one-step estimate I had developed to obtain asymptotic analogues of linear combinations of order statistics in regression, I might have preceded at least some of the lovely work of Koenker and Basset on quantile regression. However, unexpected questions such as adaptation came up in talking to Peter Huber and influenced my subsequent work greatly.

The moral of these stories is that it is very important to expose yourself to new work when you're young and starting out, but that their effects may not be felt for a long time if one is, as I am, basically conservative.

6.3.3 Collaboration and students

Most of my best work in the pre-1979 period was with collaborators who often deserve much more credit than I do, including Yossi Yahav, W.R. van Zwet,

M. Rosenblatt, my advisor and friend, E.L. Lehmann, and his favorite collaborator who brought me into statistics, J.L. Hodges, Jr. This trend continued with additional characters throughout my career. I should note that I have an unfair advantage over these friends since, with 'B' starting my name, I am almost always listed first in the list of authors, though at that time much less fuss was made about this in the mathematical sciences than now.

I acquired PhD students rather quickly, partly for selfish reasons. I always found that I could think more clearly and quickly in conversation than in single-handedly batting my head against a brick wall. More significantly, I like to interact with different minds whose foci and manner of operation are quite different from mine and whose knowledge in various directions is broader and deeper.

Thus my knowledge of invariance principles, concentration inequalities and the like which led to the work on distribution-free multivariate tests came in part out of working with Hira Koul on confidence regions for multivariate location based on rank tests.

There are, of course, students who have had a profound effect on my research directions. Some of them became lifelong collaborators. I will name six in advance. Their roles will become apparent later. There are others, such as Jianqing Fan and Jeff Wu, who have played and are playing very important roles in the field but whose interests have only occasionally meshed with mine after their doctorates, though I still hope for more collaborations with them, too.

(a) Ya'acov Ritov (PhD in Statistics from Hebrew University, Jerusalem, supervised during a 1979–80 sabbatical)

(b) Elizaveta Levina (PhD in Statistics, UC Berkeley, 2002)

(c) Katerina Kechris (PhD in Statistics, UC Berkeley, 2003)

(d) Aiyou Chen (PhD in Statistics, UC Berkeley, 2004)

(e) Bo Li (PhD in Statistics, UC Berkeley, 2006)

(f) James (Ben) Brown (PhD in Applied Science and Technology, College of Engineering, UC Berkeley, 2008).

6.4 My work after 1979

In most fields the amount, types and complexity of data have increased on an unprecedented scale. This, of course, originated from the increasing impact of computers and the development of refined sensing equipment. The rise in computing capabilities also increased greatly the types of analysis we could make. This was first perceived in statistics by Brad Efron, who introduced Monte Carlo in the service of inference by inventing the bootstrap. David Freedman and I produced some of the first papers validating the use of the bootstrap in a general context.

6.4.1 Semiparametric models

In the 1980s new types of data, arising mainly from complex clinical trials, but also astronomy and econometrics, began to appear. These were called semiparametric because they needed both finite- and infinite-dimensional parameters for adequate description.

Semiparametric models had been around for some time in the form of classical location and regression models as well as in survival analysis and quality control, survey sampling, economics, and to some extent, astronomy. There had been theoretical treatments of various aspects by Ibragimov and Khasminskii, Pfanzagl, and Lucien LeCam at a high level of generality. The key idea for their analysis was due to Charles Stein. Chris Klaassen, Ya'acov Ritov, Jon Wellner and I were able to present a unified viewpoint on these models, make a key connection to robustness, and develop methods both for semiparametric performance lower bounds and actual estimation. Our work, of which I was and am still proud, was published in book form in 1993. Much development has gone on since then through the efforts of some of my coauthors and others such as Aad van der Vaart and Jamie Robins. I worked with Ritov on various aspects of semiparametrics throughout the years, and mention some of that work below.

6.4.2 Nonparametric estimation of functions

In order to achieve the semiparametric lower bounds that we derived it became clear that restrictions had to be placed on the class of infinite dimensional "nuisance parameters." In fact, Ritov and I were able to show in a particular situation, the estimation of the integral of the square of a density, that even though the formal lower bounds could be calculated for all densities, efficient estimation of this parameter was possible if and only if the density obeyed a Lipschitz condition of order larger than $1/4$ and $\sqrt{n}$ estimation was possible if and only if the condition had an exponent greater than or equal to $1/4$.

In the mid-'80s and '90s David Donoho, Iain Johnstone and their collaborators introduced wavelets to function estimation in statistics; see, e.g., Donoho et al. (1996). With this motivation they then exploited the Gaussian white noise model. This is a generalization of the canonical Gaussian linear model, introduced by Ibragimov and Khasminskii, in which one could quantitatively study minimax analysis of estimation in complex function spaces whose definition qualitatively mimics properties of functions encountered in the real world. Their analysis led rather naturally to regularization by thresholding, a technique which had appeared in some work of mine on procedures which could work well in the presence of bias (Bickel, 1981) although I was far from putting this work in its proper context. Their work in this area, earlier work on detecting sparse objects (Donoho et al., 1992), and earlier work of Stone (1977) made it apparent that, without much knowledge of a statistical problem, minimax bounds indicated that nothing much could be achieved in even moderate dimensions.

On the other hand, a branch of computer science, machine learning, had developed methodology such as neural nets, and on the statistics side, Leo Breiman and Jerry Friedman, working with Richard Olshen and Charles Stone, developed CART. Both of these methods of classification use very high dimensional predictors and, relative to the number of predictors, small training sets. These methods worked remarkably well, far better than the minimax theory would lead us to believe. These approaches and a plethora of other methods developed in the two communities, such as Boosting, Random Forests, and above all "lasso" driven methods involve, implicitly or explicitly, "regularization," which pulls solutions of high dimensional optimization problems towards low dimensional spaces.

In many situations, while we know little about the problem, if we can assume that, in an appropriate representation, only a relatively few major factors matter, then theorists can hope to reconcile the "Curse of Dimensionality" minimax results with the observed success of prediction methods based on very high dimensional predictor sets.

Under the influence of Leo Breiman I became very aware of these developments and started to contribute, for instance, to the theory of boosting in Bickel et al. (2006). I particularly liked a simple observation with Bo Li, growing out of my Rietz lecture (Bickel and Li, 2007). If predictors of dimension p are assumed to lie on an unknown smooth d-dimensional manifold of $\mathbb{R}^p$ with $d \ll p$, then the difficulty of the nonparametric regression problem is governed not by p but by d, provided that regularization is done in a suitably data-determined way; that is, bandwidth selection is done after implicit or explicit estimation of d.

6.4.3 Estimating high dimensional objects

My views on the necessary existence of low dimensional structure were greatly strengthened by working with Elizaveta Levina on her thesis. We worked with Jitendra Malik, a specialist in computer vision, and his students, first in analyzing an algorithm for texture reconstruction developed by his then student, Alexei Effros, and then in developing some algorithms for texture classification. The first problem turned out to be equivalent to a type of spatial bootstrap. The second could be viewed as a classification problem based on samples of 1000+ dimensional vectors (picture patches) where the goal was to classify the picture from which the patches were taken into one of several classes.

We eventually developed methods which were state-of-the-art in the field but the main lesson we drew was that just using patch marginal distributions (a procedure sometimes known as naive Bayes) worked better than trying to estimate joint distributions of patch pixels.

The texture problem was too complex to analyze further, so we turned to a simpler problem in which explicit asymptotic comparisons could be made: classifying a new p-dimensional multivariate observation into one of two unknown Gaussian populations with equal covariance matrices, on the basis of a sample of n observations from each of the two populations (Bickel and Levina, 2004). In this context we compared the performance of

(i) Fisher's linear discriminant function

(ii) Naive Bayes: Replace the empirical covariance matrix in Fisher's function by the diagonal matrix of estimated variances, and proceed as usual.

We found that if the means and covariance matrices range over a sparsely approximable set and we let p increase with n, so that $p/n \to \infty$, then Fisher's rule (using the Moore–Penrose inverse) performed no better than random guessing while naive Bayes performed well, though not optimally, as long as $n^{-1} \log p \to 0$.

The reason for this behavior was that, with Fisher's rule, we were unnecessarily trying to estimate too many covariances. These results led us — Levina and I, with coworkers (Bickel and Levina, 2008) — to study a number of methods for estimating covariance matrices optimally under sparse approximation assumptions. Others, such as Cai et al. (2010), established minimax bounds on possible performance.

At the same time as this work there was a sharp rise of activity in trying to understand sparsity in the linear model with many predictors, and a number of important generalizations of the lasso were proposed and studied, such as the group lasso and the elastic net. I had — despite appearing as first author — at most a supporting part in this endeavor on a paper with Ritov and Tsybakov (Bickel et al., 2009) in which we showed the equivalence of a procedure introduced by Candès and Tao, the Danzig selector, with the more familiar lasso.

Throughout this period I was (and continue to be) interested in semiparametric models and methods. An example I was pleased to work on with my then student, Aiyou Chen, was Independent Component Analysis, a methodology arising in electrical engineering, which had some clear advantages over classical PCA (Chen and Bickel, 2006). Reconciling ICA and an extension with sparsity and high dimension is a challenge I'm addressing with another student.

A more startling and important analysis is one that is joint with Bin Yu, several students, and Noureddine El Karoui (Bean et al., 2013; El Karoui et al., 2013), whose result appears in PNAS. We essentially studied robust regression when $p/n \to c$ for some $c \in (0, 1)$, and showed that, contrary to what is suggested in a famous paper of Huber (Huber, 1973), the asymptotic normality and $1/\sqrt{n}$ regime carries over, with a few very significant exceptions. However, limiting variances acquire a new Gaussian factor. As Huber discovered, contrasts with coefficients dependent on the observed predictors may behave quite differently. Most importantly, the parameters of the Gaussian limits depend intricately on the nature of the design, not simply on the covariance matrix of the predictors. This work brought out that high dimension really presents us with novel paradigms. For p very large, garden variety models exhibit strange properties. For instance, symmetric p-variate Gaussian distributions put almost all of their mass on a thin shell around the border of the sphere of radius $\sqrt{p}$ that is centered at the mean. It was previously noted that for inference to be possible one would like mass to be concentrated on low dimensional structures. But finding these structures and taking the search process into account in inference poses very new challenges.

6.4.4 Networks

My latest interest started around 2008 and is quite possibly my last theoretical area of exploration. It concerns inference for networks, a type of data arising first in the social sciences, but which is now of great interest in many communities, including computer science, physics, mathematics and, last but not least, biology.

Probabilistically, this is the study of random graphs. It was initiated by Erdős and Rényi (1959). If you concentrate on unlabeled graphs, which is so far essentially the only focus of the probability community, it is possible to formulate a nonparametric framework using work of Aldous and Hoover which permits the identification of analogues of i.i.d. variables and hence provides the basis of inference with covariates for appropriate asymptotics (Bickel and Chen, 2009). Unfortunately, fitting even the simplest parametric models by maximum likelihood is an NP-hard problem. Nevertheless, a number of simple fitting methods based on spectral clustering of the adjacency or Laplacian matrices (Rohe et al., 2011; Fishkind et al., 2013), combined with other ideas, seem to work well both in theory and practice. An interesting feature of this theory is that it ties into random matrix theory, an important and very active field in probability and mathematics with links to classical Gaussian multivariate theory which were recently discovered by Iain Johnstone and his students (Johnstone, 2008).

In fact, more complex types of models and methods, all needing covariates, are already being used in an ad hoc way in many applications. So there's lots to do.

6.4.5 Genomics

Following work initiated in 2005 with my then student, Katerina Kechris, and a colleague in molecular biology, Alex Glazer, I began to return to a high school interest, biology. After various missteps I was able to build up a group with a new colleague, Haiyan Huang, supported initially by an NSF/NIGMS grant, working on problems in molecular biology with a group at the Lawrence Berkeley Lab. Through a series of fortunate accidents our group became the only statistics group associated with a major multinational effort, the ENCODE (Encyclopaedia of DNA) project. The end product of this effort, apart from many papers in *Nature*, *Science*, *Genome Research* and the like, was a terabyte of data (Birney et al., 2007).

I was fortunate enough to acquire a student, James (Ben) Brown, from an engineering program at Berkeley, who had both an intense interest in, and knowledge of, genomics and also the critical computational issues that are an integral part of such a collaboration. Through his participation, I, and to a considerable extent, Haiyan, did not need to immerse ourselves fully in the critical experimental issues underlying a sensible data analysis. Ben could translate and pose old and new problems in terms we could understand.

The collaboration went on for more than five years, including a pilot project. During this time Ben obtained his PhD, continued as a postdoc and is now beset with job offers from computational biology groups at LBL and all over. Whether our group's participation in such large scale computational efforts can continue at the current level without the kind of connection to experimentalists provided by Ben will, I hope, not be tested since we all wish to continue to collaborate.

There have been two clearly measurable consequences of our participation.

(a) Our citation count has risen enormously as guaranteed by participation in high-visibility biology journals.

(b) We have developed two statistical methods, the GSC (Genome Structural Correction) and the IDR (the Irreproducible Discovery Rate) which have appeared in *The Annals of Applied Statistics* (Bickel et al., 2010; Li et al., 2011) and, more significantly, were heavily used by the ENCODE consortium.

6.5 Some observations

One of the things that has struck me in writing this is that "old ideas never die" and they may not fade away. Although I have divided my interests into coherent successive stages, in fact, different ideas frequently reappeared.

For instance, second-order theory and early papers in Bayes procedures combined in a paper with J.K. Ghosh in 1990 (Bickel and Ghosh, 1990) which gave what I still view as a neat analysis of a well-known phenomenon called the Bartlett correction. Theoretical work on the behavior of inference in Hidden Markov Models with Ritov and Ryden (Bickel et al., 1998) led to a study which showed how difficult implementation of particle filters is in the atmospheric sciences (Snyder et al., 2008). The work in HMM came up again in the context of traffic forecasting (Bickel et al., 2007) and some work in astrophysics (Meinshausen et al., 2009). Both papers were close collaborations with John Rice and the second included Nicolai Meinshausen as well. The early bootstrap work with Freedman eventually morphed into work on the m out of n bootstrap with Götze and van Zwet (Bickel et al., 1997) and finally into the Genome Structural Correction Method (Bickel et al., 2010).

Another quite unrelated observation is that to succeed in applications one has to work closely with respected practitioners in the field. The main reason for this is that otherwise, statistical (and other mathematical science) contributions are dismissed because they miss what practitioners know is the essential difficulty. A more pedestrian reason is that without the imprimatur and ability to translate of a respected scientist in the field of application, statistical papers will not be accepted in the major journals of the science and hence ignored.

Another observation is that high-order computing skills are necessary to successfully work with scientists on big data. From a theoretical point of view, the utility of procedures requires not only their statistical, but to an equal extent, their computational efficiency. Performance has to be judged through simulations as well as asymptotic approximations.

I freely confess that I have not subscribed to the principle of honing my own computational skills. As a member of an older generation, I rely on younger students and collaborators for help with this. But for people starting their careers it is essential. The greater the facility with computing, in addition to R and including Matlab, C++, Python or their future versions, the better you will succeed as a statistician in most directions.

As I noted before, successful collaboration requires the ability to really understand the issues the scientist faces. This can certainly be facilitated by direct study in the field of application.

And then, at least in my own career, I've found the more mathematics I knew, from probability to functional analysis to discrete mathematics, the better. And it would have been very useful to have learned more information theory, statistical physics, etc., etc.

Of course I'm describing learning beyond what can be done or is desirable in a lifetime. (Perhaps with the exception of John von Neumann!) We all specialize in some way. But I think it's important to keep in mind that statistics should be viewed as broadly as possible and that we should glory in this time when statistical thinking pervades almost every field of endeavor. It is really a lot of fun.

References

Bean, D., Bickel, P.J., El Karoui, N., and Yu, B. (2013). Optimal M-estimation in high-dimensional regression. *Proceedings of the National Academy of Sciences*, 110:14563–14568.

Bickel, P.J. (1974). Edgeworth expansions in non parametric statistics. *The Annals of Statistics*, 2:1–20.

Bickel, P.J. (1981). Minimax estimation of the mean of a normal distribution when the parameter space is restricted. *The Annals of Statistics*, 9:1301–1309.

Bickel, P.J., Boley, N., Brown, J.B., Huang, H., and Hang, N.R. (2010). Subsampling methods for genomic inference. *The Annals of Applied Statistics*, 4:1660–1697.

Bickel, P.J. and Chen, A. (2009). A nonparametric view of network models and Newman–Girvan and other modularities. *Proceedings of the National Academy of Sciences*, 106:21068–21073.

Bickel, P.J., Chen, C., Kwon, J., Rice, J., van Zwet, E., and Varaiya, P. (2007). Measuring traffic. *Statistical Science*, 22:581–597.

Bickel, P.J. and Ghosh, J.K. (1990). A decomposition for the likelihood ratio statistic and the Bartlett correction — a Bayesian argument. *The Annals of Statistics*, 18:1070–1090.

Bickel, P.J., Götze, F., and van Zwet, W.R. (1997). Resampling fewer than n observations: Gains, losses, and remedies for losses. *Statistica Sinica*, 7:1–31.

Bickel, P.J., Hammel, E.A., O'Connell, J.W. (1975). Sex bias in graduate admissions: Data from Berkeley. *Science*, 187:398–404.

Bickel, P.J. and Levina, E. (2004). Some theory of Fisher's linear discriminant function, 'naive Bayes', and some alternatives when there are many more variables than observations. *Bernoulli*, 10:989–1010.

Bickel, P.J. and Levina, E. (2008). Regularized estimation of large covariance matrices. *The Annals of Statistics*, 36:199–227.

Bickel, P.J. and Li, B. (2007). Local polynomial regression on unknown manifolds. In *Complex Datasets and Inverse Problems*. IMS Lecture Notes vol. 54, Institute of Mathematical Statistics, Beachwood, OH, pp. 177–186.

Bickel, P.J., Ritov, Y., and Rydén, T. (1998). Asymptotic normality of the maximum-likelihood estimator for general hidden Markov models. *The Annals of Statistics*, 26:1614–1635.

Bickel, P.J., Ritov, Y., and Tsybakov, A.B. (2009). Simultaneous analysis of lasso and Dantzig selector. *The Annals of Statistics*, 37:1705–1732.

Bickel, P.J., Ritov, Y., and Zakai, A. (2006). Some theory for generalized boosting algorithms. *Journal of Machine Learning Research*, 7:705–732.

Birney, E., Stamatoyannopoulos, J.A., Dutta, A., Guigó, R., Gingeras, T.R., Margulies, E.H., Weng, Z., Snyder, M., Dermitzakis, E.T. et al. (2007). Identification and analysis of functional elements in 1% of the human genome by the encode pilot project. *Nature*, 447:799–816.

Cai, T.T., Zhang, C.-H., and Zhou, H.H. (2010). Optimal rates of convergence for covariance matrix estimation. *The Annals of Statistics*, 38:2118–2144.

Chen, A. and Bickel, P.J. (2006). Efficient independent component analysis. *The Annals of Statistics*, 34:2825–2855.

Donoho, D.L., Johnstone, I.M., Hoch, J.C., and Stern, A.S. (1992). Maximum entropy and the nearly black object (with discussion). *Journal of the Royal Statistical Society, Series B*, 54:41–81.

Donoho, D.L., Johnstone, I.M., Kerkyacharian, G., and Picard, D. (1996). Density estimation by wavelet thresholding. *The Annals of Statistics*, 24:508–539.

El Karoui, N., Bean, D., Bickel, P.J., Lim, C., and Yu, B. (2013). On robust regression with high-dimensional predictors. *Proceedings of the National Academy of Sciences*, 110:14557–14562.

Erdős, P. and Rényi, A. (1959). On random graphs. I. *Publicationes Mathematicae Debrecen*, 6:290–297.

Fishkind, D.E., Sussman, D.L., Tang, M., Vogelstein, J.T., and Priebe, C.E. (2013). Consistent adjacency-spectral partitioning for the stochastic block model when the model parameters are unknown. *SIAM Journal of Matrix Analysis and Applications*, 34:23–39.

Hodges, J.L., Jr. and Lehmann, E.L. (1970). Deficiency. *The Annals of Mathematical Statistics*, 41:783–801.

Huber, P.J. (1973). Robust regression: Asymptotics, conjectures and Monte Carlo. *The Annals of Statistics*, 1:799–821.

Johnstone, I.M. (2008). Multivariate analysis and Jacobi ensembles: Largest eigenvalue, Tracy–Widom limits and rates of convergence. *The Annals of Statistics*, 36:2638–2716.

Li, Q., Brown, J.B., Huang, H., and Bickel, P.J. (2011). Measuring reproducibility of high-throughput experiments. *The Annals of Applied Statistics*, 5:1752–1779.

Meinshausen, N., Bickel, P.J., and Rice, J. (2009). Efficient blind search: Optimal power of detection under computational cost constraints. *The Annals of Applied Statistics*, 3:38–60.

Rohe, K., Chatterjee, S., and Yu, B. (2011). Spectral clustering and the high-dimensional stochastic blockmodel. *The Annals of Statistics*, 39:1878–1915.

Snyder, C., Bengtsson, T., Bickel, P.J., and Anderson, J. (2008). Obstacles to high-dimensional particle filtering. *Monthly Weather Review*, 136:4629–4640.

Stone, C.J. (1977). Consistent nonparametric regression (with discussion). *The Annals of Statistics*, 5:595–645.

7

The accidental biostatistics professor

Donna J. Brogan
Department of Biostatistics and Bioinformatics
Emory University, Atlanta, GA

Several chapters in this book summarize the authors' career paths after completion of graduate school. My chapter includes significant childhood and early adult experiences that coalesced into eventually completing a PhD degree in statistics. I also summarize some highlights of my biostatistics academic career post PhD at two universities over 37 years. My educational and career paths had twists and turns and were not planned in advance, but an underlying theme throughout was my strong interest and ability in mathematics and statistics.

7.1 Public school and passion for mathematics

I grew up in a working class neighborhood in Baltimore and loved school from the moment I entered. I was a dedicated and conscientious student and received encouragement from many teachers. My lifelong interest in math and my perseverance trait developed at an early age, as the following vignettes illustrate.

As a nine year old in 1948 I rode the public bus each month to a local bank, clutching tightly in my fist $50 in cash and a passbook, in order to make the mortgage payment for a row house in which I lived. I asked the tellers questions over time about the passbook entries. Once I grasped some basic ideas, I did some calculations and then asked why the mortgage balance was not reduced each month by the amount of the mortgage payment. A teller explained to me about interest on loans of money; it sounded quite unfair to me.

Two years later my sixth grade teacher, Mr. Loughran, noted my mathematical ability and, after school hours, taught me junior high and high school mathematics, which I loved. He recommended me for admission to the only accelerated junior high school in Baltimore where I completed in two years the work for 7th, 8th, and 9th grades.

My maternal grandfather, whose public education ended after 8th grade, fanned my passion for math and analysis by showing me math puzzles and tricks, how to calculate baseball statistics like RBI, and how to play checkers and chess, in which he was a local champion.

I chose the unpopular academic track in my inner city working class high school simply because it offered the most advanced math courses. I had no plans to go to college.

I decided that typing would be a useful skill but was denied enrollment because I was not in the commercial track. However, I persisted and was enrolled. When personal computers appeared a few decades later, I was fast and accurate on the keyboard, unlike most of my male academic colleagues.

A gifted math teacher in 11th and 12th grades, Ms. Reese, gave a 10-minute drill (mini-test) to students at the beginning of each daily class. She encouraged my math interest and challenged me daily with a different and more difficult drill, unknown to other students in the class.

A female high school counselor, Dr. Speer, strongly advised me to go to college, a path taken by few graduates of my high school and no one in my immediate family. I applied to three schools. I won substantial scholarships to two state schools (University of Maryland and Western Maryland) but chose to attend Gettysburg College, in Pennsylvania, with less financial aid, because it was smaller and seemed less intimidating to me.

7.2 College years and discovery of statistics

College was my first exposure to middle class America. I majored in math and planned to be a high school math teacher of the caliber of Ms. Reese. However, I rashly discarded this goal in my sophomore year after disliking intensely my first required education course. In my junior year I became aware of statistics via two math courses: probability and applied business statistics. However, two courses during my senior year solidified my lifelong interest in statistics: mathematical statistics and abnormal psychology.

A new two-semester math-stat course was taught by a reluctant Dr. Fryling, the only math (or college) faculty member who had studied statistical theory. He commented frequently that he felt unqualified to teach the course, but I thought he did a great job and I was wildly excited about the topic. I worked all assigned problems in the textbook and additional ones out of general interest. When midterm exam time approached, Dr. Fryling stated that he did not know how to construct an exam for the course. Without thinking, and not yet having learned the social mores of college life, my hand shot up, and I said that I could construct a good exam for the course. The other students noticeably groaned. After class Dr. Fryling discussed with me my unorthodox suggestion and took me up on my offer. After reviewing my prepared exam and answer key, he accepted it. I did not take the exam, of course, and he graded the students' answers. We continued this arrangement for the rest of the year.

In my abnormal psychology course that same year we were assigned to read in the library selected pages from the Kinsey books on human sexual behavior (Kinsey et al., 1948, 1953). The assigned readings were not all that interesting, but I avidly read in each book the unassigned methods chapter that discussed, among other things, statistical analysis strategy and sampling issues (i.e., difficulty in obtaining a representative sample of people who were willing to answer sensitive questions about their sexual behavior). This was my first exposure to sampling theory applications, which eventually evolved into my statistical specialty. Sometime later I read a critique of the 1948 Kinsey book methodology by Cochran et al. (1954); this book is a real education in sampling and data analysis, and I highly recommend it.

Dr. Fryling, noting my blossoming fascination with statistics, asked me about my career plans. I had none, since giving up secondary school teaching, but mentioned physician and actuary as two possibilities, based on my science and statistics interests. He advised that a medicine career was too hard for a woman to manage with family life and that the actuarial science field was not friendly to women. I accepted his statements without question. Neither one of us had a strong (or any) feminist perspective at the time; the second wave of feminism in the United States was still ten years into the future.

A fellow male math major had suggested that I go into engineering since I was good at math. I did not know what engineering was and did not investigate it further; I thought an engineer was the person who drove the train. Even though I was passionate about math and statistics and performed well in them, there were obvious gaps in my general education and knowledge; some family members and friends say this is still true today.

Dr. Fryling strongly recommended that I apply for a Woodrow Wilson National Fellowship, with his nomination, and pursue a doctoral degree in statistics or math. These competitive fellowships were prestigious and provided full graduate school funding for persons who planned a college teaching career. I had not considered such a career, nor did it appeal to me, perhaps because I never saw a female faculty member at Gettysburg College except for girls' physical education. Although Dr. Fryling indicated that I would not be legally bound to teach college by accepting a Wilson fellowship, I felt that it would not be appropriate to apply when I had no intention of becoming a college teacher. Other Wilson applicants may not have been so scrupulous about "the rules." Looking back now, the Wilson fellowship was excellent advice, but limited self awareness of my own talents and interests prevented me from taking this opportunity.

7.3 Thwarted employment search after college

Having discarded high school and college teaching, actuarial science, and medicine, I sought employment after college graduation in 1960. I was aware of only two methods to find a job: look in the newspapers' "Help Wanted" sections and talk with employers at job fairs on campus.

The newspaper route proved fruitless. Younger readers may not be aware that newspapers had separate "Help Wanted Female" and "Help Wanted Male" sections until the late 1960s or early 1970s when such practice eventually was ruled to be illegal sex discrimination. In 1960 advertised positions using math skills and interest were in "Help Wanted Male," and I assumed that it would be futile to apply. Job interviews on campus with employers played out similarly; all positions were segregated by gender and all technical positions were for males. One vignette, among many, illustrates the employment culture for women in the US in 1960.

When I registered for an interview on campus with IBM, I was required to take a math aptitude test. The IBM interviewer commented that he had never seen such a high score from any applicant and offered me either a secretarial or an entry sales position. I countered that I was interested in their advertised technical positions that required a math background, especially given my score on their math aptitude test, but he simply said that those positions were for males. End of conversation.

7.4 Graduate school as a fallback option

Although it is hard for me to believe now, I did not view my failed employment search in 1960 to be the result of systematic societal sex discrimination against women in employment. Rather, I concluded that if I were more qualified, I would be hired even though I was female.

Thus, I decided to pursue a Master's degree in statistics. Looking back, my search for graduate schools seems naive. I scanned available college catalogs at the Gettysburg College library, identified schools that had a separate statistics department, and applied to three that somehow appealed to me. All three accepted me and offered financial aid: University of Chicago, Columbia University, and Purdue University. I chose Purdue because it was the least expensive for me after credit from financial aid.

7.5 Master's degree in statistics at Purdue

Upon arriving at the Purdue Statistics Department in fall of 1960, three new graduate students (including me) chose the M.S. applied statistics track while the many remaining new stat graduate students chose the mathematical statistics track. After one semester I noted that more than half of the math stat track students switched to the applied statistics track. I had a fleeting concern that I might have chosen a "flunky" track, but I loved it and continued on. At the end of first semester I married a Purdue graduate student in English.

As part of my financial aid, I assisted in teaching undergraduate calculus courses and gained valuable instructor experience which supplemented my extensive math tutoring experience begun as an undergraduate. I began to think that teaching college might not be a bad idea after all.

7.6 Thwarted employment search after Master's degree

After my husband and I completed our Purdue Master's degrees in 1962, we moved to Ames, Iowa, where he began a faculty position in the English Department at Iowa State University (ISU). I visited the ISU Statistics Department to inquire about employment opportunities and was offered a technical typist position. My interviewer was enthusiastic because I would understand many formulas and thus make fewer typing errors. Upon inquiring about positions using my statistical skills, I was told that no statistical staff positions were available. Ames was a small town, so I searched in Des Moines, about 35 miles away. I was able to find only clerical or secretarial positions; all technical positions were reserved for males.

7.7 Graduate school again as a fallback option

Since I was living in Ames, home to one of the best statistics departments in the country, I decided to take additional courses to become more qualified for a statistical position. I was not allowed to take ISU courses unless I was a degree seeking student; thus I applied for the statistics doctoral program. Not only was I accepted by the same department that had offered me a statistical typist position, but I was awarded a prestigious and competitive university-wide doctoral fellowship for one year that paid all expenses and an attractive stipend. My daughter Jennifer was born at the end of my first year at ISU.

For my second and subsequent years at ISU the department appointed me to a National Institute of Health (NIH) biostatistics traineeship that paid all expenses and an attractive stipend. I had never heard the word biostatistics.

I especially enjoyed my ISU sampling courses, building upon my initial interest in this topic from the Kinsey et al. (1948, 1953) reports and the Cochran et al. (1954) critique. Most of the other doctoral students disliked sampling: boring topic and too many formulas. I found sampling fascinating, but I frequently have been known for being out of the mainstream.

During the summer following my second ISU year my traineeship paid for me to take courses in biostatistics and epidemiology at the School of Public Health at University of North Carolina at Chapel Hill, since ISU did not offer these courses. I began to understand the scope of biostatistics. The application of statistical theory and methods to public health and medicine appealed to me, combining my then current interests in statistics and psychology with my earlier interest in medicine.

Now that I had taken all of the required coursework for a statistics doctoral degree, fulfilling my limited objective of learning more about statistics to become more employable, I decided to take the scheduled doctoral exams. If I did well, I would continue on to finish the work for a PhD, i.e., write a dissertation. To my surprise, I received the George Snedecor Award for the most outstanding PhD candidate that year, based on doctoral exam performance, and shared the award with another student because we were tied.

7.8 Dissertation research and family issues

Completing my dissertation took longer than anticipated due to academic and family issues. I began my dissertation research a few months after my doctoral exams and the birth of my son Jeffrey. Unfortunately, he was diagnosed with stomach cancer shortly thereafter and had a limited life expectancy. After one year's work on a dissertation topic that had been chosen for me, I discarded my limited research results, feeling that I was not a good match for the topic or for the dissertation advisor. I took a six-month leave of absence from graduate school to spend more time with my two children.

Upon returning to school I requested, and was granted, permission by the department to change my dissertation advisor and topic, an unusual occurrence. I felt this strategy was the only way I would ever finish my degree. I began working with Dr. Joseph Sedransk on a sampling problem of interest to me and, with his expert guidance and assistance, completed my dissertation in a little over one year in summer of 1967. My son died during the middle of this dissertation work, a few days before his second birthday.

This clearly was a difficult time period for me and my family, and I appreciate very much the support given to me by the ISU Department of Statistics.

7.9 Job offers — finally!

I planned to move to Chapel Hill after finishing my PhD because my husband had been accepted at the University of North Carolina (UNC) as a linguistics doctoral student for fall of 1967. Early that year I contacted the UNC Biostatistics Department and the Duke University Medical Center to inquire about available positions. Surprisingly, each school invited me for an interview. I visited both schools and gave a seminar about my dissertation results to date.

Within the next several weeks I received from each school an attractive offer of a tenure-track Assistant Professor position. After much deliberation I chose UNC, primarily because of an interesting and unique opportunity there. Dr. Bernard Greenberg, Biostatistics Chair, offered to appoint me as the director of an already funded training grant in the department from National Institute of Mental Health (NIMH) to develop and implement an MSPH program in mental health statistics. This offer combined my interests in statistics and psychology; I had minored in psychology for both graduate degrees, including some psychometrics during my doctoral studies.

7.10 Four years at UNC-Chapel Hill

Upon arrival at UNC another newly hired Assistant Professor and I met with a human resources specialist to review our fringe benefits. At the end of the meeting, the specialist informed the other faculty member that he would receive an attractive disability insurance policy paid for by UNC. When I inquired if I would receive this fringe benefit, the male specialist answered no, explaining that women don't need disability insurance since their husbands take care of them financially. It did not matter to him, or the university, that I was the wage earner for the family since my husband was a full-time graduate student.

Finally I began to recognize these frequent occurrences as sex discrimination. I joined a women's liberation group in Chapel Hill, and my feminist consciousness was raised indeed. I became an activist on women's barriers to employment and education, primarily within the American Statistical Association (ASA) but also at UNC. With others I founded the Caucus for Women in Statistics in 1971 and served as its president the first three years. Concurrently I spearheaded the formation of the ASA Committee on Women in Statistics (COWIS) and served as a member in its early days. These and later actions were the basis for my receiving the COPSS Elizabeth Scott Award in 1994.

At UNC I worked with collaborators in mental health and psychiatry to develop, implement and administer the MSPH training program in mental health statistics, and I created and taught three new courses in this track (Brogan and Greenberg, 1973). Looking back, it seems to have been an unusual responsibility to be given to a brand new assistant professor just one month after PhD completion. However, Dr. Greenberg and my mental health colleagues seemed confident that I could do it, and I enjoyed the challenge of creating a new MSPH track.

During my fourth year at UNC, I wrote a grant application to NIMH to continue the MSPH program in mental health statistics but also to extend it to a doctoral level training program. NIMH funded this training grant, and my salary support within the department was covered for another five years.

However, I had a few concerns about what appeared to be the opportunity for a potentially stellar academic future. First, my departmental teaching was restricted to the specialized mental health statistics courses that I created since I was the only faculty person who could (or would) teach them. I felt that I wanted more variety in my teaching. Second, although the extension of the training program to the doctoral level was a fantastic opportunity to develop further the niche into which I had fortuitously fallen, and hopefully to make substantial and needed contributions therein, I began to feel that I was in a niche. For some reason I did not like the feeling of being so specialized. Finally, I had tired of living in small college towns for the past 15 years and was interested in locating to a metropolitan area, especially since my husband and I had recently divorced.

7.11 Thirty-three years at Emory University

In what might have seemed to be irrational behavior to some of my UNC-Biostatistics colleagues, I accepted a position in fall of 1971 at Emory University School of Medicine in the small and fledgling Department of Statistics and Biometry, its first ever female faculty member. Emory transformed itself over subsequent decades into a world destination university, including the formation in 1990 of the Rollins School of Public Health (RSPH), currently one of the top-tier public health schools in the country. I was one of only a few female faculty members in RSPH upon its formation and the only female Full Professor.

At Emory I had ample opportunity to be a biostatistical generalist by conducting collaborative research with physicians and other health researchers in different disciplines. My collaborative style was involvement with almost all aspects of the research project rather than only the purely biostatistical components, primarily because I was interested in the integrity of the data that

I would be analyzing later. I enjoyed working with a few collaborators over extended time periods, including a medical sociologist colleague for thirty years.

In the early 1980s, I used my sample survey skills in an NHLBI funded multi-site collaborative contract where I designed and implemented area probability samples of adults in Georgia in order to estimate hypertension related parameters. I learned several nitty-gritty applied sampling techniques not in textbooks from the sample survey statisticians at the other sites and first used the SUDAAN software for analysis of complex survey data.

In the mid 1980s, I was diagnosed with breast cancer. My personal experience and my biostatistics background combined to make me a useful contributor to the founding group of the breast cancer advocacy movement, culminating in the formation of the National Breast Cancer Coalition (NBCC) and similar organizations.

I served as Biostatistics chair in RSPH in the early 1990s, the first ever female chair of the school. The so-called power of the position (money and space, primarily) did not interest me. Rather, I attempted to maintain a collegial department that was successful in the typical academic arenas and was supportive for each of its members (faculty, students, and staff). My best training for the chair position was a few years that I had spent in group therapy in earlier decades (another way of saying that my training was minimal). After three years I resigned as chair because academic administration took me away from what I really loved: being a practicing biostatistician.

During the early 1990s, I began to teach continuing education workshops on analysis of complex survey data at summer programs in biostatistics and epidemiology (e.g., University of Michigan), at government agencies such as CDC and at annual meetings of health researchers. I continue this teaching today, even after retirement, because I enjoy it. To date I have taught about 130 of these workshops to over 3000 participants.

Upon my retirement from Emory in 2004 the Biostatistics Department and the RSPH sponsored a gala celebration with 140 guests, an exquisite sit-down dinner, and a program with many speakers who reviewed aspects of my professional life. I felt quite honored and much loved.

7.12 Summing up and acknowledgements

I enjoyed immensely my unintended academic career in biostatistics and highly recommend the discipline to those who are interested and qualified. I liked the diverse areas in which I worked as biostatistical collaborator, in essence acquiring a mini medical education. I found teaching for very different audiences to be great fun: graduate students in biostatistics and the health sciences, health professionals, and health researchers. It took a while to find my enjoyable statistical niche: sample survey statistician. I was able to combine some major aspects of my personal life, feminism and breast cancer history, with collaborative research and activism. I regret having had less enthusiasm for biostatistical methodological research and was not as productive in this area as I would have liked.

I am grateful to many people and institutions for helping me to prepare for and navigate my career, some mentioned above. There are too many people to mention individually here, but one must be recognized. I am indebted to my ex-husband Dr. Charles Ruhl for his strong support and encouragement of my educational and career goals; for his crucial role in our family life, especially during Jeffrey's illness; for living in Iowa longer than he wanted so that I could finish my PhD degree; for encouraging me to join a women's liberation group; and for being a feminist long before I knew what the word meant.

References

Brogan, D.R. and Greenberg, B.G. (1973). An educational program in mental health statistics. *Community Mental Health Journal*, 9:68–78.

Cochran, W.G., Mosteller, F., and Tukey, J.W. (1954). *Statistical Problems of the Kinsey Report on Sexual Behavior in the Human Male*, American Statistical Association, Washington, DC.

Kinsey, A.C., Pomeroy, W.B., and Martin, C.E. (1948). *Sexual Behavior in the Human Male.* W.B. Saunders Company, Philadelphia, PA.

Kinsey, A.C., Pomeroy, W.B., Martin, C.E., and Gebhard, P.H. (1953). *Sexual Behavior in the Human Female.* W.B. Saunders Company, Philadelphia, PA.

8

Developing a passion for statistics

Bruce G. Lindsay
Department of Statistics
Pennsylvania State University, University Park, PA

This chapter covers the major milestones of the early years of my career in statistics. It is really the story of the transitions I made, from early uncertainty about my choice of career and my level of talent, up to crossing the tenure line and realizing that I had not only been deemed a success, I had a passion for the subject. The focus will be on aspects of those adventures that seem most relevant to young people who are partway along the same journey.

8.1 Introduction

I have had a long career in the academic world of statistics, something like 40 years. I have seen the whole process from many points of view, including eight years as a department head. I have supervised 30 PhD students. I would hope that from all that experience I might have forged something worthwhile to say here, something not found in a research paper. I have chosen to focus on the early part of my career, as those are the days of major transitions.

For those of you early in your career, there are many choices to make as you navigate this world. It starts with the choice of statistics for your education, then a graduate school, then an advisor, then a topic for the thesis, then a place of employment. This is done while clearing a series of hurdles meant to separate the qualified from the unqualified, starting with entrance exams and ending with tenure. In this essay I will review some of these critical moments in my early career. With each of these milestones, I gained some wisdom about statistics and myself, and went from being an unsure young man to being a passionate scholar of statistics.

Since I joined Penn State in 1979 I have been paid to do cutting edge research that makes my university famous. I therefore have welcomed this rare opportunity to look backward instead of forward, and think about the roots of my career. One aspect of academic life that has been frustrating to me is its ruthless vitality, always rushing forward, often ending up looking like a garden sadly in need of weeding. I wish there were more reflection, more respect for the past. The intellectual rewards, however, have always been largest for creativity, for those who till new soil, and so that is where most of the energy is spent.

And to be fair, I too deserve criticism, as I have too rarely taken on the role of oversight, the role of putting an order to what is important, and saying why. One of my passions is for discovering something new. It is like being Christopher Columbus, discovering a New World. My discoveries have sometimes involved basic understanding of scientific phenomena, but the big magic for me comes from the beautiful way that statistics, through mathematics, can find the signals in the midst of noise. In whatever way I can add something to this, by discovery of new ways to build models, or compute statistics, or generate mathematical understanding of scientific questions: that is part of what makes me feel valuable.

However, I also have a passion for mentoring young people. After all, why else 30 PhD students? I therefore take on the career counselor role here.

Before I describe my early career-changing events, let me touch on a couple of personal perspectives.

There is an important element of philosophy to statistics, epitomized by the frequentist/Bayesian schism. I was fortunate to be trained by a pair of powerful statistical thinkers: Norman Breslow and David Cox. Their frequentist thinking definitely colors my perspective on the philosophy of statistics to this day. However, I am not passionate about the distinction between Bayes and frequency. Although I am interested in the basic logic behind statistics, it will be a small part of my essay. This will be more about the process of becoming excited about the entire statistics culture, and what to do with that excitement.

I am also someone who has learned to collaborate, and loves it. It is a key part of maintaining my passion. For the first seven or so years of my career, I only wrote solo papers. About 1985 or so, though, I had an eye-opening research discussion with my colleague Clifford Clogg. We were both young men then, about at the point of tenure. He had a joint appointment in the Departments of Statistics and Sociology. In our discussion we realized that we, from very different backgrounds and points of view, had just found the exact same result by two completely different means (Lindsay et al., 1991). His was computational, mine was geometric. He concluded our meeting by saying, with wonder, "I can't believe that I get paid to do this!" I wholeheartedly agreed with him, but I must say that a lot of my joy came because I was doing it with him. We became fast friends and collaborators.

Sad to say, Cliff died in 1995, at the age of 45, from a sudden heart attack. It was working with him that I first learned, in a deep sense, that the biggest joys in statistical work are those that are shared. Nowadays, I often think about problem solving alone, but I very rarely work alone.

8.2 The first statistical seeds

So now let me set the scene for my first contacts with statistics. I started out with a mathematics degree from University of Oregon in 1969, but I had virtually no statistics training there. I then went to Yale graduate school in mathematics. I only lasted one year because I was drafted into the US military in 1970. I had no probability or statistics at Yale, rather a shame given the eminent faculty members there.

I took my first basic statistics course while in the US Coast Guard, about 1972. It was a night course at Berkeley, taught by an adjunct. Frankly, like many other "Stat 100's," it was not very inspiring. My impression from it was that statistics was a collection of strange recipes that had been generated by a foreign culture. Surely this was not mathematics, but what was it?!

On the plus side, however, I did my first real statistical analysis during those Coast Guard years. I had done poorly in an Armed Services exam that the military used to screen applicants to their training schools. The particular exam involved repeatedly looking at two long numbers side by side and saying if they were identical or not. (I suspect my poor performance on that exam is now reflected in my poor memory of phone numbers.)

As a result of my exam results, I had to get a waiver to get into Yeoman School. Yeomen are the clerk typists of the Navy and Coast Guard. In the end I did very well in the school, and was convinced that the screening exam was worthless. And I knew that I would need statistics to prove it! My opportunity arose because I had been assigned to the same Yeoman School as its secretary. I analyzed the school's data to show that there was zero correlation between the screening exam result and performance in the school. However my letter to the Coast Guard Commandant was never answered.

I must confess that at this time I was still a long ways from being a fan of statistics. It seemed like a messy version of mathematics constructed from a variety of disconnected black boxes. I could calculate a correlation, and look up a significance level, but why? The fact that there were multiple ways to measure correlation only made it less satisfactory. But the seeds of change had been planted in me.

8.3 Graduate training

By the time I left the Coast Guard in 1974, I had decided to drop out of Yale and out of pure mathematics. I wanted something closer to life on this planet. This led me to switch to graduate school at University of Washington (UW). It was an excellent choice.

Their Biomathematics degree offered lots of room for me to explore my applied mathematical interests. I took courses in ecology, fisheries, epidemiology, and population genetics as I looked about for interesting applications and interesting applied mathematical areas. Certainly I have no regrets about my second choice of graduate schools. It had highly talented faculty and a broad range of possibilities. I now always recommend that prospective graduate students select schools based on these characteristics.

As I took the required Biomathematics courses, I began to find statistics, and its applications, more deeply interesting. In particular, Bob Smythe's mathematical statistics course, where I first saw the magic of maximum likelihood, had me intrigued. And the more courses I took, the more I liked the subject. It is not a cohesive subject, but it is a powerful one.

A very important part of my education was not conventional coursework. I was in a consulting class in the Center for Quantitative Science unit, which meant sitting in a room and seeing clients. My very positive experience there has left me a proponent of graduate consulting classes all my life.

One of my clients brought in a fisheries problem that did not seem to fit any of the traditional models we had learned in applied classes. If it was not regression or ANOVA or discrete data, what could it be?

Salmon are fish with a complex life cycle. It starts when they are born in a home river, but they soon leave this river to mature in the open ocean. They return to their home river at the end of their lives in order to lay and fertilize eggs and die. The set of fish coming from a single river are thus a distinct subpopulation with its own genetic identity. This was important in fisheries management, as many of the fish were caught in the open ocean, but with some diagnostic measurements, one could learn about the river of origin.

In the problem I was asked to consult upon, the salmon were being caught in the Puget Sound, a waterway that connects to both American and Canadian rivers. Since each country managed its own stock of fish, the fisheries managers wanted to know how many of the caught fish came from "Canadian" rivers and how many "American."

The data consisted of electrophoretic measurements, an early form of DNA analysis, made on a sample of fish. It was important that the salmon from various rivers were physically mixed together in this sample. It was also important that the scientists also had previously determined the genetic profile of salmon from each river system. However, these genetic "fingerprints" did not provide a 100% correct diagnosis of the river that each fish came from. That would have made the problem a simple one of decoding the identities, and labelling the salmon, creating a multinomial problem.

I now know that a very natural way to analyze such data is to build an appropriate mixture model, and then use maximum likelihood or a Bayesian solution. At the time, having never seen a mixture model in my coursework, I was quite clueless about what to do. However, I did my due diligence as a consultant, talked to a population geneticist Joseph Felsenstein, and found a relevant article in the wider genetics literature — it had a similar structure. In the so-called admixture problem, the goal was to determine the racial components of a mixed human population; see, e.g., Wang (2003).

This whole process was a great discovery for me. I found the ability of statistics to ferret out the hidden information (the home rivers of each salmon) to be the most fascinating thing I had seen to date. It was done with the magic of likelihood. The fact that I could find the methods on my own, in the literature, and make some sense of them, also gave me some confidence. My lesson was learned, statistics could be empowering. And I had ignited a passion.

In retrospect, I was also in the right place at the right time. As I now look back at the fisheries literature, I now see that the scientific team that approached me with this problem was doing very cutting edge research. The decade of the 80s saw considerable development of maximum likelihood methods for unscrambling mixed populations. Indeed, I later collaborated on a paper that identified in detail the nature of the maximum likelihood solutions, as well as the identifiability issues involved (Roeder et al., 1989). In this case, the application was a different biological problem involving plants and the fertility of male plants as it depended on the distance from the female plants. In fact, there are many interesting applications of this model.

A consultant needs to do more than identify a model, he or she also needs to provide an algorithm for computation. In providing a solution to this problem, I also first discovered the EM algorithm in the literature. Mind you, there existed no algorithm called the "EM" until 1977, a year or two after this project (Dempster et al., 1977). But like many other discoveries in statistics, there were many prequels. The version I provided to the client was called the "gene-counting" algorithm, but the central idea of the EM, filling in missing data by expectation, was already there (Ott, 1977).

This algorithm became its own source of fascination to me. How and why did it work? Since that period the EM algorithm has become a powerful tool for unlocking hidden structures in many areas of statistics, and I was fortunate to be an early user, advocate, and researcher. Its key feature is its reliability in complex settings, situations where other methods are likely to fail. Whenever I teach a mixture models course, one of my first homework assignments is for the student to understand and program the EM algorithm.

So there you have it. Through my choice of graduate education, by taking a consulting class, and by drawing the right consulting client, I had entered at an early stage into the arenas of mixture models and the EM algorithm, both of which were to display considerable growth for the next thirty years. I got in on the ground floor, so to speak. I think the message for young people is to be open to new ideas, and be ready to head in surprising directions, even if they are not popular or well known.

In many ways the growth of mixture models and the EM algorithm came from a shift in computing power. As an old-timer, I feel some obligation to offer here a brief side discussion on the history of computing in statistics during the first years of my career. There was a revolution underway, and it was to change completely how we thought about "feasible" ways to do statistics.

As an undergraduate, I had used punch cards for computing in my computer science course. All computing on campus was done through a single "mainframe" computer. First one would "punch" the desired program onto cards, one program line being one card, on a special machine. One would then take this pile of cards to a desk in the computer center for submission. Sometime later one would get the output, usually with some errors that needed fixing. Turnaround to a completed successful program was very, very slow.

Computing was still in the "punch card" era when I went to graduate school at UW in 1974. However, during my years there, the shift to "terminals" and personal computing was starting. It was so much more attractive that the focus on mainframe computing quickly faded out. At that point, efficiency in programming was proving to be much more valuable than speed of the machine.

This efficiency had immediate benefits in statistics. Previously there had been a great emphasis on methods that could be computed explicitly. I happened, by chance, to be an observer at one of the most important events of the new era. I attended a meeting in Seattle in 1979 when Bradley Efron gave one the first talks on the bootstrap, a subject that blossomed in the 1980s (Efron, 1979). Statistics was waking up to the idea that computing power might drive a revolution in methodology.

The Bayesian revolution was to come a little later. In 1986 I attended an NSF–CBMS workshop by Adrian Smith on Bayesian methods — he gave a lot of emphasis to techniques for numerical integration, but the upper limit was seven dimensions, as I recall (Naylor and Smith, 1982). All this hard work on integration techniques was to be swept away in the 90s by MCMC methods (Smith and Gelfand, 1992). This created a revolution in access to Bayes, although most of the rigor of error bounds was lost. In my mind, the answers are a bit fuzzy, but then so is much of statistics based on asymptotics.

8.4 The PhD

Returning to my graduate education, after two years my exams were taken and passed, and research about to begin. Like most young people I started by examining projects that interested my possible PhD mentors. The UW Biomathematics program was tremendously rich in opportunities for choosing a thesis direction. It was clear to me that I preferred statistics over the other possible biomathematics areas. In addition, it was clear from my qualifying exam results that I was much better at mathematical statistics than applied statistics. I liked the scientific relevance of applied statistics, but also I felt more of a research curiosity about mathematical statistics.

My first PhD investigation was with Ron Pyke of the UW math department. The subject related to two-dimensional Brownian motion, but it generated little fascination in me — it was too remote from applications. I therefore had to go through the awkward process of "breaking off" with Ron. Ever since then I have always told students wishing to do research with me that they should not be embarrassed about changing advisors to suit their own interests.

The best applied people at UW were in the Biostatistics group, and after talking to several of them, I ended up doing my dissertation with Norm Breslow, who also had a pretty strong theoretical bent and a Stanford degree.

My first contact with Norm was not particularly auspicious. Norm taught a course in linear models in about 1975 that had the whole class bewildered. The textbook was by Searle (2012), which was, in that edition, doggedly matrix oriented. However, for the lectures Norm used material from his graduate days at Stanford, which involved new ideas from Charles Stein about "coordinate free" analysis using projections. The mismatch of textbook and lecture was utterly baffling to all the students — go ask anyone in my class. I think we all flunked the first exam. But I dug in, bought and studied the book by Scheffé (1999), which was also notoriously difficult. In the end I liked the subject matter and learned quite a bit about how to learn on my own. I am sure the geometric emphasis I learned there played a later role in my development of geometric methods in likelihood analyses.

Fortunately my second class with Norm, on categorical variables, went rather better, and that was where I learned he was quite involved in epidemiology and cancer studies. I later learned that his father Lester Breslow was also something of a celebrity in science, being Dean of the School of Public Health at UCLA.

Although he was an Associate Professor, my years of military service meant that Norm was only a few years older than me. He had already made a name for himself, although I knew nothing about that. I found him inspiring through his statistical talent and biological knowledge, but mainly his passion for statistics.

I sometimes wonder if there was not some additional attraction because of his youth. The mathematics genealogy website shows me to be his second PhD student. Going up my family tree, Norm Breslow was Brad Efron's first student, and Brad Efron was Rupert Millers' second. Going back yet further, Rupert Miller was fifth of Samuel Karlin's 43 offspring. Going down the tree, Kathryn Roeder, a COPSS award winner, was my first PhD student. This "first-born" phenomenon seems like more than chance. At least in my line of descent, youthful passion and creativity created some sort of mutual attraction between student and advisor.

One of the most important challenges of graduate life is settling on a research topic with the advisor. This will, after all, set the direction for your career, if you go into research. I would like to discuss my graduate experience in some detail here because of combination of chance and risk taking that a research career entails.

In my experience, most students pick from a small list of suggestions by their chosen advisor. Indeed, that is the way I started my research. Norm had suggested the development of a sequential testing method based on partial likelihood (Cox, 1975). I am sure Norm foresaw applications in clinical trials. He probably had some confidence I could do the hard math involved.

Being a conscientious student, I started off on this problem but I was very soon sidetracked onto related problems. I am not sure if I would recommend this lack of focus to others. However, I do think you have to love what you are doing, as there is no other way to succeed in the long run. I did not love the problem he gave me. It seemed to be technically difficult without being deep, and not really a chance to grow intellectually.

In the end I wrote a thesis on my own topic, about three steps removed from Norm Breslow's proposed topic. I started by studying partial likelihood, which was then a very hot topic. But as I read the papers I asked myself this — what is the justification for using this partial likelihood thing beyond its ease of computation? A better focused student would have hewn to Norm's original suggestion, but I was already falling off the track. Too many questions in my mind.

I assure you my independence was not derived from high confidence. On the contrary, no matter my age, I have always felt inferior to the best of my peers. But I am also not good at following the lead of others — I guess I like marching to the beat of my own drummer. It does not guarantee external success, but it gives me internal rewards.

One risk with research on a hot topic is that you will be scooped. As it turns out, the justification, on an efficiency basis, for Cox's partial likelihood in the proportional hazards model was on Brad Efron's research plate about that time. So it was a lucky thing for me that I had already moved on to an older and quieter topic.

The reason was that the more I read about the efficiency of likelihood methods, the less that I felt like I understood the answers being given. It all started with the classic paper by Neyman and Scott (1948) which demonstrated severe issues with maximum likelihood when there were many nuisance parameters and only a few parameters of interest. I read the papers that followed up on Neyman and Scott, working forward to the current time. I have to say that I found the results to that time rather unsatisfying, except for models in which there was a conditional likelihood that could be used.

My early exposure to mixture models provided me with a new way to think about consistency and efficiency in nuisance parameter problems, particularly as it related to the use of conditional and partial likelihoods. I put these models in a semiparametric framework, where the nuisance parameters were themselves drawn from a completely unknown "mixing distribution." In retrospect, it seems that no matter how much I evolved in my interests, I was still drawing strength from that 1975 consulting project.

My research was mostly self-directed because I had wandered away from Norm's proposed topic. I would report to Norm what I was working on, and he would humor me. Kung Yee Liang, who followed me as a Breslow student, once told me that Norm had asked him to read my thesis and explain it to him.

I tell this story not because I advise students to follow this kind of independent course. The fact that Norm barely understood what I was doing and where I was headed was something of a handicap to me. I spent years figuring out the relevant literature. The real problem is that a graduate student just does not know the background, has not seen the talks, and cannot know whether the statistical community will think the research is important. I was taking serious risks, and consider myself fortunate that things worked out in the end.

After settling on a topic and doing the research, another big hurdle must be crossed. I must say that at the beginning I found it very difficult to write up my statistical research, and so I am very sympathetic to my PhD students when they struggle with the organization, the motivation, the background, and reporting the results. I keep telling them that they are simply telling a story, just like they do when they give an oral presentation. If you can give a good talk, you can write a good paper. At Penn State these days many of the graduate students give multiple talks at meetings and in classes. I am sure this must help them immensely when it comes to writing and defending their dissertations, and going on job interviews. I had no such opportunities at Washington, and I am sure it showed in my early writing.

At any rate, Norm returned my thesis drafts with lots of red marks. In the beginning I felt like I had failed, but then bit by bit my writing became clearer and more fitting to the statistical norm. I still had a problem with figuring out the distinction between what was important and what was merely interesting. In the end, I wrote a very long thesis titled "Efficiency in the presence of nuisance parameters." It was a long ways from being publishable.

I must say that in those days proper editing was very difficult. I would start the process by turning my handwritten drafts over to a mathematical typist. In those days mathematical results were usually typed on an IBM Selectric typewriter. There were little interchangeable balls that had the various symbols and fonts. Typing in math meant stopping to change the ball, maybe more than once for each equation. This slow typing system very distinctly discouraged editing manuscripts. Mistakes could mean redoing entire pages, and revisions could mean retyping the whole manuscript. Changes in an introduction would alter everything thereafter.

Thank goodness those days also disappeared with the advent of personal computing. Now one can spend time refining a manuscript without retyping it, and I am sure we have all benefited from the chance to polish our work.

At last the finish line was reached, my thesis submitted and approved in 1978.

8.5 Job and postdoc hunting

I was ready to move on, a bright-eyed 31 year old. I had enjoyed doing research, and having received positive feedback about it, I was ready to try the academic job market. I submitted applications to many schools. While I would have preferred to stay on the West Coast, the list of job opportunities seemed pretty limiting. With Norm's encouragement, I also applied for an NSF–NATO postdoctoral fellowship.

To my somewhat shocked surprise, I had six job interviews. I can only infer that I must have had some good letters from well known scholars. In the end I had interviews at UC Berkeley, UC Davis, Purdue, Florida State, Princeton, and Penn State. I was quite awestruck about being paid to fly around the country, as at that point in my life I had never flown anywhere. It was rather nice to be treated like a celebrity for a couple of months.

Of course, the interviews could sometimes be intimidating. One colorful character was Herman Rubin of Purdue, who was notorious for cross-examining job candidates in his office. At the end of my seminar, he raised his hand and stated that my results could not possibly be correct. It was a bit disconcerting. Another place that was frightening, mostly by the fame of its scholars, was Berkeley. Peter Bickel, not much older than I, was already Department Head there. Another place with some intellectual firepower was Princeton, where John Tukey sat in the audience.

Wherever I did an interview, I told the school that I would like to take the NATO postdoc if it became available, and that being able to do so would be a factor in my decision. In the end, several did make me offers with an open start date, and after considerable deliberation, and several coin tosses, I accepted Penn State's offer over Princeton's. Since the Princeton department closed soon thereafter, I guess I was right.

8.6 The postdoc years

In the end, I did garner the postdoc. With it, I went to Imperial College in London for 1978–79, where my supervisor was the famous Sir David Cox. His paper (Cox, 1972) was already on its way to being one of the most cited works ever (Ryan and Woodall, 2005). My thanks to Norm for opening this door. All those early career choices, like University of Washington and Breslow, were paying off with new opportunities.

In London I went back to work on my dissertation topic. When I had visited Berkeley, I learned that Peter Bickel had a student working on related problems, and Peter pointed out some disadvantages about my approach to asymptotics. I had drawn heavily on a monograph by Bahadur (1971). Neither

I nor my advisors knew anything about the more attractive approaches coming out of Berkeley. So my first agenda in London was to repair some of my work before submitting it.

It was extremely inspiring to be around David Cox for a year. Although David seemed to fall asleep in every seminar, his grasp of the problems people were working on, as evidenced by his piercing questions and comments, was astounding. He single-handedly ran *Biometrika*. He was the master of the whole statistical domain.

David was very kind to me, even though he did not have a lot to say to me about my research. I think it was not really his cup of tea. He did, however, make one key link for me. He suggested that I read the paper by Kiefer and Wolfowitz (1956) about a consistent method of estimation for the Neyman–Scott problem. That paper was soon to pull me into the nascent world of nonparametric mixture modelling. An article by Laird (1978) had just appeared, unbeknownst to me, but for the most part the subject had been dead since the Kiefer and Wolfowitz paper.

My postdoc year was great. It had all the freedom of being a grad student, but with more status and knowledge. After a great year in London, I returned to the US and Penn State, ready to start on my tenure track job.

8.7 Starting on the tenure track

Going back to my early career, I confess that I was not sure that I was up to the high pressure world of publish-or-perish, get tenure or move on. In fact, I was terrified. I worried about my mental health and about my ability to succeed. I am sure that I was not the first or last to feel these uncertainties, and have many times talked in sympathy with people on the tenure track, at Penn State and elsewhere.

It took a number of years for the thesis research to end its stumbling nature, and crystallize. I published several papers in *The Annals of Statistics* that would later be viewed as early work on efficiency in semiparametric models. I also made some contributions to the problem of estimating a mixing distribution nonparametrically by maximum likelihood. In the process I learned a great deal about convex optimization and inverse problems.

I often tell my students that I had plenty of doubts about my success when I was an Assistant Professor. My publication list was too short. The first paper from my 1978 thesis was not written until 1979, and appeared in 1980. It was not even a statistics journal, it was *The Philosophical Transactions of the Royal Society of London* (Lindsay, 1980).

My first ten papers were solo-authored, so I was definitely flying on my own. It must have been a kinder era, as I don't recall any rejections in that period. And that certainly helps with confidence. (You might find it reassuring to know that I have had plenty of rejections since.)

I had six papers on my CV when I came up for tenure in 1984–85. And I had published nothing at all in 1984. I know why: I had spent most of a year trying to build a new EM theory, and then giving up on it. I was a bit scared. Nowadays, and even then, my number would be considered below average. Indeed, many of our recent Assistant Professor candidates at Penn State University seem to have had that many when they arrived at Penn State. Just the same, the people who wrote the external letters for me were very supportive, and I was promoted with tenure.

Back then it was still not obvious to me that statistics was the right place for me. My wife Laura likes to remind me that some time in the 1980s I told a graduate student that "Statistics is dead." I can understand why I said it. The major conceptual and philosophical foundations of statistics, things like likelihood, Bayes, hypothesis testing, multivariate analysis, robustness, and more, had already been developed and investigated. A few generations of ingenious thought had turned statistics into an academic subject in its own right, complete with Departments of Statistics. But that highly energetic creative era seemed to be over. Some things had already fossilized, and the academic game contained many who rejected new or competing points of view. It seemed that the mathematical-based research of the 1970s and 1980s had in large part moved on to a refinement of ideas rather than fundamentally new concepts.

But the granting of tenure liberated me from most of these doubts. I now had a seal of approval on my research. With this new confidence, I realized that, in a larger sense, statistics was hardly dead. In retrospect, I should have been celebrating my participation in a subject that, relative to many sciences, was a newborn baby. In particular, the computer and data revolutions were about to create big new and interesting challenges. Indeed, I think statistics is much livelier today than it was in my green age. It is still a good place for discovery, and subject worthy of passion.

For example, multivariate analysis is now on steroids, probing ever deeper into the mysteries of high-dimensional data analysis, big p and little n, and more. New techniques, new thinking, and new theory are arising hand in hand. Computational challenges that arise from complex models and enormous data abound, and are often demanding new paradigms for inference. This is exciting! I hope my experiences have shed some light on your own passionate pursuits in the new statistics.

References

Bahadur, R.R. (1971). *Some Limit Theorems in Statistics.* Society for Industrial and Applied Mathematics, Philadelphia, PA.

Breslow, N.E. and Day, N.E. (1980). *Statistical Methods in Cancer Research. Vol. 1. The Analysis of Case-control Studies*, vol. 1. Distributed for IARC by the World Health Organization, Geneva, Switzerland.

Cox, D.R. (1972). Regression models and life-tables (with discussion). *Journal of the Royal Statistical Society, Series B*, 34:187–220.

Cox, D.R. (1975). Partial likelihood. *Biometrika*, 62:269–276.

Dempster, A.P., Laird, N.M., and Rubin, D.B. (1977). Maximum likelihood from incomplete data via the EM algorithm. *Journal of the Royal Statistical Society, Series B*, 39:1–38.

Efron, B. (1977). The efficiency of Cox's likelihood function for censored data. *Journal of the American Statistical Association*, 72:557–565.

Efron, B. (1979). Bootstrap methods: Another look at the jackknife. *The Annals of Statistics*, 7:1–26.

Kiefer, J. and Wolfowitz, J. (1956). Consistency of the maximum likelihood estimator in the presence of infinitely many incidental parameters. *The Annals of Mathematical Statistics*, 27:887–906.

Laird, N.M. (1978). Nonparametric maximum likelihood estimation of a mixing distribution. *Journal of the American Statistical Association*, 73:805–811.

Lindsay, B.G. (1980). Nuisance parameters, mixture models, and the efficiency of partial likelihood estimators. *Philosophical Transactions of the Royal Society of London, Series A*, 296:639.

Lindsay, B.G. (1983). The geometry of mixture likelihoods: A general theory. *The Annals of Statistics*, 11:86–94.

Lindsay, B.G., Clogg, C.C., and Grego, J. (1991). Semiparametric estimation in the Rasch model and related exponential response models, including a simple latent class model for item analysis. *Journal of the American Statistical Association*, 86:96–107.

Naylor, J.C. and Smith, A.F.M. (1982). Applications of a method for the efficient computation of posterior distributions. *Applied Statistics*, 31:214–225.

Neyman, J. and Scott, E.L. (1948). Consistent estimates based on partially consistent observations. *Econometrica*, 16:1–32.

Ott, J. (1977). Counting methods (EM algorithm) in human pedigree analysis: Linkage and segregation analysis. *Annals of Human Genetics*, 40:443–454.

Roeder, K., Devlin, B., and Lindsay, B.G. (1989). Application of maximum likelihood methods to population genetic data for the estimation of individual fertilities. *Biometrics*, 45:363–379.

Ryan, T.P. and Woodall, W.H. (2005). The most-cited statistical papers. *Journal of Applied Statistics*, 32:461–474.

Scheffé, H. (1999). *The Analysis of Variance.* Wiley, New York.

Searle, S.R. (2012). *Linear Models.* Wiley, New York.

Smith, A.F.M. and Gelfand, A.E. (1992). Bayesian statistics without tears: A sampling-resampling perspective. *The American Statistician*, 46:84–88.

Wang, J. (2003). Maximum-likelihood estimation of admixture proportions from genetic data. *Genetics*, 164:747–765.

9

Reflections on a statistical career and their implications

R. Dennis Cook
School of Statistics
University of Minnesota, Minneapolis, MN

This chapter recounts the events that steered me to a career in statistics and describes how my research and statistical temperament were set by my involvement in various applications. The discussion encompasses the historical and contemporary role of statistical diagnostics in practice and reflections on the importance of applications in the professional life of a statistician.

9.1 Early years

It was mostly serendipity that led me to a career in statistics.

My introduction to statistics started between my Sophomore and Junior years in high school. At the time I was looking for summer employment so I could earn money to customize my car — neat cars elevated your social standing and attracted the girls. I was fortunate to secure summer and eventually after-school employment with the Agronomy Department at Fort Assiniboine, an agriculture experimentation facility located just outside Havre, Montana. The surrounding area is largely devoted to wheat production, thousands and thousands of acres of spring and winter wheat. The overarching goal of the Agronomy Department was to develop contour maps of suggested fertilization regimes for use by wheat farmers along the High Line, a run of about 130 miles between Havre and Cut Bank, Montana, and to occasionally develop targeted recommendations for specific tracts of land at the request of individual farmers.

I continued to work at Fort Assiniboine until I graduated from high school, at which point I enlisted in the military to avoid the uncertainty of the draft. After fulfilling my military obligation, which ended just before the buildup to the Vietnam war, I returned to full-time employment at Fort Assiniboine while pursuing an undergraduate degree at Northern Montana College. In order to meet the needs of the surrounding community, Northern offered four degree programs — nursing, education, liberal arts and modern farming methods. Given the choices, I decided that education was my best bet, although I was ambivalent about a speciality. While in the military I developed a strong distaste for standing in line, so on the first day of registration when I encountered long lines everywhere except for mathematics education, my choice was clear. I continued to work at Fort Assiniboine for four more years until I completed my undergraduate degree in mathematics education with a minor in biology.

My duties during the seven years of employment at Fort Assiniboine focused on statistics at one level or another, the same cycle being repeated year after year. Starting in the late winter, we would prepare the fertilizer combinations to be tested in the next cycle and lay out the experimental designs on paper. We typically used randomized complete block designs, but had to be prepared with completely randomized and Latin square designs, since we never knew what the experimental location was like before arriving with the planting crew. Split plot and split block designs were also used from time to time. Experimental plots would be planted in the spring (or the late fall in the case of winter wheat), and tended throughout the summer by keeping the alleys between the plots free of weeds. Plots were harvested in the fall, followed by threshing and weighing the wheat. Most of the winter was spent constructing analysis of variance tables with the aid of large desktop Monroe calculators and drawing conclusions prior to the next cycle of experimentation.

During my first year or so at Fort Assiniboine, I functioned mostly as a general laborer, but by the time I finished high school I had developed an appreciation for the research. I was fortunate that, from the beginning, the Department Head, who had a Master's degree in agronomy with a minor in statistics from a Canadian university, encouraged me to set aside time at work to read about experimental design and statistical methods. This involved studying Snedecor's text on *Statistical Methods* and Fisher's monograph on *The Design of Experiments*, in addition to other references. The material came quickly for me, mostly because nearly all that I read corresponded to something we were actually doing. But I recall being a bit baffled by the need to select a significance level and the role of p-values in determining recommendations. The possibility of developing a formal cost function to aid our recommendations did not arise until graduate school some years later. My undergraduate education certainly helped with the mathematics, but was little help with statistics since the only directly relevant offering was a course in probability with a cursory treatment of introductory statistics.

I was eventually given responsibility for nearly all aspects of the experimental trials at the Fort. I had learned to assess the experimental location, looking for hollows and moisture gradients, and to select and arrange an appropriate design. I learned that mis-entering a number could have costly consequences. A yield of 39 bushels mis-entered as 93 bushels per acre could make a non-significant factor seem highly significant, resulting in an unjustified costly recommendation to the wheat farmers. It is for this reason that I instituted "parallel computing." Two of us would sit at adjacent desks and simultaneously construct analysis of variance tables, checking that our results matched at each step of the analysis. A mismatch meant that we had to repeat the calculation in full, since there was no way of recovering what we had entered. We would occasionally lose an experiment at a remote location because the grain was too wet to harvest during the window of opportunity. In an effort to recover some information, I came up with the idea of estimating the number of seed heads per foot of row. That operation required only counting and the moisture content of the grain was irrelevant. A few pilot experiments showed that the count was usefully correlated with the grain weight, so we were able to gain some information from experiments that would be otherwise lost.

During my final year at Northern Montana College, I was required to spend six months student teaching at the local high school where I taught junior algebra and sophomore biology. I found actual teaching quite rewarding, but my overall experience was a disappointment because colorless non-teaching duties dominated my days. The Department Head at the Fort had been encouraging me to pursue a graduate degree in statistics or perhaps mathematics and, after my experience student teaching, I decided to follow his advice. I applied to four universities, two in mathematics and two in statistics, that did not require a fee to process my application because finances were extremely tight. My decision rule was to accept the first that offered a fellowship or assistantship. The following fall I began my graduate studies in statistics at Kansas State University, aided by a traineeship from the National Institutes of Health and subsequently a fellowship under the National Defense Education Act, which was enacted in response to the Soviet Union's successful launch of Sputnik and President Kennedy's moon initiative.

Although my degree from Kansas State was in statistics, my dissertation was in genetics; it was entitled "The Dynamics of Finite Populations: The Effects of Variable Selection Intensity and Population Size on the Expected Time to Fixation and the Ultimate Probability of Fixation of an Allele." I enjoyed seeing genetic theory in action and many hours were spent during my graduate career conducting laboratory experiments with *Drosophila melanogaster*. My first paper was on Bayes' estimators of gene frequencies in natural populations. My background and fellowships enabled me to complete my PhD degree in three years, at which point I joined the then nascent School of Statistics at the University of Minnesota, with an appointment consisting of intramural consulting, teaching and research, in roughly equal proportions. I continued my genetics research for about four years until I had achieved tenure and then began a transition to statistical research, which was largely stimulated and guided by my consulting experiences.

9.2 Statistical diagnostics

In his path-breaking 1922 paper "On the mathematical foundations of theoretical statistics," R.A. Fisher established the contemporary role of a statistical model and anticipated the development of diagnostic methods for model assessment and improvement. Diagnostics was a particularly active research area from the time of Fisher's death in 1962 until the late 1980s, and the area is now an essential ingredient in Fisher modeling.

9.2.1 Influence diagnostics

My involvement with diagnostics began early in my career at Minnesota. A colleague from the Animal Science Department asked me to review a regression because his experiment had apparently produced results that were diametrically opposed to his prior expectation. The experiment consisted of injecting a number of rats with varying doses of a drug and then measuring the fraction of the doses, which were the responses, that were absorbed by the rats' livers. The predictors were various measurements on the rats plus the actual dose. I redid his calculations, looked at residual plots and performed a few other checks that were standard for the time. This confirmed his results, leading to the possibilities that either there was something wrong with the experiment, which he denied, or his prior expectations were off. All in all, this was not a happy outcome for either of us.

I subsequently decided to use a subset of the data for illustration in a regression course that I was teaching at the time. Astonishingly, the selected subset of the data produced results that clearly supported my colleague's prior expectation and were opposed to those from the full data. This caused some anxiety over the possibility that I had made an error somewhere, but after considerable additional analysis I discovered that the whole issue centered on one rat. If the rat was excluded, my colleague's prior expectations were sustained; if the rat was included his expectations were contradicted. The measurements on this discordant rat were accurate as far as anyone knew, so the ball was back in my now quite perplexed colleague's court.

The anxiety that I felt during my exploration of the rat data abated but did not disappear completely because of the possibility that similar situations had gone unnoticed in other regressions. There were no methods at the time that would have identified the impact of the one unusual rat; for example, it was not an outlier as judged by the standard techniques. I decided that I needed a systematic way of finding such influential observations if they were to occur in future regressions, and I subsequently developed a method that easily identified the irreconcilable rat. My colleagues at Minnesota encouraged me to submit my findings for publication (Cook, 1977), which quickly took on a life of their own, eventually becoming known as *Cook's Distance*, although no one sought my acquiescence. In 1982 I coauthored a fairly comprehensive research monograph on the state of diagnostic methods (Cook and Weisberg, 1982).

Encouraged by the wide acceptance of *Cook's Distance* and my other diagnostic contributions, and aided by a year-long fellowship from the Mathematics Research Center at the University of Wisconsin, I continued working in diagnostics with the goal of developing local differential geometric measures that might detect various influential characteristics of a generic likelihood-based analysis. In 1986 I read before the Royal Statistical Society a paper on a local likelihood-based technique for the development of diagnostics to detect influential aspects of an analysis (Cook, 1986).

Today models can be and often are much more complicated than those likely entertained by Fisher or in common use around the time that I was earnestly working on influence diagnostics. As a consequence, the methods developed prior to the 1990s are generally not applicable in more complicated contemporary contexts, and yet these contexts are no less affected by influential observations. Intricate models are prone to instability and the lack of proper influence diagnostics can leave a cloud of doubt about the strength of an analysis. While influence diagnostics have been keeping pace with model development largely through a series of important papers by Hongtu Zhu and his colleagues (Zhu et al., 2007, 2012), methods to address other diagnostic issues, or issues unique to a particular modeling environment, are still lagging far behind. Personally, I am reluctant to accept findings that are not accompanied by some understanding of how the data and model interacted to produce them.

9.2.2 Diagnostics more generally

A substantial battery of diagnostic methods for regression was developed during the 1970s and 1980s, including transformation diagnostics, various graphical diagnostics like residual plots, added variable plots (Cook and Weisberg, 1982), partial residual plots and CERES plots for predictor transformations (Cook, 1993), methods for detecting outliers and influential observations, and diagnostics for heteroscedasticity (Cook and Weisberg, 1983). However, it was unclear how these methods should be combined in a systematic way to aid an analysis, particularly since many of them addressed one issue at a time. For instance, diagnostics for heteroscedasticity required that the mean function be correct, regardless of the fact that an incorrect mean function and homoscedastic errors can manifest as heteroscedasticity. Box's paradigm (Box, 1980) for model criticism was the most successful of the attempts to bring order to the application of diagnostic methods and was rapidly adopted by many in the field. It consists essentially of iteratively improving a model based on diagnostics: an initial model is posited and fitted to the data, followed by applications of a battery of diagnostic methods. The model is then modified to correct the most serious deficiencies detected, if any. This process is then iterated until the model and data pass the selected diagnostic checks.

Diagnostic methods guided by Box's paradigm can be quite effective when the number of predictors is small by today's standards, say less than 20, but in the early 1990s I began encountering many regressions that had too many predictors to be addressed comfortably in this way. I once spent several days analyzing a data set with 80 predictors (Cook, 1998, p. 296). Box's paradigm was quite useful and I was pleased with the end result, but the whole process was torturous and not something I would look forward to doing again. A different diagnostic paradigm was clearly needed to deal with regressions involving a relatively large number of predictors.

9.2.3 Sufficient dimension reduction

Stimulated by John Tukey's early work on computer graphics and the revolution in desktop computing, many dynamic graphical techniques were developed in the late 1980s and 1990s, including linking, brushing, scatterplot matrices, three-dimensional rotation and its extensions to grand tours, interactive smoothing and plotting with parallel coordinates. My first exposure to dynamic graphics came through David Andrews' Macintosh program called McCloud. At one point I thought that these tools might be used effectively in the context of diagnostics for regressions with a relatively large number of predictors, but that proved not to be so. While dynamic graphical techniques allow many plots to be viewed in relatively short time, most low-dimensional projective views of data can be interesting and ponderable, but at the same time do not necessarily provide useful information about the higher dimensional data. In regression for example, two-dimensional plots of the response against various one-dimensional projections of the predictors can be interesting as individual univariate regressions but do not necessarily provide useful information about the overarching multiple regression employing all predictors simultaneously. Many projective views of a regression seen in short time can quickly become imponderable, leaving the viewer with an array of disconnected facts about marginal regressions but little substantive knowledge about the full regression.

My foray into dynamic computer graphics was methodologically unproductive, but it did stimulate a modest epiphany in the context of regression that is reflected by the following question: Might it be possible to construct a low-dimensional projective view of the data that contains all or nearly all of the relevant regression information without the need to pre-specify a parametric model? If such a view could be constructed then we may no longer need to inspect many diagnostic plots and Box's paradigm could be replaced with a much simpler one, requiring perhaps only a single low-dimensional display as a guide to the regression. Stated more formally, can we find a low-dimensional subspace $\mathcal{S}$ of the predictor space with the property that the response Y is independent of the predictor vector X given the projection $P_{\mathcal{S}}X$ of X onto $\mathcal{S}$;

that is, $Y \perp\!\!\!\perp X | P_{\mathcal{S}} X$? Subspaces with this property are called dimension reduction subspaces. The smallest dimension reduction subspace, defined as the intersection of all dimension reduction subspaces when it is itself a dimension reduction subspace, is called the central subspace $\mathcal{S}_{Y|X}$ (Cook, 1994, 1998). The name "central subspace" was coined by a student during an advanced topics course in regression that I was teaching in the early 1990s. This area is now widely know as sufficient dimension reduction (SDR) because of the similarity between the driving condition $Y \perp\!\!\!\perp X | P_{\mathcal{S}_{Y|X}} X$ and Fisher's fundamental notion of sufficiency. The name also serves to distinguish it from other approaches to dimension reduction.

The central subspace turned out to be a very effective construct, and over the past 20 years much work has been devoted to methods for estimating it; the first two methods being sliced inverse regression (Li, 1991) and sliced average variance estimation (Cook and Weisberg, 1991). These methods, like nearly all of the subsequent methods, require the so-called linearity and constant covariance conditions on the marginal distribution of the predictors. Although these conditions are largely seen as mild, they are essentially uncheckable and thus a constant nag in application. Ma and Zhu (2012) recently took a substantial step forward by developing a semi-parametric approach that allows modifications of previous methods so they no longer depend on these conditions. The fundamental restriction to linear reduction $P_{\mathcal{S}_{Y|X}} X$ has also been long recognized as a limitation. Lee et al. (2013) recently extended the foundations of sufficient dimension reduction to allow for non-linear reduction. This breakthrough, like that from Ma and Zhu, opens a new frontier in dimension reduction that promises further significant advances. Although SDR methods were originally developed as comprehensive graphical diagnostics, they are now serviceable outside of that context.

Technological advances resulted in an abundance of applied regressions that Box's paradigm could no longer handle effectively, and SDR methods were developed in response to this limitation. But technology does not stand still. While SDR methods can effectively replace Box's paradigm in regressions with many predictors, they seem ill suited for high-dimensional regressions with many tens or hundreds of predictors. Such high-dimensional regressions were not imagined during the rise of diagnostic or SDR methods, but are prevalent today. We have reached the point where another diagnostic template is needed.

9.2.4 High-dimensional regressions

High-dimensional regressions often involve issues that were not common in the past. For instance, they may come with a sample size n that is smaller than the number of predictors p, leading to the so called "$n < p$" and "$n \ll p$" problems. Some type of specialized structure is needed for the analysis of high-dimensional regressions since they cannot be addressed typically by using traditional methods.

One favored framework imposes a sparsity condition — only a few of the many predictors are relevant for the regression — which reduces the regression goal to finding the relevant predictors. This is now typically done by assuming a model that is (generalized) linear in the predictors and then estimating the relevant predictors by optimizing a penalize objective function. An analysis of a high-dimensional regression based on this approach involves two acts of faith.

The first act of faith is that the regression is truly sparse. While there are contexts where sparsity is a driving concept, some seem to view sparsity as akin to a natural law. If you are faced with a high-dimensional regression then naturally it must be sparse. Others have seen sparsity as the only recourse. In the logic of Bartlett et al. (2004), the bet-on-sparsity principle arose because, to continue the metaphor, there is otherwise little chance of a reasonable payoff. In contrast, it now seems that reasonable payoffs can be obtained also in abundant regressions where many predictors contribute useful information on the response, and prediction is the ultimate goal (Cook et al., 2012).

The second and perhaps more critical act of faith involves believing the data and initial model are flawless, apart from the statistical variation that is handled through the objective function. In particular, there are no outliers or influential observations, any curvature in the mean function is captured adequately by the terms in the model, interactions are largely absent, the response and predictors are in compatible scales and the errors have constant variation. It has long been recognized that regressions infrequently originate in such an Elysian condition, leading directly to the pursuit of diagnostic methods. I can think of no compelling reason these types of considerations are less relevant in high-dimensional regressions. Diagnostic methods can and perhaps should be used after elimination of the predictors that are estimated to be unrelated with the response, but this step alone may be inadequate. Failings of the types listed here will likely have their greatest impact during rather than after penalized fitting. For instance, penalized fitting will likely set the coefficient β of a standard normal predictor X to zero when the mean function in fact depends on X only through a quadratic term βX^2. Findings that are not accompanied by an understanding of how the data and model interacted to produce them should ordinarily be accompanied by a good dose of skepticism.

9.3 Optimal experimental design

My interest in an optimal approach to experimental design arose when designing a comprehensive trial to compare poultry diets at six universities. Although the experimental diets came from a common source, the universities had different capabilities and facilities which made classical Box–Fisher–Yates variance reduction designs difficult to employ, particularly since the underlying non-linear model called for an unbalanced treatment design.

Optimal experimental design was for many years regarded as primarily a mathematical subject. While applications were encountered from time to time, it was seen as largely a sidelight. Few would have acknowledged optimal design as having a secure place in statistical practice because the approach was too dependent on knowledge of the model and because computing was often an impediment to all but the most straightforward applications. During the 1970s and most of the 1980s, I was occasionally a party to vigorous debates on the relative merits of classical design versus optimal design, pejoratively referred to by some as "alphabetic design" in reference to the rather unimaginative design designations like D-, A- and G-optimality. Today classical and optimal design are no longer typically seen as distinct approaches and the debate has largely abated. The beginning of this coalescence can be traced back to technological advances in computing and to the rise of unbalanced experimental settings that were not amenable to classical design (Cook and Nachtsheim, 1980, 1989).

9.4 Enjoying statistical practice

Statistics has its tiresome aspects, to be sure, but for me the practice of statistics has also been the source of considerable pleasure and satisfaction, and from time to time it was even thrilling.

For several years I was deeply involved with the development of aerial survey methods. This included survey methods for snow geese on their molting grounds near Arviat on the west shore of Hudson Bay, moose in northern Minnesota, deer in southern Manitoba and wild horses near Reno, Nevada. It became apparent early in my involvement with these studies that the development of good survey methods required that I be actively involved in the surveys themselves. This often involved weeks in the field observing and participating in the surveys and making modifications on the fly.

The moose and deer surveys were conducted in the winter when foliage was largely absent and the animals stood out against a snowy background. Nevertheless, it soon became clear from my experience that aerial observers would inevitably miss some animals, leading to underestimation of the population size. This visibility bias would be a constant source of uncertainty unless a statistical method could be developed to adjust the counts. I developed different adjustment methods for moose and deer. Moose occur in herds, and it seemed reasonable to postulate that the probability of seeing an animal is a function of the size of its herd, with solitary animals being missed the most frequently. Adding a stable distribution for herd size then led to an adjustment method that resulted in estimates of population size that were in qualitative agreement with estimates from other sources (Cook and Martin,

1974). A different adjustment method for deer censuses was developed based on a design protocol that involved having two observers on the same side of the aircraft. The primary observer in the front seat called out and recorded all the deer that he saw. The secondary observer in the rear seat recorded only deer that the primary observer missed. The resulting data plus a few reasonable assumptions on the generation process led directly to adjusted population counts (Cook and Jacobson, 1979).

A version of mark-capture was developed for estimating population sizes of wild horses. The horses were marked by a tethered shooter leaning out the right side of a helicopter flying near tree-top level above the then running animals. The shooter's demanding task was to use a fancy paint-ball gun to mark the animal on its left rear quarter. I was the primary shooter during the development phase, and I still recall the thrill when the helicopter pulled up sharply to avoid trees or other obstacles.

9.5 A lesson learned

Beginning in my early days at Fort Assiniboine, my statistical perspectives and research have been driven by applications. My work in diagnostic methods originated with a single rat, and my attitude toward inference and diagnostics was molded by the persistent finding that plausible initial models often do not hold up when contrasted against the data. The development of SDR methods was stimulated by the inability of the then standard diagnostic methods to deal effectively with problems involving many variables. And, as mentioned previously, we are now at a point where a new diagnostic paradigm is needed to deal with the high-dimensional regressions of today. My interest in optimal design arose because of the relative rigidity of classical design. My contributions to aerial surveys would have been impossible without imbedding myself in the science. This has taught me a lesson that may seem retrospectively obvious but was not so for me prospectively.

Statistics is driven by applications which are propelled by technological advances, new data types and new experimental constructs. Statistical theory and methods must evolve and adapt in response to technological innovation that give rise to new data-analytic issues. High-dimensional data, which seems to dominate the pages of contemporary statistics journals, may now be overshadowed by "Big Data," a tag indicating a data collection so large that it cannot be processed and analyzed with contemporary computational and statistical methods. Young statisticians who are eager to leave a mark may often find themselves behind the curve when too far removed from application. The greatest statistical advances often come early in the growth of a new area, to be followed by a fleshing out of its nooks and crannies. Immersing oneself in an application can bring a type of satisfaction that may not otherwise be possible.

References

Bartlett, P.L., Bickel, P.J., Bühlmann, P., Freund, Y., Friedman, J., Hastie, T., Jiang, W., Jordan, M.J., Koltchinskii, V., Lugosi, G., McAuliffe, J.D., Ritov, Y., Rosset, S., Schapire, R.E., Tibshirani, R.J., Vayatis, N., Yu, B., Zhang, T., and Zhu, J. (2004). Discussions of boosting papers. *The Annals of Statistics*, 32:85–134.

Box, G.E.P. (1980). Sampling and Bayes' inference in scientific modelling and robustness (with discussion). *Journal of the Royal Statistical Society, Series A*, 143:383–430.

Cook, R.D. (1977). Detection of influential observation in linear regression. *Technometrics*, 19:15–18.

Cook, R.D. (1986). Assessment of local influence (with discussion). *Journal of the Royal Statistical Society, Series B*, 48:133–169.

Cook, R.D. (1993). Exploring partial residual plots. *Technometrics*, 35:351–362.

Cook, R.D. (1994). Using dimension-reduction subspaces to identify important inputs in models of physical systems. In *Proceedings of the Section on Physical Engineering Sciences*, American Statistical Association, Washington, DC, pp. 18–25.

Cook, R.D. (1998). *Regression Graphics*. Wiley, New York.

Cook, R.D., Forzani, L., and Rothman, A.J. (2012). Estimating sufficient reductions of the predictors in abundant high-dimensional regressions. *The Annals of Statistics*, 40:353–384.

Cook, R.D. and Jacobson, J.O. (1979). A design for estimating visibility bias in aerial surveys. *Biometrics*, 34:735–742.

Cook, R.D. and Martin, F. (1974). A model for quadrant sampling with "visibility bias." *Journal of the American Statistical Association*, 69:345–349.

Cook, R.D. and Nachtsheim, C.J. (1980). A comparison of algorithms for constructing exact D-optimal designs. *Technometrics*, 22:315–324.

Cook, R.D. and Nachtsheim, C.J. (1989). Computer-aided blocking of factorial and response surface designs. *Technometrics*, 31:339–346.

Cook, R.D. and Weisberg, S. (1982). *Residuals and Influence in Regression.* Chapman & Hall, London.

Cook, R.D. and Weisberg, S. (1983). Diagnostics for heteroscedasticity in regression. *Biometrika*, 70:1–10.

Cook, R.D. and Weisberg, S. (1991). Comment on "Sliced inverse regression for dimension reduction" (with discussion). *Journal of the American Statistical Association*, 86:316–342.

Lee, K.-Y., Li, B., and Chiaromonte, F. (2013). A general theory for nonlinear sufficient dimension reduction: Formulation and estimation. *The Annals of Statistics*, 41:221–249.

Li, K.-C. (1991). Sliced inverse regression for dimension reduction (with discussion). *Journal of the American Statistical Association*, 86:328–332.

Ma, Y. and Zhu, L. (2012). A semiparametric approach to dimension reduction. *Journal of the American Statistical Association*, 107:168–179.

Zhu, H., Ibrahim, J.G., and Cho, H. (2012). Perturbation and scaled Cook's distance. *The Annals of Statistics*, 40:785–811.

Zhu, H., Ibrahim, J.G., Lee, S., and Zhang, H. (2007). Perturbation selection and influence measures in local influence analysis. *The Annals of Statistics*, 35:2565–2588.

10

Science mixes it up with statistics

Kathryn Roeder
Department of Statistics
Carnegie Mellon University, Pittsburgh, PA

I have many people to thank for encouraging me to write this essay. Indeed I believe I am among the very last group of stragglers to complete the task. My biggest problem was deciding who was the likely audience. My wonderful thesis advisor, Bruce Lindsay, also an author in this volume, told me to pick my own audience. So, while I welcome any reader, I hope that the story of my collaborative work might provide some insights for young researchers.

10.1 Introduction

An early inspiration for my career was a movie shown in an introductory biology class "The Story of Louis Pasteur." Paul Muni won an Academy Award playing Pasteur, the renowned scientist who revolutionized microbiology. Filmed in 1936, the movie was dark, creepy, and melodramatic. Some people might have taken inspiration from the contributions Pasteur made to mankind, but what struck me was that he was portrayed as a real person — vain, egotistical, and driven by his ideas. It resonated with me and gave me a glimpse of a future that I could not have imagined when I was growing up on a farm in rural Kansas. It provided a clue that the crazy intensity I felt could be put to good use. My next realization was that while I felt driven, I was not a great scientist. After working for some years as a research assistant, it was apparent that the life of a mediocre scientist would be dreary indeed; however, I liked the mathematical and statistical stuff the other science majors found dull. And so an academic statistician was born.

10.2 Collaborators

Good collaborators make all the difference, both in terms of happiness and productivity. But how does one find them? Good collaborators, like good friends, cannot be found by direct search. They appear when you are pursuing mutual interests. I've been lucky to collaborate with many great statistical colleagues, post docs and graduate students. Here I'll focus on scientific collaborators, because such bonds are not likely to happen by chance. To gain entrance into good scientific collaborations requires a substantial investment of time and effort. In some cases, even when you are an established researcher you have to work for several years as part of the research team before you get regular access to the leading scientist on a project. I have spent many an hour talking with brilliant graduate students and post docs working for leading researchers. This is a great way to participate in big science projects. I find that if I provide them with statistical insights and guidance, they will help me with the data. Sharing expertise and effort in this way is productive and fun.

A successful applied project requires good data and typically a statistician cannot produce data on her own. Invariably, scientists have invested years of their lives to procure the data to which we want to gain access. Hence, it is traditional for the lab director to be last author, a position of honor, on any papers involving the initial publication of these data. In addition, typically a post doc or graduate student who has also played a substantial role in getting the data is the first author of such papers. Because these data are presented to us electronically, it is easy to forget the tremendous investment others have made. We too want to have a leading role for the statistics team, and the authorship rules can be frustrating. But I have found that it is immensely worthwhile to participate in such projects and contribute where possible. Having made such an investment, it is usually possible to do more involved statistical analysis in a follow-up paper. Naturally, this is where the statisticians get key authorship roles.

Collaboration requires a tremendous amount of sharing and trust, so it is not surprising that it can be challenging to succeed. Just as good collaborations buoy our spirits, bad collaborations wear us down. My mother never went to college, but she was fond of the maxims of economics: "Time is money" and "Don't throw good money after bad" were her favorites. Both of these shed light on the dilemma of what to do about a bad collaboration. None of us likes to invest a lot of effort and get nothing in return, and yet, an unsatisfying or unhappy collaboration does not ultimately lead to good research. I have had many experiences where I've walked away from a project after making substantial investments of effort. I have never regretted getting out of such projects. This leaves more time for other great collaborations.

10.3 Some collaborative projects

I started my search for collaborators as a graduate student. Because I had studied so much biology and chemistry as an undergraduate, it struck me that all those years of training should not be ignored. I decided to hang out with the evolutionary biology graduate students, attending their seminars and social events. In time I found an opportunity to collaborate on a project. The research involved plant paternity. Little did I know that this would be the first joint paper in a long collaborative venture with Bernie Devlin. Years later we married and to date we have co-published 76 papers. Since the early years Bernie's interests have evolved to human genetics and statistical genetics, dovetailing very nicely with my own. So while I can't recommend everyone marry a collaborator, it has benefitted me immensely.

Genetic diversity has been a theme for much of our joint research. To provide an example of how an initial investment in a research topic can lead from one paper to another, I will explain this line of research. But first, to promote understanding of this section, I will provide a very brief primer on genetics. An allele is an alternative form of DNA, located at a specific position on a chromosome. The allele frequency is the probability distribution of the alleles among individuals in the population. Frequently this distribution is estimated using a sample of alleles drawn from the reference database. For example, sickle cell anemia is due to a single base pair change (A to a T) in the beta-globin gene. The particular allelic form with a T is extremely rare in Caucasian populations, but more common in African populations. The reason for this difference in allele frequencies is that the T form provides some benefit in resisting malaria. Thus the selective advantage of the T allele would be felt more strongly in African populations, causing a shift in frequency distribution. Finally, to complete the genetics primer, at each location, a pair of alleles is inherited, one from each parent (except, of course, on chromosome X). By Mendel's law, a parent passes on half of their genetic material to an offspring. It is through these simple inheritance rules that numerous genetic relationships can be inferred (such as paternity).

In our first project we needed to infer the paternal source of inheritance for all the seeds produced by a plant. The maternal source is obvious, because the seeds are produced on the maternal plant. While plants don't pay alimony, paternity is interesting for other reasons. Plants are obviously stationary, but the paternal genes are transmitted via pollen by natural vectors (butterflies and such), so genetic material moves much more widely than expected. It is important to know how far genes move naturally so that we can predict the consequences of genetic engineering. From a statistical point of view, for plants, paternity is inferred just as in humans. When a child matches the alleged father at half her alleles, then he is not excluded for paternity. And if the chance of matching these alleles by chance is very low, then paternity is inferred.

My work with plant paternity was about sex, but it could not be considered sexy. I enjoyed this project very much, but it was of interest to a specialized community. It was my next project that attracted the interest of other statisticians. In the late 1980s and early 1990s, DNA forensic inference was in its infancy. One of the earliest uses of this technique occurred in England. Based on circumstantial evidence a boy was accused of the rape and murder of two girls in his village. During his interrogation he asked for a blood test. As luck would have it, Alec Jeffries, who had just developed a method of DNA fingerprinting, was located just six miles from the village. The boy's blood was tested and he was found to be innocent. This was done by comparing the observed alleles at several genetic locations between the DNA left at the crime scene to the DNA of the suspect. The choice of genetic locations was made so that there were a large number of possible alleles at each location. Consequently the probability of two people matching by chance was extremely low.

This method of garnering evidence was tremendously powerful and as such was highly controversial. Genetic evidence had been used in paternity cases for quite some time, but the impact of using DNA to convict people of serious crimes was much more compelling and there was an obvious need for serious statistical inquiry. Very early on, our colleague, Professor Neil Risch, was invited to work on the problem by a private company, LIFECODES, and also by the FBI. Neil invited us into collaboration. This was the beginning of an exciting period of investigation. The three of us published several papers and were occasionally quoted in the *New York Times*. Although the setting and the technology were new, many statistical questions were familiar from the paternity project. Although we did not set out with a plan to get involved in this hot topic, we would never have had this opportunity if we hadn't gained expertise in the original botanical project.

There were many aspects to this controversy. Here I'll discuss one issue — the suitability of available reference databases for calculating the probability of a match between a suspect and the perpetrator. The question at issue was how much people vary in their allele frequencies across populations. For instance, if an Asian reference sample is available, will it be suitable if the suspect is Korean? Naturally, controversy always rages most keenly when there are little data available from which to definitively answer the questions. Let's examine this one guided by Sewell Wright. If we divide the world up into populations (continental groups) and subpopulations (ethnic groups within a continent), then we can begin to examine this question. We can partition the variance into various levels, among populations, among subpopulations within a population, among individuals within a subpopulation, and finally, layered over all of this is sampling error. Moderate sized samples were immediately available at the population level and it was apparent that populations did not differ strongly. But at the subpopulation level there was considerable sampling error, making it impossible to determine empirically if subpopulations varied strongly. Finally, we could clearly see that individuals varied tremendously, at least if we looked at individuals at the population level. The question remained, could individuals in subpopulations be quite similar and hence be falsely accused? I don't want to rehash the controversy, but suffice it to say that it was a profitable debate and it was great for statisticians to be involved on both sides of the argument. By the end of the decade DNA forensics became a model for solid forensic evidence. Not to mention that, in the meantime, I earned tenure.

The impact of variability in allele frequencies also arises when looking for associations between genetic variants and disease status. Assume two populations vary in allele frequency and they differ in disease prevalence. If the data are pooled over populations, there will be an association between the locus and disease that falsely implies that one allelic variant increases the risk of disease. Indeed this is Simpson's paradox. In most applications, confounding is a frustration about which little can be done beyond taking care not to draw causal inferences for association studies. However, in the genetic context there is a fascinating option that allows us to get a little closer to making causal inferences. It all goes back to population genetics where the variability in allele frequencies is called population substructure.

Suppose we plan to test for association at many genetic locations (SNPs) across the genome using a simple χ^2 test for association at each SNP. Relying on some statistical and population genetic models we can show that in the presence of population substructure the test statistic approximately follows an inflated χ^2-distribution. And, under certain reasonable assumptions, the inflation factor is approximately constant across the genome. Consequently we can estimate this quantity and determine the severity of the confounding. Based on this principle, the approach we developed is called Genomic Control (GC). Applying GC is such a routine part of most tests of genetic association that the original paper is no longer cited.

Recently large platforms of SNPs became available as part of genome wide association studies, or GWAS. From this immense source of genetic information, a very good proxy was discovered that creates an approximate map of genetic ancestry. Using this as a covariate, one can essentially remove the confounding effect of population substructure. The map is generated using dimension reduction techniques such as principal components, or spectral analysis. It is remarkable that while the subtle differences in allele frequency among SNPs is small enough to allow forensic inference to be sufficiently accurate, and yet when accumulated over a huge number of alleles, the combined effect is quite informative about local ancestry.

More recently I've been involved in DNA sequence studies to find genes that increase the risk of autism and other disorders and diseases. This type of research involves large consortiums. Indeed, some papers have hundreds of authors. In big science projects it may seem that there is no place for young researchers to play a role, but this is not the case. Almost all of the established researchers in our group are eager to support young researchers. Moreover, they understand that young researchers need to have primary authorship on these ventures.

10.4 Conclusions

One theme in much of my work has been a willingness and desire to work on things that are new and somewhat controversial. Not surprisingly, these topics often garner more attention than others, and certainly more attention than the statistical contributions would merit on their own. I personally find that such topics excite my curiosity and push me to work more intensely. While working on such topics has the benefit of leading to publication in top journals, it also has drawbacks. When you've been on the opposite side of an argument, tempers can flare, and in the process grants and papers can be rejected. The best policy in this regard is to maintain good communication and try to respect the people on the other side of the argument, even while disagreeing with their opinions. Regarding another controversial topic we published on — heritability of IQ — we encountered some stiff criticism at a leading journal. Our finding was that the so-called maternal effect had a big impact on the IQ of a child. As a consequence it could be argued that it was important that a mother experience a good environment during pregnancy, which hardly seems controversial. Nevertheless, a top researcher from psychometrics said we would damage an entire field of scientific inquiry if we published this paper. Fortunately, Professor Eric Lander argued for our paper and we managed to publish it in *Nature.* Just last year David Brooks, a *New York Times* columnist, mentioned the paper in one of his columns. For me, knowing that an idea could stay alive for nearly two decades and rise to the attention of someone like Brooks was enormously satisfying.

While I have been blessed with singular good luck in my life, writing only about the successes of a career can leave the wrong impression. I wouldn't want young readers to think mine was a life of happy adventures, filled with praise. Just before my first job interview a beloved faculty member told me I was only invited because I was a girl. Nonplussed, I asked him if he could help me choose my interview shoes. As a young faculty member the undergraduate students insisted on calling me Mrs. Roeder until I finally listed my proper name on the syllabus as Professor Roeder, offering helpfully — if you have trouble pronouncing my last name, you can just call me Professor. More recently I had three papers rejected in a month, and one was rejected within four hours of submission. I believe it took me longer to upload it to the journal web site than it took them to review it. But truth be told, I have not found rejections hard to bear. Oddly it is success that can fill me with dread. The year I won the COPSS award I was barely able to construct an adequate acceptance speech. Any field that felt I was a winner was surely a sorry field.

Since that year (1997) I have had many stretches of time wherein I received no praise whatsoever. This gave me ample opportunity to regret my lack of joy at such a wonderful gift. Yet, with or without external validation, I continue to feel the greatest happiness when I see that an idea of mine has worked out. It is at these times I still think of Louis Pasteur and his wild passion to try the rabies vaccine (illegally) and the thrill he must have felt when it worked.

So many years later I cannot help but marvel on the randomness of life and the paths we take in our careers. That biology professor, probably up late writing a grant proposal, may have shown the Pasteur movie because he didn't have time to prepare a lecture. And in some small way it launched my ship. Just thinking about that movie inspired me to look for it on Wikipedia. Imagine my surprise when I discovered that it was a successful Hollywood venture. Indeed it was nominated for best picture. Had I known that this was an award-winning movie, I never would have put so much stock in the personal message I felt it was conveying to me. And if I hadn't, would I have had the satisfying adventures of my career? Life is a mystery.

11

Lessons from a twisted career path

Jeffrey S. Rosenthal
Department of Statistical Sciences
University of Toronto, Toronto, ON

I reflect upon my academic career path that ultimately led to receiving the COPSS Presidents' Award, with the hopes of providing lessons and insights for younger researchers.

11.1 Introduction

On a chilly Toronto evening in February, 2007, my wife and I returned home from a restaurant. My wife went into the kitchen to put some leftovers in the fridge, while I flopped onto the couch and absent-mindedly picked up a laptop computer to check my email. A minute later my wife heard a dazed and confused "Oh my god!" and rushed back in to see what was wrong. I was barely able to mutter that, to my amazement, I had just been selected to receive that year's COPSS Presidents' Award.

The email message talked mostly about boring details, like the importance of my keeping my award "STRICTLY CONFIDENTIAL" until the official announcement (over five months later!). And the award's web page focused more on its sponsorship and eligibility requirements than on its actual meaning and value. But none of that mattered to me: I knew full well that this award was a biggie, generally regarded as the world's top academic prize in statistics. I couldn't believe that they had chosen me to receive it.

Six years later, I still can't.

I was struck then, as I often am, by my career's twists and turns: how some of the most interesting developments were also the least expected, and how unlikely it would have seemed that I would ever win something like the COPSS. In fact, I never set out to be a statistician at all.

Many young statisticians picture COPSS winners as having clear, linear career paths, in which their statistical success always appeared certain. In my case, nothing could be further from the truth. So, in this chapter, I will reflect upon some of the twists and turns of my academic career to date, with the hopes of providing lessons and insights (written in *italics*) for younger researchers.

11.2 Student days

I was an undergraduate student at the University of Toronto from 1984 to 1988. What I remember most from those years is the huge excitement that I felt at being surrounded by so much knowledge and learning. I would run enthusiastically to lectures and meetings, unable to wait for what I would learn next. In addition to my regular classes, I took or audited courses in other subjects of interest (astronomy, chemistry, philosophy, linguistics), joined various clubs and activities, socialized a great deal, played music with friends, developed my spoken French, went on fantastic camping and canoeing trips, and discussed everything with everyone. Around that time, a high school acquaintance (in fact, the young lady that I had taken to my high school prom) remarked that she saw me on campus from time to time, but never managed to talk to me, since I was always rushing off to somewhere else.

Subsequent years of pressure and deadlines have somewhat dulled that initial sense of excitement, but I can still feel and remember it well, and it has carried me through many difficult times. Indeed, if I could give just one piece of advice to students and young academics, it would be this: *maintain your enthusiasm about learning as much as you can about everything. With enough excitement and passion, everything else will follow.*

In my undergraduate studies, I concentrated primarily on pure mathematics and physics, with some computer science on the side. You will notice that "statistics" has not been mentioned here. Indeed, I am *a COPSS winner who never took a single statistics course.* I did, however, benefit tremendously from the rigorous mathematical training that I received instead.

11.2.1 Applying to graduate school

When my undergraduate studies were coming to an end, I was excited to apply to graduate programs. All around me, students were rambling on about being unsure what they wanted to study or what they would do next. I scoffed at them, since I already "knew" what I wanted to study: mathematical analysis with applications to physics! (Statistics never even crossed my mind.)

Despite my successful undergraduate years, I fretted enormously over my grad school applications, applying to loads of programs, wondering what my professors would write about me, thinking I wouldn't get accepted, and so on. That's right: *even future COPSS winners worry about succeeding in academics.*

My math professors advised me that, while there were many good mathematics graduate programs, the best one was at Princeton University. So, I was amazed and delighted to receive a letter accepting me into their PhD program! They even offered a bit of money to help me visit their campus before deciding. So, although I "knew" that I was planning to accept their offer, I found myself on a flight to Newark to visit the famous Princeton campus.

And then a funny thing happened. My visit made me very depressed. It did reinforce the amazing research depth of the Princeton math faculty. But none of the PhD students there seemed happy. They felt a lot of pressure to write very deep doctoral theses, and to finish in four years. They admitted that there wasn't much to "do" at Princeton, and that everyone spent all their time on work with little time for fun. (I asked one of them if there were clubs to go hear music, but they didn't seem to even understand my question.)

I returned to Toronto feeling worried about my choice, and fearing that I might be miserable at Princeton. At the same time, I wondered, did it really make sense to consider such intangible factors when making important academic decisions? I finally decided that the answer was yes, and I stand by that conclusion today: *it is perfectly reasonable to balance personal preferences against academic priorities.*

So, I decided to consider other graduate schools too. After some more travel and much agonizing, I enrolled in the Harvard University Mathematics PhD Program. Harvard also had incredible mathematical research depth, including in mathematical physics, and in addition it was in a fun-seeming city (Boston) with students who seemed to find at least a bit of time to enjoy themselves.

I had made a decision. I had even, I think, made the right decision. Unfortunately, I wasn't sure I had made the right decision. Now, it should be obvious that: *once you have made a decision, stick with it and move on; don't waste time and effort worrying about whether it was correct.* But I didn't follow that advice. For several years, I worried constantly, and absurdly, about whether I should have gone to Princeton instead.

11.2.2 Graduate school beginnings

And so it was that I began my PhD in the Harvard Mathematics Department. I struggled with advanced mathematics courses about strange-seeming abstract algebraic and geometric concepts, while auditing a physics course about the confusing world of quantum field theory. It was difficult, and stressful, but exciting too.

My first big challenge was the PhD program's comprehensive examination. It was written over three different afternoons, and consisted of difficult questions about advanced mathematical concepts. New PhD students were encouraged to take it "on a trial basis" just months after beginning their program. I did my best, and after three grueling days I thought I was probably "close" to the passing line. The next week I nervously went to the graduate secretaries' office to learn my result. When she told me that I passed (unconditionally), I was so thrilled and amazed that I jumped up and down, patted various office staff on their shoulders, raced down to the departmental library, and danced in circles around the tables there. I couldn't believe it.

Passing the comps had the added bonus that I was henceforth excused from all course grades. Three months after arriving at Harvard, "all" I had left to do was write my PhD thesis. Easy, right?

No, not right at all. I was trying to learn enough about state-of-the-art mathematical physics research to make original contributions. But the research papers on my little desk were so difficult and abstract, using technical results from differential geometry and algebraic topology and more to prove impenetrable theorems about 26-dimensional quantum field theories. I remember looking sadly at one such paper, and estimating that I would have to study for about two more years to understand its first sentence.

I got worried and depressed. I had thought that applications of mathematics to physics would be concrete and intuitive and fun, not impossibly difficult and abstract and intangible. It seemed that I would have to work so hard for so many years to even have a chance of earning a PhD. Meanwhile, I missed my friends from Toronto, and all the fun times we had had. I didn't see the point of continuing my studies, and considered moving back to Toronto and switching to something more "practical" like computer programming. That's right: *a COPSS winner nearly dropped out of school.*

11.2.3 Probability to the rescue

While beating my head against the wall of mathematical physics, I had been casually auditing a course in probability theory given by Persi Diaconis. In contrast to all of the technical mathematics courses and papers I was struggling with, probability with Persi seemed fun and accessible. He presented numerous open research problems which could be understood (though not solved) in just a few minutes. There were connections and applications to other subjects and perhaps even to the "real world." I had little to lose, so I nervously asked Persi if I could switch into probability theory. He agreed, and there I was.

I started a research project about random rotations in high dimensions — more precisely, random walks on the compact Lie group $SO(n)$. Although today that sounds pretty abstract to me, at the time it seemed relatively concrete. Using group representation theory, I got an initial result about the mixing time of such walks. I was excited, and told Persi, and he was excited too. I hoped to improve the result further, but for a few weeks I mostly just basked in the glory of success after so much frustration.

And then a horrible thing happened. I realized that my result was wrong! In the course of doing extensive calculations on numerous scraps of paper, I had dropped an "unimportant" constant multiplier. One morning it suddenly occurred to me that this constant couldn't be neglected after all; on the contrary, it nullified my conclusion. In short: *a COPSS winner's first research result was completely bogus.*

I felt sick and ashamed as I informed Persi of the situation, though fortunately he was very kind and understanding. It did teach me a lesson, that I don't always follow but always should: *when you think you have a result, write it down very* carefully *to make sure it is correct.*

After that setback, I worked very hard for months. I wrote out long formulas for group representation values. I simplified them using subtle calculus tricks, and bounded them using coarse dominating integrals. I restricted to a particular case (where each rotation was 180 degrees through some hyperplane) to facilitate computations. Finally, hundreds of pages of scrap paper later, I had actually proved a theorem. I wrote it up carefully, and finally my first research paper was complete. (All of the author's research papers mentioned here are available at `www.probability.ca`.) I knew I had a long road ahead — how long I could not estimate — but I now felt that I was on my way. I enthusiastically attended lots of research seminars, and felt like I was becoming part of the research community.

Over the next couple of years, I worked on other related research projects, and slowly got a few other results. One problem was that I couldn't really judge how far along I was towards my PhD. Did I just need a few more results to finish, or was I still years away? I was mostly too shy or nervous to ask my supervisor, and he didn't offer any hints. I finally asked him if I should perhaps submit my random rotations paper for publication in a research journal (a new experience for me), but he demurred, saying it was "too much of a special case of a special case," which naturally discouraged me further. (As it happens, after I graduated I submitted that very same random rotations paper to the prestigious *Annals of Probability*, and it was accepted essentially without change, leading me to conclude: *PhD students should be encouraged to submit papers for publication.* But I didn't know that then.)

I started to again despair for the future. I felt that if only I could finish my PhD, and get tenure at a decent university, then life would be good. But I wondered if that moment would ever come. Indeed, I was *a future COPSS winner who thought he would never graduate.*

A few weeks later, I was lifted out of my funk by a rather awkward occurrence. One Friday in November 1990, as I was leaving a research meeting with Persi, he casually mentioned that perhaps I should apply for academic jobs for the following year. I was speechless. Did this mean he thought I was already nearly finished my PhD, even while I was despairing of graduating even in the years ahead? I left in a daze, and then spent the weekend puzzled and enthusiastic and worried about what this all meant. When Monday finally came, I sought out Persi to discuss details. In a quick hallway conversation, I told him that if he really did think that I should apply for academic jobs, then I should get on it right away since some of the deadlines were already approaching. Right before my eyes, he considered for several seconds, and then changed his mind! He said that it might be better for me to wait another year instead.

This was quite a roller coaster for me, and I've tried to remember to *be as clear as possible with PhD students about expectations and prognoses.* Nevertheless, I was delighted to know that at least I would (probably) graduate the following year, i.e., after a total of four years of PhD study. I was thrilled to see light at the end of the tunnel.

Finally, the next year, I did graduate, and did apply for academic jobs. The year 1992 was a bad time for mathematical employment, and I felt pessimistic about my chances. I didn't even think to include my full contact information in my applications, since I doubted anyone would bother to contact me. Indeed, when the photocopier added huge ink splotches to some of my application materials, I almost didn't bother to recopy them, since I figured no one would read them anyway. Yes, *a future COPSS winner barely even considered the possibility that anyone would want to offer him a job.*

11.3 Becoming a researcher

To my surprise, I did get job offers after all. In fact, negotiating the job interviews, offers, terms, and acceptances turned out to be quite stressful in and of itself — I wasn't used to discussing my future with department chairs and deans!

Eventually I arranged to spend 1.5 years in the Mathematics Department at the University of Minnesota. They had a large and friendly probability group there, and I enjoyed talking with and learning from all of them. *It is good to be part of a research team.*

I also arranged that I would move from Minnesota to the Statistics Department at my *alma mater*, the University of Toronto. I was pleased to return to the city of my youth with all its fond memories, and to the research-focused (though administration-heavy) university. On the other hand, I was joining a Statistics Department even though *I had never taken a statistics course.*

Fortunately, my new department did not try to "mold" me into a statistician; they let me continue to work as a mathematical probabilist. I applaud them for this, and have come to believe that *it is always best to let researchers pursue interests of their own choosing.*

Despite the lack of pressure, I did hear more about statistics (for the first time) from my new colleagues. In addition, I noticed something interesting in the way my research papers were being received. My papers that focused on technical/mathematical topics, like random walks on Lie groups, were being read by a select few. But my papers that discussed the theory of the newly-popular Markov chain Monte Carlo (MCMC) computer algorithms, which Persi with his usual foresight had introduced me to, were being cited by lots of statistical researchers. This caused me to focus more on MCMC issues, and ultimately on other statistical questions too. Of course, research should never be a popularity contest. Nevertheless, *it is wise to work more on research questions which are of greater interest to others.*

In my case, these reactions led me to focus primarily on the theory of MCMC, which served me very well in building my initial research career. I still considered myself a probabilist (indeed, I recall someone referring to me as "a statistician" around that time, and me feeling uncomfortable with the designation), but my research more and more concerned applications to statistical algorithms. I was publishing papers, and working very hard — my initial small office was right off the main hallway, and colleagues commented about seeing my light still on at 10:30 or 11:00 many evenings. Indeed, *research success always requires lots of hard work and dedication.*

11.3.1 Footprints in the sand

My interactions with the research community developed a slightly odd flavor. MCMC users were aware of my work and would sometimes cite it in general terms ("for related theoretical issues, see Rosenthal"), but hardly anyone would read the actual *details* of my theorems. Meanwhile, my department was supportive from a distance, but not closely following my research. My statistics colleagues were working on questions that I didn't have the background to consider. And probability colleagues wouldn't understand the statistical/MCMC motivation and thus wouldn't see the point of my research direction. So, despite my modest research success, I was becoming academically somewhat isolated.

That was to change when I met Gareth Roberts. He was a young English researcher who also had a probability background, and was also studying theoretical properties of MCMC. The summer of 1994 featured three consecutive conferences that we would both be attending, so I looked forward to meeting him and exploring common interests. Our first encounter didn't go well: I finally cornered him at the conference's opening reception, only to hear him retort "Look, I've just arrived from England and I'm tired and jet-lagged; I'll talk to you tomorrow." Fortunately the next day he was in better form, and we quickly discovered common interests not only in research, but also in music, sports, chess, bridge, jokes, and more. Most importantly, he had a similar (though more sophisticated) perspective about applying probability theory to better understand the nature and performance of MCMC. We developed a fast friendship which has now lasted through 19 years and 33 visits and 38 joint research papers (and counting). Gareth has improved my research career and focus immeasurably; *social relationships often facilitate research collaborations.*

My career still wasn't all smooth sailing. Research projects always seemed to take longer than they should, and lead to weaker results than I'd hoped. *Research, by its very nature, is a slow and frustrating process.* Around that time, one of my PhD students had a paper rejected from a journal, and shyly asked me if that had ever happened to me. I had to laugh; of course it did! *Yes, even COPSS winners get their papers rejected. Often.* Nevertheless, I was getting papers published and doing okay as a researcher — not making any huge impact, but holding my own. I was honored to receive tenure in 1997, thus fulfilling my youthful dream, though that did lead to a depressing few months of drifting and wondering "what should I do next." A very unexpected answer to that question was to come several years later.

11.3.2 The general public

Like many mathematical researchers, sometimes I felt frustrated that I couldn't easily explain my work to non-academics (joking that I was the guy no one wanted to talk to at a party), but I had never pursued this further. In 2003, some writers and journalists in my wife's family decided that I should write a probability book for the general public. Before I knew it, they had put me in touch with a literary agent, who got me to write a few sample chapters, which quickly scored us an actual publishing contract with HarperCollins Canada. To my great surprise, and with no training or preparation, I had agreed to write a book for a general audience about probabilities in everyday life, figuring that *it is good to occasionally try something new and different.*

The book took two years to write. I had to constantly remind myself that writing for a general audience was entirely different from writing a research paper or even a textbook. I struggled to find amusing anecdotes and catchy examples without getting bogged down in technicalities. Somehow I pulled it off: "Struck by Lightning: The Curious World of Probabilities" was published in sixteen editions and ten languages, and was a bestseller in Canada. This in turn led to numerous radio/TV/newspaper interviews, public lectures, appearances in several documentaries, and invitations to present to all sorts of different groups and organizations. Completely unexpectedly, I became a little bit of a "public persona" in Canada. This in turn led to several well-paid consulting jobs (including one involving computer parsing of pdf files of customers' cell phone bills to compare prices), assisting with a high-profile media investigation of a lottery ticket-swapping scandal and publishing about that in the *RCMP Gazette*, serving as an expert witness in a brief to Supreme Court of Canada, and more. *You just can't predict what twists your career will take.*

11.3.3 Branching out: Collaborations

In a different direction, I have gradually started to do more interdisciplinary work. As I have become slightly better known due to my research and/or book and interviews, academics from a variety of departments have started asking me to collaborate on their projects. I have found that it is impossible to "prepare" for such collaborations — rather, you have to *listen carefully and be open to whatever research input your partners require.* Nevertheless, due to some combination of my mathematical and computer and social skills, I have managed to be more helpful than I would have predicted, leading to quite a number of different joint papers. (I guess I am finally a statistician!)

For example, I provided mathematical analysis about generators of credit rating transition processes for a finance colleague. I worked on several papers with computer science and economics colleagues (one of which led to a tricky probability problem, which in turn led to a nice probability paper with Robin Pemantle). I was also introduced to some psychologists working on analyzing youth criminal offender data, which began a long-term collaboration which continues to this day. Meanwhile, an economics colleague asked me to help investigate temperature and population changes in pre-industrial Iceland. And a casual chat with some philosophy professors led to a paper about the probability-related philosophical dilemma called *the Sleeping Beauty problem.*

Meanwhile, I gradually developed a deeper friendship with my department colleague Radu Craiu. Once again, social interaction led to discovering common research interests, in this case concerning MCMC methodology. Radu and I ended up co-supervising a PhD student, and publishing a joint paper in the top-level *Journal of the American Statistical Association* (JASA), with two more papers in preparation. Having a longer-term collaborator within my own department has been a wonderful development, and has once again reminded me that *it is good to be part of a research team.*

More recently, I met a speech pathologist at a lecture and gave her my card. She finally emailed me two years later, asking me to help her analyze subjects' tongue positions when producing certain sounds. Here my undergraduate linguistic course — taken with no particular goal in mind — was suddenly helpful; *knowledge can provide unexpected benefits.* Our resulting collaboration led to a paper in the *Journal of the Acoustical Society of America*, a prestigious journal which is also my second one with the famous initials "JASA."

I was also approached by a law professor (who was the son-in-law of a recently-retired statistics colleague). He wanted to analyze the text of supreme court judgments, with an eye towards determining their authorship: did the judge write it directly, or did their law clerks do it? After a few false starts, we made good progress. I submitted our first methodological paper to *JASA*, but they rejected it quickly and coldly, saying it might be more appropriate for an educational journal like *Chance.* That annoyed me at the time, but made its later acceptance in *The Annals of Applied Statistics* all the more sweet. A follow-up paper was published in the *Cornell Law Review*, and later referred to in the *New York Times*, and more related publications are on the way.

These collaborations were all very different, in both content and process. But each one involved a personal connection with some other researcher(s), which after many discussions eventually led to worthwhile papers published in high-level research journals. I have slowly learned to always be on the lookout for such connections: *Unexpected encounters and social interactions can sometimes lead to major new research collaborations.*

11.3.4 Hobbies to the fore

Another surprise for me has been the extent to which my non-research interests and hobbies have in turn fed my academic activities in unexpected ways.

As a child I did a lot of computer programming of games and other silly things. When email and bulletin boards first came out, I used them too, even though they were considered unimportant compared to "real" computer applications like numerical computations. I thought this was just an idle pastime. Years later, computer usage has become very central to my research and consulting and collaborations: from Monte Carlo simulations to text processing to internet communications, I couldn't function without them. And I've been helped tremendously by the skills acquired through my "silly" childhood hobby.

I'd always played a lot of music with friends, just for fun. Later on, music not only cemented my friendship with Gareth Roberts, it also allowed me to perform at the infamous Bayesian conference "cabarets" and thus get introduced to more top researchers. In recent years, I even published an article about the mathematical relationships of musical notes, which in turn gave me new material for my teaching. Not bad for a little "fun" music jamming.

In my late twenties I studied improvisational comedy, eventually performing in small local comedy shows. Unexpectedly, improv's attitude of "embracing the unexpected" helped me to be a more confident and entertaining teacher and presenter, turning difficult moments into humorous ones. This in turn made me better at media interviews when promoting my book. Coming full circle, I was later asked to perform musical accompaniment to comedy shows, which I continue to this day.

I'd always had a strong interest in Canadian electoral politics. I never dreamed that this would impact my research career, until I suddenly found myself using my computer skills to analyze polling data and projections from the 2011 Canadian federal election, leading to a publication in *The Canadian Journal of Statistics.*

Early in my teaching career, I experimented with alternative teaching arrangements such as having students work together in small groups during class time. (Such practices are now more common, but back in the early 1990s I was slightly ahead of my time.) To my surprise, that eventually led to a publication in the journal *Studies in Higher Education.*

In all of these cases, topics I had pursued on their own merits without connection to my academic career, turned out to be useful in my career after all. So, *don't hesitate to pursue diverse interests — they might turn out to be useful in surprising ways.*

11.4 Final thoughts

Despite my thinking that I "had it all planned out," my career has surprised me many times over. I never expected to work in statistics. I had no idea that MCMC would become such a central part of my research. I never planned to write for the general public, or appear in the media. And I certainly never dreamed that my music or improv or political interests would influence my research profile in any way.

Nevertheless, I have been very fortunate: to have strong family and educational foundations, to attend top-level universities and be taught by top-level professors, and to have excellent opportunities for employment and publishing and more. I am very grateful for all of this. And, it seems that my ultimate success has come *because* of all the twists and turns along the way, not in *spite* of them. Perhaps that is the real lesson here — that, like in improv, we should not fear unexpected developments, but rather embrace them.

My career, like most, has experienced numerous research frustrations, rejected papers, and dead ends. And my university's bureaucratic rules and procedures sometimes make me want to scream. But looking back, I recall my youthful feeling that if only I could get tenure at a decent university, then life would be good. I was right: it has been.

12

Promoting equity

Mary W. Gray
Department of Mathematics and Statistics
American University, Washington, DC

12.1 Introduction

"I'm not a gentleman." This phrase should alert everyone to the fact that this is not a conventional statistics paper; rather it is about the issues of equity that the Elizabeth Scott Award addresses. But that was the phrase that marked my entry onto the path for which I received the award. It had become obvious to me early in my career as a mathematician that women were not playing on a level field in the profession. The overt and subtle discrimination was pervasive. Of course, it was possible to ignore it and to get on with proving theorems. But I had met Betty Scott and other women who were determined not to remain passive. Where better to confront the issues but the Council of the American Mathematical Society (AMS), the group that considered itself the guardian of the discipline. I arrived at one of its meetings, eager to observe the operation of the august group. "This meeting is open only to Council members," I was told. Secure in having read the AMS By-laws, I quoted the requirement that the Council meetings be open to all AMS members. "Oh," said the president of the society, "it's a gentlemen's agreement that they be closed." I uttered my open-sesame and remained.

However, the way the "old boys' network" operated inspired many mathematicians to raise questions at Council meetings and more generally. Results were often discouraging. To the notion of introducing blind-refereeing of AMS publications, the response from our distinguished colleagues was, "But how would we know the paper was any good if we didn't know who wrote it?" Outside Council premises at a national mathematics meeting a well-known algebraist mused, "We once hired a woman, but her research wasn't very good." Faced with such attitudes about women in mathematics, a group of us founded the Association for Women in Mathematics (AWM). In the 40+ years of its existence, AWM has grown to thousands of members from around the world (about 15% of whom are men) and has established many travel grant and fellowship programs to encourage women and girls to study mathematics and to support those in the profession. Prizes are awarded by AWM for the research of established mathematicians and undergraduates and for teaching. Through grant support from NSF, NSA, other foundations and individuals the organization supports "Sonya Kovalevskaya Days" for high school and middle school students at institutions throughout the country, events designed to involve the participants in hands-on activities, to present a broad view of the world of mathematics, and to encourage a sense of comradeship with others who might be interested in mathematics.

And things got better, at least in part due to vigorous advocacy work. The percentage of PhDs in math going to women went from 6% when I got my degree to 30% before settling in the mid-twenties (the figure is more than 40% in statistics). Even "at the top" there was progress. The most prestigious university departments discovered that there were women mathematicians worthy of tenure-track positions. A woman mathematician was elected to the National Academy of Sciences, followed by several more. In its 100+ year history, the AMS finally elected two women presidents (the first in 1983, the last in 1995), compared with five of the seven most recent presidents of the American Statistical Association (ASA). However, a combination of the still chilly climate for women in math and the fact that the gap between very abstract mathematics and the work for political and social rights that I considered important led me to switch to applied statistics.

12.2 The Elizabeth Scott Award

To learn that I had been selected for the Elizabeth Scott Award was a particular honor for me. It was Betty Scott who was responsible in part for my deciding to make the switch to statistics. Because I came to know of her efforts to establish salary equity at Berkeley, when a committee of the American Association of University Professors decided to try to help other institutions accomplish a similar task, I thought of asking Betty to develop a kit (Scott, 1977) to assist them.

Using regression to study faculty salaries now seems an obvious technique; there are probably not many colleges or universities who have not tried to do so in the forty years since Title VII's prohibition of discrimination in employment based on sex became applicable to college professors. Few will remember that when it first was enacted, professors were exempted on the grounds that the qualifications and judgments involved were so subjective and specialized as to be beyond the requirement of equity and that in any case, discrimination would be too difficult to prove. It still is difficult to prove, and unfortunately it still exists. That, through litigation or voluntary action, salary inequities have generally diminished is due partly to the cascade of studies based on the kit and its refinements. As we know, discrimination cannot be proved by statistics alone, but extreme gender-disproportionate hiring, promotion and pay are extremely unlikely to occur by chance.

Professor Scott was not happy with the "remedies" that institutions were using to "fix" the problem. Once women's salaries were fitted to a male model, administrators appeared to love looking at a regression line and circling the observations below the line. "Oh look, if we add $2000 to the salaries of a few women and $1000 to the salaries of a few more, we will fix the problem," they exclaimed. Such an implementation relied on a common misunderstanding of variation. Sure, some women will be below the line and, of course, some women (although probably not many) may be above the male line. But what the regression models generally show is that the overall effect is that, on average, women are paid less than similarly qualified men. What is even worse, all too frequently administrators then engage in a process of showing that those "circled" women really "deserve" to be underpaid on some subjective basis, so that the discrimination continues.

Not only does the remedy described confuse the individual observation with the average, but it may be rewarding exactly the wrong women. No matter how many variables we throw in, we are unlikely entirely to account for legitimate, objective variation. Some women are "better" than other women, or other men, with ostensibly similar qualifications, no matter what metric the institution uses: research, teaching, or service (Gray, 1988). But systematically their salaries and those of all other women trail those of similarly qualified men.

What is appropriate for a statistically identified problem is a statistically based remedy. Thus Betty and I wrote an article (Gray and Scott, 1980), explaining that if the average difference between men's and women's salaries as shown by a regression model is $2000, then the salary of each woman should be increased by that amount. Sorry to say, this is not an idea that has been widely accepted. Women faculty are still paid less on the whole, there are still occasional regression-based studies, there are spot remedies, and often the very best women faculty continue to be underpaid.

The great interest in statistical evaluation of salary inequities, particularly in the complex setting of academia, led to Gray (1993), which expanded on the methods we used. Of course, statistical analysis may fail to show evidence of inequity, but even when it does, a remedy may not be forthcoming because those responsible may be disinclined to institute the necessary changes. If the employers refuse to remedy inequities, those aggrieved may resort to litigation, which is a long, expensive process, often not successful and almost always painful.

Experience as an expert witness teaches that however convincing the statistical evidence might be, absent anecdotal evidence of discrimination and a sympathetic plaintiff, success at trial is unlikely. Moreover, one should not forget the frequent inability of courts to grasp the significance of statistical evidence. Juries are often more willing to make the effort to understand and evaluate such evidence than are judges (or attorneys on both sides). My favorite example of the judicial lack of comprehension of statistics was a US District Court judge's inability to understand that if women's initial salaries were less than men's and yearly increases were across-the-board fixed percentages, the gap would grow progressively larger each year.

A 2007 Supreme Court decision (Ledbetter, 2007) made achieving pay equity for long-term victims of discrimination virtually impossible. The Court declared that inequities in salary that had existed — and in many cases increased — for many years did not constitute a continuing violation of Title VII. Litigation would have had to be instituted within a short time after the very first discriminatory pay check in order for a remedy to be possible. Fortunately this gap in coverage was closed by the passage of the Lilly Ledbetter Fair Pay Act (Ledbetter, 2009), named to honor the victim in the Supreme Court case; the path to equity may prove easier as a result.

12.3 Insurance

Salary inequities are not the only obstacle professional women face. There continues to exist discrimination in fringe benefits, directly financial as well as indirectly through lab space, assignment of assistants, and exclusionary actions of various sorts. Early in my career I received a notice from Teachers Insurance and Annuity Association (TIAA), the retirement plan used at most private and many public universities including American University, listing what I could expect in retirement benefits from my contribution and those of the university in the form of x dollars per $100,000 in my account at age 65. There were two columns, one headed "women" and a second, with amounts 15% higher, headed "men." When I contacted the company to point out that Title VII prohibited discrimination in fringe benefits as well as in salary, I was informed that the figures represented discrimination on the basis of "longevity," not on the basis of sex.

When I asked whether the insurer could guarantee that I would live longer than my male colleagues, I was told that I just didn't understand statistics. Learning that the US Department of Labor was suing another university that had the same pension plan, I offered to help the attorney in charge, the late Ruth Weyand, an icon in women's rights litigation. At first we concentrated on gathering data to demonstrate that the difference in longevity between men and women was in large part due to voluntary lifestyle choices, most notably smoking and drinking. In a settlement conference with the TIAA attorneys, one remarked, "Well, maybe you understand statistics, but you don't understand the law."

This provided inspiration for me to sign up for law courses, thinking I would learn a little. However, it turned out that I really loved the study of law and it was easy (relatively). While I was getting a degree and qualifying as an attorney, litigation in the TIAA case and a parallel case involving a state employee pension plan in Arizona (Arizona Governing Committee, 1983) continued. The latter reached the US Supreme Court first, by which time I was also admitted to the Supreme Court Bar and could not only help the appellee woman employee's lawyer but could write an *amicus curiæ* brief on my own.

Working with a distinguished feminist economist, Barbara Bergmann, we counteracted the legal argument of the insurer, namely that the law required that similarly situated men and women be treated the same, but men and women were not so situated because of differences in longevity. To show that they are similarly situated, consider a cohort of 1000 men and 1000 women at age 65. The death ages of 86% of them can be matched up. That is, a man dies at 66 and a woman dies at 66, a man dies at 90 and a woman dies at 90, etc. The remaining 14% consists of 7% who are men who die early unmatched by the deaths of women and 7% who are women who live longer, unmatched by long-lived men. But 86% of the cohort match up, i.e., men and women are similarly situated and must be treated the same (Bergmann and Gray, 1975).

Although the decision in favor of equity mandated only prospective equal payments, most plans, including TIAA, equalized retirement benefits resulting from past contributions as well. Women who had been doubly disadvantaged by discriminatorily low salaries and gender-based unjustly low retirement income told of now being able to have meat and fresh fruit and vegetables a few times a week as well as the security of a phone (this being before the so-called "Obama phone").

Whereas retirement benefits and employer-provided life insurance are covered by Title VII, the private insurance market is not. Insurance is state-regulated and only a few states required sex equity; more, but not all, required race equity. A campaign was undertaken to lobby state legislatures, state insurance commissions, and governors to establish legal requirements for sex equity in all insurance. Of course, sex equity cuts both ways. Automobile insurance and life insurance in general are more expensive for males than for females. In fact an argument often used to argue against discriminatory rates is that in some states a young male driver with a clean record was being charged more than an older driver (of either sex) with two DUI convictions.

Today women are still underrepresented in the study of mathematics, but thirty years ago, the disparity was even more pronounced. As a result, few activists in the women's movement were willing and able to dig into the rating policies of insurance companies to make the case for equity in insurance on the basis of statistics. One time I testified in Helena, Montana, and then was driven in a snow storm to Great Falls to get the last flight out to Spokane and then on to Portland to testify at a legislative session in Salem, leaving the opposition lobbyists behind. The insurance industry, having many resources in hand, had sent in a new team from Washington that was very surprised to see in Salem the next day that they were confronted once again by arguments for equity. But with little or no money for political contributions, the judicial victory was unmatched with legislative ones. It has been left to the Affordable Care Act (2009) to guarantee non-discrimination in important areas like health insurance. In the past, women were charged more on the ostensibly reasonable grounds that childbirth is expensive and that even healthy women seek health care more often than do men. However, eventually childbirth expenses were no longer a major factor and men begin to accrue massive health care bills, in particular due to heart attacks — in part, of course, because they fail to visit doctors more frequently — but under the existing private systems, rates were rarely adjusted to reflect the shift in the cost of benefits.

12.4 Title IX

Experience with the insurance industry led to more awareness of other areas of sex discrimination and to work with the Women's Equity Action League (WEAL), a lobbying organization concentrating primarily on economic issues and working for the passage and implementation of Title IX, which makes illegal a broad range of sex discrimination in education. The success of American women in sports is the most often cited result of Title IX, but the legislation also led to about half of the country's new doctors and lawyers being women, once professional school admission policies were revised. Statistics also played an important role in Title IX advances (Gray, 1996). *Cohen versus Brown University* (Cohen, 1993) established that women and men must be treated equitably with regard to opportunities to participate in and expenditures for collegiate sports, relying on our statistical evidence of disparities. My favorite case, however, involved Temple University, where the sports director testified that on road trips women athletes were housed three to a room and men two to a room because "women take up less space" (Haffer, 1982). As noted, anecdotal evidence is always useful. In another Philadelphia case a course in Italian was offered at Girls High as the "equal" of a calculus course at the males-only Central High. The US Supreme Court let stand a lower court decision that this segregation by sex was not unconstitutional, but girls were admitted to Central High when the practice was later found unconstitutional under Pennsylvania law (Vorchheimer, 1976).

12.5 Human rights

The Elizabeth Scott Award cites my work exposing discrimination and encouraging political action, which in fact extends beyond work for women in the mathematical sciences to more general defense of human rights. Hands-on experience began when I represented the AMS in a delegation to Montevideo that also included members of the French Mathematical Society, the Mexican Mathematical Society, and the Brazilian Applied Mathematics Society. The goal was to try to secure the release of José Luis Massera, a prominent Uruguayan mathematician who had been imprisoned for many years. We visited prisons, officials, journalists and others, ending up with the colonel who had the power to release the imprisoned mathematician. We spoke of Massera's distinguished mathematics, failing health, and international concern for his situation. The colonel agreed that the prisoner was an eminent personage and added, "He will be released in two months." To our astonishment, he was.

Success is a great motivator; it led to work with Amnesty International (AI) and other organizations on cases, not only of mathematicians, but of many others unjustly imprisoned, tortured, disappeared, or murdered. Once at the Council of the AMS, the issue of the people who had "disappeared" in Argentina, several of them mathematicians known to some on the Council, arose. Then another name came up, one that no one recognized, not surprising because at the time she was a graduate student in mathematics. One of my fellow Council members suggested that as such she was not a "real" mathematician and thus not worthy of the attention of the AMS. Fortunately, that view did not prevail.

Of the cases we worked on, one of the most memorable was that of Moncef Ben Salem, a differential geometer, who had visited my late husband at the University of Maryland before spending more than 20 years in prison or under house arrest in his home country of Tunisia. As a result of the revolution in 2011, he became Minister of Higher Education and Scientific Research there. Meeting in Tunis several times recently, we recalled the not-so-good old days and focused on improvements in higher education and the mathematics education of young people. Much of my work in international development and human rights has come through my association with the American Middle East Education Foundation (Amideast) and Amnesty International, where I served for a number of years as international treasurer — someone not afraid of numbers is always welcome as a volunteer in non-profit organizations. Integrating statistics into human rights work has now become standard in many situations.

An opportunity to combine human rights work with statistics arose in the aftermath of the Rwanda genocide in the 90s. As the liberating force moved across Rwanda from Tanzania in the east, vast numbers of people were imprisoned; two years later essentially the only way out of the prisons was death. The new government of Rwanda asserted quite correctly that the number of prisoners overwhelmed what was left, or what was being rebuilt, of the judicial system. On the other hand, major funders, including the US, had already built some new prisons and could see no end in sight to the incarceration of a substantial portion of the country's population. A basic human rights principle is the right to a speedy trial, with certain due process protections. The

Assistant Secretary for Democracy, Human Rights and Labor in the US State Department, a former AI board member, conceived a way out of the impasse. Begin by bringing to trial a random sample of the internees. Unfortunately the Rwandan government was unhappy that the sample selected did not include their favorite candidates for trial and the scheme was not adopted. It took a decade to come up with a system including village courts to resolve the problem.

A few months after the invasion of Iraq, it became imperative to identify the location and needs of schools in the country. Initial efforts at repair and rehabilitation had been unsuccessful because the schools could not be located — no street addresses were available in Iraq. Using teachers on school holiday for the survey, we managed within two weeks to find the locations (using GPS) and to gather information on the status and needs of all but a handful of the high schools in the country. A later attempt to do a census of the region in Iraq around Kirkuk proved impossible as each interested ethnic group was determined to construct the project in such a way as to inflate the proportion of the total population that came from their group.

Other projects in aid of human rights include work with Palestinian universities in the Occupied Territories, as well as with universities elsewhere in the Middle East, North Africa, Latin America, and the South Pacific on curriculum, faculty governance, and training. A current endeavor involves working the American Bar Association to survey Syrian refugees in Jordan, Lebanon, and Turkey in order to document human rights abuses. Designing and implementing an appropriate survey has presented a huge challenge as more than half of the refugees are not in camps and thus are difficult to locate as well as often reluctant to speak of their experiences.

12.6 Underrepresented groups

Women are by no means the only underrepresented group among mathematicians. In the spirit of the Scott Award is my commitment to increasing the participation of African Americans, Latinos, American Indians and Alaskan Natives, Native Hawaiians and other Pacific Islanders. including National Science Foundation and Alfred P. Sloan funded programs for young students (Hayden and Gray, 1990), as well as support of PhD students, many of whom are now faculty at HBCUs. But certainly much remains to be done, here and abroad, in support of the internationally recognized right to enjoy the benefits of scientific progress (ICESCR, 2013).

Efforts in these areas are designed to support what Elizabeth Scott worked for — equity and excellence in the mathematical sciences — and I am proud to have been given the award in her name for my work.

References

Affordable Health Care Act, Pub. L. 111–148 (2009).

Arizona Governing Committee vs. Norris, 464 US 808 (1983).

Article 15 of the International Covenant on Economic, Social and Cultural Rights (ICESCR).

Bergmann, B. and Gray, M.W. (1975). Equality in retirement benefits. *Civil Rights Digest*, 8:25–27.

Cohen vs. Brown University, 991 F. 2d 888 (1st Cir. 1993), cert. denied, 520 US 1186 (1997).

Gray, M.W. (1988). Academic freedom and nondiscrimination — enemies or allies? *University of Texas Law Review*, 6:1591–1615.

Gray, M.W. (1993). Can statistics tell the courts what they do not want to hear? The case of complex salary structures. *Statistical Science*, 8:144–179.

Gray, M.W. (1996). The concept of "substantial proportionality" in Title IX athletics cases. *Duke Journal of Gender and Social Policy*, 3:165–188.

Gray, M.W. and Scott, E.L. (1980). A "statistical" remedy for statistically identified discrimination. *Academe*, 66:174–181.

Haffer vs. Temple University, 688 F. 2d 14 (3rd Cir. 1982).

Hayden, L.B. and Gray, M.W. (1990). A successful intervention program for high ability minority students. *School Science and Mathematics*, 90:323–333.

Ledbetter vs. Goodyear Tire & Rubber, 550 US 618 (2007).

Lilly Ledbetter Fair Pay Act, Pub. L. 112–2 (2009).

Scott, E.L. (1977). *Higher Education Salary Evaluation Kit.* American Association of University Professors, Washington, DC.

Vorchheimer vs. School District of Philadelphia, 532 F. 2d 880 (3rd Cir. 1976).

Part III

Perspectives on the field and profession

13

Statistics in service to the nation

Stephen E. Fienberg
Department of Statistics
Carnegie Mellon University, Pittsburgh, PA

13.1 Introduction

Let me begin with a technical question:

> Who first implemented large-scale hierarchical Bayesian models, when, and why?

I suspect the answer will surprise you. It was none other than John Tukey, in 1962, with David Wallace and David Brillinger, as part of the NBC Election Night team. What I am referring to is their statistical model for predicting election results based on early returns; see Fienberg (2007).

The methods and election night forecasting model were indeed novel, and are now recognizable as hierarchical Bayesian methods with the use of empirical Bayesian techniques at the top level. The specific version of hierarchical Bayes in the election night model remains unpublished to this day, but Tukey's students and his collaborators began to use related ideas on "borrowing strength," and all of this happened before the methodology was described in somewhat different form by I.J. Good in his 1965 book (Good, 1965) and christened as "hierarchical Bayes" in the classic 1970 paper by Dennis Lindley and Adrian Smith (Lindley and Smith, 1972); see also Good (1980).

I was privileged to be part of the team in 1976 and 1978, and there were close to 20 PhD statisticians involved in one form or another, working in Cherry Hill, New Jersey, in the RCA Lab which housed a large mainframe computer dedicated to the evening's activities (as well as a back-up computer), and a few in New York interacting with the NBC "decision desk." The analysts each had a computer terminal and an assignment of states and political races. A summary of each run of the model for a given race could be read easily from the terminal console but full output went to a nearby line printer and was almost immediately available for detailed examination. Analysts worked with the model, often trying different prior distributions (different past elections chosen as "models" for the ones for which they were creating forecasts) and checking on robustness of conclusions to varying specifications. This experience was one among many that influenced how I continue to think about my approach to Bayesian hierarchical modeling and its uses to the present day; see, e.g., Airoldi et al. (2010).

All too often academic statisticians think of their role as the production of new theory and methods. Our motivation comes in large part from the theoretical and methodological work of others. Our careers are built on, and judged by, our publications in prestigious journals. And we often build our careers around such research and publishing activities.

In this contribution, I want to focus on a different role that many statisticians have and should play, and how this role interacts with our traditional role of developing new methods and publishing in quality journals. This is the role we can fulfill in support of national activities and projects requiring statistical insights and rigor. Much of my story is autobiographical, largely because I know best what has motivated my own efforts and research. Further, I interpret the words "to the nation" in my title quite liberally, so that it includes election night forecasts and other public uses of statistical ideas and methods.

In exploring this theme, I will be paying homage to two of the most important influences on my view of statistics: Frederick Mosteller, my thesis advisor at Harvard, and William Kruskal, my senior colleague at the University of Chicago, where I went after graduation; see Fienberg et al. (2007).

One of the remarkable features of Mosteller's autobiography (Mosteller, 2010) is that it doesn't begin with his early life. Instead, the volume begins by providing chapter-length insider accounts of his work on six collaborative, interdisciplinary projects: evaluating the pre-election polls of 1948, statistical aspects of the Kinsey report on sexual behavior in the human male, mathematical learning theory, authorship of the disputed Federalist papers, safety of anesthetics, and a wide-ranging examination of the Coleman report on equality of educational opportunity. With the exception of mathematical learning theory, all of these deal with important applications, where new theory and methodology or adaption of standard statistical thinking was important. Mosteller not only worked on such applications but also thought it important to carry the methodological ideas back from them into the mainstream statistical literature.

A key theme I emphasize here is the importance of practical motivation of statistical theory and methodology, and the iteration between application and theory. Further, I want to encourage readers of this volume, especially the students and junior faculty, to get engaged in the kinds of problems I'll describe, both because I'm sure you will find them interesting and also because they may lead to your own professional development and advancement.

13.2 The National Halothane Study

I take as a point of departure The National Halothane Study (Bunker et al., 1969). This was an investigation carried out under the auspices of the National Research Council. Unlike most NRC studies, it involved data collection and data analysis based on new methodology. The controversy about the safety of the anesthetic halothane was the result of a series of published cases involving deaths following the use of halothane in surgery. Mosteller offers the following example:

> "A healthy young woman accidentally slashed her wrists on a broken windowpane and was rushed to the hospital. Surgery was performed using the anesthetic halothane with results that led everyone to believe that the outcome of the treatment was satisfactory, but a few days later the patient died. The cause was traced to massive hepatic necrosis — so many of her liver cells died that life could not be sustained. Such outcomes are very rare, especially in healthy young people." (Mosteller, 2010, p. 69)

The NRC Halothane Committee collected data from 50,000 hospital records that were arrayed in the form of a very large, sparse multi-way contingency table, for 34 hospitals, 5 anesthetics, 5 years, 2 genders, 5 age groups, 7 risk levels, type of operation, etc., and of course survival. There were 17,000 deaths. A sample of 25 cases per hospital to estimate the denominator made up the residual 33,000 cases.

When we say the data are sparse we are talking about cells in a contingency table with an average count less than 1! The common wisdom of the day, back in the 1960s, was that to analyze contingency tables, one needed cell counts of 5 or more, and zeros in particular were an anathema. You may even have read such advice in recent papers and books.

The many statisticians involved in the halothane study brought a number of standard and new statistical ideas to bear on this problem. One of these was the use of log-linear models, work done largely by Yvonne Bishop, who was a graduate student at the time in the Department of Statistics at Harvard. The primary theory she relied upon was at that time a somewhat obscure paper by an Englishman named Birch (1963), whose theorem on the existence of maximum likelihood estimates assumed that all cell counts are positive. But she needed a way to actually do the computations, at a time when we were still carrying boxes of punched cards to the computer center to run batch programs!

The simple version of the story — see Fienberg (2011) for a more technical account — was that Yvonne used Birch's results (ignoring the condition on positive cell counts) to derive connections between log-linear and logit models, and she computed maximum likelihood estimates (MLEs) using a version of the method of iterative proportional fitting (IPF), developed by Deming and

Stephan in 1940 for a different but related problem. She applied this new methodology to the halothane study. Because the tables of interest from this study exceeded the capacity of the largest available computers of the day, she was led to explore ways to simplify the IPF calculations by multiplicative adjustments to the estimates for marginal tables — an idea related to models with direct multiplicative estimates such as conditional independence. The amazing thing was that the ideas actually worked and the results were a crucial part of the 1969 published committee report and formed the heart of her 1967 PhD thesis (Bishop, 1967).

Now let me step back to the summer of 1966, when Mosteller suggested a pair of different research problems to me involving contingency tables, one of which involving Bayesian smoothing — the smoothing problem utilized the idea of hierarchical models and thus linked in ways I didn't understand at the time to the election night forecasting model from NBC. Both of these problems were motivated by the difficulties Mosteller had encountered in his work on the halothane study and they ended up in my 1968 PhD thesis (Fienberg, 1968).

It was only later that I began to think hard about Yvonne's problem of zeros and maximum likelihood estimation. This work involved several of my students (and later colleagues), most notably Shelby Haberman (1974) and Mike Meyer. Today the computer programs for log-linear model methods are rooted in code originally developed by Yvonne and Shelby; as for the theoretical question of when zeros matter and when they do not for log-linear models, it was finally resolved by my former Carnegie Mellon student and now colleague, Alessandro Rinaldo (Rinaldo, 2005; Fienberg and Rinaldo, 2012).

The log-linear model work took on a life of its own, and it led to the book with Yvonne and Paul Holland, "Discrete Multivariate Analysis" (1975). Others refer to this book as "the jolly green giant" because of the color of the original cover, and it included many new applications involving extensions of the contingency table ideas. But, over my career, I have found myself constantly returning to problems that treat this log-linear model work as a starting point. I'd like to mention some of these, but only after introducing some additional chronological touch-points.

13.3 The President's Commission and CNSTAT

In 1970, when I was a junior faculty member at the University of Chicago, my senior colleague, Bill Kruskal, seemed to be headed to Washington with enormous frequency. After a while, I learned about his service on the President's Commission on Federal Statistics, led by W. Allan Wallis, who was then President of the University of Rochester, and Fred Mosteller. When the Commission reported in 1971, it included many topics that provide a crosswalk between the academic statistics community and statisticians in the federal government.

The release of the two-volume Report of the President's Commission on Federal Statistics was a defining moment not only for the Federal Statistical System, but also for the National Academy of Sciences. It had many recommendations to improve aspects of the Federal statistical system and its co-ordination. One topic explored at length in the report was privacy and confidentiality — I'll return to this shortly. For the moment I want to focus on the emphasis in the report on the need for outside advice and assessment for work going on the federal government:

> **Recommendation 5–4**: The Commission recommends that a National Academy of Sciences–National Research Council committee be established to provide an outside review of federal statistical activities.

That committee was indeed established a few years later as the Committee on National Statistics (CNSTAT) and it has blossomed to fulfill not only the role envisioned by the Commission members, but also to serve as a repository of statistical knowledge, both about the system and statistical methodology for the NRC more broadly. The agenda was well set by the committee's first chair, William Kruskal, who insisted that its focus be "national statistics" and not simply "federal statistics," implying that its mandate reaches well beyond the usual topics and problems associated with the federal statistics agencies and their data series.

I joined the committee in 1978 and served as Chair from 1981 through 1987. CNSTAT projects over the past 35 years serve as the backdrop for the other topics I plan to cover here.

13.4 Census-taking and multiple-systems estimation

One of the most vexing public controversies that has raged for the better part of the last 40 years, has been the accuracy of the decennial census. As early as 1950 the Census Bureau carried out a post enumeration survey to gauge the accuracy of the count. NRC committees in 1969, and again in 1979, addressed the topic of census accuracy and the possibility that the census counts be adjusted for the differential undercount of Blacks. Following the 1980 census, New York City sued the Bureau, demanding that it use a pair of surveys conducted at census time to carry out an adjustment. The proposed adjustment methodology used a Bayesian version of something known as dual-system estimation, or capture-recapture methodology. For those who have read my 1972 paper (Fienberg, 1972) or Chapter 6 of Bishop, Fienberg and Holland (1975), you will know that one can view this methodology as a variant of a special case of log-linear model methodology.

In the 1980s and again in the 1990s, I was a member of a CNSTAT panel addressing this and other methodological issues. Several authors during this period wrote about multiple-system estimation in the census context. I was one of these authors, a topic to which I contributed.

Political pressure and lawsuits have thwarted the use of this methodology as a formal part of the census. Much of this story is chronicled in my 1999 book with Margo Anderson, "Who Counts? The Politics of Census-Taking in Contemporary America" (Anderson and Fienberg, 1999). In conjunction with the 2000 decennial census, the Bureau used related log-linear methodology to produce population estimates from a collection of administrative lists. This work was revived following the 2010 census. In the meantime, I and others have produced several variants on the multiple-recapture methodology to deal with population heterogeneity, and I have a current project funded by the Census Bureau and the National Science Foundation that is looking once again at ways to use these more elaborate approaches involving multiple lists for both enumeration and census accuracy evaluation.

I'm especially proud of the fact that the same tools have now emerged as major methodologies in epidemiology in the 1990s (IWG, 1995a,b) and in human rights over the past decade (Ball and Asher, 2002). This is an amazing and unexpected consequence of work begun in a totally different form as a consequence of the National Halothane Study and the methodology it spawned.

13.5 Cognitive aspects of survey methodology

One of the things one learns about real sample surveys is that the measurement problems, especially those associated with questionnaire design, are immense. I learned this firsthand while working with data from the National Crime Survey and as a technical advisor to the National Commission on Employment and Unemployment Statistics, both in the 1970s. These matters rarely, if ever, show up in the statistics classroom or in textbooks, and they had long been viewed as a matter of art rather than science.

Triggered by a small 1980 workshop on victimization measurement and cognitive psychology, I proposed that the NRC sponsor a workshop on cognitive aspects of survey measurement that would bring together survey specialists, government statisticians, methodologically oriented statisticians, and cognitive scientists. My motivation was simple: the creative use of statistical thinking could suggest new ways of carrying out interviews that ultimately could improve not only specific surveys, but the field as a whole. Under the leadership of Judy Tanur and Miron Straf, the Committee on National Statistics hosted such a workshop in the summer of 1983 and it produced a widely-cited volume (Jabine et al., 1984), as well as a wide array of unorthodox ideas that have now been instantiated in three major US statistical agencies and are part of the training of survey statisticians. Census Bureau surveys and census forms are now regularly developed using cognitive principles. Today, few students or practitioners understand the methodological roots of this enterprise. Tanur and I also linked these ideas to our ongoing work on combining experiments and surveys (Fienberg and Tanur, 1989).

13.6 Privacy and confidentiality

I mentioned that the President's Commission explored the topic of privacy and confidentiality in depth in its report. But I failed to tell you that most of the discussion was about legal and other protections for statistical databases. Indeed, as I participated in discussions of large government surveys throughout the 1970s and 1980s, the topic was always present but rarely in the form that you and I would recognize as statistical. That began to change in the mid-1980s with the work of my then colleagues George Duncan and Diane Lambert (1986, 1989) interpreting several rules for the protection of confidentiality in government practice using the formalism of statistical decision theory.

I was finally drawn into the area when I was asked to review the statistics literature on the topic for a conference in Dublin in 1992; see Fienberg (1994). I discovered what I like to refer to as a statistical gold-mine whose veins I have been working for the past 21 years. There has been a major change since the President's Commission report, linked to changes in the world of computing and the growth of the World Wide Web. This has produced new demands for access to statistical data, and new dangers of inappropriate record linkage and statistical disclosures. These are not simply national American issues, but rather they are international ones, and they have stimulated exciting technical statistical research.

The Committee on National Statistics has been actively engaged in this topic with multiple panels and workshops on different aspects of privacy and confidentiality, and there is now substantial technical statistical literature on the topic, and even a new online *Journal of Privacy and Confidentiality*; see `http://repository.cmu.edu/jpc/`.

How does this work link to other topics discussed here? Well, much of my research has dealt with the protection of information in large sparse contingency tables, and it will not surprise you to learn that it ties to theory on log-linear models. In fact, there are also deep links to the algebraic geometry literature and the geometry of 2×2 contingency tables, one of those problems Fred Mosteller introduced me to in 1966. See Dobra et al. (2009) for some details.

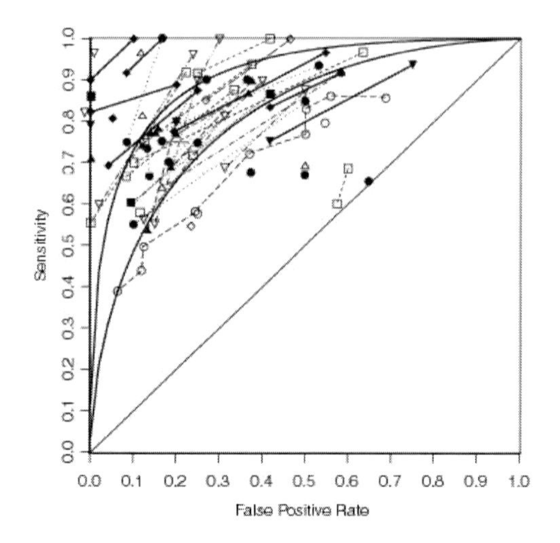

FIGURE 13.1
Sensitivity and false positive rates in 52 laboratory datasets on polygraph validity. Reprinted with permission from *The Polygraph and Lie Detection* by the National Academy of Sciences (NRC, 2003). Courtesy of the National Academies Press, Washington, DC.

13.7 The accuracy of the polygraph

In 2000, I was asked to chair yet another NRC committee on the accuracy of polygraph evidence in the aftermath of the "Wen-Ho Lee affair" at Los Alamos National Laboratory, and the study was in response to a congressional request. My principal qualification for the job, beyond my broad statistical background and my research and writing on forensic topics, was ignorance — I had never read a study on the polygraph, nor had I been subjected to a polygraph exam. I will share with you just one figure from the committee's report (NRC, 2003).

Figure 13.1 takes the form of a receiver operating characteristic plot (or ROC curve plot) that is just a "scatterplot" showing through connected lines the sensitivity and specificity figures derived from each of the 52 laboratory studies that met the committee's minimal quality criteria. I like to refer to this as our "show me the data" plot. Each study has its own ROC curve, points connected by dotted lines. Each point comes from a 2×2 contingency table. You can clearly see why we concluded that the polygraph is better than chance but far from perfect! Further, there are two smooth curves on the graph representing the accuracy scores encompassed by something like the interquartile range of the experimental results. (Since there does not exist a natural definition for quartiles for multivariate data of this nature, the committee first computed the area under the curve for each study, A, rank-ordered the values of A, and then used symmetric ROC curves corresponding to the quartiles for these values. These curves also happen to enclose approximately 50% of the data points as well.) The committee chose this scatterplot, which includes essentially all of the relevant data on accuracy from the 52 studies, rather than a single summary statistic or one with added standard error bounds because we judged it important to make the variability in results visible, and because our internal analyses of the characteristics of the studies left us suspicious that the variability was non-random. Polygraph accuracy likely depends on unknown specifics of the test situation, and we did not want to create the impression that there is a single number to describe polygraph accuracy appropriately across situations.

Although I thought this was going to be a one-of-a-kind activity, I guess I should have known better. About a year after the release of the report, I testified at a senate hearing on the Department of Energy's polygraph policy (Fienberg and Stern, 2005), and continue to be called upon by the media to comment on the accuracy of such methods for the detection of deception — as recently as this month, over 10 years after the publication of our report.

13.8 Take-home messages

What are the lessons you might leave this chapter having learned?

(a) First, it's fun to be a statistician, especially because we can ply our science in a diverse set of ways. But perhaps most of you already knew that.

(b) Second, big problems, especially those confronting the nation, almost always have a statistical component and working on these can be rewarding, both personally and professionally. Your work can make a difference.

(c) Third, working on even small aspects of these large national problems can stimulate the creation of new statistical methodology and even theory. Thus if you engage in such activities, you will still have a chance to publish in the best journals of our field.

(d) Fourth, such new methods often have unplanned-for applications in other fields. This will let you, as a statistician, cross substantive boundaries in new and exciting ways.

Who would have thought that working on a few problems coming out of the Halothane Study would lead to a new integrated set of models and methods that would have impact in many fields and in different forms?

References

Airoldi, E.M., Erosheva, E.A., Fienberg, S.E., Joutard, C., Love, T.M., and Shringarpure, S. (2010). Reconceptualizing the classification of PNAS articles. *Proceedings of the National Academy of Sciences*, 107:20899–20904.

Anderson, M. and Fienberg, S.E. (1999). *Who Counts? The Politics of Census-Taking in Contemporary America*. Russell Sage Foundation, New York.

Ball, P. and Asher, J. (2002). Statistics and Slobodan: Using data analysis and statistics in the war crimes trial of former President Milosevic. *Chance*, 15(4):17–24.

Birch, M.W. (1963). Maximum likelihood in three-way contingency tables. *Journal of the Royal Statistical Society, Series B*, 25:220–233.

Bishop, Y.M.M. (1967). *Multi-Dimensional Contingency Tables: Cell Estimates*. PhD thesis, Department of Statistics, Harvard University, Cambridge, MA.

Bishop, Y.M.M., Fienberg, S.E., and Holland, P.W. (1975). *Discrete Multivariate Analysis: Theory and Practice*. MIT Press, Cambridge MA.

Bunker, J.P., Forrest, W.H., Mosteller, F., and Vandam, L.D., Eds. (1969). *The National Halothane Study*. US Government Printing Office, Washington DC.

Dobra, A., Fienberg, S.E., Rinaldo, A., Slavkovic, A., and Zhou, Y. (2009). Algebraic statistics and contingency table problems: Log-linear models, likelihood estimation, and disclosure limitation. In *Emerging Applications of Algebraic Geometry* (M. Putinar and S. Sullivant, Eds.). *IMA Volumes in Mathematics and its Applications*, vol. 148. Springer, New York, pp. 63–88.

Duncan, G.T. and Lambert, D. (1986). Disclosure-limited data dissemination (with discussion). *Journal of the American Statistical Association*, 81:10–28.

Duncan, G.T. and Lambert, D. (1989). The risk of disclosure for microdata. *Journal of Business & Economic Statistics*, 7:207–217.

Fienberg, S.E. (1968). *The Estimation of Cell Probabilities in Two-Way Contingency Tables*. PhD thesis, Department of Statistics, Harvard University, Cambridge, MA.

Fienberg, S.E. (1972). The multiple recapture census for closed populations and incomplete 2^k contingency tables. *Biometrika*, 59:591–603.

Fienberg, S.E. (1994). Conflicts between the needs for access to statistical information and demands for confidentiality. *Journal of Official Statistics*, 10:115–132.

Fienberg, S.E. (2007). Memories of election night predictions past: Psephologists and statisticians at work. *Chance*, 20(4):8–17.

Fienberg, S.E. (2011). The analysis of contingency tables: From chi-squared tests and log-linear models to models of mixed membership. *Statistics in Biopharmaceutical Research*, 3:173–184.

Fienberg, S.E. and Rinaldo, A. (2012). Maximum likelihood estimation in log-linear models. *The Annals of Statistics*, 40:996–1023.

Fienberg, S.E. and Stern, P.C. (2005). In search of the magic lasso: The truth about the polygraph. *Statistical Science*, 20:249–260.

Fienberg, S.E., Stigler, S.M., and Tanur, J.M. (2007). The William Kruskal legacy: 1919–2005 (with discussion). *Statistical Science*, 22:255–278.

Fienberg, S.E. and Tanur, J.M. (1989). Combining cognitive and statistical approaches to survey design. *Science*, 243:1017–1022.

Good, I.J. (1965). *The Estimation of Probabilities.* MIT Press, Cambridge, MA.

Good, I.J. (1980). Some history of the hierarchical Bayesian methodology (with discussion). In *Bayesian Statistics: Proceedings of the First International Meeting in Valencia (Spain)* (J.M. Bernardo, M.H. DeGroot, D.V. Lindley, and A.F.M. Smith, Eds.), Universidad de Valencia, pp. 489–519.

Haberman, S.J. (1974). *Analysis of Frequency Data.* University of Chicago Press, Chicago, IL.

International Working Group for Disease Monitoring and Forecasting (1995). Capture-recapture and multiple-record systems estimation I: History and theoretical development. *American Journal of Epidemiology*, 142:1047–1058.

International Working Group for Disease Monitoring and Forecasting. (1995). Capture-recapture and multiple-record systems estimation II: Applications in human diseases. *American Journal of Epidemiology*, 142:1059–1068.

Jabine, T.B., Straf, M.L., Tanur, J.M., and Tourangeau, R., Eds. (1984). *Cognitive Aspects of Survey Methodology: Building a Bridge Between Disciplines.* National Academies Press, Washington, DC.

Lindley, D.V. and Smith, A.F.M. (1972). Bayes estimates for the linear model (with discussion). *Journal of the Royal Statistical Society, Series B*, 34:1–41.

Mosteller, F. (2010). *The Autobiography of Frederick Mosteller* (S.E. Fienberg, D.C. Hoaglin, and J.M. Tanur, Eds.). Springer, New York, 2010.

National Research Council Committee to Review the Scientific Evidence on the Polygraph. (2003). *The Polygraph and Lie Detection.* National Academies Press, Washington DC.

President's Commission on Federal Statistics (1971). Federal Statistics. Volumes I and 2. US Government Printing Office, Washington, DC.

Rinaldo, A. (2005). *On Maximum Likelihood Estimates for Contingency Tables.* PhD thesis, Department of Statistics, Carnegie Mellon University, Pittsburgh, PA.

14

Where are the majors?

Iain M. Johnstone
Department of Statistics
Stanford University, Stanford, CA

We present a puzzle in the form of a plot. No answers are given.

14.1 The puzzle

Figure 14.1 suggests that the field of statistics in the US is spectacularly and uniquely unsuccessful in producing Bachelor's degree graduates when compared with every other undergraduate major for which cognate Advanced Placement (AP) courses exist. In those same subject areas, statistics also produces by far the fewest Bachelor's degrees when the normalization is taken as the number of doctoral degrees in that field. We are in an era in which demand for statistical/data scientists and analytics professionals is exploding. The puzzle is to decide whether this plot is spur to action or merely an irrelevant curiosity.

14.2 The data

The "Subjects" are, with some grouping, those offered in AP courses by the College Board in 2006. The number of students taking these courses varies widely, from over 500,000 in English to under 2,000 in Italian. In the 2006 data, the statistics number was about 88,000 (and reached 154,000 in 2012).

The National Center for Education Statistics (NCES) publishes data on the number of Bachelor's, Master's and doctoral degrees conferred by degree granting institutions.

Bachelor-to-AP ratio is, for each of the subjects, the ratio of the number of Bachelor's degrees in 2009–10 to the number of AP takers in 2006. One

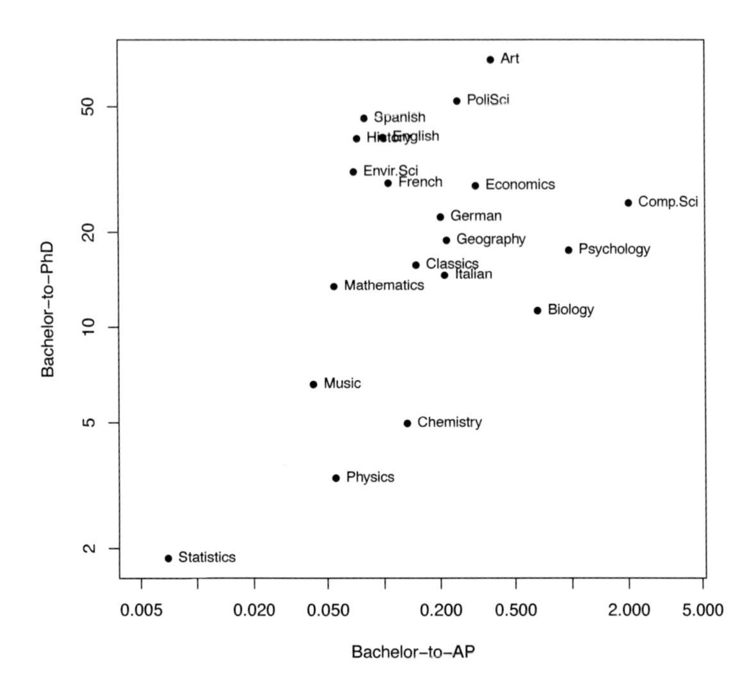

FIGURE 14.1
Normalized Bachelor's degree productivity by discipline.

perhaps should not think of it as a "yield" of Bachelor's degrees from AP takers, but rather as a measure of the number of Bachelor's degrees normalized by Subject size.

Bachelor-to-PhD ratio, for each of the subjects, is the ratio of the number of Bachelor's degrees in 2009–10 to the number of doctoral degrees. Again it is not a measure of the "yield" of doctoral degrees from Bachelor's degrees.

The data (and sources) are in file `degree.xls`. This and other files, including R code, are in directory `Majors` on the page that accompanies this volume: `http://www.crcpress.com/product/isbn/9781482204964`.

14.3 Some remarks

1. It is likely that the number of Bachelors' degrees in statistics is undercounted, as many students may be getting the equivalent of a statistics major in a Mathematics Department. The undercount would have to be by a factor of five to ten to put statistics into the bulk of the plot.

The NCES data shows almost three times as many Master's degrees in Statistics for 2009–10 as Bachelor's degrees (1602 versus 593). This is significant, and may in part reflect the undercount just cited. However, a focus on the Bachelor's level still seems important, as this is seen as the flagship degree at many institutions.

2. There have always been many routes to a career in statistics. For a mathematically inclined student, it often used to be said that an undergraduate focus on statistics was too narrow, and that a broad training in mathematics was a better beginning. My own department did not offer an undergraduate degree in statistics for this reason.

   In 2013, it seems that young people who like data and mathematics should study at least computing, mathematics, and statistics (the order is alphabetical) and that Statistics Departments might design and promote majors that encourage that breadth.

3. An Academic Dean spends a lot of time reading letters of evaluation of teaching, research, etc. In my own experience, the most consistently outstanding student letters about teaching were in philosophy. Why? Here is a conjecture rather than an explanation: in the US, philosophy is not a high school subject — and there is no AP exam — and so its professors have to "start from scratch" in winning converts to the philosophy major. They appear to do this by dint of exemplary teaching.

   What is the relevance of this to statistics? I'm not sure, but statistics likewise has not traditionally been a high school subject (this is changing, as the advent of AP statistics shows), and it seems that statisticians in research universities have not in the past felt the same sense of mission about recruiting to an undergraduate major in statistics.

4. These are data aggregated at a *national* level, and — perhaps — point to a national phenomenon. The good news is that there are outstanding examples of individual departments promoting and growing statistics majors.

Acknowledgements

Thanks to Robert Gould, Nicholas Horton, and Steve Pierson for comments on an earlier draft. After much procrastination, this short piece was written while not preparing a conference talk, in conformance with the Ig Nobel Prize winning article by Perry (1996). The title echoes that of the sometimes forgotten classic Schwed Jr. (1940), without any other claimed similarity.

References

Perry, J. (1996). How to procrastinate and still get things done. *Chronicle of Higher Education.* http://chronicle.com/article/How-to-ProcrastinateStill/93959

Schwed, F. Jr. (1940). *Where Are the Customers' Yachts? Or a Good Hard Look at Wall Street.* Simon and Schuster, Toronto, Canada.

15

We live in exciting times

Peter G. Hall
Department of Mathematics and Statistics
University of Melbourne, Australia
and
Department of Statistics
University of California, Davis, CA

15.1 Introduction

15.1.1 The beginnings of computer-intensive statistics

My involvement in statistics research started at about the time that significant interactive computing power began to become available in university statistics departments. Up until that point, those of us using powerful electronic computers in universities generally were at the mercy of punch cards operating main frame computers, typically at relatively distant locations. This severely hindered the use of computers for assessing the performance of statistical methodology, particularly for developing new approaches. However, once computational experiments could be performed from one's desk, and parameter settings adjusted as the results came in, vast new horizons opened up for methodological development.

The new statistical approaches to which this led were able, by virtue of powerful statistical computing, to do relatively complex things to data. For many of us, Cox's regression model (Cox, 1972), and Efron's bootstrap (Efron, 1979a), became feasible only in the late 1970s and early 1980s. Efron (1979b) gave a remarkably prescient account of the future relationship between theory and computation in modern statistics, noting that:

> "The need for a more flexible, realistic, and dependable statistical theory is pressing, given the mountains of data now being amassed. The prospect for success is bright, but I believe the solution is likely to lie along the lines suggested in the previous sections — a blend of traditional mathematical thinking combined with the numerical and organizational aptitude of the computer."

Critically, Efron saw theory and computing working together to ensure the development of future statistical methodology, meeting many different demands. And, of course, that is what happened, despite the concerns of some (see, e.g., Hall, 2003, p. 165) that advances in computing threatened to replace theoretical statistical argument.

15.1.2 Computer-intensive statistics and me

In this short paper I shall give a personal account of some of my experiences at this very exciting time. My intention is to focus particularly on a 15-year period, from the late 1970s to the early 1990s. However, it will be necessary from time to time to stray to the present, in order to make a didactic point, and to look back at the past, so as to see how much we still have in common with our forebears like R.A. Fisher and E.J.G. Pitman.

I feel particularly fortunate to have been able to work on the development of statistical methodology during such an era of seminal change. I'll explain how, as in many of the most important things in life, I came to this role largely by accident, while looking for a steady job in probability rather than statistics; and how the many advances in computing that have taken place since then have actually created an increasingly high demand for theoretical analysis, when some of my colleagues had predicted there would actually be much less.

I'm going to begin, in Section 15.2, by trying to capture some of the recent concerns I have heard about the directions being taken by statistics today, and by indicating why I generally do not share the anxiety. To some extent, I feel, the concerns exist only if we try to resist change that we should accept as exciting and stimulating, rather than as a threat.

Also in Section 15.2 I'll try to set this disquiet against the background of the many changes that stimulated my work, particularly from the late 1970s to the early 1990s, and argue that the changes we are seeing today are in part a continuation of the many technological advances that have taken place over a long period. The changes add to, rather than subtract from, our field.

In Section 15.3, with the reader's indulgence I'll focus more sharply on my own experience — on how I came to be a theoretical statistician, and how theory has always guided my intuition and led, somewhat indirectly, to my work on computer-intensive statistical methods. I know that, for many of my colleagues in Australia and elsewhere, my "cart" of theoretical analysis comes before their "horse" of statistical computing. The opportunity to write this short chapter gives me a chance of explaining to them what I've been doing all these years, and why.

15.2 Living with change

15.2.1 The diversification of statistics

Just a few years ago I had a conversation with a colleague who expressed grave concern for the future of statistics. He saw it being taken over by, or subsumed into, fields as diverse and disparate as computer science and bioinformatics, to name only two. He was worried, and wondered what we could do to stop the trend.

Similar disquiet has been articulated by others, and not just recently. The eminent British statistician D.J. Finney expressed apprehensions similar to those of my colleague, although at a time when my colleague had not detected much that was perceptibly wrong. Writing in the newsletter of the Royal Statistical Society (*RSS News*) in 2000, Professor Finney argued that:

> "... professional statisticians may be losing control of — perhaps even losing concern for — what is done in the name of our discipline. I began [this article] by asking 'Whither... [statistics]?' My answer is 'Downhill!' Any road back will be long and tortuous, but unless we find it we fail to keep faith with the lead that giants gave us 75 years ago."

I'm not sure whether it is pragmatism or optimism that keeps me from worrying about these issues — pragmatism because, even if these portents of calamity were well founded, there would not be much that we could do about them, short of pretending we had the powers of Cnut the Great and commanding the tide of statistical change to halt; or optimism, on the grounds that these changes are actually healthy, and more likely to enrich statistics than destroy it.

In 1986 the UCLA historian Theodore Porter wrote that:

> "Statistics has become known in the twentieth century as the mathematical tool for analysing experimental and observational data. Enshrined by public policy as the only reliable basis for judgements such as the efficacy of medical procedures or the safety of chemicals, and adopted by business for such uses as industrial quality control, it is evidently among the products of science whose influence on public and private life has been most pervasive."

The years since then have only deepened the involvement of statistics in science, technology, social science and culture, so that Porter's comments about the 20th Century apply with even greater force in the early 21st. Hal Varian's famous remark in *The McKinsey Quarterly*, in January 2009, that "the sexy job in the next ten years will be statisticians," augments and reinforces Porter's words almost 30 years ago. Statistics continues to be vibrant and vital, I think because of, not despite, being in a constant state of change.

In general I find that the changes lamented by Finney, and applauded by Varian, invigorate and energize our field. I'm not troubled by them, except for the fact that they can make it more challenging to get funding for positions in statistics, and more generally for research and teaching in the field. Indeed, it is not just in Australia, but across the globe, that the funding pool that is notionally allocated to statistics is being challenged by many different multi-disciplinary pressures, to such an extent that financial support for core research and teaching in statistics has declined in many cases, at least relative to the increasing size of our community.

Funding is flowing increasingly to collaborative research-centre type activities, where mathematical scientists (including statisticians) are often not involved directly at all. If involved, they are often present as consultants, rather than as true collaborators sharing in the funding. This is the main danger that I see, for statisticians, in the diversification of statistics.

15.2.2 Global and local revolutions

The diversification has gone hand in hand with a revolution, or rather several revolutions, that have resulted in the past from the rapid development of inexpensive computing power since the 1970s, and from technologies that have led to an explosion in machine recorded data. Indeed, we might reasonably think that massive technological change has altered our discipline mainly through the ways our data are generated and the methods we can now use to analyse them.

However, while those changes are very important, they are perhaps minor when set against the new questions that new sorts of data, and new computational tools, enable scientists to ask, and the still newer data types, and data analyses, that they must address in order to respond to those questions. Statisticians, and statistical methods, are at the heart of exploring these new puzzles and clarifying the new directions in which they point.

Some aspects of the revolutions are "global," in the sense that, although the motivation might come from a particular area of application, the resulting changes influence many fields. Others are more local; their impact is not so widespread, and sometimes does not stray terribly far from the area of application that first motivated the new developments.

During the last 30 years or so we have seen examples of both global and local revolutions in statistical methodology. For example, Efron's (1979a) bootstrap was definitely global. Although it arguably had its origins in methodology for sample survey data, for example in work of Hubback (1946), Mahalanobis (1946), Kish (1957), Guerney (1963), and McCarthy (1966, 1969), it has arguably touched all fields of statistics. [Hall (2003) gave a brief account of the prehistory of the bootstrap.] Statistical "revolutions" that are more local, in terms of influence, include work during the 1980s on image analysis, and much of today's statistical research on very high-dimensional data analysis.

Some of this work is having as much influence on statistics itself as on the fields of science and technology that motivated it, and some of that influence is particularly significant. For example, we have benefited, and are still benefiting, from appreciating that entropy is a metric that can be used to assess the performance of statistical smoothing in general, non-imaging contexts (Donoho et al., 1992). And we are using linear models to select variables in relatively low-dimensional settings, not just for high-dimensional data. (The linear model is, after all, a wonderful parametric surrogate for monotonicity in general contexts.) Some of these new methodologies have broken free from statistical gravity and risen above the turmoil of other developments with which they are associated. They include, in the setting of modern high-dimensional data analysis, the lasso and basis pursuit; see, e.g., Tibshirani (1996, 2014a,b) and Chen et al. (1998).

15.3 Living the revolution

15.3.1 A few words to the reader

A great deal of Section 15.3, particularly Section 15.3.2, is going to be about me, and for that I must apologize. Please feel free to skip Section 15.3.2, and come back in at Section 15.3.3.

15.3.2 A little background

Einstein's (1934) advocacy of mathematics as a pillar for creative reasoning in science, or at least in physics, has been cited widely:

> "... experience of course remains the sole criterion of the serviceability of a mathematical construction for physics, but the truly creative principle resides in mathematics."

I'm not going to suggest for a moment that all statisticians would agree that Einstein's argument is valid in their field. In fact, I know some who regard theory as a downright hindrance to the way they do research, although it works for me. However, let's begin at the beginning.

I spent my high school days happily, at a very good government school in Sydney. However, the school's science curriculum in the 1960s effectively permitted no experimental work for students. This saved money, and since there were no state-wide exams in practical work then the effects of that privation were not immediately apparent. It meant that my study of science was almost entirely theoretical, and that suited me just fine. But it didn't prepare me well for my first year at university, where in each of biology, chemistry and physics I had several hours of practical work each week.

I entered the University of Sydney as an undergraduate in early 1970, studying for a science degree and convinced I would become a theoretical physicist, but I hadn't bargained for all the lab work I would have to do before I reached that goal. Moreover, the mathematics in my physics classes was not rigorous enough for me. Asymptotic mathematical formulae abounded, but they were never proved "properly." I quickly came to the conclusion that it would be better for all concerned, including those around me (I once flooded the physics lab), if I devoted my energies to mathematics.

However, for my second year I needed three subjects. I was delighted to be able to study pure mathematics and applied mathematics in separate programs, not just one as had been the case in first year. But I had to find a third subject, one that didn't require any lab work. I saw in the student handbook that there was a subject called mathematical statistics, which in those days, at the University of Sydney, started as a major in second year — so I hadn't missed anything in my first year. And it had "mathematical" in its title, so I felt it would probably suit.

Indeed it did, particularly the course on probability from Feller's wonderful Volume I (Feller, 1968). In second and third years the mathematical statistics course was streamed into "pass" and "honors" levels, and for the latter I had to take extra lectures, which were invariably at a high mathematical level and which I found fascinating. I even enjoyed the classes that led to Ramsey's theorem, although I could not any longer reproduce the proof!

I took a course on measure theory in the third year pure mathematics curriculum, and it prepared me very well for a fourth year undergraduate mathematical statistics course in probability theory, based on Chung's graduate text (Chung, 1968). That course ran for a full year, three lectures a week, and I loved both it and Chung's book.

I appreciate that the book is not for everyone. Indeed, more than a few graduate students have confided to me how difficult they have found it to get into that text. But for me it had just the right mix of intuition, explanation, and leave-it-up-to-the-reader omissions to keep my attention. Chung's style captivated me, and I'm pleased to see that I'm still not alone. (I've just read the five reviews of the third edition on Amazon, and I'm delighted to say that each of them gives the book five stars.)

Another attraction of the course was that I could give every third lecture myself. It turned out I was the only student in the course, although a very amiable logician, Gordon Monro from the Department of Pure Mathematics, also attended. The two of us, and our assigned lecturer Malcolm Quine, shared the teaching among us. It was a wonderful experience. It gave me a lifelong love of probability, and also of much of the theory that underpins statistics.

Now let's fast forward to late 1976, when I returned to Australia from the UK after completing my doctorate in probability. I had a short-term contract job at the University of Melbourne, with no opportunity in 1976 for anything more permanent. In particular, I was having significant difficulty finding a longer-term position in probability. So I applied for any job that had some connection to probability or theoretical statistics, including one at the Australian National University that was advertised with a preference for a biometrician.

This was in 1977, and I assume that they had no plausible applicants in biometrics, since they offered the position to me. However, the Head of Department, Chip Heathcote, asked me (quite reasonably) to try to migrate my research interests from probability to statistics. (Biometrics wasn't required.) He was very accommodating and nice about it, and in particular there was no deadline for making the switch.

I accepted the position and undertook to make the change, which I found that I quite enjoyed. On my reckoning, it took me about a decade to move from probability to statistics, although some of my colleagues, who perhaps wouldn't appreciate that Einstein's remark above, about physics, might apply to statistics as well, would argue that I have still got a way to go. I eased myself into statistics by taking what I might call the "contemporary nonparametric" route, which I unashamedly admit was much easier than proceeding along a parametric path.

At least in the 1970s and 1980s, much of nonparametric statistics (function estimation, the jackknife, the bootstrap, etc) was gloriously ad hoc. The basic methodology, problems and concepts (kernel methods, bias and variance estimation, resampling, and so forth) were founded on the fact that they made good intuitive sense and could be justified theoretically, for example in terms of rates of convergence. To undertake this sort of research it was not necessary to have at your fingertips an extensive appreciation of classical statistical foundations, based for example on sufficiency and efficiency and ancillarity and completeness and minimum variance unbiasedness. Taking a nonparametric route, I could start work immediately. And it was lots of fun.

I should mention, for the benefit of any North American readers who have got this far, that in Australia at that time there was virtually no barrier between statistics and probability. Practitioners of both were in the same department, typically a Department of Statistics, and a Mathematics Department was usually devoid of probabilists (unless the department also housed statisticians). This was one of many structures that Australian universities inherited from the United Kingdom, and I have always found it to be an attractive, healthy arrangement. However, given my background you'd probably expect me to have this view.

The arrangement persists to a large extent today, not least because many Australian Statistics Departments amalgamated with Mathematics Departments during the budget crises that hit universities in the mid to late 1990s. A modern exception, more common today than thirty years ago, is that strong statistics groups exist in some economics or business areas in Australian universities, where they have little contact with probability.

15.3.3 Joining the revolution

So, this is how I came to statistics — by necessity, with employment in mind and having a distinct, persistently theoretical outlook. A paper on nonparametric density estimation by Eve Bofinger (1975), whom I'd met in 1974 while I was an MSc student, drew a connection for me between the theory of order statistics and nonparametric function estimation, and gave me a start there in the late 1970s.

I already had a strong interest in rates of convergence in the central limit theorem, and in distribution approximations. That gave me a way, in the 1980s, of accessing theory for the bootstrap, which I found absolutely fascinating.

All these methodologies — function estimation, particularly techniques for choosing tuning parameters empirically, and of course the bootstrap — were part of the "contemporary nonparametric" revolution in the 1970s and 1980s. It took off when it became possible to do the computation. I was excited to be part of it, even if mainly on the theoretical side. In the early 1980s a senior colleague, convinced that in the future statistical science would be developed through computer experimentation, and that the days of theoretical work in statistics were numbered, advised me to discontinue my interests in theory and focus instead on simulation. However, stubborn as usual, I ignored him.

It is curious that the mathematical tools used to develop statistical theory were regarded firmly as parts of probability theory in the 1970s, 80s and even the 90s, whereas today they are seen as statistical. For example, recent experience serving on an IMS committee has taught me that methods built around empirical processes, which were at the heart of wondrous results in probability in the 1970s (see, e.g., Komlós et al., 1975, 1976), are today seen by more than a few probabilists as distinctly statistical contributions. Convergence rates in the central limit theorem are viewed in the same light. Indeed, most results associated with the central limit theorem seem today to be seen as statistical, rather than probabilistic. (So, I could have moved from probability to statistics simply by standing still, while time washed over me!)

This change of viewpoint parallels the "reinterpretation" of theory for special functions, which in the era of Whittaker and Watson (1902), and indeed also of the more widely used fourth edition in 1927, was seen as theoretical mathematics, but which, by the advent of Abramowitz and Stegun (1964) (planning for that volume had commenced as early as 1954), had migrated to the realm of applied mathematics.

Throughout all this work in nonparametric statistics, theoretical development was my guide. Using it hand in hand with intuition I was able to go much further than I could have managed otherwise. This has always been my approach — use theory to augment intuition, and allow them to work together to elucidate methodology.

Function estimation in the 1970s and 1980s had, to a theoretician like myself, a fascinating character. Today we can hardly conceive of constructing a nonparametric function estimator without also estimating an appropriate smoothing parameter, for example a bandwidth, from the data. But in the 1970s, and indeed for part of the 80s, that was challenging to do without using a mainframe computer in another building and waiting until the next day to see the results. So theory played a critical role.

For example, Mike Woodroofe's paper (Woodroofe, 1970) on asymptotic properties of an early plug-in rule for bandwidth choice was seminal, and was representative of related theoretical contributions over the next decade or so. Methods for smoothing parameter choice for density estimation, using cross-validation and suggested by Habemma et al. (1974) in a Kullback–Leibler setting, and by Rudemo (1982) and Bowman (1984) for least squares, were challenging to implement numerically at the time they were introduced, especially in Monte Carlo analysis. However, they were explored enthusiastically and in detail using theoretical arguments; see, e.g., Hall (1983, 1987) and Stone (1984).

Indeed, when applied to a sample of size n, cross-validation requires $O(n^2)$ computations, and even for moderate sample sizes that could be difficult in a simulation study. We avoided using the Gaussian kernel because of the sheer computational labour required to compute an exponential. Kernels based on truncated polynomials, for example the Bartlett–Epanechnikov kernel and the biweight, were therefore popular.

In important respects the development of bootstrap methods was no different. For example, the double bootstrap was out of reach, computationally, for most of us when it was first discussed (Hall, 1986; Beran, 1987, 1988). Hall (1986, p. 1439) remarked of the iterated bootstrap that "it could not be regarded as a general practical tool." Likewise, the computational challenges posed by even single bootstrap methods motivated a variety of techniques that aimed to provide greater efficiency to the operation of sampling from a sample, and appeared in print from the mid 1980s until at least the early 1990s. However, efficient methods for bootstrap simulation are seldom used today, so plentiful is the computing power that we have at our disposal.

Thus, for the bootstrap, as for problems in function estimation, theory played a role that computation really couldn't. Asymptotic arguments pointed authoritatively to the advantages of some bootstrap techniques, and to the drawbacks associated with others, at a time when reliable numerical corroboration was hard to come by. The literature of the day contains muted versions of some of the exciting discussions that took place in the mid to late 1980s on this topic. It was an extraordinary time — I feel so fortunate to have been working on these problems.

I should make the perhaps obvious remark that, even if it had been possible to address these issues in 1985 using today's computing resources, theory still would have provided a substantial and unique degree of authority to the development of nonparametric methods. In one sweep it enabled us to address issues in depth in an extraordinarily broad range of settings. It allowed us to diagnose and profoundly understand many complex problems, such as the high variability of a particular method for bandwidth choice, or the poor coverage properties of a certain type of bootstrap confidence interval. I find it hard to believe that numerical methods, on their own, will ever have the capacity to deliver the level of intuition and insightful analysis, with such breadth and clarity, that theory can provide.

Thus, in those early developments of methodology for function estimation and bootstrap methods, theory was providing unrivaled insights into new methodology, as well as being ahead of the game of numerical practice. Methods were suggested that were computationally impractical (e.g., techniques for bandwidth choice in the 1970s, and iterated bootstrap methods in the 1980s), but they were explored because they were intrinsically attractive from an intuitive viewpoint. Many researchers had at least an impression that the methods would become feasible as the cost of computing power decreased, but there was never a guarantee that they would become as easy to use as they are today. Moreover, it was not with a view to today's abundant computing resources that those computer-intensive methods were proposed and developed. To some extent their development was unashamedly an intellectual exercise, motivated by a desire to push the limits of what might sometime be feasible.

In adopting this outlook we were pursuing a strong precedent. For example, Pitman (1937a,b, 1938), following the lead of Fisher (1935, p. 50), suggested general permutation test methods in statistics, well in advance of computing technology that would subsequently make permutation tests widely applicable. However, today the notion that we might discuss and develop computer-intensive statistical methodology, well ahead of the practical tools for implementing it, is often frowned upon. It strays too far, some colleagues argue, from the practical motivation that should underpin all our work.

I'm afraid I don't agree, and I think that some of the giants of the past, perhaps even Fisher, would concur. The nature of revolutions, be they in statistics or somewhere else, is to go beyond what is feasible today, and devise something remarkable for tomorrow. Those of us who have participated in some of the statistics revolutions in the past feel privileged to have been permitted free rein for our imaginations.

Acknowledgements

I'm grateful to Rudy Beran, Anirban Dasgupta, Aurore Delaigle, and Jane-Ling Wang for helpful comments.

References

Abramowitz, M. and Stegun, I.A., Eds. (1964). *Handbook of Mathematical Functions with Formulas, Graphs, and Mathematical Tables*, National Bureau of Standards, Washington, DC.

Beran, R. (1987). Prepivoting to reduce level error in confidence sets. *Biometrika*, 74:457–468.

Beran, R. (1988). Prepivoting test statistics: A bootstrap view of asymptotic refinements. *Journal of the American Statistical Association*, 83:687–697.

Bofinger, E. (1975). Estimation of a density function using order statistics. *Australian Journal of Statistics*, 17:1–7.

Bowman, A.W. (1984). An alternative method of cross-validation for the smoothing of density estimates. *Biometrika*, 71:353–360.

Chen, S., Donoho, D., and Saunders, M. (1998). Atomic decomposition by basis pursuit. *SIAM Journal of Scientific Computing*, 20:33–61.

Chung, K.L. (1968). *A Course in Probability Theory*. Harcourt, Brace & World, New York.

Cox, D.R. (1972). Regression models and life-tables (with discussion). *Journal of the Royal Statistical Society, Series B*, 34:187–220.

Donoho, D., Johnstone, I.M., Hoch, J.C., and Stern, A.S. (1992). Maximum entropy and the nearly black object (with discussion). *Journal of the Royal Statistical Society, Series B*, 54:41–81.

Efron, B. (1979a). Bootstrap methods: Another look at the jackknife. *The Annals of Statistics*, 7:1–26.

Efron, B. (1979b). Computers and the theory of statistics: Thinking the unthinkable. *SIAM Review*, 21:460–480.

Einstein, A. (1934). On the method of theoretical physics. *Philosophy of Science*, 1:163–169.

Feller, W. (1968). *An Introduction to Probability Theory and its Applications*, Vol. I, 3rd edition. Wiley, New York.

Fisher, R.A. (1935). *The Design of Experiments*. Oliver & Boyd, Edinburgh.

Guerney, M. (1963). *The Variance of the Replication Method for Estimating Variances for the CPS Sample Design*. Unpublished Memorandum, US Bureau of the Census, Washington, DC.

Habemma, J.D.F., Hermans, J., and van den Broek, K. (1974). A stepwise discriminant analysis program using density estimation. In *Proceedings in Computational Statistics, COMPSTAT'1974*, Physica Verlag, Heidelberg, pp. 101–110.

Hall, P. (1983). Large sample optimality of least squares cross-validation in density estimation. *The Annals of Statistics*, 11:1156–1174.

Hall, P. (1986). On the bootstrap and confidence intervals. *The Annals of Statistics*, 14:1431–1452.

Hall, P. (1987). On Kullback–Leibler loss and density estimation. *The Annals of Statistics*, 15:1491–1519.

Hall, P. (2003). A short prehistory of the bootstrap. *Statistical Science*, 18:158–167.

Hubback, J. (1946). Sampling for rice yield in Bihar and Orissa. *Sankhyā*, 7:281–294. (First published in 1927 as Bulletin 166, Imperial Agricultural Research Institute, Pusa, India).

Kish, L. (1957). Confidence intervals for clustered samples. *American Sociological Review*, 22:154–165.

Komlós, J., Major, P., and Tusnády, G. (1975). An approximation of partial sums of independent RV's and the sample DF. I. *Zeitschrift für Wahrscheinlichkeitstheorie und Verwandte Gebiete*, 32:111–131.

Komlós, J., Major, P., and Tusnády, G. (1976). An approximation of partial sums of independent RV's and the sample DF. II. *Zeitschrift für Wahrscheinlichkeitstheorie und Verwandte Gebiete*, 34:33–58.

Mahalanobis, P. (1946). Recent experiments in statistical sampling in the Indian Statistical Institute (with discussion). *Journal of the Royal Statistical Society*, 109:325–378.

McCarthy, P. (1966). *Replication: An Approach to the Analysis of Data From Complex Surveys. Vital Health Statistics.* Public Health Service Publication 1000, Series 2, No. 14, National Center for Health Statistics, Public Health Service, US Government Printing Office.

McCarthy, P. (1969). Pseudo-replication: Half samples. *Review of the International Statistical Institute*, 37:239–264.

Pitman, E.J.G. (1937a). Significance tests which may be applied to samples from any population. *Royal Statistical Society Supplement*, 4:119–130.

Pitman, E.J.G. (1937b). Significance tests which may be applied to samples from any population, II. *Royal Statistical Society Supplement*, 4:225–232.

Pitman, E.J.G. (1938). Significance tests which may be applied to samples from any population. Part III. The analysis of variance test. *Biometrika*, 29:322–335.

Rudemo, M. (1982). Empirical choice of histograms and kernel density estimators. *Scandinavian Journal of Statistics*, 9:65–78.

Stone, C.J. (1984). An asymptotically optimal window selection rule for kernel density estimates. *The Annals of Statistics*, 12:1285–1297.

Tibshirani, R.J. (1996). Regression shrinkage and selection via the lasso. *Journal of the Royal Statistical Society, Series B*, 58:267–288.

Tibshirani, R.J. (2014a). In praise of sparsity and convexity. *Past, Present, and Future of Statistical Science* (X. Lin, C. Genest, D.L. Banks, G. Molenberghs, D.W. Scott, and J.-L. Wang, Eds.). Chapman & Hall, London, pp. 497–505.

Tibshirani, R.J. (2014b). Lasso and sparsity in statistics. *Statistics in Action: A Canadian Outlook* (J.F. Lawless, Ed.). Chapman & Hall, London, pp. 79–91.

Whittaker, E.T. and Watson, G.N. (1902). *A Course of Modern Analysis.* Cambridge University Press, Cambridge, UK.

Woodroofe, M. (1970). On choosing a delta-sequence. *The Annals of Mathematical Statistics*, 41:1665–1671.

16

The bright future of applied statistics

Rafael A. Irizarry

Department of Biostatistics and Computational Biology,
Dana-Farber Cancer Institute
and
Department of Biostatistics
Harvard School of Public Health, Boston, MA

16.1 Introduction

When I was asked to contribute to this book, titled *Past, Present, and Future of Statistical Science*, I contemplated my career while deciding what to write about. One aspect that stood out was how much I benefited from the right circumstances. I came to one clear conclusion: it is a great time to be an applied statistician. I decided to describe the aspects of my career that I have thoroughly enjoyed in the *past* and *present* and explain why this has led me to believe that the *future* is bright for applied statisticians.

16.2 Becoming an applied statistician

I became an applied statistician while working with David Brillinger on my PhD thesis. When searching for an advisor I visited several professors and asked them about their interests. David asked me what I liked and all I came up with was "I don't know. Music?" to which he responded, "That's what we will work on." Apart from the necessary theorems to get a PhD from the Statistics Department at Berkeley, my thesis summarized my collaborative work with researchers at the Center for New Music and Audio Technology. The work involved separating and parameterizing the harmonic and non-harmonic components of musical sound signals (Irizarry, 2001). The sounds had been digitized into data. The work was indeed fun, but I also had my first glimpse into the incredible potential of statistics in a world becoming more and more data-driven.

Despite having expertise only in music, and a thesis that required a CD player to hear the data, fitted models and residuals (`http://www.biostat.jhsph.edu/~ririzarr/Demo/index.html`), I was hired by the Department of Biostatistics at Johns Hopkins School of Public Health. Later I realized what was probably obvious to the school's leadership: that regardless of the subject matter of my thesis, my time series expertise could be applied to several public health applications (Crone et al., 2001; DiPietro et al., 2001; Irizarry, 2001). The public health and biomedical challenges surrounding me were simply too hard to resist and my new department knew this. It was inevitable that I would quickly turn into an applied biostatistician.

16.3 Genomics and the measurement revolution

Since the day that I arrived at Johns Hopkins University 15 years ago, Scott Zeger, the department chair, fostered and encouraged faculty to leverage their statistical expertise to make a difference and to have an immediate impact in science. At that time, we were in the midst of a measurement revolution that was transforming several scientific fields into data-driven ones. Being located in a School of Public Health and next to a medical school, we were surrounded by collaborators working in such fields. These included environmental science, neuroscience, cancer biology, genetics, and molecular biology. Much of my work was motivated by collaborations with biologists that, for the first time, were collecting large amounts of data. Biology was changing from a data poor discipline to a data intensive one.

A specific example came from the measurement of gene expression. Gene expression is the process in which DNA, the blueprint for life, is copied into RNA, the templates for the synthesis of proteins, the building blocks for life. Before microarrays were invented in the 1990s, the analysis of gene expression data amounted to spotting black dots on a piece of paper (Figure 16.1, left). With microarrays, this suddenly changed to sifting through tens of thousands of numbers (Figure 16.1, right). Biologists went from using their eyes to categorize results to having thousands (and now millions) of measurements per sample to analyze. Furthermore, unlike genomic DNA, which is static, gene expression is a dynamic quantity: different tissues express different genes at different levels and at different times. The complexity was exacerbated by unpolished technologies that made measurements much noisier than anticipated. This complexity and level of variability made statistical thinking an important aspect of the analysis. The biologists that used to say, "If I need statistics, the experiment went wrong" were now seeking out our help. The results of these collaborations have led to, among other things, the development of breast can-

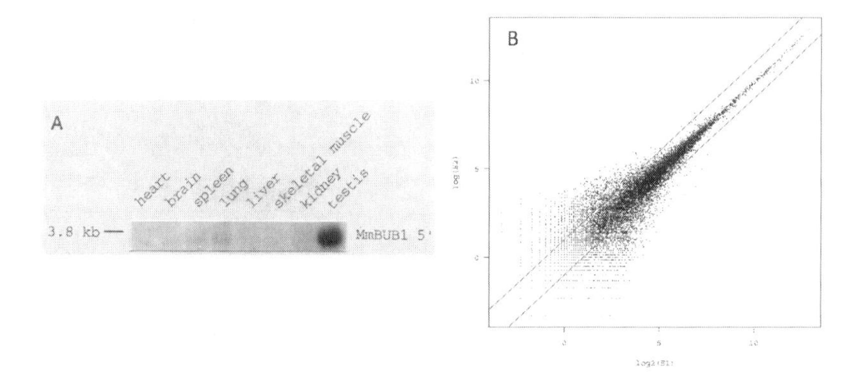

FIGURE 16.1
Illustration of gene expression data before and after microarrays.

cer recurrence gene expression assays making it possible to identify patients at risk of distant recurrence following surgery (van't Veer, 2002).

When biologists at Johns Hopkins first came to our department for help with their microarray data, Scott put them in touch with me because I had experience with (what was then) large datasets (digitized music signals are represented by 44,100 points per second). The more I learned about the scientific problems and the more data I explored, the more motivated I became. The potential for statisticians having an impact in this nascent field was clear and my department was encouraging me to take the plunge. This institutional encouragement and support was crucial as successfully working in this field made it harder to publish in the mainstream statistical journals; an accomplishment that had traditionally been heavily weighted in the promotion process. The message was clear: having an immediate impact on specific scientific fields would be rewarded as much as mathematically rigorous methods with general applicability.

As with my thesis applications, it was clear that to solve some of the challenges posed by microarray data I would have to learn all about the technology. For this I organized a sabbatical with Terry Speed's group in Melbourne where they helped me accomplish this goal. During this visit I reaffirmed my preference for attacking applied problems with simple statistical methods, as opposed to overcomplicated ones or developing new techniques. Learning that discovering clever ways of putting the existing statistical toolbox to work was good enough for an accomplished statistician like Terry gave me the necessary confidence to continue working this way. More than a decade later this continues to be my approach to applied statistics. This approach has been instrumental for some of my current collaborative work. In particular, it led to important new biological discoveries made together with Andy Feinberg's lab (Irizarry, 2009).

During my sabbatical we developed preliminary solutions that improved precision and aided in the removal of systematic biases from microarray data (Irizarry, 2003). I was aware that hundreds, if not thousands, of other scientists were facing the same problematic data and were searching for solutions. Therefore I was also thinking hard about ways in which I could share whatever solutions I developed with others. During this time I received an email from Robert Gentleman asking if I was interested in joining a new software project for the delivery of statistical methods for genomics data. This new collaboration eventually became the Bioconductor project (`http://www.bioconductor.org`), which to this day continues to grow its user and developer base (Gentleman et al., 2004). Bioconductor was the perfect vehicle for having the impact that my department had encouraged me to seek. With Ben Bolstad and others we wrote an R package that has been downloaded tens of thousands of times (Gautier et al., 2004). Without the availability of software, the statistical method would not have received nearly as much attention. This lesson served me well throughout my career, as developing software packages has greatly helped disseminate my statistical ideas. The fact that my department and school rewarded software publications provided important support.

The impact statisticians have had in genomics is just one example of our field's accomplishments in the 21st century. In academia, the number of statisticians becoming leaders in fields such as environmental sciences, human genetics, genomics, and social sciences continues to grow. Outside of academia, sabermetrics has become a standard approach in several sports (not just baseball) and inspired the Hollywood movie *Money Ball.* A PhD statistician led the team that won the Netflix million dollar prize (`http://www.netflixprize.com/`). Nate Silver (`http://mashable.com/2012/11/07/nate-silver-wins/`) proved the pundits wrong by once again using statistical models to predict election results almost perfectly. R has become a widely used programming language. It is no surprise that statistics majors at Harvard have more than quadrupled since 2000 (`http://nesterko.com/visuals/statconcpred2012-with-dm/`] and that statistics MOOCs are among the most popular (`http://edudemic.com/2012/12/the-11-most-popular-open-online-courses/`).

The unprecedented advance in digital technology during the second half of the 20th century has produced a measurement revolution that is transforming science. Scientific fields that have traditionally relied upon simple data analysis techniques have been turned on their heads by these technologies. Furthermore, advances such as these have brought about a shift from hypothesis-driven to discovery-driven research. However, interpreting information extracted from these massive and complex datasets requires sophisticated statistical skills as one can easily be fooled by patterns that arise by chance. This has greatly elevated the importance of our discipline in biomedical research.

16.4 The bright future

I think that the data revolution is just getting started. Datasets are currently being, or have already been, collected that contain, hidden in their complexity, important truths waiting to be discovered. These discoveries will increase the scientific understanding of our world. Statisticians should be excited and ready to play an important role in the new scientific renaissance driven by the measurement revolution.

References

Crone, N.E., Hao, L., Hart, J., Boatman, D., Lesser, R.P., Irizarry, R., and Gordon, B. (2001). Electrocorticographic gamma activity during word production in spoken and sign language. *Neurology*, 57:2045–2053.

DiPietro, J.A., Irizarry, R.A., Hawkins, M., Costigan, K.A., and Pressman, E.K. (2001). Cross-correlation of fetal cardiac and somatic activity as an indicator of antenatal neural development. *American Journal of Obstetrics and Gynecology*, 185:1421–1428.

Gautier, L., Cope, L., Bolstad, B.M., and Irizarry, R.A. (2004). Affyanalysis of affymetrix genechip data at the probe level. *Bioinformatics*, 20:307–315.

Gentleman, R.C., Carey, V.J., Bates, D.M., Bolstad, B., Dettling, M., Dudoit, S., Ellis, B., Gautier, L., Ge, Y., Gentry, J. et al. (2004). Bioconductor: Open software development for computational biology and bioinformatics. *Genome Biology*, 5:R80.

Irizarry, R.A. (2001). Local harmonic estimation in musical sound signals. *Journal of the American Statistical Association*, 96:357–367.

Irizarry, R.A., Hobbs, B., Collin, F., Beazer-Barclay, Y.D., Antonellis, K.J., Scherf, U., and Speed, T.P. (2003). Exploration, normalization, and summaries of high density oligonucleotide array probe level data. *Biostatistics*, 4:249–264.

Irizarry, R.A., Ladd-Acosta, C., Wen, B., Wu, Z., Montano, C., Onyango, P., Cui, H., Gabo, K., Rongione, M., Webster, M. et al. (2009). The human colon cancer methylome shows similar hypo- and hypermethylation at conserved tissue-specific CpG island shores. *Nature Genetics*, 41:178–186.

Irizarry, R.A., Tankersley, C., Frank, R., and Flanders, S. (2001). Assessing homeostasis through circadian patterns. *Biometrics*, 57:1228–1237.

van't Veer, L.J., Dai, H., Van De Vijver, M.J., He, Y.D., Hart, A.AM., Mao, M., Peterse, H.L., van der Kooy, K., Marton, M.J., Witteveen, A.T. et al. (2002). Gene expression profiling predicts clinical outcome of breast cancer. *Nature*, 415:530–536.

17

The road travelled: From statistician to statistical scientist

Nilanjan Chatterjee

Biostatistics Branch, Division of Cancer Epidemiology and Genetics
National Cancer Institute
National Institutes of Health
Department of Health and Human Services

In this chapter, I will attempt to describe how a series of problems in statistical genetics, starting from a project about estimation of risk for BRCA1/2 mutation carriers, have driven a major part of my research at the National Cancer Institute over the last fourteen years. I try to share some of the statistical and scientific perspectives in this field that I have developed over the years during which I myself have transformed from a theoretical statistician to a statistical scientist. I hope my experience would draw the attention of other statisticians, especially young researchers who have perceived strength in statistical theory but are keen on using their knowledge to advance science, about the tremendous opportunity that lies ahead.

17.1 Introduction

I have been wondering for a while now what would I like to share with the readers about my experience as a statistician. Suddenly, a news event that has captivated the attention of the world in the last few days gave me the impetus to start on this piece. Angelina Jolie, the beautiful and famous Hollywood actress, also well-known for her humanitarian work across the globe, has made the bold decision to opt for bilateral mastectomy after knowing that she carries a mutation in the gene called BRCA1 which, according to her doctors, gives her a risk of 87% for developing breast cancers and 50% for developing ovarian cancer (Jolie, May 14, 2013). The print and online media, blogosphere and social media sites are all buzzing with the discussion about her courage of not only making the drastic decision to take control of her own health, but also to help increase awareness about the issues by sharing the deeply personal story with the general public. At the same time, a debate is also raging about whether she got the right estimates of her risks and whether she could have made alternative, less drastic choices, to minimize and manage her long-term risks of these two cancers. And of course, all of these are embedded in a much broader debate in the community about the appropriate use of comprehensive genome sequencing information for personalized medicine (treatment and prevention) in the future.

17.2 Kin-cohort study: My gateway to genetics

This debate about Angelina Jolie's decision is taking me back through the memory lane to May of 1999 when I joined the Division of Cancer Epidemiology and Genetics of the National Cancer Institute as a post-doctoral fellow after finishing my PhD in Statistics from the University of Washington, Seattle. The very first project my mentor, Sholom Wacholder, introduced me to involved estimation of risk of certain cancers, including those of breast and ovary, associated with mutations in genes BRCA1 and BRCA2 from the Washington Ashkenazi Study (WAS) (Struewing et al., 1997). My mentor and his colleagues had previously developed the novel "kin-cohort" approach that allowed estimation of age-specific cumulative risk of a disease associated with a genetic mutation based on the disease history of the set of relatives of genotyped study participants (Wacholder et al., 1998). This approach, when applied to the WAS study, estimated the lifetime risk or penetrance of breast cancer to be between 50–60%, substantially lower than the estimates of penetrance around 90–100% that have been previously obtained from analysis of highly disease enriched families. It was thought that WAS, which were volunteer-based and not as prone to ascertainment bias as family-studies, provided more unbiased estimate of risk for BRCA1/2 mutation carriers in the general population. Other studies, that have employed "un-selected" designs, have estimated the penetrance to be even lower.

The approach my mentor and colleagues had previously developed was very simple and elegant. It relied on the observation that since approximately half of the first-degree relatives of BRCA1/2 mutation carriers are expected to be carriers by themselves due to Mendelian laws of transmission, the risks of the disease in this group of relatives should be approximately 50:50 mixture of the risk of the disease associated with carriers and non-carriers. Further, for a rare mutation, since very few of the first degree relatives of the non-carriers are expected to be carriers themselves, the risk of the disease in the group of relatives should be approximately the same as that for non-carriers themselves. Thus they employed a simple method-of-moment approach to estimate the age-specific cumulative risks associated with carriers and non-carriers using the Kaplan–Meyer estimators of risk for the first-degree relative of carriers and those for non-carriers.

I attempted to formulate the problem in terms of a composite likelihood framework (Chatterjee and Wacholder, 2001) so that the resulting estimator has desirable statistical properties such as monotonicity of the age-specific cumulative risk curve and are robust to strong assumptions about residual familial correlation of disease among family members. The likelihood-based framework was quite attractive due to its flexibility for performing various additional analyses and I was happy that I was able to make a quick methodologic contribution, even being fairly novice to the field. However, the actual application of the method to WAS hardly changed the estimate of penetrance for BRCA1/2 mutation compared to the method of moment estimates previously available.

In retrospect, I learned several lessons from my first post-doctoral project. First, it is often hard to beat a simple but sensible statistical method. Although this may be obvious to many seasoned applied statisticians, this first-hand experience was an important lesson for me, fresh out of graduate school where my PhD thesis involved understanding of semi-parametric efficient estimation methodology, the purpose of which is to get the last drop of information from the data with minimal assumption about "nuisance" parameters. Second, although the substantive contribution of my first project was modest, it was an invaluable exercise for me as it opened my gateway to the whole new area of statistical genetics. To get a solid grasp of the problem without having any knowledge of genetics *a priori*, I had to teach myself concepts of population as well as molecular genetics. Self-teaching and my related struggles were an invaluable experience for me that help me to date to think about each problem in my own way.

17.3 Gene-environment interaction: Bridging genetics and theory of case-control studies

As I was wrapping up my work on kin-cohort studies, one day Sholom asked me for some help to analyze data from a case-control study of Ovarian Cancer to assess interaction between BRCA1/2 mutations and certain reproductive factors, such as oral contraceptive use, which are known to reduce the risk for the disease in the general population (Modan et al., 2001). Because BRCA1/2 mutations are very rare in the general population, standard logistic regression analysis of interaction would have been very imprecise. Instead, the investigators in this study were pursuing an alternative method that uses a log-linear modeling approach that can incorporate the reasonable assumption that reproductive factors are distributed independently of genetic mutation status in the general population. Earlier work has shown that incorporation of the gene-environment independence assumption can greatly enhance the efficiency of interaction analysis in case-control studies (Piegorsch et al., 1994; Umbach and Weinberg, 1997).

While I was attempting to analyze the study using the log-linear model framework, I realized it is a bit of a cumbersome approach that requires creating massive contingency tables by categorizing all of the variables under study, carefully tracking different sets of parameters (e.g., parameters related to disease odds-ratios and exposure frequency distributions) and then constraining specific parameters to zero for incorporation of the gene-environment independence assumption. I quickly realized that all of these details can be greatly simplified by some of the techniques I had learned from my advisors Norman Breslow and Jon Wellner during my PhD thesis regarding semi-parametric analysis of case-control and other types of studies that use complex outcome dependent sampling designs (Chatterjee et al., 2003). In particular, I was able to derive a profile-likelihood technique that simplifies fitting of logistic regression models to case-control data under the gene-environment independence assumption(Chatterjee and Carroll, 2005). Early in the development, I told Ray Carroll, who has been a regular visitor at NCI for a long time, about some of the results I have derived, and he got everyone excited because of his own interest and earlier research in this area. Since then Ray and I have been partners in crime in many papers related to inference on gene-environment interactions from case-control studies.

This project also taught me a number of important lessons. First, there is tremendous value to understanding the theoretical underpinning of standard methods that we routinely use in practice. Without the understanding of the fundamentals of semi-parametric inference for analysis of case-control data that I developed during my graduate studies, I would never have made the connection of this problem to profile-likelihood, which is essentially the backbone for many standard methods, such as Cox's partial likelihood analysis of lifetime data. The approach not only provided a simplified framework for exploiting the gene-environment independence assumption for case-control studies, but also led to a series of natural extensions of practical importance so that the method is less sensitive to the violation of the critical gene-environment independence assumption. My second lesson was that it is important not to lose the applied perspective even when one is deeply involved in the development of theory. Because we paid close attention to the practical limitation of the original methods and cutting-edge developments in genetic epidemiology, my collaborators, trainee and I (Ray Carroll, Yi-Hau Chen, Bhramar Mukherjee, Samsiddhi Bhattacharjee to name a few) were able to propose robust extensions of the methodology using conditional logistic regression, shrinkage estimation techniques and genomic control methods (Chatterjee et al., 2005; Mukherjee and Chatterjee, 2008; Bhattacharjee et al., 2010).

17.4 Genome-wide association studies (GWAS): Introduction to big science

BRCA1/2 mutations, which pose high-risks for breast and ovarian cancer but rare in the general population, were originally discovered in the early 1990s through linkage studies that involve analysis of the co-segregation of genetic mutations and disease within highly affected families (Hall et al., 1990). From the beginning of the 21st century, after the human genome project got completed and large scale genotyping technologies evolved, the genetic community started focusing on genome-wide association studies (GWAS). The purpose of these studies was to identify genetic variants which may pose more modest risk of diseases, like breast cancer, but are more common in the general populations. Early in this effort, the leadership of our Division decided to launch two such studies, one for breast cancer and one for prostate cancer, under the rubric of the Cancer Genetics Marker of Susceptibility of Studies (CGEMS) (Hunter et al., 2007; Yeager et al., 2007).

I participated in these studies as a four-member team of statisticians who provided the oversight of the quantitative issues in the design and analysis aspect of these studies. For me, this was my first exposure to large "team science," where progress could only be made through collaborations of a team of researchers with diverse background, such as genomics, epidemiology, bioinformatics, and statistics. Getting into the nitty-gritty of the studies gave me an appreciation of the complexities of large scale genomic studies. I realized that while we statisticians are prone to focus on developing an "even more optimal" method of analysis, some of the most fundamental and interesting quantitative issues in these types of studies lies elsewhere, in particular in the areas of study design, quality control and characterization following discovery (see next section for more on the last topic).

I started thinking seriously about study design when I was helping one of my epidemiologic collaborators put together a proposal for conducting a genome-wide association study for lung cancer. As a principled statistician, I felt some responsibility to show that the proposed study is likely to make new discoveries beyond three GWAS of lung cancer that were just published in high-profile journals such as *Nature* and *Nature Genetics*. I realized that standard power calculations, where investigators typically show that the study has 80–90% power to detect certain effect sizes, is not satisfactory for ever-growing GWA studies. I realized if I had to do a more intelligent power calculation, I first needed to make an assessment of what might be the underlying genetic architecture of the trait, in particular how many genetic variants might be associated with the trait and what are their effect-sizes.

I made a very simple observation that the discoveries made in an existing study can be thought of as a random sample from the underlying "population" of susceptibility markers where the probability of discovery of any given marker is proportional to the power of the study at the effect-size associated with that marker. My familiarity with sample-survey theory, which I developed during my PhD thesis on two-phase study design, again came very handy here. I worked with a post-doctoral Fellow, JuHyun Park, to develop an "inverse-power-weighted" method, similar to the widely used "inverse-probability-weighted" (IPW) methods for analysis of survey data, for inferring the number of underlying susceptibility markers for a trait and their effect-size distribution using published information on known discoveries and the study design of the underlying GWA studies (Park et al., 2010). We inferred genetic architecture for several complex traits using this method and made projections about the expected number of discoveries in GWAS of these traits. We have been very pleased to see that our projections were quite accurate when results from larger and larger GWA studies have come out for these traits since the publication of our report (Allen et al., 2010; Anderson et al., 2011; Eeles et al., 2013; Michailidou et al., 2013).

Realizing how optimal study design is fundamentally related to the underlying genetic architecture of traits, both JuHyun and I continued to delve into these related issues. Again using known discoveries from published studies and information on design of existing studies, we showed that there is very modest or no evidence of an inverse relationship between effect-size and allele frequency for genetic markers, a hypothesis in population genetics postulated from a selection point of view and one that often has been used in the past by scientists to motivate studies of less common and rare variants using sequencing technologies (Park et al., 2011). From the design point of view, we conjectured that lack of strong relationship between allele frequency and effect-size implies future studies for less common and rare variants will require even larger sample sizes than current GWAS to make comparable numbers of discoveries for underlying susceptibility loci.

Understanding its implications for discoveries made us question the implication of genetic architecture for risk-prediction, another hotly debated topic. Interestingly, while the modern statistical literature is very rich regarding optimal algorithms for building models, very little attention is given to more fundamental design questions, such as how our ability to predict a trait is inherently limited by sample-size of the training datasets and the genetic architecture of the trait, or more generally the etiologic architecture that may involve both genetic and non-genetic factors. This motivated us to develop a mathematical approximation for the relationship between expected predictive performance of models, sample size of training datasets and genetic architecture of traits. Based on these formulations, we projected that highly polygenic nature of complex traits implies future GWAS will require extremely large sample sizes, possibly of a higher order magnitude than even some of the largest GWAS to date, for substantial improvement of risk-prediction based on genetic information (Chatterjee et al., 2013).

Although the study of genetic architecture and its implications for study designs is now a significant part of my research portfolio, it was not by design by any means. I just stumbled upon the problem when I was attempting to do basic power calculations for a collaborator. Looking back, it was a risky thing to undertake as I was not sure where my effort was going to lead to, other than maybe I could advise my collaborators a little more intelligently about study designs. It was more tempting to focus, like many of my peers have done, sometimes very successfully, on so-called "hot" problems such as developing an optimal association test. Although I have put some effort in those areas as well, today I am really glad that instead of chasing more obvious problems, I gave myself the freedom to venture into unknown territories. The experimentation has certainly helped me, and hopefully the larger scientific community, to obtain some fresh insights into study designs, statistical power calculations and risk-prediction in the context of modern high-throughput studies.

17.5 The post-GWAS era: What does it all mean?

It is quite amazing that even more than two decades after the BRCA1/2 mutations were discovered, there is so much ambiguity about what are the true risks associated with these genes for various cancers. In the literature, available estimates of lifetime-risk of breast cancer, for example, vary from 20–90%. As noted above, while estimates available from highly-enriched cancer families tend to reside at the higher range, their counterparts from population-based studies tend to be more at the lower range. Risk for an individual carrier would also depend on other information, such as the specific mutation type, cancer history among family members and information on other risk-factors. The problem of estimation of risk associated with rare highly penetrant mutations poses many interesting statistical challenges and has generated a large volume of literature.

Discoveries from genome-wide association studies are now fueling the debate how discovery of low penetrant common variants can be useful for public health. Some researchers argue that common variants, irrespective of how modest their effects are, can individually or collectively highlight interesting biologic pathways that are involved in the pathogenesis of a disease and hence potentially be useful for development of drug targets. Although this would be a highly desirable outcome, skepticism exists given that discoveries of even major genes like BRCA1/2 have seldom led to successful development of drug targets. Utility of common variants for genetic risk prediction is also now a matter of great debate. While a number of early studies painted mostly a negative picture, large numbers of discoveries from the most recent very large GWAS suggests that there is indeed potential for common variants to improve risk-prediction.

I am an avid follower of this debate. While the focus of the genetics community is quickly shifting towards what additional discoveries are possible using future whole-exome or genome sequencing studies, I keep wondering about how the knowledge from GWAS could be further refined and ultimately applied to improve public health. In general, I see the pattern that whenever a new technology emerges, there is tremendous interest about making new "discovery," but the effort is not proportionate when it comes to following-up these discoveries for better "characterization" of risk or/and the underlying causal mechanism. Interestingly, while in the discovery effort statisticians face stiff competition from researchers in other quantitative disciplines, like geneticists, bioinformaticians, computer scientists and physicists, statisticians have the potential to develop a more exclusive niche in the "characterization" steps where the underlying inferential issues are often much more of complex nature than simple hypothesis testing.

17.6 Conclusion

In the last fourteen years, the biggest change I observe within myself is how I think about a problem. When I started working on refinement of kin-cohort methods, I focused on developing novel methods but was not very aware of all the underlying very complex clinical and epidemiologic subject-matter issues. Now that I am struggling with the question of what would the discoveries from current GWAS and future sequencing studies mean for personalized medicine and public health, I feel I have a better appreciation of those pertinent scientific issues and the related debate. For this, I owe much to the highly stimulating scientific environment of our Division, created and fostered over decades by our recently retired director Dr. Joseph Fraumeni Jr. and a number of other leaders. The countless conversations and debates I had with my statistician, epidemiologist and geneticist colleagues in the meeting rooms, corridors and cafeteria of DCEG about cutting-edge issues for cancer genetics, epidemiology and prevention had a major effect on me. At the same time, my training in theory and methods always guides my thinking about these applied problems in a statistically rigorous way. I consider myself to be fortunate to inherit my academic "genes" through training in statistics and biostatistics from the Indian Statistical Institute and the University of Washington, and then be exposed to the great "environment" of DCEG for launching my career as a statistical scientist.

Congratulations to COPSS for sustaining and supporting such a great profession as ours for 50 years! It is an exciting time to be a statistician in the current era of data science. There are tremendous opportunities for our profession which also comes with tremendous responsibility. While I was finishing up this chapter, the US Supreme Court ruled that genes are not "patentable," implying genetic tests would become more openly available to consumers in the future. The debate about whether Angelina Jolie got the right information about her risk from BRCA1/2 testing is just a reminder about the challenge that lies ahead for all of us to use genetic and other types of biomedical data to create objective "knowledge" that will benefit, and not misguide or harm, medical researchers, clinicians and most importantly the public.

References

Allen, H.L. et al. (2010). Hundreds of variants clustered in genomic loci and biological pathways affect human height. *Nature*, 467:832–838.

Anderson, C.A., Boucher, G., Lees, C.W., Franke, A., D'Amato, M., Taylor, K.D., Lee, J.C., Goyette, P., Imielinski, M., Latiano, A. et al. (2011). Meta-analysis identifies 29 additional ulcerative colitis risk loci, increasing the number of confirmed associations to 47. *Nature Genetics*, 43:246–252.

Bhattacharjee, S., Wang, Z., Ciampa, J., Kraft, P., Chanock, S., Yu, K., and Chatterjee, N. (2010). Using principal components of genetic variation for robust and powerful detection of gene-gene interactions in case-control and case-only studies. *The American Journal of Human Genetics*, 86:331–342.

Chatterjee, N., and Carroll, R.J. (2005). Semiparametric maximum likelihood estimation exploiting gene-environment independence in case-control studies. *Biometrika*, 92:399–418.

Chatterjee, N., Chen, Y.-H., and Breslow, N.E. (2003). A pseudoscore estimator for regression problems with two-phase sampling. *Journal of the American Statistical Association*, 98:158–168.

Chatterjee, N., Kalaylioglu, Z., and Carroll, R.J. (2005). Exploiting gene-environment independence in family-based case-control studies: Increased power for detecting associations, interactions and joint effects. *Genetic Epidemiology*, 28:138–156.

Chatterjee, N. and Wacholder, S. (2001). A marginal likelihood approach for estimating penetrance from kin-cohort designs. *Biometrics*, 57:245–252.

Chatterjee, N., Wheeler, B., Sampson, J., Hartge, P., Chanock, S.J., and Park, J.-H. (2013). Projecting the performance of risk prediction based on polygenic analyses of genome-wide association studies. *Nature Genetics*, 45:400–405.

Eeles, R.A., Al Olama, A.A., Benlloch, S., Saunders, E.J., Leongamornlert, D.A., Tymrakiewicz, M., Ghoussaini, M., Luccarini, C., Dennis, J.,

Jugurnauth-Little, S. et al. (2013). Identification of 23 new prostate cancer susceptibility loci using the icogs custom genotyping array. *Nature Genetics*, 45:385–391.

Hall, J.M., Lee, M.K., Newman, B., Morrow, J.E., Anderson, L.A., Huey, B., King, M.-C. et al. (1990). Linkage of early-onset familial breast cancer to chromosome 17q21. *Science*, 250:1684–1689.

Hunter, D.J., Kraft, P., Jacobs, K.B., Cox, D.G., Yeager, M., Hankinson, S.E., Wacholder, S., Wang, Z., Welch, R., Hutchinson, A. et al. (2007). A genome-wide association study identifies alleles in FGFR2 associated with risk of sporadic postmenopausal breast cancer. *Nature Genetics*, 39:870–874.

Jolie, A. (May 14, 2013). My Medical Choice. *The New York Times*.

Michailidou, K., Hall, P., Gonzalez-Neira, A., Ghoussaini, M., Dennis, J., Milne, R.L., Schmidt, M.K., Chang-Claude, J., Bojesen, S.E., Bolla, M.K. et al. (2013). Large-scale genotyping identifies 41 new loci associated with breast cancer risk. *Nature Genetics*, 45:353–361.

Modan, B., Hartge, P., Hirsh-Yechezkel, G., Chetrit, A., Lubin, F., Beller, U., Ben-Baruch, G., Fishman, A., Menczer, J., Struewing, J.P. et al. (2001). Parity, oral contraceptives, and the risk of ovarian cancer among carriers and noncarriers of a BRCA1 or BRCA2 mutation. *New England Journal of Medicine*, 345:235–240.

Mukherjee, B. and Chatterjee, N. (2008). Exploiting gene-environment independence for analysis of case–control studies: An empirical Bayes-type shrinkage estimator to trade-off between bias and efficiency. *Biometrics*, 64:685–694.

Park, J.-H., Gail, M.H., Weinberg, C.R., Carroll, R.J., Chung, C.C., Wang, Z., Chanock, S.J., Fraumeni, J.F., and Chatterjee, N. (2011). Distribution of allele frequencies and effect sizes and their interrelationships for common genetic susceptibility variants. *Proceedings of the National Academy of Sciences*, 108:18026–18031.

Park, J.-H., Wacholder, S., Gail, M.H., Peters, U., Jacobs, K.B., Chanock, S.J., and Chatterjee, N. (2010). Estimation of effect size distribution from genome-wide association studies and implications for future discoveries. *Nature Genetics*, 42:570–575.

Piegorsch, W.W., Weinberg, C.R., and Taylor, J.A. (1994). Non-hierarchical logistic models and case-only designs for assessing susceptibility in population-based case-control studies. *Statistics in Medicine*, 13:153–162.

Struewing, J.P., Hartge, P., Wacholder, S., Baker, S.M., Berlin, M., McAdams, M., Timmerman, M.M., Brody, L.C., and Tucker, M.A. (1997).

The risk of cancer associated with specific mutations of BRCA1 and BRCA2 among Ashkenazi Jews. *New England Journal of Medicine*, 336:1401–1408.

Umbach, D.M. and Weinberg, C.R. (1997). Designing and analysing case-control studies to exploit independence of genotype and exposure. *Statistics in Medicine*, 16:1731–1743.

Wacholder, S., Hartge, P., Struewing, J.P., Pee, D., McAdams, M., Brody, L., and Tucker, M. (1998). The kin-cohort study for estimating penetrance. *American Journal of Epidemiology*, 148:623–630.

Yeager, M., Orr, N., Hayes, R.B., Jacobs, K.B., Kraft, P., Wacholder, S., Minichiello, M.J., Fearnhead, P., Yu, K., Chatterjee, N. et al. (2007). Genome-wide association study of prostate cancer identifies a second risk locus at 8q24. *Nature Genetics*, 39:645–649.

18

A journey into statistical genetics and genomics

Xihong Lin
Department of Biostatistics
Harvard School of Public Health, Boston, MA

This chapter provides personal reflections and lessons I learned through my journey into the field of statistical genetics and genomics in the last few years. I will discuss the importance of being both a statistician and a scientist; challenges and opportunities in analyzing massive genetic and genomic data, and training the next generation statistical genetic and genomic scientists in the 'omics era.

18.1 The 'omics era

The human genome project in conjunction with the rapid advance of high throughput technology has transformed the landscape of health science research in the last ten years. Scientists have been assimilating the implications of the genetic revolution, characterizing the activity of genes, messenger RNAs, and proteins, studying the interplay of genes and the environment in causing human diseases, and developing strategies for personalized medicine. Technological platforms have advanced to a stage where many biological entities, e.g., genes, transcripts, and proteins, can be measured on the whole genome scale, yielding massive high-throughput 'omics data, such as genetic, genomic, epigenetic, proteomic, and metabolomic data. The 'omics era provides an unprecedented promise of understanding common complex diseases, developing strategies for disease risk assessment, early detection, and prevention and intervention, and personalized therapies.

The volume of genetic and genomic data has exploded rapidly in the last few years. Genome-wide association studies (GWAS) use arrays to genotype 500,000–5,000,000 common Single Nucleotide Polymorphisms (SNPs) across the genome. Over a thousand of GWASs have been conducted in the last few years. They have identified hundreds of common genetic variants that are associated with complex traits and diseases (`http://www.genome.gov/gwastudies/`). The emerging next generation sequencing technology offers an exciting new opportunity for sequencing the whole genome, obtaining information about both common and rare variants and structural variation. The next generation sequencing data allow to explore the roles of rare genetic variants and mutations in human diseases. Candidate gene sequencing, whole exome sequencing and whole genome sequencing studies are being conducted. High-throughput RNA and epigenetic sequencing data are also becoming rapidly available to study gene regulation and functionality, and the mechanisms of biological systems. A large number of public genomic databases, such as the HapMap Project (`http://hapmap.ncbi.nlm.nih.gov/`), the 1000 genomes project (`www.1000genomes.org`), are freely available. The NIH database of Genotypes and Phenotypes (dbGaP) archives and distributes data from many GWAS and sequencing studies funded by NIH freely to the general research community for enhancing new discoveries.

The emerging sequencing technology presents many new opportunities. Whole genome sequencing measures the complete DNA sequence of the genome of a subject at three billion base-pairs. Although the current cost of whole genome sequencing prohibits conducting large scale studies, with the rapid advance of biotechnology, the "1000 dollar genome" era will come in the near future. This provides a new era of predictive and personalized medicine during which the full genome sequencing for an individual or patient costs only $1000 or lower. Individual subject's genome map will facilitate patients and physicians with identifying personalized effective treatment decisions and intervention strategies.

While the 'omics era presents many exciting research opportunities, the explosion of massive information about the human genome presents extraordinary challenges in data processing, integration, analysis and result interpretation. The volume of whole genome sequencing data is substantially larger than that of GWAS data, and is in the magnitude of tens or hundreds of terabites (TBs). In recent years, limited quantitative methods suitable for analyzing these data have emerged as a bottleneck for effectively translating rich information into meaningful knowledge. There is a pressing need to develop statistical methods for these data to bridge the technology and information transfer gap in order to accelerate innovations in disease prevention and treatment. As noted by John McPherson, from the Ontario Institute for Cancer Research,

> "There is a growing gap between the generation of massively parallel sequencing output and the ability to process and analyze the resulting data. Bridging this gap is essential, or the coveted $1000 genome will come with a $20,000 analysis price tag." (McPherson, 2009)

This is an exciting time for statisticians. I discuss in this chapter how I became interested in statistical genetics and genomics a few years ago, lessons

I learned while making my journey into this field. I will discuss a few open and challenging problems to demonstrate statistical genetics and genomics is a stimulating field with many opportunities for statisticians to make methodological and scientific contributions. I will also discuss training the next generation quantitative scientists in the 'omics era.

18.2 My move into statistical genetics and genomics

Moving into statistical genetics and genomics is a significant turn in my career. My dissertation work was on GLMMs, i.e., generalized linear mixed models (Breslow and Clayton, 1993). In the first twelve years out of graduate school, I had been primarily working on developing statistical methods for analysis of correlated data, such as mixed models, measurement errors, and nonparametric and semiparametric regression for longitudinal data. When I obtained my PhD degree, I had quite limited knowledge about nonparametric and semiparametric regression using kernels and splines. Learning a new field is challenging. One is more likely to be willing to invest time and energy to learn a new field when stimulated by problems in an open new area and identifying niches. I was fascinated by the opportunities of developing nonparametric and semiparametric regression methods for longitudinal data as little work had been done in this area and there were plenty of open problems. Such an experience can be rewarding if timing and environment are right and good collaborators are found. One is more likely to make unique contributions to a field when it is still at an early stage of development. This experience also speaks well of the lessons I learned in my journey into statistical genetics and genomics.

After I moved to the Harvard School of Public Health in 2005, I was interested in exploring new areas of research. My collaborative projects turned out to be mainly in genetic epidemiological studies and environmental genetic studies, a field I had little knowledge about. In the next several years, I was gradually engaged in several ongoing genome-wide association studies, DNA methylation studies, and genes and environment studies. I was fascinated by the challenges in the analysis of large genetic and genomic data, and rich methodological opportunities for addressing many open statistical problems that are likely to facilitate new genetic discovery and advance science. At the same time, I realized that to make contributions in this field, one has to understand genetics and biology well enough in order to identify interesting problems and develop methods that are practically relevant and useful. In my sabbatical year in 2008, I decided to audit a molecular biology course, which was very helpful for me to build a foundation in genetics and understand the genetic jargon in my ongoing collaborative projects. This experience prepared me to get started working on methodological research in statistical genetics and genomics. Looking back, a good timing and a stimulating collaborative environment made my transition easier. In the mean time, moving into a new field with limited background requires patience, courage, and willingness to sacrifice, e.g., having a lower productivity in the first few years, and more importantly, identifying a niche.

18.3 A few lessons learned

Importance to be a scientist besides a statistician: While working on statistical genetics in the last few years, an important message I appreciate more and more is to be a scientist first and then a statistician. To make a quantitative impact in the genetic field, one needs to be sincerely interested in science, devote serious time to learn genetics well enough to identify important problems, and closely collaborate with subject-matter scientists. It is less a good practice to develop methods first and then look for applications in genetics to illustrate the methods. By doing so, it would be more challenging to make such methods have an impact in real world practice, and it is more likely to follow the crowd and work on a problem at a later and more matured stage of the area. Furthermore, attractive statistical methods that are likely to be popular and advance scientific discovery need to integrate genetic knowledge well in method development. This will require a very good knowledge of genetics, and identifying cutting-edge scientific problems that require new method development, and developing a good sense of important and less important problems.

Furthermore, the genetic field is more technology-driven than many other health science areas, and technology moves very fast. Statistical methods that were developed for data generated by an older technology might not be applicable for data generated by new technology. For example, normalization methods that work well for array-based technology might not work well for sequencing-based technology. Statistical geneticists hence need to closely follow technological advance.

Simple and computationally efficient methods carry an important role: To make analysis of massive genetic data feasible, computationally efficient and simple enough methods that can be easily explained to practitioners are often more advantageous and desirable. An interesting phenomenon is that simple classical methods seem to work well in practice. For example, in GWAS, simple single SNP analysis has been commonly used in both the discovery phase and the validation phase, and has led to discovery of hundreds of SNPs that are associated with disease phenotypes. This presents a significant challenge to statisticians who are interested in developing more advanced and sophisticated methods that can be adopted for routine use and outperform these simple methods in practical settings, so that new scientific discoveries can be made that are missed by the simple methods.

Importance of developing user-friendly open-access software: As the genetic field moves fast especially with the rapid advance of biotechnology, timely development of user-friendly and computationally efficient open access software is critical for a new method to become popular and used by practitioners. The software is more likely to be used in practice if it allows data to be input using the standard formats of genetic data. For example, for GWAS data, it is useful to allow to input data in the popular Plink format. Furthermore, a research team is likely to make more impact if a team member has a strong background in software engineering and facilitates software development.

Importance of publishing methodological papers in genetic journals: To increase the chance for a new statistical method to be used in the genetic community, it is important to publish the method in leading genetic journals, such as *Nature Genetics, American Journal of Human Genetics*, and *Plos Genetics,* and make the paper readable to this audience. Thus strong communication skills are needed. Compared to statistical journals, genetic journals not only have a faster review time, but also and more importantly have the readership that is more likely to be interested in these methods and be immediate users of these methods. By publishing methodological papers in genetic journals, the work is more likely to have an impact in real-world practice and speed up scientific discovery; also, as a pleasant byproduct a paper gets more citations. It is a tricky balance in terms of publishing methodological papers in genetic journals or statistical journals.

18.4 A few emerging areas in statistical genetics and genomics

To demonstrate the excitement of the field and attract more young researchers to work on this field, I provide in this section a few emerging areas in statistical genetics that require advanced statistical method developments.

18.4.1 Analysis of rare variants in next generation sequencing association studies

GWAS has been successful in identifying susceptible common variants associated with complex diseases and traits. However, it has been found that disease associated common variants only explain a small fraction of heritability (Manolio et al., 2009). Taking lung cancer as an example, the relative risks of the genetic variants found to be associated with lung cancer in GWAS (Hung et al., 2008) are much smaller than those from traditional epidemiological or environmental risk factors, such as cigarette smoking, radon and asbestos exposures. Different from the "common disease, common variants" hypothesis behind GWAS, the hypothesis of "common disease, multiple rare variants" has been proposed (Dickson et al., 2010; Robinson, 2010) as a complementary approach to search for the missing heritability.

The recent development of Next Generation Sequencing (NGS) technologies provides an exciting opportunity to improve our understanding of complex diseases, their prevention and treatment. As shown by the 1000 Genome Project (The 1000 Genomes Project Consortium, 2010) and the NHLBI Exome Sequencing Project (ESP) (Tennessen et al., 2012), a vast majority of variants on the human genome are rare variants. Numerous candidate genes, whole exome, and whole genome sequencing studies are being conducted to identify disease-susceptibility rare variants. However, analysis of rare variants in sequencing association studies present substantial challenges (Bansal et al., 2010; Kiezun et al., 2012) due to the presence of a large number of rare variants.

Individual SNP based analysis commonly used in GWAS has little power to detect the effects of rare variants. SNP set analysis has been advocated to improve power by assessing the effects of a group of SNPs in a set, e.g., using a gene, a region, or a pathway. Several rare variant association tests have been proposed recently, including burden tests (Morgenthaler and Thilly, 2007; Li and Leal, 2008; Madsen and Browning, 2009), and non-burden tests (Lin and Tang, 2011; Neale et al., 2011; Wu et al., 2011; Lee et al., 2012). A common theme of these methods is to aggregate individual variants or individual test statistics within a SNP set. However, these tests suffer from power loss when a SNP set has a large number of null variants. For example, a large gene often has a large number of rare variants, with many of them being likely to be null. Aggregating individual variant test statistics is likely to introduce a large amount of noises when the number of causal variants is small.

To formulate the problem in a statistical framework, assume n subjects are sequenced in a region, e.g., a gene, with p variants. For the ith subject, let Y_i be a phenotype (outcome variable), $\mathbf{G}_i = (G_{i1}, \ldots, G_{ip})^\top$ be the genotypes of p variants ($G_{ij} = 0, 1, 2$) for 0, 1, or 2 copies of the minor allele in a SNP set, e.g., a gene/region, $\mathbf{X}_i = (x_{i1}, \ldots, x_{iq})^\top$ be a covariate vector. Assume the Y_i are independent and follow a distribution in the exponential family with $\mathrm{E}(y_i) = \mu_i$ and $\mathrm{var}(y_i) = \phi v(\mu_i)$, where v is a variance function. We model the effects of p SNPs $\mathbf{G}_i$ in a set, e.g., a gene, and covariates $\mathbf{X}_i$ on a continuous/categorical phenotype using the generalized linear model (GLM) (McCullagh and Nelder, 1989),

$$g(\mu_i) = \mathbf{X}_i^\top \boldsymbol{\alpha} + \mathbf{G}_i^\top \beta, \tag{18.1}$$

where g is a monotone link function, $\boldsymbol{\alpha} = (\alpha_1, \ldots, \alpha_q)^\top$ and $\beta = (\beta_1, \ldots, \beta_p)^\top$ are vectors of regression coefficients for the covariates and the genetic variants, respectively. The $n \times p$ design matrix $\mathbf{G} = (\mathbf{G}_1, \ldots, \mathbf{G}_n)^\top$ is very sparse, with each column containing only a very small number of 1 or 2 and the rest being

0's. The association between a region consisting of the p rare variants $\mathbf{G}_i$ and the phenotype Y can be tested by evaluating the null hypothesis that $\mathcal{H}_0$: $\beta = (\beta_1, \ldots, \beta_p)^\top = \mathbf{0}$. As the genotype matrix $\mathbf{G}$ is very sparse and p might be moderate or large, estimation of β is difficult. Hence the standard p-DF Wald and LR tests are difficult to carry out and also lose power when p is large. Further, if the alternative hypothesis is sparse, i.e., only a small fraction of β's are non-zero but one does not know which ones are non-zeros, the classical tests do not effectively take the knowledge of the sparse alternative and the sparse design matrix into account.

18.4.2 Building risk prediction models using whole genome data

Accurate and individualized prediction of risk and treatment response plays a central role in successful disease prevention and treatment. GWAS and Genome-wide Next Generation Sequencing (NGS) studies present rich opportunities to develop a risk prediction model using massive common and rare genetic variants across the genome and well known risk factors. These massive genetic data hold great potential for population risk prediction, as well as improving prediction of clinical outcomes and advancing personalized medicine tailored for individual patients. It is a very challenging statistical task to develop a reliable and reproducible risk prediction model using millions or billions of common and rare variants, as a vast majority of these variants are likely to be null variants, and the signals of individual variants are often weak.

The simple strategy of building risk prediction models using only the variants that are significantly associated with diseases and traits after scanning the genome miss a substantial amount of information. For breast cancer, GWASs have identified over 32 SNPs that are associated with breast cancer risk. Although a risk model based on these markers alone can discriminate cases and controls better than risk models incorporating only non-genetic factors (Hüsing et al., 2012), the genetic risk model still falls short of what should be possible if all the genetic variants driving the observed familial aggregation of breast cancer were known: the AUC is .58 (Hüsing et al., 2012) versus the expected maximum of .89 (Wray et al., 2010). Early efforts of including a large number of non-significant variants from GWAS in estimating heritability models show encouraging promises (Yang et al., 2010).

The recent advancement in NGS holds great promises in overcoming such difficulties. The missing heritability could potentially be uncovered by rare and uncommon genetic variants that are missed by GWAS (Cirulli and Goldstein, 2010). However, building risk prediction models using NGS data present substantial challenges. First, there are a massive number of rare variants cross the genome. Second, as variants are rare and the data dimension is large, their effects are difficult to be estimated using standard MLEs. It is of substan-

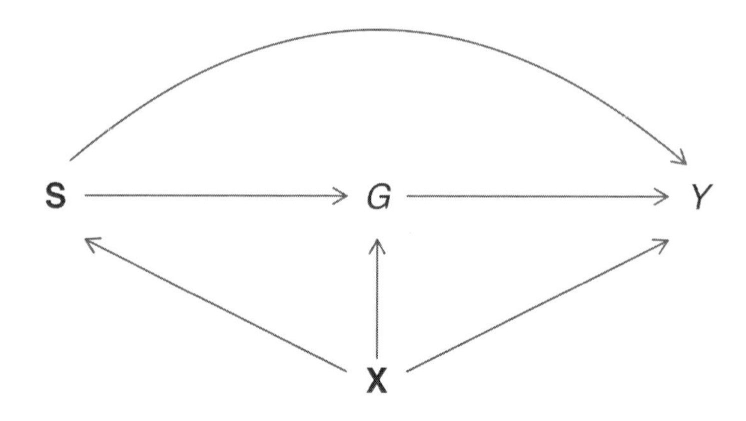

FIGURE 18.1
Causal mediation diagram: S is a SNP; G is a mediator, e.g., gene expression; Y is an outcome; and $\mathbf{X}$ a vector of covariates.

tial interest to develop statistical methods for risk prediction using massive NGS data.

18.4.3 Integration of different 'omics data and mediation analysis

An important emerging problem in genetic and genomic research is how to integrate different types of genetic and genomic data, such as SNPs, gene expressions, DNA methylation data, to improve understanding of disease susceptibility. The statistical problem of jointly modeling different types of genetic and genomic data and their relationships with a disease can be described using a causal diagram (Pearl, 2001) and be framed using a causal mediation model (VanderWeele and Vansteelandt, 2010) based on counterfactuals. For example, one can jointly model SNPs, gene expressions and a disease outcome using the causal diagram in Figure 18.1, with gene expression serving as a potential mediator.

To formulate the problem, assume for subject $i \in \{1, \ldots, n\}$, an outcome of interest Y_i is dichotomous (e.g., case/control), whose mean is associated with q covariates ($\mathbf{X}_i$), p SNPs ($\mathbf{S}_i$), mRNA expression of a gene (G_i) and possibly interactions between the SNPs and the gene expression as

$$\text{logit}\{\Pr(Y_i = 1|\mathbf{S}_i, G_i, \mathbf{X}_i)\} = \mathbf{X}_i^\top \beta_X + \mathbf{S}_i^\top \beta_S + G_i \beta_G + \mathbf{S}_i^\top G_i \beta_{GS}, \quad (18.2)$$

where $\beta_X, \beta_S, \beta_G, \beta_{GS}$ are the regression coefficients for the covariates, the SNPs, the gene expression, and the interactions of the SNPs and the gene expression, respectively. The gene expression G_i (i.e., the mediator) depends on the q covariates ($\mathbf{X}_i$) and the p SNPs ($\mathbf{S}_i$) through a linear model, as

$$G_i = \mathbf{X}_i^\top \boldsymbol{\alpha}_X + \mathbf{S}_i^\top \boldsymbol{\alpha}_S + \epsilon_i, \tag{18.3}$$

where $\boldsymbol{\alpha}_X$ and $\boldsymbol{\alpha}_S$ are the regression coefficients for the covariates and the SNPs, respectively. Here, ϵ_i follows a Normal distribution with mean 0 and variance σ_G^2.

The total effect (TE) of SNPs of the disease outcome Y can be decomposed into the Direct Effect (DE) and the Indirect Effect (IE). The Direct Effect of SNPs is the effect of the SNPs on the disease outcome that is not through gene expression, whereas the Indirect Effect of the SNPs is the effect of the SNPs on the disease outcome that is through the gene expression. Under no unmeasured confounding assumptions (VanderWeele and Vansteelandt, 2010), the TE, DE and IE can be estimated from the joint causal mediation models (18.2)–(18.3). In genome-wide genetic and genomic studies, the numbers of SNPs ($\mathbf{S}$) and gene expressions (G) are both large. It is of interest to develop mediation analysis methods in such settings.

18.5 Training the next generation statistical genetic and genomic scientists in the 'omics era

To help expedite scientific discovery in the 'omics era and respond to the pressing quantitative needs for handling massive 'omics data, there is a significant need to train a new generation of quantitative genomic scientists through an integrative approach designed to meet the challenges of today's biomedical science. The traditional biostatistical training does not meet the need. We need to train a cadre of interdisciplinary biostatisticians with strong quantitative skills and biological knowledge to work at the interface of biostatistics, computational biology, molecular biology, and population and clinical science genomics. They will be poised to become quantitative leaders in integrative and team approaches to genetic research in the public health and medical arenas. Trainees are expected to (1) have strong statistical and computational skills for development of statistical and computational methods for massive 'omics data and for integration of large genomic data from different sources; (2) have sufficient biological knowledge and understanding of both basic science and population science; (3) work effectively in an interdisciplinary research environment to conduct translation research from basic science to population and clinical sciences; (4) play a quantitative leadership role in contributing to frontier scientific discovery and have strong communication skills to be able to engage in active discussions of the substance of biomedical research.

Recent advances in genomic research have shown that such integrated training is critical. First, biological systems are complex. It is crucial to understand how the biological systems work. Such knowledge facilitates result interpretation and new scientific discovery in population sciences and clinical sciences. Computational biology plays a pivotal role in understanding the complexity of the biological system and integrating sources of different genomic data. Statistical methods provide systematic and rigorous tools for analyzing complex biological data and allow for making statistical inference accounting for randomness in data. Second, many complex diseases are likely to be governed by the interplay of genes and environment. The 2003 IOM Committee on "Assuring the Public's Health" has argued that advances in health will require a population health perspective that integrates understanding of biological and mechanistic science, human behavior, and social determinants of health. Analysis of GWAS and whole genome sequencing (WGS) data requires development of advanced biostatistical, computational, and epidemiological methods for big data. The top SNPs identified from the GWAS and WGS scan often have unknown functions. Interpretation of these findings requires bioinformatics tools and data integration, e.g., connecting SNP data with gene expression or RNA-seq data (eQTL data). Furthermore, to increase analysis power, integration with other genomic information, such as pathways and networks, in statistical analysis is important.

Ground breaking research and discovery in the life sciences in the 21st century are more interdisciplinary than ever, and students studying within the life sciences today can expect to work with a wider range of scientists and scholars than their predecessors could ever have imagined. One needs to recognize this approach to scientific advancement when training the next generation of quantitative health science students. Rigorous training in the core statistical theory and methods remains important. In addition, students must have a broad spectrum of quantitative knowledge and skills, especially in the areas of statistical methods for analyzing big data, such as statistical and machine learning methods, more training in efficient computational methods for large data, programming, and information sciences. Indeed, analysis of massive genomic data requires much stronger computing skills than what is traditionally offered in biostatistics programs. Besides R, students are advantageous to learn other programming languages, such as `scripts`, `python` and `perl`.

The next generation statistical genetic and genomic scientists should use rigorous statistical methods to analyze the data, interpret results, harness the power of computational biology to inform scientific hypotheses, and work effectively as leading quantitative scientists with subject-matter scientists engaged in genetic research in basic sciences, population science and clinical science. To train them, we need to develop an interdisciplinary curriculum, foster interactive research experiences in laboratory rotations ranging from wet labs on biological sciences to dry labs (statistical genetics, computational biology, and genetic epidemiology), developing leadership and communication skills through seminars, workshops and projects to enable our trainees to meet modern challenges for conducting translational genomic research.

18.6 Concluding remarks

We are living in an exciting time of genetic and genomic science, where massive 'omics data present statisticians with many opportunities and challenges. To take a full advantage of the opportunities and meet the challenges, we need to strategically broaden our roles and quantitative and scientific knowledge, so that we can play a quantitative leadership role as statistical genetic and genomic scientists in both method development and scientific discovery. It is important to develop new strategies to train our students along these lines so they can succeed in the increasingly interdisciplinary research environment with massive data. With the joint effort of our community, we can best position ourselves and the younger generation as quantitative leaders for new scientific discovery in the 'omics era.

References

Bansal, V., Libiger, O., Torkamani, A., and Schork, N.J. (2010). Statistical analysis strategies for association studies involving rare variants. *Nature Reviews Genetics*, 11:773–785.

Breslow, N.E. and Clayton, D.G. (1993). Approximate inference in generalized linear mixed models. *Journal of the American Statistical Association*, 88:9–25.

Cirulli, E.T. and Goldstein, D.B. (2010). Uncovering the roles of rare variants in common disease through whole-genome sequencing. *Nature Reviews Genetics*, 11:415–425.

Dickson, S.P., Wang, K., Krantz, I., Hakonarson, H., and Goldstein, D.B. (2010). Rare variants create synthetic genome-wide associations. *PLoS Biology*, 8:e1000294, doi:10.1371/journal.pbio.1000294.

Hung, R.J., McKay, J.D., Gaborieau, V., Boffetta, P., Hashibe, M., Zaridze, D., Mukeria, A., Szeszenia-Dabrowska, N., Lissowska, J., Rudnai, P., Fabianova, E., Mates, D., Bencko, V., Foretova, L., Janout, V., Chen, C., Goodman, G., Field, J.K., Liloglou, T., Xinarianos, G., Cassidy, A. et al (2008). A subunit genes on 15q25. *Nature*, 452:633–637.

Hüsing, A., Canzian, F., Beckmann, L., Garcia-Closas, M., Diver, W.R., Thun, M.J., Berg, C.D., Hoover, R.N., Ziegler, R.G., Figueroa, J.D. et al. (2012). Prediction of breast cancer risk by genetic risk factors, overall and by hormone receptor status. *Journal of Medical Genetics*, 49:601–608.

Kiezun, A., Garimella, K., Do, R., Stitziel, N.O., Neale, B.M., McLaren, P.J., Gupta, N., Sklar, P., Sullivan, P.F., Moran, J.L., Hultman, C.M., Lichtenstein, P., Magnusson, P., Lehner, T., Shugart, Y.Y., Price, A.L., de Bakker, P.I.W., Purcell, S.M., and Sunyaev, S.R. (2012). Exome sequencing and the genetic basis of complex traits. *Nature Genetics*, 44:623–630.

Lee, S., Wu, M.C., and Lin, X. (2012). Optimal tests for rare variant effects in sequencing association studies. *Biostatistics*, 13:762–775.

Li, B. and Leal, S.M. (2008). Methods for detecting associations with rare variants for common diseases: Application to analysis of sequence data. *The American Journal of Human Genetics*, 83:311–321.

Lin, D.Y. and Tang, Z.Z. (2011). A general framework for detecting disease associations with rare variants in sequencing studies. *The American Journal of Human Genetics*, 89:354–367.

Madsen, B.E. and Browning, S.R. (2009). A groupwise association test for rare mutations using a weighted sum statistic. *PLoS Genet*, 5:e1000384.

Manolio, T.A., Collins, F.S., Cox, N.J., Goldstein, D.B., Hindorff, L.A., Hunter, D.J., McCarthy, M.I., Ramos, E.M., Cardon, L.R., Chakravarti, A., Cho, J.H., Guttmacher, A.E., Kong, A., Kruglyak, L., Mardis, E., Rotimi, C.N., Slatkin, M., Valle, D., Whittemore, A.S., Boehnke, M., Clark, A.G., Eichler, E.E., Gibson, G., Haines, J.L., Mackay, T.F.C., McCarroll, S.A., and Visscher, P.M. (2009). Finding the missing heritability of complex diseases. *Nature*, 461:747–753.

McCullagh, P. and Nelder, J.A. (1989). *Generalized Linear Models.* Chapman & Hall, London.

McPherson, J.D. (2009). Next-generation gap. *Nature Methods*, 6:S2–S5.

Morgenthaler, S. and Thilly, W.G. (2007). A strategy to discover genes that carry multi-allelic or mono-allelic risk for common diseases: A cohort allelic sums test (CAST). *Mutation Research/Fundamental and Molecular Mechanisms of Mutagenesis*, 615:28–56.

Neale, B.M., Rivas, M.A., Voight, B.F., Altshuler, D., Devlin, B., Orho-Melander, M., Kathiresan, S., Purcell, S.M., Roeder, K. and Daly, M.J. (2011). Testing for an unusual distribution of rare variants. *PLoS Genetics*, 7:e1001322, doi:10.1371/journal.pgen.1001322.

Pearl, J. (2001). Direct and indirect effects. In *Proceedings of the Seventeenth Conference on Uncertainty in Artificial Intelligence*. Morgan Kaufmann Publishers, Burlington, MA, pp. 411–420.

Robinson, R. (2010). Common disease, multiple rare (and distant) variants. *PLoS Biology*, 8:e1000293, doi:10.1371/journal.pbio.1000293.

Tennessen, J.A., Bigham, A.W., O'Connor, T.D., Fu, W., Kenny, E.E., Gravel, S., McGee, S., Do, R., Liu, X., Jun, G., Kang, H.M., Jordan, D., Leal, S.M., Gabriel, S., Rieder, M.J., Abecasis, G., Altshuler, D., Nickerson, D.A., Boerwinkle, E., Sunyaev, S., Bustamante, C.D., Bamshad, M.J., Akey, J.M., Broad, G.O., Seattle, G.O., and the Exome Sequencing Project. (2012). Evolution and functional impact of rare coding variation from deep sequencing of human exomes. *Science*, 337:64–69.

The 1000 Genomes Project Consortium (2010). A map of human genome variation from population scale sequencing. *Nature*, 467:1061–1073.

VanderWeele, T.J. and Vansteelandt, S. (2010). Odds ratios for mediation analysis for a dichotomous outcome. *American Journal of Epidemiology*, 172:1339–1348.

Wray, N.R., Yang, J., Goddard, M.E., and Visscher, P.M. (2010). The genetic interpretation of area under the ROC curve in genomic profiling. *PLoS Genetics*, 6:e1000864.

Wu, M.C., Lee, S., Cai, T., Li, Y., Boehnke, M., and Lin, X. (2011). Rare-variant association testing for sequencing data with the sequence kernel association test. *The American Journal of Human Genetics*, 89:82–93.

Yang, J., Benyamin, B., McEvoy, B.P., Gordon, S., Henders, A.K., Nyholt, D.R., Madden, P.A., Heath, A.C., Martin, N.G., Montgomery, G.W., Goddard, M.E., and Visscher, P.M. (2010). Common SNPs explain a large proportion of the heritability for human height. *Nature Genetics*, 42:565–569.

19

Reflections on women in statistics in Canada

Mary E. Thompson
Department of Statistics and Actuarial Science
University of Waterloo, Waterloo, ON

Having been a recipient of the Elizabeth L. Scott Award, I have chosen the subject of women in statistics in Canada. The chapter is a selective history, combined with a series of reflections, and is intended as a tribute to women in statistics in Canada — past, present, and future.

Elizabeth L. Scott touches the story herself in a couple of ways, though fleetingly. I am fortunate to have met her in Delhi in December 1977, at the 41st Session of the International Statistical Institute. I had studied some of her work in modeling and inference — and had recently become aware of her pioneering approaches to the study of gender disparities in faculty salaries.

I have written about some individuals by name, mainly senior people. Because there are many women in the field in Canada, I have mentioned few whose last degree is more recent than 1995. Even so, I am conscious of omissions, and hope that this essay may inspire others to tell the story more completely.

19.1 A glimpse of the hidden past

The early history of statistics in Canada is not unrecorded, but apart from a few highlights, has never been processed into a narrative. Quite possibly women were involved in statistics from the beginning. Before the Statistical Society of Canada brought statisticians together and began to record their contributions, there were no chronicles of the profession. There may be other stories like the following, told by historian of statistics David Bellhouse:

> "Last summer on a family holiday in Winnipeg, we had dinner with my father-in-law at his seniors' apartment building in the downtown area. Seated at our table in the dining room were four sisters. Since

mealtime conversations at any dinner party I have ever attended rarely turn to the topic of statistics, I was surprised by the turn of events. For some reason that I cannot recall one of the sisters, a diminutive and exuberant nonagenarian, stated that she was a statistician. After asking where she had worked and when, and where, when and with whom she had studied, I came to the conclusion that I was talking not only to the oldest surviving statistician in the province of Manitoba, but also to one of the first women, if not the first, to work professionally as a statistician in Canada." (Bellhouse, 2002)

The statistician was Isobel Loutit, who was born in Selkirk, Manitoba, in July of 1909. She obtained a BA in mathematics with a minor in French from the University of Manitoba in 1929. She was taught statistics by Professor Lloyd Warren, using textbooks by Gavett and Yule. She started out as a teacher — in those days, school teaching, nursing and secretarial work were virtually the only career paths open to women on graduation — but with the World War II, as positions for women in industry opened up, she began a career in statistical quality control. In 1969, at a time where it was still rare for women to hold leadership positions, Loutit became chair of the Montréal Section of the American Society for Quality Control (ASQC).

She was made an Honorary Member of the Statistical Society of Canada in 2009, and the Business and Industrial Statistics Section instituted the Isobel Loutit Invited Address in her honor. She died in April 2009 at the age of 99.

19.2 Early historical context

Although the first census in North America was conducted in New France by Intendant Jean Talon in 1665–66, data gathering in Canada more generally seems to have become serious business only in the 19th century. Canadians of the time were avidly interested in science. Zeller (1996) credits Alexander von Humboldt's encouragement of worldwide studies of natural phenomena and the culture of the Scottish Enlightenment with leading to the Victorian "tradition of collecting 'statistics'... using detailed surveys to assess resources and quality of life in various districts." In Canada, such surveys for agricultural potential began as early as 1801 in Nova Scotia. Natural history societies were founded in the 1820s and later, while scientific organizations began to be formed in the 1840s. Statistics was used in analyses related to public health in Montréal as early as the 1860s (Bellhouse and Genest, 2003), a few years after the pioneering work of Florence Nightingale and John Snow in the 1850s.

In both Britain and North America there were substantial increases in opportunity for education of women in the late 19th and early 20th centuries, and women began to enter the scientific professions. At the same time, statistics as a mathematical discipline came into being, through the efforts of Karl

Pearson in England. Always on the lookout for references to the practice of statistics in Canada, Dr. Bellhouse has recently come across a 1915 letter in the files of Karl Pearson from one of his assistants in the field of craniometry, New Zealander Kathleen Ryley, who had by 1915 emigrated to Canada. We find that she had first visited Canada on a trip to Winnipeg in August 1909, among many women from abroad attending the meetings of the British Association for the Advancement of Science at the University of Manitoba. When she moved to Canada, she undertook the running of the Princess Patricia Ranch in Vernon, British Columbia — a fascinating story in itself (Yarmie, 2003). Pearson's biographer notes that the women who studied or worked with Karl Pearson in the early days generally had trouble finding positions in which they could continue statistical work (Porter, 2004).

Official statistics came into its own in Québec with the founding in 1912 of the Bureau de la statistique du Québec (Beaud and Prévost, 2000), and the Dominion Bureau of Statistics (now Statistics Canada) came about through the Statistics Act in 1918; for a detailed account, see Worton (1998). Data from censuses conducted at least every ten years since 1851 can now be studied to provide a picture of the evolution of the status of women in the country, particularly following digitization of samples from the census beginning with 1911; see Thomas (2010).

Statistics was slow to enter the academy in North America, in both the United States and Canada (Huntington, 1919). According to Watts (Watts, 1984), statistics courses were taught at the University of Manitoba beginning in 1917 by Lloyd A.H. Warren — Isobel Loutit's professor — who was developing curricula for actuarial science and commerce (Rankin, 2011). Watts cites a 1918 article by E.H. Godfrey as saying that no Canadian university was then teaching statistics as "a separate branch of science." However, there were at that time in the École des hautes études commerciales in Montréal (now HEC Montréal) a Chair of Statistics and "a practical and comprehensive curriculum"; at the University of Toronto, statistics was a subject in the second year of the course in Commerce and Finance.

The first teachers of statistics as a subject in its own right in mathematics departments included E.S. Keeping in Alberta, and George L. Edgett at Queen's, who taught a statistics course in 1933. Daniel B. DeLury, who taught my first course in statistics in 1962 at the University of Toronto, had begun teaching at the University of Saskatchewan in the period 1932–35. Some statistics was taught by biologists, particularly those involved in genetics. Geneticist and plant breeder Cyril H. Goulden is said to have written the first North American textbook in biostatistics in 1937, for the students he was teaching at the University of Manitoba. Yet as late as 1939, Dominion Statistician Robert H. Coats lamented that "five of our twenty-two universities do not know the word in their curricula" (Coats, 1939).

In 1950, the University of Manitoba was the first to recognize statistics in the name of a department (the Department of Actuarial Mathematics and Statistics). The two first Departments of Statistics were formed at the University of Waterloo and the University of Manitoba in 1967. There are now four other separate statistics or statistics and actuarial science departments, at the University of Toronto, Western University in Ontario, the University of British Columbia, and Simon Fraser University in Burnaby, British Columbia.

19.3 A collection of firsts for women

When I was a university student in the years 1961–65, there were few women professors in departments of mathematics in Canada. One of the exceptions was Constance van Eeden, who had come in the early 1960s (Clarke, 2003). Dr. van Eeden was born in 1927 in the Netherlands. She finished high school in 1944, and when the Second World War was over, she attended university, graduating in 1949 with a first degree in mathematics, physics and astronomy. She then entered a new actuarial program for her second degree, and in 1954 began as a part-time assistant at the Statistics Department at the Math Research Center in Amsterdam. She received her PhD cum laude in 1958. After some time at Michigan State and some time in Minneapolis, she and her husband (Charles Kraft, also a statistician) moved to the Université de Montréal — where, in contrast to the situation at their previous two institutions, there was no regulation against both members of a couple having tenure in the same department.

A specialist in mathematical statistics and in particular estimation in restricted parameter spaces, van Eeden was the first woman to receive the Gold Medal of the Statistical Society of Canada, in 1990. She has an extensive scientific "family tree"; two of her women students went on to academic careers in Canada: Louise Dionne (Memorial University of Newfoundland) and Sorana Froda (Université du Québec à Montréal).

There are, or were, other women statisticians in Canada born in the 1920s, but most have had their careers outside academia. The 34th Session of the International Statistical Institute was held in Ottawa fifty years ago, in 1963 — the only time the biennial meeting of the ISI has been held in Canada. The Proceedings tell us the names of the attendees. (Elizabeth L. Scott, from the University of California, Berkeley, was one.) Of the 136 Canadian "guests," 14 were women. Their names are listed in Table 19.1.

One of the two from universities in this list is the first and so far the only woman to hold the position of Chief Statistician of Canada: Winnipeg-born Sylvia Ostry, CC OM FRSC, an economist (PhD 1954) who was Chief Statistician from 1972 to 1975. In 1972, as she began her mandate, Dr. Ostry

TABLE 19.1
The 14 Canadian women who attended the 34th Session of the International Statistical Institute held in Ottawa, Ontario, in 1963.

Marjorie M. Baskerville	Dominion Foundries and Steel Ltd, Hamilton
M. Anne Corbet	Department of Secretary of State, Ottawa
I. June Forgie	Department of Transport, Ottawa
Geraldine E. Fulton	Sir George Williams University, Montréal
Irene E. Johnson	Department of Labour, Ottawa
Elma I. Kennedy	Department of Forestry, Ottawa
Pamela M. Morse	Department of Agriculture, Ottawa
Monique Mousseau	Radio-Canada, Montréal
Sylvia Ostry	Université de Montréal
Dorothy J. Powell	Bank of Nova Scotia, Toronto
Margaret R. Prentis	Department of Finance, Ottawa
Jean R. Proctor	Department of Agriculture, Ottawa
Joan Grace Sloman	Ontario Department of Health, Toronto
Dorothy Walters	National Energy Board, Ottawa

was the first woman working in Canada to be elected a Fellow of the American Statistical Association.

Among statistical researchers in Canada who were born in the 1930s and early 1940s, many (both men and women) have worked in the area of the design of experiments. One such is Agnes M. Herzberg, who was a student of Dr. Norman Shklov at the University of Saskatchewan in Saskatoon. Dr. Herzberg went to England on an Overseas Fellowship in 1966, soon after obtaining her PhD, and stayed as a member of the Department of Mathematics at Imperial College until 1988, when she moved back to Canada and a professorship at Queen's University. She was the first woman to serve as President of the Statistical Society of Canada, in 1991–92, and the first to be awarded the SSC Distinguished Service Award in 1999. In recent years, in addition to her research, Dr. Herzberg has focused much of her energy on organizing a series of annual conferences on Statistics, Science and Public Policy, held at Herstmonceux Castle in England. These meetings are attended by a wide variety of participants, from science, public service and the press (Lawless, 2012).

Canada is the adopted home of Priscilla E. (Cindy) Greenwood of the University of British Columbia (PhD 1963, University of Wisconsin, Madison), a distinguished probabilist who has also worked in mathematical statistics and efficient estimation in stochastic processes. Her work and love of science have been celebrated in a special Festschrift volume of *Stochastics* published in 2008 (vol. 80). In 1997, she was awarded a grant of $500,000 from the Peter Wall Institute of Advanced Studies for their first topic study: "Crisis Points and Models for Decision."

A rise in consciousness of the status of women in the late 1960s and early 1970s was marked by several initiatives in Canada; a landmark was the tabling of the Royal Commission on the Status of Women in 1970, which recommended that gender-based discrimination in employment be prohibited across the country (as it had been in several provinces). Women began to be elected in greater numbers to positions in the learned societies in those years. In 1975, Audrey Duthie of the University of Regina and Kathleen (Subrahmaniam) Kocherlakota of the University of Manitoba were the first women to be elected to the Board of Directors of the Statistical Science Association of Canada, one of the ancestors of the Statistical Society of Canada (SSC); this story is chronicled in Bellhouse and Genest (1999). Gail Eyssen (now of the University of Toronto) was elected to the SSC Board in 1976. Since that year, there has always been at least one woman serving on the Board; beginning in about 1991, there have been several each year.

The second woman to become President of the SSC was Jane F. Gentleman, whom I met first when I came to Waterloo in 1969. She was a fine role model and supportive mentor. Dr. Gentleman moved to Statistics Canada in 1982, and in 1999 she became the Director of the Division of Health Interview Statistics at the National Center for Health Statistics in Maryland. She was the winner of the first annual Janet L. Norwood Award in 2002 for Outstanding Achievement by a Woman in the Statistical Sciences.

I met K. Brenda MacGibbon-Taylor of the Université du Québec à Montréal in 1993, when I was appointed to the Statistical Sciences Grant Selection Committee by the Natural Sciences and Engineering Research Council of Canada (NSERC). Dr. MacGibbon was Chair of the Committee that year, the first woman to fill that position, and I came to have a great admiration for her expertise and judgment. She obtained her PhD at McGill in 1970, working in K-analytic spaces and countable operations in topology under the supervision of Donald A. Dawson. During her career, Dr. MacGibbon has worked in many areas of statistics, with a continuing interest in minimax estimation in restricted parameter spaces.

Another student of Dawson, probabilist Gail Ivanoff, was the first woman to fill the position of Group Chair of the Mathematical and Statistical Sciences Evaluation Group at NSERC, from 2009 to 2012. Dr. Ivanoff, who works on stochastic processes indexed by sets, as well as point processes and asymptotics, is a key member of the very strong group of probabilists and mathematical statisticians in the Ottawa–Carleton Institute for Mathematics and Statistics.

A third Dawson student, Colleen D. Cutler, was the first woman to be awarded the CRM–SSC Prize, awarded jointly by the SSC and the Centre de recherches mathématiques (Montréal) in recognition of accomplishments in research by a statistical scientist within 15 years of the PhD. The award, which she received in the third year of its bestowing, recognizes her work at the interface of non-linear dynamics and statistics, and in particular non-linear time series, the study of determinism in time series, and the computation of fractal dimension.

Professor Nancy M. Reid of the University of Toronto, another former President of the SSC, is the owner of "firsts" in several categories. Most notably, she was the first woman, and the first statistician working in Canada, to receive the COPSS Presidents' Award, in 1992. She is the only Canadian-based statistician to have served as President of the Institute of Mathematical Statistics (IMS), in 1997. In 1995 she was awarded the first Krieger–Nelson Prize of the Canadian Mathematical Society — a prize "in recognition of an outstanding woman in mathematics." She was also the first woman in Canada to be awarded a Canada Research Chair in the statistical sciences, and is the only woman to have served as Editor of *The Canadian Journal of Statistics.*

In this International Year of Statistics, for the first time ever, the Program Chair and the Local Arrangements Chair of the 41st SSC Annual Meeting are both women, respectively Debbie Dupuis of HEC Montréal and Rhonda Rosychuk of the University of Alberta. The meeting was held in Edmonton, Alberta, May 26–29, 2013.

19.4 Awards

I have always had conflicted feelings about awards which, like the Elizabeth L. Scott Award, are intended to recognize the contributions of women. Why do we have special awards for women? Is it to compensate for the fact that relatively few of us win prestigious research awards in open competition? Or is it rather to recognize that it takes courage to be in the field as a woman, and that those of us who are here now should be mindful of the sacrifices and difficulties faced by our forerunners.

In general, the awards in an academic or professional field recognize a few in the context of achievements by the many. Perhaps it is as well to remember that awards are really mainly pretexts for celebrations. I will never forget the toast of a student speaker at a Waterloo awards banquet: "Here's to those who have won awards, and to those who merely deserve them."

The SSC has two awards named after women, and open to both men and women. Besides the Isobel Loutit Invited Address, there is the biennial Lise Manchester Award. Lise Manchester of Dalhousie University in Halifax was the first woman to receive *The Canadian Journal of Statistics* Award, in 1991, the second year of its bestowing, for a paper entitled "Techniques for comparing graphical methods." The community was saddened at the passing of this respected young researcher, teacher, and mother. The Lise Manchester Award, established in 2007, commemorates her "abiding interest in making use of statistical methods to provide insights into matters of relevance to society at large."

The other awards of the Statistical Society of Canada are also open to both men and women. One is the afore-mentioned *The Canadian Journal of Statistics* Award, for the best paper appearing in a volume of the journal. Of 23 awards from 1990 to 2012, women have been authors or co-authors of 10; Nancy E. Heckman of the University of British Columbia has received it twice, the first time (with John Rice, 1997) for a paper on line transects of two-dimensional random fields, and the second time (with James O. Ramsay, 2000) for penalized regression with model-based penalties.

Women are well represented among winners of the CRM–SSC Prize since its creation in 1999 (Colleen Cutler, Charmaine Dean, Grace Y. Yi). To date, three of us have received the SSC Gold Medal (Constance van Eeden, Nancy Reid and myself) since it was first awarded in 1985. In the first 27 awards of the Pierre Robillard Award for the best thesis in probability or statistics defended at a Canadian university in a given year, three of the winners were women (Maureen Tingley, Vera Huse-Eastwood, and Xiaoqiong Joan Hu) but since the year 2001 there have been eight: Grace Chiu, Rachel MacKay-Altman, Zeny Zhe-Qing Feng, Mylène Bédard, Juli Atherton, Jingjing Wu, Qian Zhou, and Bei Chen.

There are just eight women who have been elected Fellow of the Institute of Mathematical Statistics while working in Canada. The one not so far mentioned is Hélène Massam of York University, who combines mathematics and statistics in the study of hierarchical and graphical models. I count sixteen who have been elected Fellow of the American Statistical Association, including Thérèse Stukel of the Institute for Clinical Evaluative Sciences in Toronto (2007), Keumhee C. Chough of the University of Alberta (2009), Xiaoqiong Joan Hu of Simon Fraser University (2012), Sylvia Esterby of UBC Okanagan (2013), and W. Y. Wendy Lou of the University of Toronto (2013).

19.5 Builders

Since the late 1980s, women in Canada have begun to find themselves more welcome in leadership positions in academia and in the statistics profession. It is as though society suddenly realized at about that time that to consider only men for such roles was to miss out on a significant resource. In some cases, despite the ephemeral nature of "service" achievements, our leaders have left lasting legacies.

One builder in academia is Charmaine B. Dean, who came to Canada from San Fernando, Trinidad. She completed an Honours Bachelor's Degree in Mathematics at the University of Saskatchewan and her PhD at the University of Waterloo. She joined the Department of Mathematics and Statistics at Simon Fraser University in 1989, and several years later played a major role in setting up the Department of Statistics and Actuarial Science, becoming the founding Chair in 2001. She received the CRM–SSC Prize in 2003 for her work on inference for over-dispersed generalized linear models, the analysis of recurrent event data, and spatial and spatio-temporal modelling for disease mapping. In 2002, Dr. Dean was President of WNAR, the Western North American Region of the Biometric Society; she was also President of the SSC in 2006–07 and is currently Dean of the Faculty of Science at Western University, London, Ontario.

Another example is Shelley Bull, a graduate of the University of Waterloo and Western University in Ontario. As a faculty member in biostatistics in the Samuel Lunenfeld Institute of Mount Sinai General Hospital, in Toronto, she became interested in research in statistical genetics in the 1990s. When the Networks of Centres of Excellence MITACS (Mathematics of Information Technology for Complex Systems) began in 1999, Dr. Bull became the leader of a national team pursuing research at the interface of statistics and genetics, in both modeling and analysis, with emphasis on diseases such as breast cancer and diabetes. The cohesiveness of the group of statistical genetics researchers in Canada owes much of its origin to her project.

I am proud to count myself among the builders, having chaired the Department of Statistics and Actuarial Science at the University of Waterloo from 1996 to 2000. Besides Charmaine Dean and myself, other women in statistics who have chaired departments include Nancy Reid (Toronto), Nancy Heckman (UBC), Karen Campbell (Epidemiology and Biostatistics, Western), Cyntha Struthers (St. Jerome's), Sylvia Esterby (UBC Okanagan) and Christiane Lemieux (Waterloo).

Nadia Ghazzali, formerly at Université Laval, has held since 2006 the NSERC Chair for Women in Science and Engineering (Québec Region). She is the first woman statistician to become President of a university in Canada. She was appointed Rector of the Université du Québec à Trois-Rivières in 2012.

Since the late 1990s, the SSC has had an active Committee on Women in Statistics, which sponsors events at the SSC annual meetings jointly with the Canadian Section of the Caucus for Women in Statistics. Cyntha Struthers was both the founding Chair of the Caucus section, in 1987–89, and Chair of the SSC Committee on Women in Statistics in 1998–2000.

For professional leadership, an early example was Nicole P.-Gendreau of the Bureau de la statistique du Québec (which has since become the Institut de la statistique du Québec). Mme Gendreau was Public Relations Officer in the Statistical Society of Canada from 1986 to 1989. She was the founder of the Newsletter, *SSC Liaison*, the chief communication vehicle of the community in Canada, now in online and print versions, and still very much in keeping with her original vision.

The process of developing SSC Accreditation was brought to fruition in 2003–04 (when I was President of the SSC) under the dedicated leadership of Judy-Anne Chapman, now at Queen's University. (At least 30 of the 140 PStat holders to date are women.) As Chair of the SSC Education Committee,

Alison Gibbs of the University of Toronto led the takeover of responsibility by the SSC of the highly successful educational program, Census at School Canada. Shirley Mills of Carleton University is the first woman to lead the day-to-day operations of the SSC as Executive Director.

Few of us in the earlier cohorts have followed a career path in the private sector. A notable exception is Janet E.A. McDougall, PStat, the founder and President of McDougall Scientific, a leading Clinical Research Organization which has been in operation since 1984.

19.6 Statistical practice

Statistics is a profession as well as an academic subject. Some of the best statisticians I know spend much of their time in statistical practice, and research that arises from practical problems.

To name just two — Dr. Jeanette O'Hara Hines was Director of the Statistical Consulting Service (and teacher of consulting) at the University of Waterloo for many years, and specialized in working with faculty in the Department of Biology on a very diverse set of problems; Hélène Crépeau, Nancy Reid's first Master's student, who has been working at the Service de consultation statistique de l'Université Laval since 1985, has been involved in biometrics research and studies of the quantification of wildlife populations.

As in 1963, many women statisticians work in government agencies. In earlier days at Statistics Canada, Estelle Bee Dagum developed the X–11–ARIMA method, variants of which are used for seasonal adjustment of times series around the world. Survey statistician Georgia Roberts of Statistics Canada is an expert on methodology and the analytical uses of complex survey data, and has led the Data Analysis Resource Centre for several years. In the same area of data analysis is Pat Newcombe-Welch, Statistics Canada analyst at the Southwestern Ontario Research Data Centre, who is on the front lines of assisting researchers of many disciplines in the access to Statistics Canada data. Susana Rubin-Bleuer, Ioana Schiopu-Kratina and Lenka Mach of Statistics Canada have published research papers in advanced theoretical topics in the analysis of complex survey data and in sample coordination.

Women statisticians have also achieved leadership roles in statistical agencies. To name just a few: Louise Bourque was Directrice de la méthodologie for several years at the BSQ/ISQ; Nanjamma Chinnappa served as Director of the Business Survey Methods Division at Statistics Canada; Marie Brodeur is Director General of the Industry Statistics Branch of Statistics Canada; and Rosemary Bender is Assistant Chief Statistician of Analytical Studies and Methodology.

19.7 The current scene

Statistics is an important profession in Canada today. Despite a steady inflow of capable young people, the demand for talented statisticians outstrips the supply. It is fortunate that women are entering the field in greater numbers than ever before. The cohorts with PhDs in the 1990s are now mid-career and making their mark on the profession, while those just beginning are achieving things that I would never have dreamed of at the same stage. Are there still barriers that are particular to women? Is it still harder for women to enter the profession, and to rise high in its ranks?

The answer is probably "yes." When we think of it, some of the obstacles that were always there are hardly likely to disappear: The difficulties of combining career and family life, the "two-body" problem, despite the adoption by most universities of family-friendly "spousal hiring policies"; and the physical toll of long working hours and dedication. Other barriers might continue to become less pervasive over time, such as the prejudices that have contributed to the adoption of double-blind refereeing by many journals. *The Canadian Journal of Statistics* was among the first to adopt this policy, in 1990.

In academia, examples of anomalies in salary and advancement may be fewer these days. At least at the University of Waterloo, statistical approaches something like those pioneered by Elizabeth L. Scott are now taken to identify and rectify such anomalies. There are still however several departments across the country with a surprisingly small number of women faculty in statistics.

It used to be the case that in order to succeed in a field like statistics, a woman had to be resolute and determined, and be prepared to work very hard. It was often said that she would have to work twice as hard as a man to achieve the same degree of recognition. It now seems to be the case that both men and women entering the field have to be either consummately brilliant or resolute and determined, to about the same degree — but it is still an easier road for those who can afford to be single-minded. I remember the days when women who pursued a career, particularly in academia, were considered rather odd, and in some ways exceptional. We are now no longer so odd, but those who make the attempt, and those who succeed, are still exceptional, in one way or another.

Acknowledgments

I would like to thank David Bellhouse, Christian Genest, and the reviewers for very helpful suggestions.

References

Beaud, J.-P. and Prévost, J.-G. (2000). L'expérience statistique canadienne. In *The Age of Numbers: Statistical Systems and National Traditions* (J.-P. Beaud and J.-G. Prévost, Eds.). Presses de l'Université du Québec à Montréal, Montréal, Canada, pp. 61–86.

Bellhouse, D.R. (2002). Isobel Loutit: Statistician of quality. *SSC Liaison*, 16(2):14–19.

Bellhouse, D.R. and Genest, C. (1999). A history of the Statistical Society of Canada: The formative years (with discussion). *Statistical Science*, 14:80–125.

Bellhouse, D.R. and Genest, C. (2003). A public health controversy in 19th century Canada. *Statistical Science*, 20:178–192.

Clarke, B. (2003). A conversation with Constance van Eeden. *SSC Liaison*, 17(4):28–35.

Coats, R.H. (1939). Science and society. *Journal of the American Statistical Association*, 34:1–26.

Huntington, E.V. (1919). Mathematics and statistics, with an elementary account of the correlation coefficient and the correlation ratio. *American Mathematical Monthly*, 26:421–435.

Lawless, J.F. (2012). A conversation with Agnes Herzberg. *SSC Liaison*, 26(4):40–45.

Porter, T.M. (2004). *Karl Pearson: The Scientific Life in a Statistical Age.* Princeton: Princeton University Press, pp. 273–275.

Rankin, L. (2011). *Assessing the Risk: A History of Actuarial Science at the University of Manitoba.* Warren Centre for Actuarial Studies and Research, Asper School of Business, University of Manitoba, Winnipeg, MB.

Thomas, D. (2010). The Census and the evolution of gender roles in early 20th century Canada. Canadian Social Trends, Catalogue No 11–008, pp. 40–46. http://www.statcan.gc.ca/pub/11-008-x/2010001/article/11125-eng.pdf

Watts, D.G. (1984). Teaching statistics in Canada: The early days. *The Canadian Journal of Statistics*, 12:237–239.

Worton, D.A. (1998). *Dominion Bureau of Statistics: A History of Canada's Central Statistical Office and its Antecedents, 1841–1972.* McGill–Queen's University Press, Montréal, Canada.

Yarmie, A. (2003). "I had always wanted to farm." The quest for independence by English Female Emigrants at the Princess Patricia Ranch, Vernon, British Columbia: 1912–1920. *British Journal of Canadian Studies*, 16:102–125.

Zeller, S. (1996). *Land of Promise, Promised Land: The Culture of Victorian Science in Canada.* Canadian Historical Association Historical Booklet No 56.

20

"The whole women thing"

Nancy M. Reid
Department of Statistical Sciences
University of Toronto, Toronto, ON

It is an honor and a pleasure to contribute to this celebratory volume, and I am grateful to the editors for their efforts. The temptation to discuss non-technical aspects of our profession and discipline has caught me too, and I have, with some trepidation, decided to look at the past, present and future of statistical science through a gender-biased lens. In the past fifty years, a great deal has changed for the better, for the position of women in science and in statistical science, but I believe we still have some way to go.

20.1 Introduction

The title of this chapter is a quote, as I remember it, from a dear friend and colleague. The occasion was a short discussion we had while rushing in opposite directions to catch talks at a Joint Statistical Meeting, probably in the early 1990s, and he asked me if I might consider being nominated to run for election as President of the Institute of Mathematical Statistics (IMS). I was completely surprised by the question, and my immediate reactions were to be honored that we were discussing it, and to assume that the question was rhetorical. He said no, this was within the realm of possibility, and I should give it careful consideration, for all the reasons one might expect: an honor for me, a chance to influence an organization I cared about, etc. He ended by saying "plus, you know, there's the whole women thing. I guess you'd be the first." In fact, Elizabeth Scott was the first woman President of the IMS, in 1978.

For various reasons, a number of unconnected events recently got me thinking about "the whole women thing." Despite many years of on-and-off thinking about issues surrounding gender and a professional career, I find I still have a lot of questions and not many answers. I have no training in social science, nor in womens' studies, nor in psychology, and no experience of what seems to me the difficult world outside academia. I will present a handful of anecdotes, haphazardly chosen published studies, and personal musings. My vantage point is very North American, but I hope that some of the issues resonate with women in other countries as well.

20.2 "How many women are there in your department?"

In September, 2012, the small part of the blogosphere that I sometimes wander through lit up with an unusual amount of angst. A study on gender bias (Moss-Racusin et al., 2012a,b) appeared in the *Proceedings of the National Academy of Sciences* (PNAS), not in itself so unusual, but this one got us rattled. By "us" I mean a handful of internet colleagues that worry about issues of women in science, at least some of the time. I was alerted to the article through Isabella Laba's blog (Laba, 2012a, 2013), but the story was also picked up by many of the major news organizations. The PNAS paper reported on a study in which faculty members in biology, chemistry and physics departments were asked to evaluate an application for a position as a student laboratory manager, and the participants were told that this was part of a program to develop undergraduate mentoring. From Moss-Racusin et al. (2012b):

> "Following conventions established in previous experimental work..., the laboratory manager application was designed to reflect slightly ambiguous competence, allowing for variability in participant responses... if the applicant had been described as irrefutably excellent, most participants would likely rank him or her highly, obscuring the variability in responses to most students for whom undeniable competence is frequently not evident."

In other words, the applicant would likely not be at the top of anyone's short list, but was qualified for the position. Faculty were asked to evaluate the application as if they were hiring the student into their own lab. The applications were all identical, but for half the scientists the student was named "John," and for the other half, "Jennifer."

The headline story was that scientists rated applications from a male student higher than those from a female student. Scores assigned to qualities of competence, hireability, and mentoring, were systematically higher for the male student application, and

> "The mean starting salary offered the female student, \$26,507.94, was significantly lower than that of \$30,238.10 to the male student [$t = 3.42, P < 0.01$]" (Moss-Racusin et al., 2012a)

Many more details about the methods for the study, the relevant literature, and the results, are available from the original publication (Moss-Racusin et al., 2012a), and the accompanying supplementary material (Moss-Racusin et al., 2012b). A particularly concerning result was that female scientists exhibited the same gender bias as their male counterparts.

These results reverberated because they felt real, and indeed I felt I recognized myself in this study. During my many years on search committees, and during my five-year tenure as department chair, efforts to hire female research-stream faculty were not successful. It took me a surprisingly long time to come to the conclusion that it was easy to decide to make an offer to the *best* female candidate in any given hiring season, but much, much harder for females in the next tier 'down' to be ranked as highly as the men in the same tier. Our intentions were good, but our biases not easily identified. With reference to the PNAS study, Laba says:

> "The scientists were not actively seeking to discriminate... They offered similar salaries to candidates that they perceived as equally competent, suggesting that, in their minds, they were evaluating the candidate purely on merit. The problem is that the female candidate was judged to be less competent, evidently for no reason other than gender, given that the resumes were exactly identical except for the name. [$\cdots$] I'm sure that most of the participants, believing themselves unbiased, would be shocked to see the results." (Laba, 2012a)

I've presented this study in two talks, and mentioned it in a number of conversations. The reaction from women is often to note other related studies of gender bias; there are a number of these, with similar designs. An early study of refereeing (Goldberg, 1968) involved submitting identical articles for publication with the author's name either Joan or John; this study featured in the report of an IMS committee to investigate double-blind refereeing; see Cox et al. (1993). A more common reaction is to speculate more broadly on whether or not women self-select out of certain career paths, are genuinely less interested in science and so on. This deflects from the results of the study at hand, and also diffuses the discussion to such an extent that the complexity of "the women thing" can seem overwhelming. Here is Laba in a related post:

> "Let's recap what the study actually said: that given identical paperwork from two hypothetical job candidates, one male and one female, the woman was judged as less competent and offered a lower salary. This is not about whether girls, statistically speaking, are less interested in science. It's about a specific candidate who had already met the prerequisites... and was received much better when his name was John instead of Jennifer." (Laba, 2012b)

We all have our biases. The ABC News report (Little, 2012) on the paper described a "small, non-random experiment," but the study was randomized, and the authors provided considerable detail on the size of the study and the response rate. The authors themselves have been criticized for their bar chart of the salary differential. By conveniently starting the y-axis at \$25,000, the visual appearance suggests a three-fold salary differential.

That the biases are subtle, and often unconscious, is much better than what many of the pioneers faced. Elizabeth Scott's first ambition was to be an astronomer, but she soon realized that this was a hopeless career for a woman (Billard and Ferber, 1991). We've heard the jaw-dropping stories of blatant sexism from days gone by, and to the extent that this is behind us, this is progress of a sort. But biases that we don't even recognize should concern us all.

How to move forward? I hope this study will help: by highlighting biases that seem to be operating well below the surface, perhaps the next search committees will work even harder. Genuine efforts are being made at my university, and my department, and I believe at universities and departments around the world, to increase the number of women hired, and to treat them well. There is progress, but it seems to be slow and imperfect.

20.3 "Should I ask for more money?"

This might be the most common question I am asked by our graduating students who are considering job offers, in academia or not. Of course uncertainty about how to negotiate as one starts a career affects men and women, and there are many aspects to the dialogue between a candidate and prospective employer, including the hiring landscape at the time, the trade-off between salary and other aspects of the position, and so on.

When I arrived at the University of Toronto in 1986, both the government of Canada, and the government of Ontario, had passed pay equity laws, enshrining the principle of "equal pay for work of equal value." This is broader than "equal pay for equal work," which had already been in force for some years in most jurisdictions in Canada. The laws led to a flurry of work on pay equity at that time, and one of my first consulting projects, undertaken jointly with Ruth Croxford of our department, was to review the salaries of all faculty at the University of Toronto, with a view to identifying pay equity issues.

I was at the time unaware of a strong legacy for statistical analyses of faculty salaries initiated by Elizabeth Scott (Gray and Scott, 1980; Scott, 1975), who led the charge on this issue; see Billard and Ferber (1991). Lynne Billard made important follow-up contributions to the discussion in 1991 and 1994 (Billard, 1991, 1994). Lynne's 1994 paper asked: "Twenty years later: Is there parity for academic women?"; her answer was "no" (the title referred to a US government law, Title IX of the Education Amendments, enacted in 1972). This continues to be the case. For example, the University of British Columbia (UBC) recommended in October 2012 an across-the-board salary increase of

2% to all female faculty (Boyd et al., 2012). The detailed review of university practices on this issue in the UBC report noted that many universities have implemented ongoing reviews and adjustments of female faculty salaries on a regular schedule.

As with hiring, it is easy to get distracted in the salaries debate by potential factors contributing to this. One is rank; it continues to be the case that women are under-represented in the professorial rank; the London Mathematical Society has just published a report highlighting this fact in mathematics (London Mathematical Society, 2013). A point very clearly explained in Gray and Scott (1980) is that systematic differences in rank are themselves a form of gender bias, so not appropriate defence against salary remedies. Even setting this aside, however, the UBC report concluded that after adjusting for rank, departmental unit, merit pay and experience, "there remains an unexplained female disadvantage of about $3000" (Boyd et al., 2012).

It may be the case that on average, women are less aggressive in negotiating starting salaries and subsequent raises, although of course levels of skill in, and comfort with, negotiation vary widely across both genders. Other explanations, related to publication rate, lack of interest in promotion, time off for family matters, and so on, seem to need to be addressed in each equity exercise, although this seems to me to be once again "changing the subject." As just one example in support of this view, the UBC report, referring to an earlier salary analysis, concluded: "our assumptions would be supported by a more complete analysis, and... parental leave does not alter the salary disadvantage."

Over the years I have often met statistical colleagues, usually women, who were also asked by their university to consult on an analysis of female faculty salary data, and it often seemed that we were each re-inventing the wheel. The pioneering work of Gray and Scott (Gray and Scott, 1980) touches on all the main issues that are identified in the UBC report, and it would be good to have a central repository for the now quite large number of reports from individual universities, as well as some of these key references.

20.4 "I'm honored"

The Elizabeth L. Scott Award was established by COPSS in 1992 to honor individuals who have "helped foster opportunities in statistics for women." The first winner was Florence Nightingale David, and the second, Donna Brogan, said in her acceptance speech that she looked forward to the day when such awards were no longer needed. While women are well-represented in this volume, I think that day is not here yet.

I was involved in a number of honor-related activities over the past year or two, including serving on the Program Committee for a major meeting, on an *ad hoc* committee of the Bernoulli Society to establish a new prize in statistics, and as Chair of the F.N. David Award Committee.

On the Program Committee, the first task for committee members was to create a long list of potential candidates for the plenary lecture sessions. In a hurry, as usual, I jotted down my personal list of usual suspects, i.e., people whose work I admire and who I thought would give interesting and important presentations. I examined my list more critically when I realized that my usual suspects were all about the same age (my age or older), and realized I'd better have another think. Some on my list had already given plenary lectures at the same meeting in earlier years, so off they came. I'm embarrassed to admit that it took me several passes before I realized I had no women on my list; another round of revisions was called for. At that point I did some research, and discovered that for this particular meeting, there had been no women plenary speakers since 1998, which seemed a pretty long time.

Then things got interesting. I sent an email to the Program Committee pointing this out, and suggesting that we should commit to having at least one female plenary lecturer. Email is the wrong medium in which to rationalize opposing views, and it turned out there were indeed opposing, as well as supporting, views of this proposal. Extremes ranged from "I do not consider gender, race or any other non-scientific characteristics to be relevant criteria" to "I do find it important for the field and for the meeting that female researchers are well represented." Without the diplomatic efforts of the Chair, we might still be arguing.

What did I learn from this? First, we all have our biases, and it takes some effort to overcome them. The more well-known people are, the more likely they are to be suggested for honors, awards, plenary lectures, and so on. The older they are, the more likely they are to be well-known. Statistical science is aging, and we have a built-in bias in favor of established researchers, that I think makes it difficult for young people to get the opportunities and recognition that I had when I was young(er). Second, our biases are unintentional, much as they surely were for the scientists evaluating lab manager applications. We are all busy, we have a lot of demands on our time, and the first, quick, answer is rarely the best one. Third, it is important to have women on committees. I wish it were not so; I have served on far too many committees in my career, and every woman I know says the same thing. It turned out I was the only woman on this particular Program Committee, and while I had good support from many members, I found it lonely.

The Bernoulli Society recently established the Wolfgang Doeblin prize in probability (see Bernoulli Society (2012)), and I chaired an *ad hoc* committee to consider a new prize in statistics. Wolfgang Doeblin died in 1940, just 25 years old, and shortly before his death wrote a manuscript later found to contain many important ideas of stochastic calculus (Göbel, 2008). The award is thus given to a single individual with outstanding work, and intended for researchers at the beginning of their mathematical career.

A parallel to this could be a prize for statisticians at the beginning of their career, and my personal bias was to restrict the award to women. On discussion with colleagues, friends, and members of the committee, it became apparent that this was not quite as good an idea as I had thought. In particular, it seemed likely that a prize for women only might have a negative connotation — "not a 'real' prize" — not among the curmudgeonly senior colleagues of the awardees, but among the potential awardees themselves. In fact I am sure I would have felt that way myself.

On to the next challenge. We decided to recommend naming the prize after a female statistician, no longer living. Well, we already have the E.L. Scott prize, the F.N. David Award, and the Gertrude Cox scholarship. Quick, how many can you name? How many of your colleagues will recognize the name?

We discovered Ethel Newbold (1882–1933), the first woman to be awarded a Guy Medal in Silver from the Royal Statistical Society, in 1928. We also discovered that the second woman to be awarded a Guy Medal in Silver was Sylvia Richardson, in 2002. Get nominating, ladies! We needn't feel smug on this side of the Atlantic, either; see Gray and Ghosh-Dastidar (2010) and Palta (2010).

The F.N. David Award is the only international award in statistical sciences that I am aware of that is restricted to women. It was established jointly by COPSS and the Caucus for Women in Statistics in 2001. The nominees this year were amazing, with nomination letters and *vitae* that could induce strong feelings of inadequacy in any reader. But a side remark from one nominator got me thinking. The nominator pointed out that the selection criteria for the award were daunting indeed, although his nominee did indeed fulfill all the criteria, and then some. I had a more careful look at these criteria, and they are

> "Excellence in the following: as a role model to women; statistical research; leadership in multidisciplinary collaborative groups; statistics education; service to the profession."

Hello Caucus! We are *much* too hard on each other! But perhaps I'm being unfair, and the intention was "one of," rather than "all of." I can say though, that our leading female colleagues do seem to manage to excel in "all of." Here's Laba again on a similar point:

> "The other way to make progress, of course, is for women to be 'twice as good,' [$\cdots$] That's what many women in science have been doing all along. It takes a toll on us. It's not a good solution. Unfortunately, sometimes it's the only one we've got." (Laba, 2012b)

20.5 "I loved that photo"

The 1992 Joint Statistical Meetings coincided with our first daughter's first birthday, so we joined the legions of families who combine meetings with children. If you've done this you'll know how hard it can be, and if you haven't, well, be assured that it is an adventure. George Styan snapped a picture of me with Ailie in a back carrier, and this photo ended up being printed in the *IMS Bulletin.* At the time I was quite embarrassed by this — I thought that this photo would suggest that I wasn't taking the meeting, and by extension research, seriously enough. But in fact I received so many emails and notes and comments from colleagues, expressing the sentiment in the section heading, that in the end I was, and am, grateful to George for his sixth sense for a good snapshot.

For me the most difficult aspect of the discussion around women and academia is children. Decisions around having and raising children are so deeply personal, cultural, and emotional, that it often seems better to leave this genie in the bottle. It is also the most disruptive part of an academic career, and by and large still seems to be more disruptive for women than for men. It risks being a two-edged sword. If differences in opportunities are tied to child-rearing, then is there a temptation to assume that women without children face no hurdles, or that men who choose to become more involved in child care should be prepared to sacrifice advancement at work? Again it is easy to get distracted by the complexity and depth of the issues, and lose the main thread.

The main thread to me, is the number of women who ask me whether or not it is possible to have an academic career in a good department and still have time for your family. When is the 'best' time to have children — grad school? Post-doc? Pre-tenure? If I wait until I have tenure, will I have waited too long? What about promotion to full professor? I don't know the answer to any of these questions, but I do know women who have had children at each of these stages, and who have had, and are having, very successful academic careers.

I can only speak for academia, but exceptionally talented and successful women are speaking about government and industry: Anne-Marie Slaughter, (Slaughter, 2012) and Sheryl Sandberg (Sandberg, 2013), to name just two.

Slaughter, a Princeton professor who spent two years in a high profile position in Washington, DC, writes:

> "I still strongly believe that women can 'have it all' (and that men can too). I believe that we can 'have it all at the same time.' But not today, not with the way America's economy and society are currently structured. My experiences over the past three years have forced me to confront a number of uncomfortable facts that need to be widely acknowledged — and quickly changed." (Slaughter, 2012)

Happily for most readers of this article, Slaughter contrasts the flexibility and freedom she has with her academic appointment with the demands of a high-profile position in government. While the very heavy demands on her time at Princeton would make most of us weep, "I had the ability to set my own schedule most of the time. I could be with my kids when I needed to be, and still get the work done" (Slaughter, 2012). This article is a great reference to provide to your graduate students, if they start wondering whether an academic career is too difficult to combine with family life. However, much of what she describes as barriers to women in high-level government positions resonates in a thoughtful analysis of the difficulties faced by women moving into the leadership ranks of universities; see Dominici et al. (2009).

Sandberg has been criticized for implying that at least some of the barriers for advancement of women are created by the womens' own attitudes, particularly around family, and she exhorts women to make sure that they are aggressively pursuing opportunities. The position set out in her book (Sandberg, 2013) is much more nuanced than that, but the notion that women are sometimes their own worst enemies did resonate with me. In an interesting radio interview with the BBC (BBC News, 2013), Sandberg suggested that the phrases "work-life balance" and "having it all" should *de facto* be mistrusted, as they are themselves quite gender-specific.

Recently I visited the lovely new building that is now home to the Department of Statistics at North Carolina State University. As it happened, there was a career mentoring event taking place in the department at the same time, and over coffee I met an enthusiastic young woman who is completing a PhD in statistics. Her first question was about balancing career and family in a tenure-stream position; I think I relied on the rather bland "advice" mentioned above. But the most encouraging part of the day was the tour of the department. There, between the faculty offices, department lounge, seminar rooms, and banks of computer terminals, was a wonderful sight: a Baby Room! I imagine that Gertrude Cox would be surprised to see this, but I hope she would also be proud of the department she founded.

20.6 Conclusion

The position of women in academia is vastly improved from the days when Elizabeth Scott was discouraged from studying astronomy, and from the days when my probability professor could state in class that "women are not suited for mathematics." Determined and forceful pioneers through the 1950s and 1960s, followed by much larger numbers of female students from the 1970s on, has meant that women do have many opportunities to succeed in academic work, and many are succeeding on a number of levels.

I find it difficult to discuss all these issues without seeming plaintive. I write from a privileged position, and I can say without hesitation that I personally do not feel disadvantaged; my career has coincided with a concerted effort to hire and promote women in academia. And yet I've felt the energy drain from trying to tackle some of the issues described here. I've experienced the well-documented attrition through the ranks: although my undergraduate statistics class had nine women in a class of 23, and my graduating PhD class had four women and three men, I continue to be the only female research stream faculty member in my department. While I enjoy my colleagues and I love my job, I believe this stark imbalance means our department is missing out on something intangible and important.

So while I don't stress about gender issues all the time, I do find that after all these years there still are many things to discuss, to ponder, to wonder over, and with luck and determination, to solve.

Acknowledgements

I was fortunate to meet as an undergraduate three (then future) leaders of our discipline: Lynne Billard, Jane Gentleman, and Mary Thompson; this is my personal proof that role models are essential. I would like to thank them for their contributions to, and achievements for, our profession.

References

BBC News (2013). Powerful women are less liked, April. http://www.bbc.co.uk/news/business-22189754. Accessed on February 11, 2014.

Bernoulli Society (2012). Wolfgang Doeblin Prize, 2012. http://www.bernoulli-society.org/index.php/prizes/158. Accessed on February 11, 2014.

Billard, L. (1991). The past, present, and future of academic women in the mathematical sciences. *Notices of the American Mathematical Society*, 38:714–717. http://www.awm-math.org/articles/notices/199107/billard/index.html. Accessed on February 11, 2014.

Billard, L. (1994). Twenty years later: Is there parity for women in academia? *Thought and Action*, 10:114–144. http://www.stat.uga.edu/stat_files/billard/20yearslater.pdf. Accessed on February 11, 2014.

Billard, L. and Ferber, M.A. (1991). Elizabeth Scott: Scholar, teacher, administrator. *Statistical Science*, 6:206–216.

Boyd, L., Creese, G., Rubuliak, D., Trowell, M., and Young, C. (2012). Report of the gender pay equity recommendation committee. http://www.facultyassociation.ubc.ca/docs/news/GenderPayEquity_JointCommunique.pdf. Accessed on February 11, 2014.

Cox, D.R., Gleser, L., Perlman, M.D., Reid, N.M., and Roeder, K. (1993). Report of the ad hoc committee on double-blind refereeing. *Statistical Science*, 8:310–317.

Dominici, F., Fried, L.P., and Zeger, S. (2009). So few women leaders. *Academe*, 95:25–27.

Göbel, S. (2008). The mathematician Wolfgang Doeblin (1915–1940) — searching the internet. In *A Focus on Mathematics* (Prof. Dr. Bernd Wegner and Staff Unit Communications, Eds.), pp. 31–34. Fachinformationszentrum Karlsruhe, Karlsruhe, Germany. http://www.zentralblatt-math.org/year-of-mathematics/. Accessed on February 11, 2014.

Goldberg, P.A. (1968). Are women prejudiced against women? *Transactions*, 5:28–30.

Gray, M.W. and Ghosh Dasidar, B. (2010). Awards for women fall short. *AmStat News*, October. http://magazine.amstat.org/blog/2010/10/01/awardswomenfallshort/. Accessed on February 11, 2014.

Gray, M.W. and Scott, E.L. (1980). A "statistical" remedy for statistically identified discrimination. *Academe*, 66:174–181.

Laba, I. (2012a). Biased. http://ilaba.wordpress.com/2012/09/25/biased/. Accessed on February 11, 2014.

Laba, I. (2012b). The perils of changing the subject. http://ilaba.wordpress.com/2012/10/02/the-perils-of-changing-the-subject/. Accessed on February 11, 2014.

Laba, I. (2013). Gender bias 101 for mathematicians. http://ilaba.wordpress.com/2013/02/09/gender-bias-101-for-mathematicians/. Accessed on February 11, 2014.

Little, L. (2012). Women studying science face gender bias, study finds. http://abcnews.go.com/blogs/business/2012/09/women-studying-science-face-gender-bias-study-finds/. Accessed on February 11, 2014.

London Mathematical Society (2013). *Advancing Women in Mathematics: Good Practice in UK University Departments.* Technical report, London Mathematical Society. Prepared by Sean MacWhinnie and Carolyn Fox, Oxford Research and Policy.

Moss-Racusin, C.A., Dovidio, J.F., Brescoll, V.L., Graham, M.J., and Handelsman, J. (2012a). Science faculty's subtle gender biases favor male students. *Proceedings of the National Academy of Science*, 109:16474–16479.

Moss-Racusin, C.A., Dovidio, J.F., Brescoll, V.L., Graham, M.J., and Handelsman, J. (2012b). Supporting information: Moss-Racusin et al. 10.1073/pnas.1211286109, 2012. `www.pnas.org/cgi/content/short/1211286109`. Accessed on February 11, 2014.

Palta, M. (2010). Women in science still overlooked. *Amstat News*, October. `http://magazine.amstat.org/blog/2010/10/01/women-in-science/`. Accessed on February 11, 2014.

Sandberg, S. (2013). *Lean In: Women, Work and the Will to Lead.* Random House, New York.

Scott, E.L. (1975). Developing criteria and measures of equal opportunity for women. In *Women in Academia: Evolving Policies Towards Equal Opportunities* (A. Lewin, E. Wasserman, and L. Bleiweiss, Eds.). Praeger, New York, pp. 82–114.

Slaughter, A.M. (2012). Why women still can't have it all. *Atlantic*, 310:85–102.

21

Reflections on diversity

Louise M. Ryan
School of Mathematical Sciences
University of Technology, Sydney, Australia

21.1 Introduction

I recall quite vividly the start in the early 1990s of my interest in fostering diversity in higher education. Professor James Ware had just been appointed as the Academic Dean at Harvard School of Public Health and was letting go of some of his departmental responsibilities. He asked if I would take over as director of the department's training grant in environmental statistics, funded through the National Institute of Environmental Health Sciences (NIEHS). Being an ambitious young associate professor, I eagerly accepted. It wasn't long before I had to start preparing the grant's competitive renewal. These were the days when funding agencies were becoming increasingly proactive in terms of pushing Universities on the issue of diversity and one of the required sections in the renewal concerned minority recruitment and retention. Not knowing much about this, I went for advice to the associate dean for student affairs, a bright and articulate African American woman named Renee (not her true name). When I asked her what the School was doing to foster diversity, she chuckled and said "not much!" She suggested that I let her know when I was traveling to another city and she would arrange for me to visit some colleges with high minority enrollments so that I could engage with students and teachers to tell them about opportunities for training in biostatistics at Harvard.

Not long after this, I received an invitation to speak at the University of Mississippi in Oxford, Mississippi. I excitedly called Renee to tell her about my invitation, naively commenting that since I would be visiting a university in the South, there must be lots of minority students there with whom I could talk about opportunities in Biostatistics. She laughed and said "Louise, it's a bit more complicated than that..." She went on to tell me about some of the history associated with "Ole Miss," including the riots in the early 60s triggered by the brave efforts of African American, James Meredith, to enroll as a student — for a fascinating, "can't put the book down" read, try Nadine Cohodas's "The Band Played Dixie" (Cohodas, 1997). Renee encouraged me to accept the invitation, but also offered to arrange visits to a couple of other schools within driving distance of Oxford, that had high enrollments of minority students. Rust College and Lemoyne-Owen College were both members of the network of Historically Black Colleges and Universities (HBCU). This network comprises 105 schools that were originally established in the days of segregation to educate black students, but which continue today, proud of their rich heritage and passionate about educating African American students and preparing them to become tomorrow's leaders. While students of any race or ethnicity can apply for admission to a HBCU, the majority of students are of African American heritage. Some HBCUs, especially the more well-known ones such as Howard University in Washington, DC, and Spelman College in Atlanta, are well endowed and have the same atmosphere of privilege and learning that one finds on so many modern liberal arts college campuses. Others, while unquestionably providing a sound college education, were not so wealthy. Rust and Lemoyne–Owen Colleges were definitely in the latter category. My visit to those two colleges felt to be in stark contrast to the sense of wealth and privilege that I encountered at Ole Miss. The experience for me was a major eye-opener and I came away with a sense of determination to do something to open up an avenue for more minority students to pursue graduate work in biostatistics.

21.2 Initiatives for minority students

Serendipitously, the NIEHS had just announced the availability of supplementary funds for universities with existing training grants to establish summer programs for minority students. We successfully applied, and the next summer (1992) ran our first ever Summer Program in Biostatistics, with six math-majors from various HBCUs, including one student from Lemoyne–Owen. The 4-week program comprised an introductory course in biostatistics, along with a series of faculty seminars designed to expose students to the breadth of interesting applications in which biostatisticians engage. We also organized practical sessions focussed on things such as how to prepare for the Graduate Record Examination (GRE) and tips on applying for graduate school. We built in lots of time for the summer students to meet more informally with students and faculty from our department. Finally, we organized various social activities and outings in Boston, always involving department students and faculty. Our goal was to create an immersive experience, with a view to giving the students a taste of what a graduate experience might be, and especially demystifying the Harvard experience. I recall very clearly one of our earlier participants saying that without having attended the Program, she would never have even considered applying for graduate studies, particularly at a place like Harvard. This particular student did apply to Harvard, was accepted and went on to be one of the strongest students in her class. She is now a successful faculty member at a major university near Washington, DC, and has published her work in top statistical journals. This student's story is just one of many similar ones and represents a measurable successful outcome of the Program. After a few years of running the Summer Program, we succeeded in winning a so-called IMSD (Initiative for Minority Student Development) grant available at that time from the National Institute of General Medical Sciences. This grant was much larger, supporting not only an expansion of our Summer Program, but also providing funds for doctoral and postdoctoral training and expanding beyond biostatistics into other departments.

The IMSD grant had a major impact on the Department and the School as a whole. It strengthened the legitimacy of our diversity efforts by generating substantial funds and also by influencing the nature of the research that many of us were doing. The IMSD grant required us to develop a strong research theme, and we had chosen to look at the development and application of quantitative methods for community-based research, with a strong emphasis on understanding and reducing health disparities. While it would be an inappropriate generalization to expect that every minority student will be interested in the study of health disparities, the reality was that many of our students were. I'll come back to this point presently, but I believe an important element of academic success is giving students the opportunity to pursue research in areas that ignite their passion. Embracing diversity will inevitably involve being exposed to new ideas and perspectives and this was just one example of how that played out in the department. We ran a weekly seminar/discussion group that provided an opportunity to not only have formal seminars on health disparities research, but also to create a supportive community where the students could talk about the various issues, academic and other, that they were encountering.

21.3 Impact of the diversity programs

Our Diversity programs had profound impacts, over time. I think it is fair to say that the students weren't exposed to overtly racist attitudes, certainly not of the extreme kind described in Nadine Cohodas's book. However, they were most definitely affected by many subtle aspects, especially in the early days of the Program. Examples included faculty expectations of lowered performance or resentment from fellow students at a perception of special treatment. By making such observations, I am not trying to criticize or cast judgment, or even excluding myself from having stereotyped views. As discussed by Malcolm Gladwell in his excellent book entitled "Outliers" (Gladwell, 2011), virtually all of us do, and this can have strong and negative effects. Gladwell discusses extensively the impact, positive and negative, of the social environment on academic performance and general success in life. He also describes some very interesting experiments designed to measure very subtle aspects of negative racial stereotyping. Recognizing our own tendency to stereotype others is in fact an important first step towards making progress towards a more equitable society. Working closely with the students over so many years provided an opportunity for me to observe and, to at least some extent, empathize with the challenges of being a minority student, even in today's relatively enlightened educational system. Internal and external expectations of underperformance very easily turn into reality. Self-doubt can erode confidence, leading students to isolate themselves, thus cutting themselves off from the beneficial effects of being in student study groups. On the flip side, however, we saw the positive and reinforcing effects of growing numbers and student success stories. I will never forget the shock we all experienced one year when a particularly bright young African American man failed the department's doctoral qualifying exam. To his credit, he dusted himself off and developed a steely determination to succeed the following year. He did so with flying colors, winning the departmental award for the top score in the exam (an award that is assigned purely on the basis of exam performance and blind to student identity). That same year, another African American, a young woman, also failed the exam. Although devastated, she was also determined to not only try again, but to repeat the outstanding performance of her classmate and win the prize. And she did. I still get cold shivers thinking about it! These were the kinds of things, along with growing critical mass, that got things changing. It is quite awe-inspiring to think about what some of our program graduates are doing today and how through their success they are inspiring and encouraging the next generation to thrive as well.

I don't feel like I have the language or skill to describe many of the profound things that I learned and experienced through directing the minority program at Harvard for so many years. However I recently read an excellent book, "Whistling Vivaldi," (Steele, 2011) by someone who does — Claude Steele, a renowned social scientist and Dean of the Graduate School of Education at Stanford. Much of Steele's work has been on the concept of stereotype threat. The idea is that when a person is being evaluated (e.g., through a test), their performance can be significantly undermined if they believe that the evaluators will be looking at them through the lens of a stereotype. While stereotyping happens anytime where there are broad-based characterizations of a person's ability or character, based on their social standing, ethnicity or race, the ones that most readily come to mind in the educational context are gender and math/science ability as well as race and general academic performance. Steele describes some fascinating experiments where test scores can be significantly impacted according to whether or not subjects are conscious of stereotype threat. Not only a great read, "Whistling Vivaldi" is a definite eye-opener.

21.4 Gender issues

I've thought a lot over the years about the issue of gender in our field of statistics. During my almost thirty years in US academia, I was never particularly conscious of experiencing any obvious bias or discrimination because I was a woman. I was even beginning to think that the days were over where special efforts were still needed to encourage and support women in science. In fact, I even felt slightly guilty when I received the COPSS award that recognizes Elizabeth L. Scott's lifelong efforts in the furtherance of the careers of women. Since returning to Australia in early 2009, however, my thinking has changed a bit. I've found the research environment here much harder to navigate than in the US and my confidence has suffered as a result. At a meeting of the Australian Academy of Science earlier this year, I had something of a light-bulb moment talking with Terry Speed and several others who assured me that the problem wasn't just me, but rather I was experiencing the impact of working in an environment that was inherently more difficult for women than for men. A telling symptom of this was that none of the 20 new fellows elected to the Australian Academy of Science in 2013 were women! While this situation was something of an embarrassment to the Academy, it did provide an important opportunity for collective self reflection and dialogue on the issue of gender diversity in Australian science. I realized that I was struggling with some of the same challenges that I had worked so hard years earlier to help my students overcome. Because there were fewer successful academic women in Australia, I felt more isolated. Also, because the guidelines for assessing success reflected a more male perspective, I was not measuring up so well. For example, because of some family responsibilities, I was generally reluctant to accept many invitations to speak at international conferences. However, such activities were seen as very important when it came to evidence of track record for grant applications. Finally, my interests didn't quite align. In the US, I had been very fortunate to spend my career in an environment that embraced interdisciplinary research and where the model of a biostatistician combining collaborative and methodological research was not only well understood, but seen as an ideal. In Australia, the model was a more traditional one of a successful, independent academic heading up a team of students, postdocs and junior staff. For me, this model just didn't fit. For all these reasons, it made sense that I was having some difficulty in finding my place within the Australian academic environment. But instead of seeing this bigger picture, I was personalizing it and starting to believe that I simply didn't have the talent to succeed. I see now that I have an opportunity to put into practice some of the advice I used to give my students about believing in myself, keeping in mind the bigger picture and understanding that by persevering, I can help change the system. My experience also underscores why having a diverse workforce helps the whole system to be healthier and more effective. A diverse workforce means a diversity of opinions and values, and a diversity of approaches to problem solving. Diversity broadens the scope of what's important, how workplaces are organized and how people are valued. In the end, a diverse workplace is good for everyone.

References

Cohodas, N. (1997). *The Band Played Dixie: Race and the Liberal Conscience at Ole Miss.* Free Press.

Gladwell, M. (2011). *Outliers: The Story of Success.* Back Bay Books, New York.

Steele, C.M. (2011). *Whistling Vivaldi: How Stereotypes Affect Us and What We Can Do.* W.W. Norton & Co, New York.

Part IV

Reflections on the discipline

22

Why does statistics have two theories?

Donald A.S. Fraser
Department of Statistical Sciences
University of Toronto, Toronto, ON

The public image of statistics is changing, and recently the changes have been mostly for the better, as we've all seen. But occasional court cases, a few conspicuous failures, and even appeals to personal feelings suggest that careful thought may be in order. Actually, statistics itself has more than one theory, and these approaches can give contradictory answers, with the discipline largely indifferent. Saying "we are just exploring!" or appealing to mysticism can't really be appropriate, no matter the spin. In this paper for the COPSS 50th Anniversary Volume, I would like to examine three current approaches to central theory. As we will see, if continuity that is present in the model is also required for the methods, then the conflicts and contradictions resolve.

22.1 Introduction

L'Aquila and 300 deaths. The earthquake at L'Aquila, Italy on April 5, 2009 had been preceded by many small shocks, and Italy's Civil Protection Department established a committee of seismologists to address the risks of a major earthquake. The committee reported before the event that there was no particularly good reason to think that a major earthquake was coming and the Department's Deputy Head even allowed that the small shocks were reducing the seismic stresses, lowering the chances of a major quake. This gave some reassurance to many who were concerned for their lives; but the earthquake did come and more than 300 died. For some details, see Pielke (2011). Charges were then brought against the seismologists and seven were sentenced to six years in prison for manslaughter, "for falsely reassuring the inhabitants of L'Aquila." Part of the committee's role had been the communication of their findings, statistics being intrinsically involved. See Marshall (2012) and Prats (2012).

Vioxx and 40,000 deaths. The pain killer Vioxx was approved by the US Food and Drug Administration (FDA) in 1999 after a relatively short eight years in the approval process and then withdrawn by the pharmaceutical company Merck in 2004 after an acknowledged excess of cardiovascular thrombotic (CVT) events under Vioxx in a placebo controlled study. But statistical assessments as early as 2000 had indicated the heightened rate of CVT events with the use of Vioxx. Statistician David Madigan of Columbia University rose to the challenge as litigation consultant against Merck, and a five billion dollar penalty against Merck went to the injured and their survivors; some felt this was a bargain for Merck, as the company had made billions in profit from the drug. One estimate from the FDA of the number of deaths attributed to the use of the drug was 40,000. See Abraham (2009).

Challenger and 7 deaths. The space shuttle Challenger had completed nine successful flights but on its tenth take-off on January 28, 1986 disintegrated within the first two minutes. The failure was attributed to the breakdown of an O-ring on a solid rocket booster. The external temperature before the flight was well below the acknowledged tolerance for the O-rings, but the flight was given the go-ahead. The 7 crew members died. See Dalai and Fowlkes (1989) and Bergin (2007).

The preceding events involve data, data analysis, determinations, predictions, presentations, then catastrophic results. Where does responsibility fall? With the various levels of the application of statistics? Or with the statistical discipline itself with its contradicting theories? Or with the attitude of many statisticians. We are just exploring and believe in the tools we use?

Certainly the discipline of statistics has more than one theory and these can give contradictory results, witness frequency-based, Bayes-based, and bootstrap-based methodology; these provide a wealth of choice among the contraindicating methods. Here I would like to briefly overview the multiple theories with a view to showing that if continuity as present in the typical model is also required for the methods, an equivalence emerges among the frequency, the bootstrap, and partially the Bayesian approach to inference.

But also, there is attitude within the discipline that tolerates the contradictions and indeed affects within-discipline valuations of statistics and statisticians. In recent years, an important Canadian grant adjudication process had mathematicians and statisticians evaluating applications from mathematicians and statisticians using standardized criteria but with a panel from mathematics for the mathematicians and a panel from statistics for the statisticians; and it was found that mathematicians rate mathematicians much higher than statisticians rate statisticians, even though it was clear that benefits would be apportioned accordingly. For details, see Léger (2013). The contradictory theory and the contradictory attitude provide a potential for serious challenges for statistics, hopefully not at the level of L'Aquila, Vioxx and Challenger.

22.2 65 years and what's new

I did my undergraduate studies in mathematics in my home town of Toronto, Ontario. An opportunity to study analysis and algebra in the doctoral program at Princeton University arose in 1947. But then, with a side interest in actuarial things, I soon drifted to the Statistics Group led by Sam Wilks and John Tukey. A prominent theme was Neyman–Pearson theory but a persistent seminar interest focussed on Fisher's writings, particularly those on fiducial inference which had in turn triggered the Neyman (Neyman, 1937) confidence methodology. But also, a paper by Jeffreys (Jeffreys, 1946) kept reemerging in discussions; it offered a default Bayes (Bayes, 1763) approach, often but incorrectly called objective Bayes in present Bayes usage. The striking thing for me at that time was the presence of two theories for statistics that gave contradictory results: if the results were contradictory, then simple logic on theories says that one or the other, or both, are wrong. This latter view, however, was not part of the professional milieu at the time, though there was some puzzlement and vague acceptance of contradictions, as being in the nature of things; and this may even be part of current thinking! "One or the other, or both, could be wrong?" Physics manages to elicit billions in taxpayer money to assess their theories! Where does statistics stand?

With a completed thesis that avoided the frequency-Bayes contradictions, I returned to Canada and accepted a faculty position in the Department of Mathematics at the University of Toronto. The interest in the frequency-Bayes contradictions, however, remained and a conference talk in 1959 and two resulting papers (Fraser, 1961a,b) explored a broad class of statistical models for which the two approaches gave equivalent results: the location model $f(y - \theta)$, of course, and the locally-generated group extensions, the transformation-parameter models. Then an opportunity for a senior faculty position in the Mathematics Department at Princeton arose in 1963, but I was unable to accept. The concerns for the frequency-Bayes contradictions, however, remained!

Now in 2013 with COPSS celebrating its 50th anniversary, we can look about and say "What's new?" And even more we are encouraged to reminisce! There is very active frequency statistics and related data analysis; and there is very active Bayesian statistics; and they still give contradictory answers. So nothing has changed on the frequency-Bayes disconnect: what goes around comes around... Does that apply to statistical theory in the 65 years I have been in the profession? Oh, of course, there have been massive extensions to data exploration, to computer implementation, to simulations, and to algorithmic approaches. Certainly we have Precision, when sought! But what about Accuracy? I mean Accuracy beyond Precision? And what about the frequency-Bayes contradictions in the theory? And even, indeed, the fact that no one seems to care? And then L'Aquila, Vioxx, Challenger, and of course the contradictory theory? Are perceptions being suppressed? It might wind up in a court, as with L'Aquila, an inappropriate place to address a scientific issue but perhaps not to address a conflict coming from discipline contradictions.

Well yes, something has changed! Now a general feeling in the milieu is acceptance of the frequency-Bayes contradiction: it just doesn't matter, we are just exploring; our models and calculations are just approximations; and we can acquire any Precision we want, even though we may not have used the full information provided by the model, so just run the MCMC longer, even though several million cycles only give two decimal places for some wanted probability or confidence calculation. Or put together an algorithm for processing numbers. Or use personal feelings as in some Bayes methods.

But even for explorations it certainly behooves one to have calibrated tools! And more generally to know with Precision and Accuracy what a model with data implies? Know as a separate issue quite apart from the descriptive Accuracy of the model in a particular context, which of course in itself is an important but separate issue! This Accuracy is rarely addressed! Indeed, as L'Aquila, Vioxx, and Challenger indicate, a concern for Accuracy in the end products of statistics may have an elusive presence in many professional endeavours. An indictment of statistics?

22.3 Where do the probabilities come from?

(i) The starting point. The statistical model $f(y;\theta)$ with data y^0 forms the starting point for the Bayes and often the frequency approach. The Bayesian approach calculates and typically uses just the observed likelihood $L^0(\theta) = f(y^0;\theta)$, omitting other model information as part of a Bayes commitment. The frequency approach uses more than the observed likelihood function: it can use distribution functions and full model calculations, sometimes component model calculations that provide relevant precision information, and more.

(ii) The ingredients for inference. In the model-data context, y^0 is an observed value and is thus a known constant, and θ is an unknown constant. And if a distribution $\pi(\theta)$ is present, assumed, proposed or created, as the source for θ, then a second distribution is on offer concerning the unknown constant. Probabilities are then sought for the unknown constant, in the context of one or two distributional sources: one part of the given and the other objective, subjective, or appended for computational or other reasons. Should these distributions be combined, or be examined separately, or should the added distribution be ignored? No over-riding principle says that distributions of different status or quality should be combined rather than having their consequences judged separately! Familiar Bayes methodology, however, takes the combining as a given, just as the use of only the observed likelihood

function is taken as a given, essentially axioms in the Bayes methodology! For a recent discussion, see Fraser (2011).

(iii) The simple location model. Consider the location model $f(y-\theta)$. This is of course rather special in that the error, the variable minus the parameter, has a fixed known distributional shape, free of the parameter. A common added or proposed prior is the flat prior $\pi(\theta)=1$ representing the translation invariance of the model. As it stands the model almost seems too simple for consideration here; but the reality is that this simple model exists as an embedded approximation in an incredibly broad class of models where continuity of parameter effect is present and should thus have its influence acknowledged.

(iv) Location case: p-value or s-value. The generic version of the p-value from observed data y^0 is

$$p^0(\theta)=\int^{y^0} f(y-\theta)\mathrm{d}y = F^0(\theta),$$

which records just the statistical position of the data relative to the parameter. As such it is just the observed distribution function. This $p(\theta)$ function is uniform on the interval $(0,1)$, which in turn implies that any related confidence bound or confidence interval has validity in the sense that it bounds or embraces the true parameter value with the stated reliability; see Fisher (1930) and Neyman (1937). In parallel, the observed Bayes survivor value is

$$s^0(\theta)=\int_\theta f(y^0-\alpha)\mathrm{d}\alpha.$$

The two different directions of integration correspond to data left of the parameter and parameter right of the data, at least in this stochastically increasing case. The two integrals are mathematically equal as is seen from a routine calculus change of variable in the integration. Thus the Bayes survivor s-value acquires validity here, validity in the sense that it is uniformly distributed on $(0,1)$; and validity also in the sense that a Bayes quantile at a level β will have the confidence property and bound the parameter at the stated level. This validity depends entirely on the equivalence of the integrals and no reference or appeal to conditional probability is involved or invoked. Thus in this location model context, a sample space integration can routinely be replaced by a parameter space integration, a pure calculus formality. And thus in the location model context there is no frequency-Bayes contradiction, just the matter of choosing the prior that yields the translation property which in turn enables the integration change of variable and thus the transfer of the integration from sample space to parameter space.

(v) The simple scalar model. Now consider a stochastically increasing scalar model $f(y;\theta)$ with distribution function $F(y;\theta)$ and some minimum continuity and regularity. The observed p-value is

$$p^0(\theta)=F^0(\theta)=\int^{y^0} F_y(y;\theta)\mathrm{d}y=\int_\theta -F_\theta(y^0;\theta)\mathrm{d}\theta,$$

where the subscripts to F denote partial differentiation with respect to the indicated argument. Each of the integrals records an $F(y, \theta)$ value as an integral of its derivative — the fundamental theorem of calculus — one with respect to θ and the other with respect to y. This is pure computation, entirely without Bayes! And then, quite separately, the Bayes survivor value using a proffered prior $\pi(\theta)$ is

$$s^0(\theta) = \int_\theta \pi(\theta) F_y(y^0; \theta) \mathrm{d}\theta.$$

(vi) Validity of Bayes posterior: Simple scalar model. The second integral for $p^0(\theta)$ and the integral for $s^0(\theta)$ are equal if and only if the integrands are equal. In other words if and only if

$$\pi(\theta) = -\frac{F_\theta(y^0; \theta)}{F_y(y^0; \theta)} = \left.\frac{\partial y(\theta; u)}{\partial \theta}\right|_{\text{fixed } F(y;\theta); y^0}$$

with an appropriate norming constant included. The second equality comes from the total derivative of $u = F(y; \theta)$ set equal to 0, thus determining how a θ-change affects y for fixed probability position. We can also view $v(\theta) = \partial y(\theta; u)/\partial\theta$ for fixed u as being the change in y caused by a change in θ, thus giving at y^0 a differential version of the y, θ analysis in the preceding subsection.

Again, with this simple scalar model analysis, there is no frequency-Bayes contradiction; it is just a matter of getting the prior right. The correct prior does depend on the data point y^0 but this should cause no concern. If the objective of Bayesian analysis is to extract all accessible information from an observed likelihood and if this then requires the tailoring of the prior to the particular data, then this is in accord with that objective. Data dependent priors have been around for a long time; see, e.g., Box and Cox (1964). But of course this data dependence does conflict with a conventional Bayes view that a prior should be available for each model type. The realities of data analysis may not be as simple as Bayes might wish.

(vii) What's the conclusion? With a location model, Bayes and frequency approaches are in full agreement: Bayes gets it right because the Bayes calculation is just a frequency confidence calculation in mild disguise. However, with a non-location model, the Bayes claim with a percentage attached to an interval does require a data-dependent prior. But to reference the conditional probability lemma, relabeled as Bayes lemma, requires that a missing ingredient for the lemma be created, that a density not from the reality being investigated be given *objective* status in order to nominally validate the term probability: this violates mathematics and science.

22.4 Inference for regular models: Frequency

(i) Normal, exponential, and regular models. Much of contemporary inference theory is organized around Normal statistical models with side concerns for departures from Normality, thus neglecting more general structures. Recent likelihood methods show, however, that statistical inference is easy and direct for exponential models and more generally for regular models using an appropriate exponential approximation. Accordingly, let us briefly overview inference for exponential models.

(ii) Exponential statistical model. The exponential family of models is widely useful both for model building and for model-data analysis. The full exponential model with canonical parameter φ and canonical variable $u(y)$ both of dimension p is $f(y;\varphi) = \exp\{\varphi' u(y) + k(\varphi)\}h(y)$. Let y^0 with $u^0 = u(y^0)$ be observed data for which statistical inference is wanted. For most purposes we can work with the model in terms of the canonical statistic u:

$$g(u;\varphi) = \exp\{\ell^0(\varphi) + (\varphi - \hat{\varphi}^0)'(u - u^0)\}g(u),$$

where $\ell^0(\varphi) = a + \ln f(y^0;\varphi)$ is the observed log-likelihood function with the usual arbitrary constant chosen conveniently to subtract the maximum log-likelihood $\ln f(y^0;\hat{\varphi}^0)$, using $\hat{\varphi}^0$ as the observed maximum likelihood value. This representative $\ell^0(\varphi)$ has value 0 at $\hat{\varphi}^0$, and $-\ell^0(\varphi)$ relative to $\hat{\varphi}^0$ is the cumulant generating function of $u - u^0$, and $g(u)$ is a probability density function. The saddle point then gives a third-order inversion of the cumulant generating function $-\ell^0(\varphi)$ leading to the third-order rewrite

$$g(u;\varphi) = \frac{e^{k/n}}{(2\pi)^{p/2}} \exp\{-r^2(\varphi;u)/2\}|\jmath_{\varphi\varphi}(\hat{\varphi})|^{-1/2},$$

where $\hat{\varphi} = \hat{\varphi}(u)$ is the maximum likelihood value for the tilted likelihood

$$\ell(\varphi;u) = \ell^0(\varphi) + \varphi'(u - u^0),$$

$r^2(\varphi;u)/2 = \ell(\hat{\varphi};u) - \ell(\varphi;u)$ is the related log-likelihood ratio quantity,

$$\jmath_{\varphi\varphi}(\hat{\varphi}) = -\frac{\partial}{\partial\varphi\partial\varphi'}\ell(\varphi;u)|_{\hat{\varphi}(u)}$$

is the information matrix at u, and finally k/n is constant to third order. The density approximation $g(u;\varphi_0)$ gives an essentially unique third-order null distribution (Fraser and Reid, 2013) for testing the parameter value $\varphi = \varphi_0$.

Then if the parameter φ is scalar, we can use standard r^*-technology to calculate the p-value $p(\varphi_0)$ for assessing $\varphi = \varphi_0$; see, e.g., Brazzale et al. (2007). For a vector φ, a directed r^* departure is available; see, e.g., Davison et al. (2014). Thus p-values are widely available with high third-order accuracy, all with uniqueness coming from the continuity of the parameter's effect on the variable involved; see in part Fraser et al. (2010b).

(iii) Testing component parameters. Now consider more generally a component parameter $\psi(\varphi)$ of dimension d with $d < p$. If ψ is linear in φ, then a rotation of coordinates lets us write $\varphi = (\chi, \lambda)$ with χ equivalent to ψ and with say (s, t) as the corresponding canonical coordinates. Statistical inference is available from the d-dimensional conditional distribution on the profile line or plane $L^0 = \{(s, t^0)\}$ with parameter χ. This uses in an essential way the profile likelihood ratio

$$r^2(\chi; s)/2 = \ell^{\mathrm{P}}(\hat{\chi}; s) - \ell^{\mathrm{P}}(\chi; s) = \ell(s, t^0; \hat{\chi}, \hat{\lambda}) - \ell(s, t^0; \chi, \hat{\lambda}_\chi)$$

and related saddle point, but does need a norming constant dependent on χ.

But more generally when the interest parameter ψ is non-linear and thus curved in the initial φ parameterization, the conditional approach just described is effectively unavailable and a marginal approach coming from recent likelihood asymptotics is needed. This involves integrating out over a nuisance parameter variable, and gives to third order the marginal distribution for an ancillary variable under $\psi = \psi_0$, viz.

$$f_{\mathrm{m}}(s; \psi_0) = \frac{e^{k/n}}{(2\pi)^{d/2}} \exp(\tilde{\ell} - \hat{\ell}) \\ \times |\jmath_{\varphi\varphi}\{\hat{\varphi}(s, t^0)\}|^{-1/2} |\jmath_{(\lambda\lambda)}(\psi_0, \hat{\lambda}_{\psi_0}; s, t^0)|^{1/2}, \quad (22.1)$$

on $L^0 = \{(s, t^0)\}$ using rotated coordinates (χ, λ) and (s, t) having $\chi = \chi_0$ first derivative equivalent to $\psi = \psi_0$ at $\hat{\varphi}^0_{\psi_0}$. Here $\hat{\ell} - \tilde{\ell}$ is the log-likelihood ratio at (s, t^0) for the tested value ψ_0, and the nuisance information uses λ with given $\psi = \psi_0$ and λ derivatives for fixed $\psi = \psi_0$ then rescaled in terms of the φ parameterization at $\hat{\varphi}(s, t^0)$ as indicated by the parentheses and described in Brazzale et al. (2007), Fraser and Reid (1993) or Davison et al. (2014). This distribution is essentially unique if continuity of parameter effect is respected; and it is simple, involving only the log-likelihood ratio for ψ_0 and information determinants. In the linear parameter case where the conditional approach is available, this agrees with that conditional result; but here with curvature where no easily accessible conditional approach is available the present marginal approach is the reference standard. My only purpose here is to report on the availability of these unique null distributions and on the availability of p-values, for both linear and curved parameters; for details see, e.g., Fraser and Reid (2013).

(iv) Regular statistical model. Now consider a statistical model $f(y; \theta)$ with continuity in parameter effect and general regularity. For such models we can find, quite widely, a quantile representation $y = y(\theta, u)$ as discussed briefly for a simple case earlier. Such is widely used for simulations and is routinely and definitively available in cases where the model has independent scalar coordinates. Let $V(\theta, y) = \partial y(\theta; u)/\partial\theta$ be the $n \times p$ matrix giving the vectors

that record the effect on y of change in the parameter coordinates $\theta_1, \ldots, \theta_p$; and let $V = V(\hat{\theta}^0, y^0) = \hat{V}^0$ be the observed matrix. Then V records tangents to an intrinsic ancillary contour, say $a(y) = a(y^0)$, that passes through the observed y^0. Thus V represents directions in which the data can be viewed as measuring the parameter, and $\mathcal{L}V$ gives the tangent space to the ancillary at the observed data, with V having somewhat the role of a design matrix. For development details, see Fraser and Reid (1995).

From ancillarity it follows that likelihood conditionally is equal to the full likelihood $L^0(\theta)$, to an order one higher than that of the ancillary used. And it also follows that the sample space gradient of the log-likelihood in the directions V along the ancillary contour gives the canonical parameter, viz.

$$\varphi(\theta) = \frac{\partial}{\partial V} \ell(\theta; y)\bigg|_{y^0},$$

whenever the conditional model is exponential, or gives the canonical parameter of an approximating exponential model otherwise. In either case, $\ell^0(\theta)$ with the preceding $\varphi(\theta)$ provides third order statistical inference for scalar parameters using the saddle point expression and the above technology. And this statistical inference is uniquely determined provided the continuity in the model is required for the inference (Fraser and Rousseau, 2008). For further discussion and details, see Fraser et al. (2010a) and Fraser and Reid (1995).

22.5 Inference for regular models: Bootstrap

Consider a regular statistical model and the exponential approximation as discussed in the preceding section, and suppose we are interested in testing a scalar parameter $\psi(\varphi) = \psi_0$ with observed data y^0. The bootstrap distribution is $f(y; \psi_0, \hat{\lambda}^0_{\psi_0})$, as used in Fraser and Rousseau (2008) from a log-model perspective and then in DiCiccio and Young (2008) for the exponential model case with linear interest parameter.

The ancillary density in the preceding section is third-order free of the nuisance parameter λ. Thus the bootstrap distribution $f(y; \psi_0, \hat{\lambda}^0_{\psi_0})$ provides full third-order sampling for this ancillary, equivalent to that from the true sampling $f(y; \psi_0, \lambda)$, just the use of a different λ value when the distribution is free of λ.

Consider the profile line L^0 through the data point y^0. In developing the ancillary density (22.1), we made use of the presence of ancillary contours cross-sectional to the line L^0. Now suppose we have a d-dimensional quantity $t(y, \psi)$ that provides likelihood centred and scaled departure for ψ, e.g., a signed likelihood root as in Barndorff-Nielsen and Cox (1994) or a Wald quantity, thus providing the choice in DiCiccio and Young (2008). If $t(y)$ is a function of the ancillary, say $a(y)$, then one bootstrap cycle gives third order, a case of direct sampling; otherwise the conditional distribution of $y|a$ also becomes involved and with the likelihood based $t(y)$ gives third order inference as in the third cycle of Fraser and Rousseau (2008).

This means that the bootstrap and the usual higher-order calculations are third-order equivalent in some generality, and in reverse that the bootstrap calculations for a likelihood centred and scaled quantity can be viewed as consistent with standard higher-order calculations, although clearly this was not part of the bootstrap design. This equivalence was presented for the linear interest parameter case in an exponential model in DiCiccio and Young (2008), and we now have that it holds widely for regular models with linear or curved interest parameters. For a general regular model, the higher order routinely gives conditioning on full-model ancillary directions while the bootstrap averages over this conditioning.

22.6 Inference for regular models: Bayes

(i) Jeffreys prior. The discussion earlier shows that Bayes validity in general requires data-dependent priors. For the scalar exponential model, however, it was shown by Welch and Peers (1963) that the root information prior of Jeffreys (1946), viz.

$$\pi(\theta) = j_{\theta\theta}^{1/2},$$

provides full second-order validity, and is presented as a globally defined prior and indeed is not data-dependent. The Welch–Peers presentation does use expected information, but with exponential models the observed and expected informations are equivalent. Are such results then available for the vector exponential model?

For the vector regression-scale model, Jeffreys subsequently noted that his root information prior (Jeffreys, 1946) was unsatisfactory and proposed an effective alternative for that model. And for more general contexts, Bernardo (1979) proposed reference posteriors and thus reference priors, based on maximizing the Kullback–Leibler distance between prior and posterior. These priors have some wide acceptance, but can also miss available information.

(ii) The Bayes objective: Likelihood based inference. Another way of viewing Bayesian analysis is as a procedure to extract maximum information from an observed likelihood function $L^0(\theta)$. This suggests asymptotic analysis and Taylor expansion about the observed maximum likelihood value $\hat{\theta}^0$. For this we assume a p-dimensional exponential model $g(u;\varphi)$ as expressed in terms of its canonical parameter φ and its canonical variable u, either as the given model or as the higher-order approximation mentioned earlier. There are also some presentation advantages in using versions of the parameter and of the

variable that give an observed information matrix $\hat{\jmath}^0_{\varphi\varphi} = I$ equal to the identity matrix.

(iii) Insightful local coordinates. Now consider the form of the log-model in the neighborhood of the observed data $(u^0, \hat{\varphi}^0)$. And let $\mathbf{e}$ be a p-dimensional unit vector that provides a direction from $\hat{\varphi}^0$ or from u^0. The conditional statistical model along the line $u^0 + \mathcal{L}\mathbf{e}$ is available from exponential model theory and is just a scalar exponential model with scalar canonical parameter ρ, where $\varphi = \hat{\varphi}^0 + \rho\mathbf{e}$ is given by polar coordinates. Likelihood theory also shows that the conditional distribution is second-order equivalent to the marginal distribution for assessing ρ. The related prior $\jmath_{\rho\rho}^{1/2} \mathrm{d}\rho$ for ρ would use $\lambda = \hat{\lambda}^0$, where λ is the canonical parameter complementing ρ.

(iv) The differential prior. Now suppose the preceding prior $\jmath_{\rho\rho}^{1/2} \mathrm{d}\rho$ is used on each line $\hat{\varphi}^0 + \mathcal{L}\mathbf{e}$. This composite prior on the full parameter space can be called the differential prior and provides crucial information for Bayes inference. But as such it is of course subject to the well-known limitation on distributions for parameters, both confidence and Bayes; they give incorrect results for curved parameters unless the pivot or prior is targeted on the curved parameter of interest; for details, see, e.g., Dawid et al. (1973) and Fraser (2011).

(v) Location model: Why not use the location property? The appropriate prior for ρ would lead to a constant-information parameterization, which would provide a location relationship near the observed $(y^0, \hat{\varphi}^0)$. As such the p-value for a linear parameter would have a reflected Bayes survivor s-value, thus leading to second order. Such is not a full location model property, just a location property near the data point, but this is all that is needed for the reflected transfer of probability from the sample space to the parameter space, thereby enabling a second-order Bayes calculation.

(vi) Second-order for scalar parameters? But there is more. The conditional distribution for a linear parameter does provide third order inference and it does use the full likelihood but that full likelihood needs an adjustment for the conditioning (Fraser and Reid, 2013). It follows that even a linear parameter in an exponential model needs targeting for Bayes inference, and a local or global prior cannot generally yield second-order inference for linear parameters, let alone for the curved parameters as in Dawid et al. (1973) and Fraser (2013).

22.7 The frequency-Bayes contradiction

So is there a frequency-Bayes contradiction? Or a frequency-bootstrap-Bayes contradiction? Not if one respects the continuity widely present in regular statistical models and then requires the continuity to be respected for the frequency calculations and for the choice of Bayes prior.

Frequency theory of course routinely leaves open the choice of pivotal quantity which provides the basis for tests, confidence bounds, and related intervals and distributions. And Bayes theory leaves open the choice of the prior for extracting information from the likelihood function. And the bootstrap needs a tactical choice of initial statistic to succeed in one bootstrap cycle. Thus on the surface there is a lot of apparent arbitrariness in the usual inference procedures, with a consequent potential for serious contradictions. In the frequency approach, however, this arbitrariness essentially disappears if continuity of parameter effect in the model is respected, and then required in the inference calculations; see Fraser et al. (2010b) and the discussion in earlier sections. And for the Bayes approach above, the arbitrariness can disappear if the need for data dependence is acknowledged and the locally based differential prior is used to examine sample space probability on the parameter space. This extracts information from the likelihood function to the second order, but just for linear parameters (Fraser, 2013).

The frequency and the bootstrap approaches can succeed without arbitrariness to third order. The Bayes approach can succeed to second order provided the parameter is linear, otherwise the prior needs to target the particular interest parameter. And if distributions are used to describe unknown parameter values, the frequency joins the Bayes in being restricted to linear parameters unless there is targeting; see Dawid et al. (1973) and Fraser (2011).

22.8 Discussion

(i) Scalar case. We began with the simple scalar location case, feeling that clarity should be present at that transparent level if sensible inference was to be available more generally. And we found at point (ii) that there were no Bayes-frequency contradictions in the location model case so long as model continuity was respected and the Bayes s-value was obtained from the location based prior. Then at point (v) in the general scalar case, we saw that the p-value retains its interpretation as the statistical position of the data and has full repetition validity, but the Bayes requires a prior determined by the form of the model and is typically data dependent. For the scalar model case this is a radical limitation on the Bayes approach; in other words inverting the distribution function as pivot works immediately for the frequency approach whereas inverting the likelihood using the conditional probability lemma as a tool requires the prior to reflect the location property, at least locally. For the scalar model context, this represents a full vindication of Fisher (1930), subject to the Neyman (1937) restriction that probabilities be attached only to the inverses of pivot sets.

(ii) Vector case. Most models however involve more than just a scalar parameter. So what about the frequency-Bayes disconnect away from the very

simple scalar case? The Bayes method arose from an unusual original example (Bayes, 1763), where at the analysis stage the parameter was retroactively viewed as generated randomly by a physical process, indeed an earlier performance of the process under study. Thus a real frequency-based prior was introduced hypothetically and became the progenitor for the present Bayes procedure. In due course a prior then evolved as a means for exploring, for inserting feelings, or for technical reasons to achieve analysis when direct methods seemed unavailable. But do we have to make up a prior to avoid admitting that direct methods of analysis were not in obvious abundance?

(iii) Distributions for parameters? Fisher presented the fiducial distribution in Fisher (1930, 1935) and in various subsequent papers. He was criticized from the frequency viewpoint because his proposal left certain things arbitrary and thus not in a fully developed form as expected by the mathematics community at that time: welcome to statistics as a developing discipline! And he was criticized sharply from the Bayes (Lindley, 1958) because Fisher proposed distributions for a parameter and such were firmly viewed as Bayes territory. We now have substantial grounds that the exact route to a distribution for a parameter is the Fisher route, and that Bayes becomes an approximation to the Fisher confidence and can even attain second-order validity (Fraser, 2011) but requires targeting even for linear parameters.

But the root problem is that a distribution for a vector parameter is inherently invalid beyond first order (Fraser, 2011). Certainly in some generality with a linear parameter the routine frequency and routine Bayes can agree. But if parameter curvature is allowed then the frequency p-value and the Bayes s-value change in *opposite directions*: the p-value retains its validity, having the uniform distribution on the interval $(0, 1)$ property, while the Bayes loses this property and associated validity, yet chooses to retain the label "probability" by discipline commitment, as used from early on. In all the Bayes cases the events receiving probabilities are events in the past, and the prior probability input to the conditional probability lemma is widely there for expediency: the lemma does not create real probabilities from hypothetical probabilities except when there is location equivalence.

(iv) Overview. Most inference contradictions disappear if continuity present in the model is required for the inference calculations. Higher order frequency and bootstrap are consistent to third order for scalar parameters. Bayes agrees but just for location parameters and then to first order for other parameters, and for this Bayes does need a prior that reflects or approximates the location relationship between variable and parameter. Some recent preliminary reports are available at `http://www.utstat.toronto.edu/dfraser/documents/` as `260-V3.pdf` and `265-V3.pdf`.

Acknowledgment

This research was funded in part by the Natural Sciences and Engineering Research Council of Canada, by the Senior Scholars Funding at York University, and by the Department of Statistical Sciences at the University of Toronto. Thanks to C. Genest and A. Wang for help in preparing the manuscript.

References

Abraham, C. (2009). Vioxx took deadly toll study. *Globe and Mail* `http://www.theglobeandmail.com/life/study-finds-vioxx-took-deadly-toll/article4114560/`

Barndorff-Nielsen, O.E. and Cox, D.R. (1994). *Inference and Asymptotics*. Chapman & Hall, London.

Bayes, T. (1763). An essay towards solving a problem in the doctrine of chances. *Philosophical Transactions of the Royal Society, London*, 53:370–418.

Bergin, C. (2007). Remembering the mistakes of Challenger. `nasaspaceflight.com`.

Bernardo, J.M. (1979). Reference posterior distributions for Bayesian inference. *Journal of the Royal Statistical Society, Series B*, 41:113–147.

Box, G.E.P. and Cox, D.R. (1964). An analysis of transformations (with discussion). *Journal of the Royal Statistical Society, Series B*, 26:211–252.

Brazzale, A.R., Davison, A.C., and Reid, N.M. (2007). *Applied Asymptotics*. Cambridge University Press, Cambridge, UK.

Dalai, S. and Fowlkes, B. (1989). Risk analysis of the space shuttle: Pre-Challenger prediction of failure. *Journal of the American Statistical Association*, 84:945–957.

Davison, A.C., Fraser, D.A.S., Reid, N.M., and Sartori, N. (2014). Accurate directional inference for vector parameters. *Journal of the American Statistical Association*, to appear.

Dawid, A.P., Stone, M., and Zidek, J.V. (1973). Marginalization paradoxes in Bayesian and structural inference. *Journal of the Royal Statistical Society, Series B*, 35:189–233.

DiCiccio, T.J. and Young, G.A. (2008). Conditional properties of unconditional parametric bootstrap procedures for inference in exponential families. *Biometrika*, 95:497–504.

Fisher, R.A. (1930). Inverse probability. *Proceedings of the Cambridge Philosophical Society*, 26:528–535.

Fisher, R.A. (1935). The fiducial argument in statistical inference. *Annals of Eugenics*, 6:391–398.

Fraser, A.M., Fraser, D.A.S., and Fraser, M.J. (2010a). Parameter curvature revisited and the Bayesian frequentist divergence. *Statistical Research: Efron Volume*, 44:335–346.

Fraser, A.M., Fraser, D.A.S., and Staicu, A.M. (2010b). Second order ancillary: A differential view with continuity. *Bernoulli*, 16:1208–1223.

Fraser, D.A.S. (1961a). The fiducial method and invariance. *Biometrika*, 48:261–280.

Fraser, D.A.S. (1961b). On fiducial inference. *The Annals of Mathematical Statistics*, 32:661–676.

Fraser, D.A.S. (2011). Is Bayes posterior just quick and dirty confidence? (with discussion). *Statistical Science*, 26:299–316.

Fraser, D.A.S. (2013). *Can Bayes inference be second-order for scalar parameters?* Submitted for publication.

Fraser, D.A.S. and Reid, N.M. (1993). Third order asymptotic models: Likelihood functions leading to accurate approximations for distribution functions. *Statistica Sinica*, 3:67–82.

Fraser, D.A.S. and Reid, N.M. (1995). Ancillaries and third order significance. *Utilitas Mathematica*, 47:33–53.

Fraser, D.A.S. and Reid, N.M. (2013). *Assessing a parameter of interest: Higher-order methods and the bootstrap.* Submitted for publication.

Fraser, D.A.S. and Rousseau, J. (2008). Studentization and deriving accurate p-values. *Biometrika*, 95:1–16.

Jeffreys, H. (1946). An invariant form for the prior probability in estimation problems. *Proceedings of the Royal Society, Series A*, 186:453–461.

Léger, C. (2013). The Statistical Society of Canada (SSC) response to the NSERC consultation on the evaluation of the Discovery Grants Program. *SSC Liaison*, 27(2):12–21.

Lindley, D.V. (1958). Fiducial distribution and Bayes theorem. *Journal of the Royal Statistical Society, Series B*, 20:102–107.

Marshall, M. (2012). Seismologists found guilty of manslaughter. *New Scientist*, October 22, 2012.

Neyman, J. (1937). Outline of a theory of statistical estimation based on the classical theory of probability. *Philosophical Transactions of the Royal Society, Series A*, 237:333–380.

Pielke, R. (2011). Lessons of the L'Aquila lawsuit. *Bridges* 31, `http://www.bbc.co.uk/news/world-europe-20025626`.

Prats, J. (2012). The L'Aquila earthquake. *Significance*, 9:13–16.

Welch, B.L. and Peers, H.W. (1963). On formulae for confidence points based on intervals of weighted likelihoods. *Journal of the Royal Statistical Society, Series B*, 25:318–329.

23

Conditioning is the issue

James O. Berger
Department of Statistical Science
Duke University, Durham, NC

The importance of conditioning in statistics and its implementation are highlighted through the series of examples that most strongly affected my understanding of the issue. The examples range from "oldies but goodies" to new examples that illustrate the importance of thinking conditionally in modern statistical developments. The enormous potential impact of improved handling of conditioning is also illustrated.

23.1 Introduction

No, this is not about *conditioning* in the sense of "I was conditioned to be a Bayesian." Indeed I was educated at Cornell University in the early 1970s, by Jack Kiefer, Jack Wolfowitz, Roger Farrell and my advisor Larry Brown, in a strong frequentist tradition, albeit with heavy use of prior distributions as technical tools. My early work on shrinkage estimation got me thinking more about the Bayesian perspective; doesn't one need to decide where to shrink, and how can that decision not require Bayesian thinking? But it wasn't until I encountered statistical conditioning (see the next section if you do not know what that means) and the Likelihood Principle that I suddenly felt like I had woken up and was beginning to understand the foundations of statistics.

Bayesian analysis, because it is completely conditional (depending on the statistical model only through the observed data), automatically conditions properly and, hence, has been the focus of much of my work. But I never stopped being a frequentist and came to understand that frequentists can also appropriately condition. Not surprisingly (in retrospect, but not at the time) I found that, when frequentists do appropriately condition, they obtain answers remarkably like the Bayesian answers; this, in my mind, makes conditioning the key issue in the foundations of statistics, as it unifies the two major perspectives of statistics.

The practical importance of conditioning arises because, when it is not done in certain scenarios, the results can be very detrimental to science. Unfortunately, this is the case for many of the most commonly used statistical procedures, as will be discussed.

This chapter is a brief tour of old and new examples that most influenced me over the years concerning the need to appropriately condition. The new ones include performing a sequence of tests, as is now common in clinical trials *and is being done badly*, and an example involving a type of false discovery rate.

23.2 Cox example and a pedagogical example

As this is more of an account of my own experiences with conditioning, I have not tried to track down when the notion first arose. Pierre Simon de Laplace likely understood the issue, as he spent much of his career as a Bayesian in dealing with applied problems and then, later in life, also developed frequentist inference. Clearly Ronald Fisher and Harold Jeffreys knew all about conditioning early on. My first introduction to conditioning was the example of Cox (1958).

A variant of the Cox example: Every day an employee enters a lab to perform assays, and is assigned an unbiased instrument to perform the assays. Half of the available instruments are new and have a small variance of 1, while the other half are old and have a variance of 3. The employee is assigned each type with probability 1/2, and knows whether the instrument is old or new.

Conditional inference: For each assay, report variance 1 or 3, depending on whether a new or an old instrument is being used.

Unconditional inference: The overall variance of the assays is $.5\times 1 + .5\times 3 = 2$, so report a variance of 2 always.

It seems silly to do the unconditional inference here, especially when noting that the conditional inference is also fully frequentist; in the latter, one is just choosing different subset of events over which to do a long run average.

The Cox example contains the essence of conditioning, but tends to be dismissed because of the issue of "global frequentism." The completely pure frequentist position is that one's entire life is a huge experiment, and so the correct frequentist average is over all possibilities in all situations involving uncertainty that one encounters in life. As this is clearly impossible, frequentists have historically chosen to condition on the experiment actually being conducted before applying frequentist logic; then Cox's example would seem irrelevant. However, the virtually identical issue can arise within an experiment, as demonstrated in the following example, first appearing in Berger and Wolpert (1984).

Pedagogical example: Two observations, X_1 and X_2, are to be taken, where

$$X_i = \begin{cases} \theta + 1 & \text{with probability } 1/2, \\ \theta - 1 & \text{with probability } 1/2. \end{cases}$$

Consider the confidence set for the unknown $\theta \in \mathbb{R}$:

$$C(X_1, X_2) = \begin{cases} \text{the singleton } \{(X_1 + X_2)/2\} & \text{if } X_1 \neq X_2, \\ \text{the singleton } \{X_1 - 1\} & \text{if } X_1 = X_2. \end{cases}$$

The frequentist coverage of this confidence set is

$$P_\theta\{C(X_1, X_2) \text{ contains } \theta\} = .75,$$

which is not at all a sensible report once the data is at hand. Indeed, if $x_1 \neq x_2$, then we know for sure that $(x_1 + x_2)/2$ is equal to θ, so that the confidence set is then actually 100% accurate. On the other hand, if $x_1 = x_2$, we do not know whether θ is the data's common value plus 1 or their common value minus 1, and each of these possibilities is equally likely to have occurred; the confidence interval is then only 50% accurate. While it is not wrong to say that the confidence interval has 75% coverage, it is obviously much more scientifically useful to report 100% or 50%, depending on the data. And again, this conditional report is still fully frequentist, averaging over the sets of data $\{(x_1, x_2) : x_1 \neq x_2\}$ and $\{(x_1, x_2) : x_1 = x_2\}$, respectively.

23.3 Likelihood and stopping rule principles

Suppose an experiment E is conducted, which consists of observing data X having density $f(x|\theta)$, where θ is the unknown parameters of the statistical model. Let x_{obs} denote the data actually observed.

Likelihood Principle (LP): *The information about θ, arising from just E and x_{obs}, is contained in the observed likelihood function*

$$L(\theta) = f(x_{\text{obs}}|\theta)\,.$$

Furthermore, if two observed likelihood functions are proportional, then they contain the same information about θ.

The LP is quite controversial, in that it effectively precludes use of frequentist measures, which all involve averages of $f(x|\theta)$ over x that are not observed. Bayesians automatically follow the LP because the posterior distribution of θ follows from Bayes' theorem (with $p(\theta)$ being the prior density for θ) as

$$p(\theta|x_{\text{obs}}) = \frac{p(\theta) f(x_{\text{obs}}|\theta)}{\int p(\theta') f(x_{\text{obs}}|\theta') \mathrm{d}\theta'}\,,$$

which clearly depends on E and $x_{\rm obs}$ only through the observed likelihood function. There was not much attention paid to the LP by non-Bayesians, however, until the remarkable paper of Birnbaum (1962), which deduced the LP as a consequence of the *conditionality principle* (essentially the Cox example, saying that one should base the inference on the measuring instrument actually used) and the *sufficiency principle*, which states that a sufficient statistic for θ in E contains all information about θ that is available from the experiment. At the time of Birnbaum's paper, almost everyone agreed with the conditionality principle and the sufficiency principle, so it was a shock that the LP was a direct consequence of the two. The paper had a profound effect on my own thinking.

There are numerous clarifications and qualifications relevant to the LP, and various generalizations and implications. Many of these (and the history of the LP) are summarized in Berger and Wolpert (1984). Without going further, suffice it to say that the LP is, at a minimum, a very powerful argument for conditioning.

Stopping Rule Principle (SRP): *The reasons for stopping experimentation have no bearing on the information about θ arising from E and $x_{\rm obs}$.*

The SRP is actually an immediate consequence of the second part of the LP, since "stopping rules" affect $L(\theta)$ only by multiplicative constants. Serious discussion of the SRP goes back at least to Barnard (1947), who wondered why thoughts in the experimenter's head concerning why to stop an experiment should affect how we analyze the actual data that were obtained.

Frequentists typically violate the SRP. In clinical trials, for instance, it is standard to "spend α" for looks at the data — i.e., if there are to be interim analyses during the trial, with the option of stopping the trial early should the data look convincing, frequentists view it to be mandatory to adjust the allowed error probability (down) to account for the multiple analyses.

In Berger and Berry (1988), there is extensive discussion of these issues, with earlier references. The complexity of the issue was illustrated by a comment of Jimmy Savage:

> "I learned the stopping rule principle from Professor Barnard, in conversation in the summer of 1952. Frankly, I then thought it a scandal that anyone in the profession could advance an idea so patently wrong, even as today I can scarcely believe that people resist an idea so patently right." (Savage et al., 1962)

The SRP does not say that one is free to ignore the stopping rule in any statistical analysis. For instance, common practice in some sciences, when testing a null hypothesis $\mathcal{H}_0$, is to continue collecting data until the p-value satisfies $p < .05$ and then report the result as if no optional stopping had been involved. This is obviously bad science in that, even if $\mathcal{H}_0$ is true, one is guaranteed to obtain $p < .05$ if one just collects enough data. This fact was noted as early as 1938 by Berkson (Berkson, 1938), who humorously observed that, since one would be sure to obtain $p < .05$ in this way, we should save everyone trouble and the cost of experimentation and just declare every hypothesis rejected with $p < .05$! The correct calculation of a p-value would have to include the stopping rule used and the problem of "sampling to a foregone conclusion" would then disappear.

What the SRP is saying is that methods of statistical inference should be used which are compatible with the SRP. Bayesian analysis is compatible with the SRP; it will ignore the stopping rule and will not suffer for doing so. As but one illustration, in testing $\mathcal{H}_0$, a Bayesian would often use a Bayes factor $B(X)$ (defined later) of $\mathcal{H}_0$ to the alternative, which will not depend on the stopping rule. But Birnbaum (1962) observed that, for any stopping rule, $\Pr\{B(X) < \epsilon | \mathcal{H}_0\} < \epsilon$, so that optional stopping cannot ensure that a small Bayes factor (small is evidence against $\mathcal{H}_0$) will be obtained. Surprisingly, there are also frequentist methods that are compatible with the SRP, but these are inevitably conditional frequentist methods. See Berger et al. (1999) and Berger et al. (1994) for examples.

23.4 What it means to be a frequentist

As we move to more complicated examples, it is necessary to define the frequentist principle of statistics.

Frequentist Principle: *In repeated practical use of a statistical procedure, the long-run average actual accuracy should not be less than (and ideally should equal) the long-run average reported accuracy.*

Suppose, for instance, that a particular statistical model and procedure are to be *repeatedly* used — for instance, a 95% classical confidence interval for a Normal mean. This procedure will, in practice, be used on a series of different problems involving a series of different Normal means with different data. In evaluating the procedure, we should simultaneously be averaging over all possible practical instances of utilization of the procedure.

Textbook statements of the frequentist principle tend to focus on fixing the value of, say, the Normal mean, and *imagining* repeatedly drawing data from the given model and utilizing the confidence procedure repeatedly on the different data draws. The word imagining is highlighted because this is solely a thought experiment. What is done in practice is to use the confidence procedure on a series of different problems — not a series of imaginary repetitions of the same problem with different data.

Neyman himself often pointed out that the motivation for the frequentist principle is in its use on differing real problems; see, e.g., Neyman (1977). Of course, the reason textbooks typically give the imaginary repetition of an experiment version is because of the mathematical fact that if, say, a confidence procedure has 95% frequentist coverage for each fixed parameter value, then it will necessarily also have 95% coverage when used repeatedly on a series of differing problems. And, if the coverage is not constant over each fixed parameter value, one can always find the minimum coverage over the parameter space, since it will follow that the real frequentist coverage in repeated use of the procedure on real problems will never be worse than this minimum coverage.

Pedagogical example continued: Reporting 50% and 100% confidence, as appropriate, is fully frequentist, in that the long run reported coverage will average .75, which is the long run actual coverage.

p-values: p-values are not frequentist measures of evidence in any long run average sense. Suppose we observe X, have a null hypothesis $\mathcal{H}_0$, and construct a proper p-value $p(X)$. Viewing the observed $p(x_{\text{obs}})$ as a conditional error rate when rejecting $\mathcal{H}_0$ is not correct from the frequentist perspective. To see this, note that, under the null hypothesis, a proper p-value will be Uniform on the interval $(0, 1)$, so that if rejection occurs when $p(X) \leq \alpha$, the average reported p-value under $\mathcal{H}_0$ and rejection will be

$$\mathrm{E}[p(X)|\mathcal{H}_0, \{p(\cdot) \leq \alpha\}] = \int_0^\alpha p \, \frac{1}{\alpha} \, \mathrm{d}p = \frac{\alpha}{2} \, ,$$

which is only half the actual long run error α. There have been other efforts to give a real frequentist interpretation of a p-value, none of them successful in terms of the definition at the beginning of this section. Note that the procedure $\{$reject $\mathcal{H}_0$ when $p(X) \leq \alpha\}$ is a fully correct frequentist procedure, but the stated error rate in rejection must be α, not the p-value.

There have certainly been other ways of defining frequentism; see, e.g., Mayo (1996) for discussion. However, it is only the version given at the beginning of the section that strikes me as being compelling. How could one want to give statistical inferences that, over the long run, systematically distort their associated accuracies?

23.5 Conditional frequentist inference

23.5.1 Introduction

The theory of combining the frequentist principle with conditioning was formalized by Kiefer in Kiefer (1977), although there were many precursors to the theory initiated by Fisher and others. There are several versions of the theory, but the most useful has been to begin by defining a conditioning statistic S which measures the "strength of evidence" in the data. Then one computes the desired frequentist measure, but does so conditional on the strength of evidence S.

Pedagogical example continued: $S = |X_1 - X_2|$ is the obvious choice, $S = 2$ reflecting data with maximal evidential content (corresponding to the situation of 100% confidence) and $S = 0$ being data of minimal evidential content. Here coverage probability is the desired frequentist criterion, and an easy computation shows that conditional coverage given S is given by

$$\begin{aligned} P_\theta\{C(X_1, X_2) \text{ contains } \theta \,|\, S = 2\} &= 1, \\ P_\theta\{C(X_1, X_2) \text{ contains } \theta \,|\, S = 0\} &= 1/2, \end{aligned}$$

for the two distinct cases, which are the intuitively correct answers.

23.5.2 Ancillary statistics and invariant models

An *ancillary statistic* is a statistic S whose distribution does not depend on unknown model parameters θ. In the pedagogical example, $S = 0$ and $S = 2$ have probability 1/2 each, independent of θ, and so S is ancillary. When ancillary statistics exist, they are usually good measures of the strength of evidence in the data, and hence provide good candidates for conditional frequentist inference.

The most important situations involving ancillary statistics arise when the model has what is called a group-invariance structure; cf. Berger (1985) and Eaton (1989). When this structure is present, the best ancillary statistic to use is what is called the maximal invariant statistic. Doing conditional frequentist inference with the maximal invariant statistic is then equivalent to performing Bayesian inference with the right-Haar prior distribution with respect to the group action; cf. Berger (1985), Eaton (1989), and Stein (1965).

Example–Location Distributions: Suppose $X_1, \ldots, X_n$ form a random sample from the location density $f(x_i - \theta)$. This model is invariant under the group operation defined by adding any constant to each observation and θ; the maximal invariant statistic (in general) is $S = (x_2 - x_1, x_3 - x_1, \ldots, x_n - x_1)$, and performing conditional frequentist inference, conditional on S, will give the same numerical answers as performing Bayesian inference with the right-

Haar prior, here simply given by $p(\theta) = 1$. For instance, the optimal conditional frequentist estimator of θ under squared error loss would simply be the posterior mean with respect to $p(\theta) = 1$, namely

$$\hat{\theta} = \frac{\int \theta \prod_{i=1}^{n} f(x_i - \theta) 1 \mathrm{d}\theta}{\int \prod_{i=1}^{n} f(x_i - \theta) 1 \mathrm{d}\theta},$$

which is also known as Pitman's estimator.

Having a model with a group-invariance structure leaves one in an incredibly powerful situation, and this happens with many of our most common statistical problems (mostly from an estimation perspective). The difficulties of the conditional frequentist perspective are (i) finding the right strength of evidence statistic S, and (ii) carrying out the conditional frequentist computation. But, if one has a group-invariant model, these difficulties can be bypassed because theory says that the optimal conditional frequentist answer is the answer obtained from the much simpler Bayesian analysis with the right-Haar prior.

Note that the conditional frequentist and Bayesian answers will have different interpretations. For instance both approaches would produce the same 95% confidence set, but the conditional frequentist would say that the frequentist coverage, conditional on S (and also unconditionally), is 95%, while the Bayesian would say the set has probability .95 of actually containing θ. Also note that it is not automatically true that analysis conditional on ancillary statistics is optimal; see, e.g., Brown (1990).

23.5.3 Conditional frequentist testing

Upon rejection of the $\mathcal{H}_0$ in unconditional Neyman–Pearson testing, one reports the same error probability α regardless of where the test statistic is in the rejection region. This has been viewed as problematical by many, and is one of the main reasons for the popularity of p-values. But as we saw earlier, p-values do not satisfy the frequentist principle, and so are not the conditional frequentist answer.

A true conditional frequentist solution to the problem was proposed in Berger et al. (1994), with modification (given below) from Sellke et al. (2001) and Wolpert (1996). Suppose that we wish to test that the data $\mathbf{X}$ arises from the simple (i.e., completely specified) hypotheses $\mathcal{H}_0 : f = f_0$ or $\mathcal{H}_1 : f = f_1$. The recommended strength of evidence statistic is

$$S = \max\{p_0(\mathbf{x}), p_1(\mathbf{x})\},$$

where $p_0(\mathbf{x})$ is the p-value when testing $\mathcal{H}_0$ versus $\mathcal{H}_1$, and $p_1(\mathbf{x})$ is the p-value when testing $\mathcal{H}_1$ versus $\mathcal{H}_0$. It is generally agreed that smaller p-values correspond to more evidence against an hypothesis, so this use of p-values in determining the strength of evidence statistic is natural. The frequentist conditional error probabilities (CEPs) are computed as

$$\begin{aligned} \alpha(s) &= \Pr(\text{Type I error}|S = s) \equiv P_0\{\text{reject } \mathcal{H}_0|S(\mathbf{X}) = s\}, \\ \beta(s) &= \Pr(\text{Type II error}|S = s) \equiv P_1\{\text{accept } \mathcal{H}_0|S(\mathbf{X}) = s\}, \end{aligned} \tag{23.1}$$

where P_0 and P_1 refer to probability under $\mathcal{H}_0$ and $\mathcal{H}_1$, respectively.

The corresponding conditional frequentist test is then

$$\begin{aligned} &\text{If } p_0 \leq p_1, \text{reject } \mathcal{H}_0 \text{ and report Type I CEP } \alpha(s); \\ &\text{If } p_0 > p_1, \text{accept } \mathcal{H}_0 \text{ and report Type II CEP } \beta(s); \end{aligned} \tag{23.2}$$

where the CEPs are given in (23.1).

These conditional error probabilities are fully frequentist and vary over the rejection region as one would expect. In a sense, this procedure can be viewed as a way to turn p-values into actual error probabilities.

It was mentioned in the introduction that, when a good conditional frequentist procedure has been found, it often turns out to be numerically equivalent to a Bayesian procedure. That is the case here. Indeed, Berger et al. (1994) shows that

$$\alpha(s) = \Pr(\mathcal{H}_0|\mathbf{x}), \quad \beta(s) = \Pr(\mathcal{H}_1|\mathbf{x}), \tag{23.3}$$

where $\Pr(\mathcal{H}_0|\mathbf{x})$ and $\Pr(\mathcal{H}_1|\mathbf{x})$ are the Bayesian posterior probabilities of $\mathcal{H}_0$ and $\mathcal{H}_1$, respectively, assuming the hypotheses have equal prior probabilities of 1/2. Therefore, a conditional frequentist can simply compute the objective Bayesian posterior probabilities of the hypotheses, and declare that they are the conditional frequentist error probabilities; there is no need to formally derive the conditioning statistic or perform the conditional frequentist computations. There are many generalizations of this beyond the simple versus simple testing.

The practical import of switching to conditional frequentist testing (or the equivalent objective Bayesian testing) is startling. For instance, Sellke et al. (2001) uses a nonparametric setting to develop the following very general lower bound on $\alpha(s)$, for a given p-value:

$$\alpha(s) \geq \frac{1}{1 - \dfrac{1}{e\, p\, \ln(p)}}. \tag{23.4}$$

Some values of this lower bound for common p-values are given in Table 23.1. Thus $p = .05$, which many erroneously think implies strong evidence against $\mathcal{H}_0$, actually corresponds to a conditional frequentist error probability at least as large as .289, which is a rather large error probability. If scientists understood that a p-value of .05 corresponded to that large a potential error probability in rejection, the scientific world would be a quite different place.

TABLE 23.1
Values of the lower bound $\alpha(s)$ in (23.4) for various values of p.

p	.2	.1	.05	.01	.005	.001	.0001	.00001
$\alpha(s)$	.465	.385	.289	.111	.067	.0184	.0025	.00031

23.5.4 Testing a sequence of hypotheses

It is common in clinical trials to test multiple endpoints but to do so sequentially, only considering the next hypothesis if the previous hypothesis was a rejection of the null. For instance, the primary endpoint for a drug might be weight reduction, with the secondary endpoint being reduction in an allergic reaction. (Typically, these will be more biologically related endpoints but the point here is better made when the endpoints have little to do with each other.) Denote the primary endpoint (null hypothesis) by $\mathcal{H}_0^1$, and the statistical analysis must first test this hypothesis. If the hypothesis is not rejected at level α, the analysis stops — i.e., no further hypotheses can be considered. However, if the hypothesis is rejected, one can go on and consider the secondary endpoint, defined by null hypothesis $\mathcal{H}_0^2$. Suppose this hypothesis is also rejected at level α.

Surprisingly, the overall probability of Type I error (rejecting at least one true null hypothesis) for this procedure is still just α — see, e.g., Hsu and Berger (1999) — even though there is the possibility of rejecting two separate hypotheses. It appears that the second test comes "for free," with rejection allowing one to claim two discoveries for the price of one. This actually seems too remarkable; how can we be as confident that both rejections are correct as we are that just the first rejection is correct?

If this latter intuition is not clear, note that one does not need to stop after two hypotheses. If the second has rejected, one can test $\mathcal{H}_0^3$ and, if that is rejected at level α, one can go on to test a fourth hypothesis $\mathcal{H}_0^4$, etc. Suppose one follows this procedure and has rejected $\mathcal{H}_0^1, \ldots, \mathcal{H}_0^{10}$. It is still true that the probability of Type I error for the procedure — i.e., the probability that the procedure will result in an erroneous rejection — is just α. But it seems ridiculous to think that there is only probability α that at least one of the 10 rejections is incorrect. (Or imagine a million rejections in a row, if you do not find the argument for 10 convincing.)

The problem here is in the use of the unconditional Type I error to judge accuracy. Before starting the sequence of tests, the probability that the *procedure* yields at least one incorrect rejection is indeed, α, but the situation changes dramatically as we start down the path of rejections. The simplest way to see this is to view the situation from the Bayesian perspective. Consider the situation in which all the hypotheses can be viewed as *a priori* independent (i.e., knowing that one is true or false does not affect perceptions of the others). If $\mathbf{x}$ is the overall data from the trial, and a total of m tests are ultimately conducted by the procedure, all claimed to be rejections (i.e., all claimed to correspond to the $\mathcal{H}_0^i$ being false), the Bayesian computes

$$\begin{aligned} \Pr\,(\text{at least one incorrect rejection }|\mathbf{x}) \\ &= 1 - \Pr(\text{no incorrect rejections}|\mathbf{x}) \\ &= 1 - \prod_{i=1}^{m}\{1 - \Pr(\mathcal{H}_0^i|\mathbf{x})\}, \end{aligned} \tag{23.5}$$

where $\Pr(\mathcal{H}_0^i|\mathbf{x})$ is the posterior probability that $\mathcal{H}_0^i$ is true given the data. Clearly, as m grows, (23.5) will go to 1 so that, if there are enough tests, the Bayesian becomes essentially sure that at least one of the rejections was wrong. From Section 23.5.3, recall that Bayesian testing can be exactly equivalent to conditional frequentist testing, so it should be possible to construct a conditional frequentist variant of (23.5). This will, however, be pursued elsewhere.

While we assumed that the hypotheses are all *a priori* independent, it is more typical in the multiple endpoint scenario that they will be *a priori* related (e.g., different dosages of a drug). This can be handled within the Bayesian approach (and will be explored elsewhere), but it is not clear how a frequentist could incorporate this information, since it is information about the prior probabilities of hypotheses.

23.5.5 True to false discovery odds

A very important paper in the history of genome wide association studies (the effort to find which genes are associated with certain diseases) was Burton et al. (2007). Consider testing $\mathcal{H}_0 : \theta = 0$ versus an alternative $\mathcal{H}_1 : \theta \neq 0$, with rejection region $\mathcal{R}$ and corresponding Type I and Type II errors α and $\beta(\theta)$. Let $p(\theta)$ be the prior density of θ under $\mathcal{H}_1$, and define the average power

$$1 - \bar{\beta} = \int \{1 - \beta(\theta)\} p(\theta) \mathrm{d}\theta.$$

Frequentists would typically just pick some value θ^* at which to evaluate the power; this is equivalent to choosing $p(\theta)$ to be a point mass at θ^*.

The paper observed that, pre-experimentally, the odds of correctly rejecting $\mathcal{H}_0$ to incorrectly rejecting are

$$O_{\text{pre}} = \frac{\pi_1}{\pi_0} \times \frac{1 - \bar{\beta}}{\alpha}, \tag{23.6}$$

where π_0 and $\pi_1 = 1 - \pi_0$ are the prior probabilities of $\mathcal{H}_0$ and $\mathcal{H}_1$. The corresponding false discovery rate would be $(1 + O_{\text{pre}})^{-1}$.

The paper went on to assess the prior odds π_1/π_0 of a genome/disease association to be $1/100,000$, and estimated the average power of a GWAS test to be .5. It was decided that a discovery should be reported if $O_{\text{pre}} \geq 10$, which from (23.6) would require $\alpha \leq 5 \times 10^{-7}$; this became the recommended standard for significance in GWAS studies. Using this standard for a large data set, the paper found 21 genome/disease associations, virtually all of which have been subsequently verified.

An alternative approach that was discussed in the paper is to use the posterior odds rather than pre-experimental odds — i.e., to condition. The posterior odds are

$$O_{\text{post}}(\mathbf{x}) = \frac{\pi_1}{\pi_0} \times \frac{m(\mathbf{x}|\mathcal{H}_1)}{f(\mathbf{x}|0)}, \tag{23.7}$$

where $m(\mathbf{x}|\mathcal{H}_1) = \int f(\mathbf{x}|\theta)p(\theta)d\theta$ is the marginal likelihood of the data $\mathbf{x}$ under $\mathcal{H}_1$. (Again, this prior could be a point mass at θ^* in a frequentist setting.) It was noted in the paper that the posterior odds for the 21 claimed associations ranged between 1/10 (i.e., evidence against the association being true) to 10^{68} (overwhelming evidence in favor of the association). It would seem that these conditional odds, based on the actual data, are much more scientifically informative than the fixed pre-experimental odds of 10/1 for the chosen α, but the paper did not ultimately recommend their use because it was felt that a frequentist justification was needed.

Actually, use of O_{post} is as fully frequentist as is use of O_{pre}, since it is trivial to show that $\text{E}\{O_{\text{post}}(\mathbf{x})|\mathcal{H}_0, \mathcal{R}\} = O_{\text{pre}}$, i.e., the average of the conditional reported odds equals the actual pre-experimental reported odds, which is all that is needed to be fully frequentist. So one can have the much more scientifically useful conditional report, while maintaining full frequentist justification. This is yet another case where, upon getting the conditioning right, a frequentist completely agrees with a Bayesian.

23.6 Final comments

Lots of bad science is being done because of a lack of recognition of the importance of conditioning in statistics. Overwhelmingly at the top of the list is the use of p-values and acting as if they are actually error probabilities. The common approach to testing a sequence of hypotheses is a new addition to the list of bad science because of a lack of conditioning. The use of pre-experimental odds rather than posterior odds in GWAS studies is not so much bad science, as a failure to recognize a conditional frequentist opportunity that is available to improve science. Violation of the stopping rule principle in sequential (or interim) analysis is in a funny position. While it is generally suboptimal (for instance, one could do conditional frequentist testing instead), it may be necessary if one is committed to certain inferential procedures such as fixed Type I error probabilities. (In other words, one mistake may require the incorporation of another mistake.)

How does a frequentist know when a serious conditioning mistake is being made? We have seen a number of situations where it is clear but, in general, there is only one way to identify if conditioning is an issue — Bayesian analysis. If one can find a Bayesian analysis for a reasonable prior that yields the same answer as the frequentist analysis, then there is probably not a conditioning issue; otherwise, the conflicting answers are probably due to the need for conditioning on the frequentist side.

The most problematic situations (and unfortunately there are many) are those for which there exists an apparently sensible unconditional frequentist analysis but Bayesian analysis is unavailable or too difficult to implement given available resources. There is then not much choice but to use the unconditional frequentist analysis, but one might be doing something silly because of not being able to condition and one will not know. The situation is somewhat comparable to seeing the report of a Bayesian analysis but not having access to the prior distribution.

While I have enjoyed reminiscing about conditioning, I remain as perplexed today as 35 years ago when I first learned about the issue; why do we still not treat conditioning as one of the most central issues in statistics?

References

Barnard, G.A. (1947). A review of 'Sequential Analysis' by Abraham Wald. *Journal of the American Statistical Association*, 42:658–669.

Berger, J.O. (1985). *Statistical Decision Theory and Bayesian Analysis.* Springer, New York.

Berger, J.O. and Berry, D.A. (1988). The relevance of stopping rules in statistical inference. In *Statistical Decision Theory and Related Topics IV, 1.* Springer, New York, pp. 29–47.

Berger, J.O., Boukai, B., and Wang, Y. (1999). Simultaneous Bayesian-frequentist sequential testing of nested hypotheses. *Biometrika*, 86:79–92.

Berger, J.O., Brown, L.D., and Wolpert, R.L. (1994). A unified conditional frequentist and Bayesian test for fixed and sequential simple hypothesis testing. *The Annals of Statistics*, 22:1787–1807.

Berger, J.O. and Wolpert, R.L. (1984). *The Likelihood Principle.* IMS Lecture Notes, Monograph Series, 6. Institute of Mathematical Statistics, Hayward, CA.

Berkson, J. (1938). Some difficulties of interpretation encountered in the application of the chi-square test. *Journal of the American Statistical Association*, 33:526–536.

Birnbaum, A. (1962). On the foundations of statistical inference. *Journal of the American Statistical Association*, 57:269–306.

Brown, L.D. (1990). An ancillarity paradox which appears in multiple linear regression. *The Annals of Statistics*, 18:471–493.

Burton, P.R., Clayton, D.G., Cardon, L.R., Craddock, N., Deloukas, P., Duncanson, A., Kwiatkowski, D.P., McCarthy, M.I., Ouwehand, W.H., Samani, N.J., et al. (2007). Genome-wide association study of 14,000 cases of seven common diseases and 3,000 shared controls. *Nature*, 447:661–678.

Cox, D.R. (1958). Some problems connected with statistical inference. *The Annals of Mathematical Statistics*, 29:357–372.

Eaton, M.L. (1989). *Group Invariance Applications in Statistics.* Institute of Mathematical Statistics, Hayward, CA.

Hsu, J.C. and Berger, R.L. (1999). Stepwise confidence intervals without multiplicity adjustment for dose-response and toxicity studies. *Journal of the American Statistical Association*, 94:468–482.

Kiefer, J. (1977). Conditional confidence statements and confidence estimators. *Journal of the American Statistical Association*, 72:789–808.

Mayo, D.G. (1996). *Error and the Growth of Experimental Knowledge.* University of Chicago Press, Chicago, IL.

Neyman, J. (1977). Frequentist probability and frequentist statistics. *Synthese*, 36: 97–131.

Savage, L.J., Barnard, G., Cornfield, J., Bross, I., Box, G.E.P., Good, I.J., Lindley, D.V., Clunies-Ross, C.W., Pratt, J.W., Levene, H. et al. (1962). On the foundations of statistical inference: Discussion. *Journal of the American Statistical Association*, 57:307–326.

Sellke, T., Bayarri, M., and Berger, J.O. (2001). Calibration of p-values for testing precise null hypotheses. *The American Statistician*, 55:62–71.

Stein, C. (1965). Approximation of improper prior measures by prior probability measures. In *Bernoulli–Bayes–Laplace Festschrift.* Springer, New York, pp. 217–240.

Wolpert, R.L. (1996). Testing simple hypotheses. In *Studies in Classification, Data Analysis, and Knowledge Organization*, Vol. 7. Springer, New York, pp. 289–297.

24

Statistical inference from a Dempster–Shafer perspective

Arthur P. Dempster
Department of Statistics
Harvard University, Cambridge, MA

24.1 Introduction

What follows is a sketch of my 2013 viewpoint on how statistical inference should be viewed by applied statisticians. The label DS is an acronym for "Dempster–Shafer" after the originators of the technical foundation of the theory. Our foundation remains essentially unchanged since the 1960s and 1970s when I and then Glenn Shafer were its initial expositors.

Present issues concern why and how the theory has the potential to develop into a major competitor of the "frequentist" and "Bayesian" outlooks. This for me is a work in progress. My understanding has evolved substantially over the past eight years of my emeritus status, during which DS has been my major focus. It was also a major focus of mine over the eight years beginning in 1961 when I first had the freedom that came with academic tenure in the Harvard Statistics Department. Between the two periods I was more an observer and teacher in relation to DS than a primary developer. I do not attempt here to address the long history of how DS got to where I now understand it to be, including connections with R.A. Fisher's controversial "fiducial" argument.

DS draws on technical developments in fields such as stochastic modeling and Bayesian posterior computation, but my DS-guided perception of the nature of statistical inference is in different ways both narrower and broader than that of its established competitors. It is narrower because it maintains that what "frequentist" statisticians call "inference" is not inference in the natural language meaning of the word. The latter means to me direct situation-specific assessments of probabilistic uncertainties that I call "personal probabilities." For example, I might predict on September 30, 2013 that with personal probability .31 the Dow Jones Industrials stock index will exceed 16,000 at the end of business on December 31, 2013.

From the DS perspective, statistical prediction, estimation, and significance testing depend on understanding and accepting the DS logical framework, as implemented through model-based computations that mix probabilistic and deterministic logic. They do not depend on frequentist properties of hypothetical ("imagined") long runs of repeated application of any defined repeatable statistical procedure, which properties are simply mathematical statements about the procedure. Knowledge of such long run properties may guide choosing among statistical procedures, but drawing conclusions from a specific application of a chosen procedure is something else again.

Whereas the logical framework of DS inference has long been defined and stable, and presumably will not change, the choice of a model to be processed through the logic must be determined by a user or user community in each specific application. It has long been known that DS logic subsumes Bayesian logic. A Bayesian instantiation of DS inference occurs automatically within the DS framework when a Bayesian model is adopted by a user. My argument for the importance of DS logic is not primarily that it encapsulates Bayes, however, but is that it makes available important classes of models and associated inferences that narrower Bayesian models are unable to represent. Specifically, it provides models where personal probabilities of "don't know" are appropriately introduced. In particular, Bayesian "priors" become optional in many common statistical situations, especially when DS probabilities of "don't know" are allowed. Extending Bayesian thinking in this way promises greater realism in many or even most applied situations.

24.2 Personal probability

DS probabilities can be studied from a purely mathematical standpoint, but when they have a role in assessing uncertainties about specific real world unknowns, they are meant for interpretation as "personal" probabilities. To my knowledge, the term "personal" was first used in relation to mathematical probabilities by Émile Borel in a book (Borel, 1939), and then in statistics by Jimmie Savage, as far back as 1950, and subsequently in many short contributions preceding his untimely death in 1971. Savage was primarily concerned with Bayesian decision theory, wherein proposed actions are based on posterior expectations. From a DS viewpoint, the presence of decision components is optional.

The DS inference paradigm explicitly recognizes the role of a user in constructing and using formal models that represent his or her uncertainty. No other role for the application of probabilities is recognized. Ordinary speech often describes empirical variation as "random," and statisticians often regard probabilities as mathematical representations of "randomness," which they are, except that in most if not all of statistical practice "random" variation is simply unexplained variation whose associated probabilities are quite properly interpretable as personal numerical assessments of specific targeted uncertainties. Such models inevitably run a gamut from objective to subjective, and from broadly accepted to proposed and adopted by a single analyst who becomes the "person" in a personalist narrative. Good statistical practice aims at the left ends of both scales, while actual practice necessarily makes compromises. As Jack Good used to say, "inference is possible."

The concept of personalist interpretation of specific probabilities is usually well understood by statisticians, but is mostly kept hidden as possibly unscientific. Nevertheless, all approaches to statistical inference imply the exercise of mature judgment in the construction and use of formal models that integrate descriptions of empirical phenomena with prescriptions for reasoning under uncertainty. By limiting attention to long run averages, "frequentist" interpretations are designed to remove any real or imagined taint from personal probabilities, but paradoxically do not remove the presence of nonprobabilistic reasoning about deterministic long runs. The latter reasoning is just as personalist as the former. Why the fear of reasoning with personal probabilities, but not a similar fear of ordinary propositional logic? This makes no sense to me, if the goal is to remove any role for a "person" performing logical analysis in establishing valid scientific findings.

I believe that, as partners in scientific inquiry, applied statisticians should seek credible models directly aimed at uncertainties through precisely formulated direct and transparent reasoning with personal probabilities. I argue that DS logic is at present the best available system for doing this.

24.3 Personal probabilities of "don't know"

DS "gives up something big," as John Tukey once described it to me, or as I now prefer to describe it, by modifying the root concepts of personal probabilities "for" and "against" that sum to one, by appending a third personal probability of "don't know." The extra term adds substantially to the flexibility of modeling, and to the expressiveness of inputs and outputs of DS probabilistic reasoning.

The targets of DS inference are binary outcomes, or equivalent assertions that the true state of some identified small world is either in one subset of the full set of possible true states, or in the complementary subset. Under what I shall refer to as the "ordinary" calculus of probability (OCP), the user is required to supply a pair of non-negative probabilities summing to one that characterize "your" uncertainty about which subset contains the true state. DS requires instead that "you" adopt an "extended" calculus of probability (ECP) wherein the traditional pair of probabilities that the true state lies or does not lie in the subset associated with a targeted outcome is supplemented by a third probability of "don't know," where all three probabilities are non-negative and sum to one. It needs to be emphasized that the small world of possible true states is characterized by binary outcomes interpretable as true/false assertions, while uncertainties about such two-valued outcomes are represented by three-valued probabilities.

For a given targeted outcome, a convenient notation for a three-valued probability assessment is (p, q, r), where p represents personal probability "for" the truth of an assertion, while q represents personal probability "against," and r represents personal probability of "don't know." Each of p, q, and r is non-negative and together they sum to unity. The outcome complementary to a given target associated with (p, q, r) has the associated personal probability triple (q, p, r). The "ordinary" calculus is recovered from the "extended" calculus by limiting (p, q, r) uncertainty assessments to the form $(p, q, 0)$, or (p, q) for short. The "ordinary" calculus permits "you" to be sure that the assertion is true through $(p, q, 0) = (1, 0, 0)$, or false through $(p, q, r) = (0, 1, 0)$, while the "extended" calculus additionally permits $(p, q, r) = (0, 0, 1)$, representing total ignorance.

Devotees of the "ordinary" calculus are sometimes inclined, when confronted with the introduction of $r > 0$, to ask why the extra term is needed. Aren't probabilities (p, q) with $p + q = 1$ sufficient to characterize scientific and operational uncertainties? Who needs probabilities of "don't know"? One answer is that every application of a Bayesian model is necessarily based on a limited state space structure (SSS) that does not assess associated (p, q) probabilities for more inclusive state space structures. Such extended state space structures realistically always exist, and may be relevant to reported inferences. In effect, every Bayesian analysis makes implicit assumptions that evidence about true states of variables omitted from an SSS is "independent" of additional probabilistic knowledge, including ECP expressions thereof, that should accompany explicitly identifiable state spaces. DS methodology makes available a wide range of models and analyses whose differences from narrower analyses can point to "biases" due to the limitations of state spaces associated with reported Bayesian analyses. Failure of narrower assumptions often accentuates non-reproducibility of findings from non-DS statistical studies, casting doubts on the credibility of many statistical studies.

DS methodology can go part way at least to fixing the problem through broadening of state space structures and indicating plausible assignments of personal probabilities of "don't know" to aspects of broadened state spaces, including the use of $(p, q, r) = (0, 0, 1)$ when no empirical basis "whatever" to quote Keynes exists for the use of "a good Benthamite calculation of a series of prospective advantages and disadvantages, each multiplied by its appropriate probability, waiting to he summed" that can be brought to bear. DS allows a wide range of possible probabilistic uncertainty assessments between complete ignorance and the fully "Benthamite" (i.e., Bayesian) models that Keynes rejected for many applications.

As the world of statistical analysis moves more and more to "big data" and associated "complex systems," the DS middle ground can be expected to become increasingly important. DS puts no restraints on making state space structures as large as judged essential for bias protection, while the accompanying increases in many probabilities of "don't know" will often require paying serious attention to the introduction of more evidence, including future research studies. Contrary to the opinion of critics who decry all dependence on mathematical models, the need is for more inclusive and necessarily more complex mathematical models that will continue to come on line as associated information technologies advance.

24.4 The standard DS protocol

Developing and carrying out a DS analysis follows a prescribed sequence of activities and operations. First comes defining the state space structure, referred to henceforth by its acronym SSS. The purpose of initializing an SSS is to render precise the implied connection between the mathematical model and a piece of the actual real world. Shafer introduced the insightful term "frame of discernment" for what I am calling the SSS. The SSS is a mathematical set whose elements are the possible true values of some "small world" under investigation. The SSS is typically defined by a vector or multi-way array of variables, each with its own known or unknown true value. Such an SSS may be very simple, such as a vector of binary variables representing the outcomes of successive tosses of a bent coin, some observed, and some such as future tosses remaining unobserved. Or, an SSS may be huge, based on a set of variables representing multivariate variation across situations that repeat across times and spatial locations, in fields such as climatology, genomics, or economics.

The requirement of an initialized SSS follows naturally from the desirability of clearly and adequately specifying at the outset the extent of admissible queries about the true state of a small world under analysis. Each such query corresponds mathematically to a subset of the SSS. For example, before the first toss of a coin in an identified sequence of tosses, I might formulate a query about the outcome, and respond by assigning a (p, q, r) personal probability triple to the outcome "head," and its reordered (q, p, r) triple to the "complementary" outcome "tail." After the outcome "head" is observed and known to "you," the appropriate inference concerning the outcome "head" is $(1, 0, 0)$, because the idealized "you" is sure about the outcome. The assertion "head on the first toss" is represented by the "marginal" subset of the SSS consisting of all possible outcomes of all the variables beyond the first toss, which has been fixed by observation. A DS inference $(1, 0, 0)$ associated with "head" signifies observed data.

The mindset of the user of DS methods is that each assignment of a (p, q, r) judgment is based on evidence. Evidence is a term of art, not a formal concept. Statistical data is one source of evidence. If the outcome of a sequence of $n =$ 10 tosses of a certain bent coin is observed to result in data HTTHHHTHTT, then each data point provides evidence about a particular toss, and queries with $(1, 0, 0)$ responses can be given with confidence concerning individual margins of the 10-dimensional SSS. The user can "combine" these marginal inferences so as to respond to queries depending on subsets of the sequence, or about any interesting properties of the entire sequence.

The DS notion of "combining" sources of evidence extends to cover probabilistically represented sources of evidence that combine with each other and with data to produce fused posterior statistical inferences. This inference process can be illustrated by revising the simple SSS of 10 binary variables to include a long sequence of perhaps $N = 10{,}000$ coin tosses, of which the observed $n = 10$ tosses are only the beginning. Queries may now be directed at the much larger set of possible outcomes concerning properties of subsets, or about any or all of the tosses, whether observed or not. We may be interested primarily in the long run fraction P of heads in the full sequence, then shrink back the revised SSS to the sequence of variables $X_1, \ldots, X_{10}$, P, whence to work with approximate mathematics that treats N as infinite so that P may take any real value on the closed interval $[0, 1]$. The resulting inference situation was called "the fundamental problem of practical statistics" by Karl Pearson in 1920 giving a Bayesian solution. It was the implicit motivation for Jakob Bernoulli writing circa 1700 leading him to introduce binomial sampling distributions. It was again the subject of Thomas Bayes's seminal posthumous 1763 note introducing what are now known as uniform Bayesian priors and associated posterior distributions for an unknown P.

My 1966 DS model and analysis for this most basic inference situation, when recast in 2013 terminology, is best explained by introducing a set of "auxiliary" variables $U_1, \ldots, U_{10}$ that are assigned a uniform personal probability distribution over the 10-dimensional unit cube. The U_i do not represent any real world quantities, but are simply technical crutches created for mathematical convenience that can be appropriately marginalized away in the end because inferences concerning the values of the U_i have no direct real world interpretation.

Each of the independent and identically distributed U_i provides the connection between a known X_i and the target unknown P. The relationships among X_i, P, and U_i are already familiar to statisticians because they are widely used to "computer-generate" a value of X_i for given P. Specifically, my suggested relations are

$$X_i = 1 \text{ if } 0 \leq U_i \leq P \quad \text{and} \quad X_i = 0 \text{ if } P < U_i \leq 1,$$

for each i, where $X_i = 1$ means "head" and $X_i = 0$ means "tail."

In practice, the above relations can be applied under DS inference logic either when P is assumed known and inferences about the X_i are sought, or when the X_i are assumed known and inferences about P are sought. The justification for the less familiar "inverse" application is tied to the fundamental DS "rule of combination" under "independence" that was the pivotal innovation of my earliest papers. The attitude here is that the X_i and P describe features of the real world, with values that may be known or unknown according to context, thereby avoiding the criticism that P and X_i must be viewed asymmetrically because P is "fixed" and X_i is "random." Under a personalist viewpoint the argument based on asymmetry is not germane. Either or both of the two independence assumptions may be assumed according as P or X_i or both have known values. (In the case of "both," the U_i become partially known according to the above formulas.)

Precise details of the concepts and operations of the "extended calculus of probability" (ECP) arise naturally when the X_i are fixed, whence the above relations do not determine P uniquely, but instead limit P to an interval associated with personal probabilities determined by the U_i. Under the ECP in its general form, personal probabilities are constructed from a distribution over subsets of the SSS that we call a "mass distribution." The mass distribution determines by simple sums the (p, q, r) for any desired subset of the SSS, according as mass is restricted to the subset in the case of p, or is restricted to the complement of the subset in the case of q, or has positive accumulated mass in both subsets in the case of r. DS combination of independent component mass distributions involves both intersection of subsets as in propositional logic, and multiplication of probabilities as in the ordinary calculus of probability (OCP). The ECP allows not only projecting a mass distribution "down" to margins, as in the OCP, but also inverse projection of a marginal mass distribution "up" to a finer margin of the SSS, or to an SSS that has been expanded by adjoining arbitrarily many new variables. DS combination takes place in principle across input components of evidence that have been projected up to a full SSS, although in practice computational shortcuts are often available. Combined inferences are then computed by projecting down to obtain marginal inferences of practical interest.

Returning to the example called the "fundamental problem of practical statistics," it can be shown that the result of operating with the inputs of data-determined logical mass distributions, together with inputs of probabilistic mass distributions based on the U_i, leads to a posterior mass distribution that in effect places the unknown P on the interval between the Rth and $(R+1)$st ordered values of the U_i, where R denotes the observed number of "heads" in the n observed trials. This probabilistic interval is the basis for significance testing, estimation and prediction.

To test the null hypothesis that $P = .25$, for example, the user computes the probability p that the probabilistic interval for P is either completely to the right or left of $P = .25$. The complementary $1 - p$ is the probability r that the interval covers $P = .25$, because there is zero probability that the interval shrinks to precisely P. Thus $(p, 0, r)$ is the triple corresponding to the assertion that the null hypothesis fails, and $r = 1-p$ replaces the controversial

"p-value" of contemporary applied statistics. The choice offered by such a DS significance test is not to either "accept" or "reject" the null hypothesis, but instead is either to "not reject" or "reject."

In a similar vein a (p, q, r) triple can be associated with any specified range of P values, such as the interval $(.25, .75)$, thus creating an "interval estimate." Similarly, if so desired, a "sharp" null hypothesis such as $P = .25$ can be rendered "dull" using an interval such as $(.24, .26)$. Finally, if the SSS is expanded to include a future toss or tosses, then (p, q, r) "predictions" concerning the future outcomes of such tosses can be computed given observed sample data.

There is no space here to set forth details concerning how independent input mass distributions on margins of an SSS are up-projected to mass distributions on the full SSS, and are combined there and used to derive inferences as in the preceding paragraphs. Most details have been in place, albeit using differing terminology, since the 1960s. The methods are remarkably simple and mathematically elegant. It is surprising to me that research on the standard protocol has not been taken up by any but an invisible sliver of the mathematical statistics community.

The inference system outlined in the preceding paragraphs can and should be straightforwardly developed to cover many or most inference situations found in statistical textbooks. The result will not only be that traditional Bayesian models and analyses can be re-expressed in DS terms, but more significantly that many "weakened" modifications of such inferences will become apparent, for example, by replacing Bayesian priors with DS mass distributions that demand less in terms of supporting evidence, including limiting "total ignorance" priors concerning "parameter" values. In the case of such $(0, 0, 1)$-based priors, traditional "likelihood functions" assume a restated DS form having a mass distribution implying stand-alone DS inferences. But when a "prior" includes limited probabilities of "don't know," the OCP "likelihood principle" no longer holds, nor is it needed. It also becomes easy in principle to "weaken" parametric forms adopted in likelihood functions, for example, by exploring DS analyses that do not assume precise normality, but might assume that cumulative distribution functions (CDFs) are within, say, .10 of a Gaussian CDF. Such "robustness" research is in its infancy, and is without financial support, to my knowledge, at the present time.

The concepts of DS "weakening," or conversely "strengthening," provide basic tools of model construction and revision for a user to consider in the course of arriving at final reports. In particular, claims about complex systems may be more appropriately represented in weakened forms with increased probabilities of "don't know."

24.5 Nonparametric inference

When I became a PhD student in the mid-50s, Sam Wilks suggested to me that the topic of "nonparametric" or "distribution-free" statistical inference had been largely worked through in the 1940s, in no small part through his efforts, implying that I might want to look elsewhere for a research topic. I conclude here by sketching how DS could introduce new thinking that goes back to the roots of this important topic.

A use of binomial sampling probabilities similar to that in my coin-tossing example arises in connection with sampling a univariate continuous observable. In a 1939 obituary for "Student" (W.S. Gosset), Fisher recalled that Gosset had somewhere remarked that given a random sample of size 2 with a continuous observable, the probability is 1/2 that the population median lies between the observations, with the remaining probabilities 1/4 and 1/4 evenly split between the two sides of the data. In a footnote, Fisher pointed out how Student's remark could be generalized to use binomial sampling probabilities to locate with computed probabilistic uncertainty any nominated population quantile in each of the $n + 1$ intervals determined by the data. In DS terms, the same ordered uniformly distributed auxiliaries used in connection with "binomial" sampling a dichotomy extend easily to provide marginal mass distribution posteriors for any unknown quantile of the population distribution, not just the quantiles at the observed data points. When the DS analysis is extended to placing an arbitrary population quantile in intervals other than exactly determined by the observations, (p, q, r) inferences arise that in general have $r > 0$, including $r = 1$ for assertions concerning the population CDF in regions in the tails of the data beyond the largest and smallest observations. In addition, DS would have allowed Student to extend his analysis to predict that a third sample draw can be predicted to lie in each of the three regions determined by the data with equal probabilities 1/3, or more generally with equal probabilities $1/(n+1)$ in the $n+1$ regions determined by n observations.

The "nonparametric" analysis that Fisher pioneered in his 1939 footnote serves to illustrate DS-ECP logic in action. It can also serve to illustrate the need for critical examination of particular models and consequent analyses. Consider the situation of a statistician faced with analysis of a sample of modest size, such as $n = 30$, where a casual examination of the data suggests that the underlying population distribution has a smooth CDF but does not conform to an obvious simple parametric form such as Gaussian. After plotting the data, it would not be surprising to see that the lengths of intervals between successive ordered data point over the middle ranges of the data vary by a factor of two or three. The nonparametric model asserts that these intervals have equal probabilities $1/(n + 1) = 1/31$ of containing the next draw, but a broker offering both sides of bets based on these probabilities would soon be losing money because takers would bet with higher and lower probabilities for longer and shorter intervals. The issue here is a version of the "multiplicity" problem of applied inferential statistics.

Several modified DS-ECP strategies come to mind. One idea is to "strengthen" the broad assumption that allows all continuous population CDFs, by imposing greater smoothness, for example by limiting consideration to population distributions with convex log probability density function. If the culprit is that invariance over all monotone continuous transforms of the scale of the observed variable is too much, then maybe back off to just linear invariance as implied by a convexity assumption. Alternatively, if it is desired to retain invariance over all monotone transforms, then the auxiliaries can be "weakened" by requiring "weakened" auxiliary "don't know" terms to apply across ranges of intervals between data points. The result would be bets with increased "don't know" that could help protect the broker against bankruptcy.

24.6 Open areas for research

Many opportunities exist for both modifying state space structures through alternative choices that delete some variables and add others. The robustness of simpler models can be studied when only weak or nonexistent personal probability restrictions can be supported by evidence concerning the effects of additional model complexity. Many DS models that mimic standard multiparameter models can be subjected to strengthening or weakening modifications. In particular, DS methods are easily extended to discount for cherry-picking among multiple inferences. There is no space here to survey a broad range of stochastic systems and related DS models that can be re-expressed and modified in DS terms.

DS versions of decision theory merit systematic development and study. Decision analysis assumes a menu of possible actions each of which is associated with a real-valued utility function defined over the state space structure (SSS). Given an evidence-based posterior mass distribution over the SSS, each possible action has an associated lower expectation and upper expectation defined in an obvious way. The lower expectation is interpreted as "your" guaranteed expected returns from choosing alternative actions, so is a reasonable criterion for "optimal" decision-making.

In the case of simple bets, two or more players compete for a defined prize with their own DS mass functions and with the same utility function on the same SSS. Here, Borel's celebrated observation applies:

> "It has been said that, given two bettors, there is always one thief and one imbecile; that is true in certain cases when one of the two bettors is much better informed than the other, and knows it; but it

can easily happen that two men of good faith, in complex matters where they possess exactly the same elements of information, arrive at different conclusions on the probabilities of an event, and that in betting together, each believes... that it is he who is the thief and the other the imbecile." (Borel, 1924)

A typical modern application involves choices among different investment opportunities, through comparisons of DS posterior lower expectations of future monetary gains for buyers or losses for sellers, and for risk-taking brokers who quote bid/ask spreads while bringing buyers and sellers together. For mathematical statisticians, the field of DS decision analysis is wide open for investigations of models and analyses, of interest both mathematically and for practical use. For the latter in particular there are many potentially useful models to be defined and studied, and to be implemented numerically by software with acceptable speed, accuracy, and cost.

A third area of potential DS topics for research statisticians concerns modeling and inference for "parametric" models, as originally formulated by R.A. Fisher in his celebrated pair of 1920s papers on estimation. The concept of a statistical parameter is plagued by ambiguity. I believe that the term arose in parallel with similar usage in physics. For example, the dynamics of physical systems are often closely approximated by classical Newtonian physical laws, but application of the laws can depend on certain "parameters" whose actual numerical values are left "to be determined." In contemporary branches of probabilistic statistical sciences, stochastic models are generally described in terms of parameters similarly left "to be determined" prior to direct application. The mathematics is clear, but the nature of the task of parameter determination for practical application is murky, and in statistics is a chief source of contention between frequentists and Bayesians.

In many stochastic sampling models, including many pioneered by Fisher, parameters such as population fractions, means, and standard deviations, can actually represent specific unknown real world population quantities. Oftentimes, however, parameters are simply ad hoc quantities constructed on the fly while "fitting" mathematical forms to data. To emphasize the distinction, I like to denote parameters of the former type by Roman capital letters such as P, M, and S, while denoting analogous "parameters" fitted to data by Greek letters π, μ and σ. The distinction here is important, because the parameters of personalist statistical science draw on evidence and utilities that can only be assessed one application at a time. "Evidence-based" assumptions, in particular, draw upon many types of information and experience.

Very little published research exists that is devoted to usable personalist methodology and computational software along the lines of the DS standard protocol. Even the specific DS sampling model for inference about a population with k exchangeable categories that was proposed in my initial 1966 paper in *The Annals of Mathematical Statistics* has not yet been implemented and analyzed beyond the trivially simple case of $k = 2$. I published a brief report in 2008 on estimates and tests for a Poisson parameter L, which is a limiting case of the binomial P. I have given eight or so seminars and lectures in recent years, more in Europe than in North America, and largely outside of the field of statistics proper. I have conducted useful correspondences with statisticians, most notably Paul Edlefsen, Chuanhai Liu, and Jonty Rougier, for which I am grateful.

Near the end of my 1966 paper I sketched an argument to the effect that a limiting case of my basic multinomial model for data with k exchangeable cells, wherein the observables become continuous in the limit, leads to defined limiting DS inferences concerning the parameters of widely adopted parametric sampling models. This theory deserves to pass from conjecture to theorem, because it bears on how specific DS methods based on sampling models generalize traditional methods based on likelihood functions that have been studied by mathematical statisticians for most of the 20th century. Whereas individual likelihood functions from each of the n data points in a random sample multiply to yield the combined likelihood function for a complete random sample of size n, the generalizing DS mass distributions from single observations combine under the DS combination rule to provide the mass distribution for DS inference under the full sample. In fact, values of likelihood functions are seen to be identical to "upper probabilities" $p+r$ obtained from DS (p, q, r) inferences for singleton subsets of the parameters. Ordinary likelihood functions are thus seen to provide only part of the information in the data. What they lack is information associated with the "don't know" feature of the extended calculus of probability.

Detailed connections between the DS system and its predecessors invite illuminating research studies. For example, mass distributions that generalize likelihood functions from random samples combine under the DS rule of combination with prior mass distributions to provide DS inferences that generalize traditional Bayesian inferences. The use of the term "generalize" refers specifically to the recognition that when the DS prior specializes to an ordinary calculus of probability (OCP) prior, the DS combination rule reproduces the traditional "posterior $=$ prior $\times$ likelihood" rule of traditional Bayesian inference. In an important sense, the DS rule of combination is thus seen to generalize the OCP axiom that the joint probability of two events equals the marginal probability of one multiplied by the conditional probability of the other given the first.

I believe that an elegant theory of parametric inference is out there just waiting to be explained in precise mathematical terms, with the potential to render moot many of the confusions and controversies of the 20th century over statistical inference.

References

Borel, É. (1924). À propos d'un traité de probabilité. *Revue philosophique* 98. English translation: Apropos of a treatise on probability. *Studies in Subjective Probability* (H.E. Kyburg Jr. and H.E. Smokler, Eds.). Wiley, New York (1964).

Borel, É. (1939). *Valeur pratique et philosophie des probabilités.* Gauthiers-Villars, Paris.

Dempster, A.P. (1966). New methods for reasoning towards posterior distributions based on sample data. *The Annals of Mathematical Statistics*, 37:355–374.

Dempster, A.P. (1967). Upper and lower probabilities induced by a multivalued mapping. *The Annals of Mathematical Statistics*, 38:325–339.

Dempster, A.P. (1968). A generalization of Bayesian inference (with discussion). *Journal of the Royal Statistical Society, Series B*, 30:205–247.

Dempster, A.P. (1990). Bayes, Fisher, and belief functions. In *Bayesian and Likelihood Methods in Statistics and Econometrics: Essays in Honor of George Barnard* (S. Geisser, J.S. Hodges, S.J. Press, and A. Zellner, Eds.). North-Holland, Amsterdam.

Dempster, A.P. (1990). Construction and local computation aspects of networked belief functions. In *Influence Diagrams, Belief Nets and Decision Analysis* (R.M. Oliver and J.Q. Smith, Eds.). Wiley, New York.

Dempster, A.P. (1998). Logicist statistics I. Models and modeling. *Statistical Science*, 13:248–276.

Dempster, A.P. (2008). Logicist Statistics II. *Inference.* (Revised version of the 1998 COPSS Fisher Memorial Lecture) Classic Works of Dempster–Shafer Theory of Belief Functions (L. Liu and R.R. Yager, Eds.).

Dempster, A.P. (2008). The Dempster–Shafer calculus for statisticians. *International Journal of Approximate Reasoning*, 48:365–377.

Dempster, A.P. and Chiu, W.F. (2006). Dempster–Shafer models for object recognition and classification. *International Journal of Intelligent Systems*, 21:283–297.

Fisher, R.A. (1922). On the mathematical foundations of theoretical statistics. *Philosophical Transactions, Series A*, 222:309–368.

Fisher, R.A. (1925). Theory of statistical estimation. *Proceedings of the Cambridge Philosophical Society*, 22:200–225.

Fisher, R.A. (1939). "Student." *Annals of Eugenics*, 9:1–9.

Keynes, J.M. (1937). The general theory of employment. *Quarterly Journal of Economics*, 51:209–223.

Pearson, K. (1920). The fundamental problem of practical statistics. *Biometrika*, 20:1–16.

Savage, L.J. (1950). The role of personal probability in statistics. *Econometrica*, 18:183–184.

Savage, L.J. (1981). *The Writings of Leonard Jimmie Savage — A Memorial Selection.* Published by the American Statistical Association and the Institute of Mathematical Statistics, Washington, DC.

Shafer, G. (1976). *A Mathematical Theory of Evidence.* Princeton University Press, Princeton, NJ.

Wilks, S.S. (1948). Order statistics. *Bulletin of the American Mathematical Society*, 54:6–48.

25

Nonparametric Bayes

David B. Dunson
Department of Statistical Science
Duke University, Durham, NC

I reflect on the past, present, and future of nonparametric Bayesian statistics. Current nonparametric Bayes research tends to be split between theoretical studies, seeking to understand relatively simple models, and machine learning, defining new models and computational algorithms motivated by practical performance. I comment on the current landscape, open problems and promising future directions in modern big data applications.

25.1 Introduction

25.1.1 Problems with parametric Bayes

In parametric Bayesian statistics, one chooses a likelihood function $L(y|\theta)$ for data y, which is parameterized in terms of a finite-dimensional unknown θ. Choosing a prior distribution for θ, one updates this prior with the likelihood $L(y|\theta)$ via Bayes' rule to obtain the posterior distribution $\pi(\theta|y)$ for θ. This framework has a number of highly appealing characteristics, ranging from flexibility to the ability to characterize uncertainty in θ in an intuitively appealing probabilistic manner. However, one unappealing aspect is the intrinsic assumption that the data were generated from a particular probability distribution (e.g., a Gaussian linear model).

There are a number of challenging questions that arise in considering, from both philosophical and practical perspectives, what happens when such an assumption is violated, as is arguably always the case in practice. From a philosophical viewpoint, if one takes a parametric Bayesian perspective, then a prior is being assumed that has support on a measure zero subset of the set of possible distributions that could have generated the data. Of course, as it is commonly accepted that all models are wrong, it seems that such a prior does not actually characterize any individual's prior beliefs, and one may question the meaning of the resulting posterior from a subjective Bayes perspective. It would seem that a rational subjectivist would assign positive prior probability to the case in which the presumed parametric model is wrong in unanticipated ways, and probability zero to the case in which the data are generated *exactly* from the presumed model. Objective Bayesians should similarly acknowledge that any parametric model is wrong, or at least has a positive probability of being wrong, in order to truly be objective. It seems odd to spend an enormous amount of effort showing that a particular prior satisfies various objectivity properties in a simple parametric model, as has been the focus of much of the Bayes literature.

The failure to define a framework for choosing priors in parametric models, which acknowledge that the "working" model is wrong, leads to some clear practical issues with parametric Bayesian inference. One of the major ones is the lack of a framework for model criticism and goodness-of-fit assessments. Parametric Bayesians assume prior knowledge of the true model which generated the data, and hence there is no allowance within the Bayesian framework for incorrect model choice. For this reason, the literature on Bayesian goodness-of-fit assessments remains under-developed, with most of the existing approaches relying on diagnostics that lack a Bayesian justification. A partial solution is to place a prior distribution over a list of possible models instead of assuming a single model is true *a priori*. However, such Bayesian model averaging/selection approaches assume that the true model is one of those in the list, the so-called M-closed viewpoint, and hence do not solve the fundamental problem.

An alternative pragmatic view is that it is often reasonable to operate under the working assumption that the presumed model is true. Certainly, parametric Bayesian and frequentist inferences often produce excellent results even when the true model deviates from the assumptions. In parametric Bayesian models, it tends to be the case that the posterior distribution for the unknown θ will concentrate at the value θ_0, which yields a sampling distribution that is as close as possible to the true data-generating model in terms of the Kullback–Leibler (KL) divergence. As long as the parametric model provides an "adequate" approximation, and this divergence is small, it is commonly believed that inferences will be "reliable." However, there has been some research suggesting that this common belief is often wrong, such as when the loss function is far from KL (Owhadi et al., 2013).

Results of this type have provided motivation for "quasi" Bayesian approaches, which replace the likelihood with other functions (Chernozhukov and Hong, 2003). For example, quantile-based substitution likelihoods have been proposed, which avoid specifying the density of the data between quantiles (Dunson and Taylor, 2005). Alternatively, motivated by avoiding specification of parametric marginal distributions in considering copula dependence models (Genest and Favre, 2007; Hoff, 2007; Genest and Nešlehová, 2012; Murray et al., 2013), use an extended rank-based likelihood. Recently, the idea of a Gibbs posterior (Jiang and Tanner, 2008; Chen et al., 2010) was introduced, providing a generalization of Bayesian inference using a loss-based pseudo likelihood. Appealing properties of this approach have been shown in various contexts, but it is still unclear whether such methods are appropriately calibrated so that the quasi posterior distributions obtained provide a valid measure of uncertainty. It may be the case that uncertainty intervals are systematically too wide or too narrow, with asymptotic properties such as consistency providing no reassurance that uncertainty is well characterized.

Fully Bayesian nonparametric methods require a full characterization of the likelihood, relying on models with infinitely-many parameters having carefully chosen priors that yield desirable properties. In the remainder of this chapter, I focus on such approaches.

25.1.2 What is nonparametric Bayes?

Nonparametric (NP) Bayes seeks to solve the above problems by choosing a highly flexible prior, which assigns positive probability to arbitrarily small neighborhoods around any true data-generating model f_0 in a large class. For example, as an illustration, consider the simple case in which $y_1, \ldots, y_n$ form a random sample from density f. A parametric Bayes approach would parameterize the density f in terms of finitely-many unknowns θ, and induce a prior for f through a prior for θ. Such a prior will in general have support on a vanishingly small subset of the set of possible densities $\mathcal{F}$ (e.g., with respect to Lebesgue measure on $\mathbb{R}$). NP Bayes instead lets $f \sim \Pi$, with Π a prior over $\mathcal{F}$ having large support, meaning that $\Pi\{f : d(f, f_0) < \epsilon\} > 0$ for some distance metric d, any $\epsilon > 0$, and any f_0 in a large subset of $\mathcal{F}$. Large support is the defining property of an NP Bayes approach, and means that realizations from the prior have a positive probability of being arbitrarily close to any f_0, perhaps ruling out some irregular ones (say with heavy tails).

In general, to satisfy the large support property, NP Bayes probability models include infinitely-many parameters and involve specifying stochastic processes for random functions. For example, in the density estimation example, a very popular prior is a Dirichlet process mixture (DPM) of Gaussians (Lo, 1984). Under the stick-breaking representation of the Dirichlet process (Sethuraman, 1994), such a prior lets

$$f(y) = \sum_{h=1}^{\infty} \pi_h \mathcal{N}(y; \mu_h, \tau_h^{-1}), \quad (\mu_h, \tau_h) \overset{\text{iid}}{\sim} P_0, \tag{25.1}$$

where the weights on the normal kernels follow a stick-breaking process, $\pi_h = V_h \prod_{\ell<h}(1 - V_\ell)$, with $V_h \sim \mathcal{B}(1, \alpha)$ independently, α is the concentration parameter in the Dirichlet process, and P_0 is the base measure. Such kernel mixture priors satisfy the large support property and can be defined so that the resulting posterior concentrates around the unknown true f_0 at the minimax optimal rate up to a log factor (de Jonge and van Zanten, 2010).

Prior (25.1) is intuitively appealing in including infinitely-many Gaussian kernels having stochastically decreasing weights. In practice, there will tend to be a small number of kernels having large weights, with the remaining having vanishingly small weights. Only a modest number of kernels will be occupied by the subjects in a sample, so that the effective number of parameters may actually be quite small and is certainly not infinite, making posterior computation and inferences tractable. Of course, prior (25.1) is only one particularly simple example (to quote Andrew Gelman, "No one is really interested in density estimation"), and there has been an explosion of literature in recent years proposing an amazing variety of NP Bayes models for broad applications and data structures.

Section 25.2 contains an (absurdly) incomplete timeline of the history of NP Bayes up through the present. Section 25.3 comments briefly on interesting future directions.

25.2 A brief history of NP Bayes

Although there were many important earlier developments, the modern view of nonparametric Bayes statistics was essentially introduced in the papers of Ferguson (1973, 1974), which proposed the Dirichlet process (DP) along with several ideal criteria for a nonparametric Bayes approach including large support, interpretability and computational tractability. The DP provides a prior for a discrete probability measure with infinitely many atoms, and is broadly employed within Bayesian models as a prior for mixing distributions and for clustering. An equally popular prior is the Gaussian process (GP), which is instead used for random functions or surfaces. A non-neglible proportion of the nonparametric Bayes literature continues to focus on theoretical properties, computational algorithms and applications of DPs and GPs in various contexts.

In the 1970s and 1980s, NP Bayes research was primarily theoretical and conducted by a narrow community, with applications focused primarily on jointly conjugate priors, such as simple cases of the gamma process, DP and GP. Most research did not consider applications or data analysis at all, but instead delved into characterizations and probabilistic properties of stochastic processes, which could be employed as priors in NP Bayes models. These developments later had substantial applied implications in facilitating computation and the development of richer model classes.

With the rise in computing power, development of Gibbs sampling and explosion in use of Markov chain Monte Carlo (MCMC) algorithms in the early 1990s, nonparametric Bayes methods started to become computationally tractable. By the late 1990s and early 2000s, there were a rich variety of inferential algorithms available for general DP mixtures and GP-based models in spatial statistics, computer experiments and beyond. These algorithms, combined with increasing knowledge of theoretical properties and characterizations, stimulated an explosion of modeling innovation starting in the early 2000s but really gaining steam by 2005. A key catalyst in this exponential growth of research activity and innovation in NP Bayes was the dependent Dirichlet process (DDP) of Steve MacEachern, which ironically was never published and is only available as a technical report. The DDP and other key modeling innovations were made possible by earlier theoretical work providing characterizations, such as stick-breaking (Sethuraman, 1994; Ishwaran and James, 2001) and the Polya urn scheme/Chinese restaurant process (Blackwell and MacQueen, 1973). Some of the circa 2005–10 innovations include the Indian buffet process (IBP) (Griffiths and Ghahramani, 2011), the hierarchical Dirichlet process (HDP) (Teh et al., 2006), the nested Dirichlet process (Rodríguez et al., 2008), and the kernel stick-breaking process (Dunson and Park, 2008).

One of the most exciting aspects of these new modeling innovations was the potential for major applied impact. I was fortunate to start working on NP Bayes just as this exponential growth started to take off. In the NP Bayes statistics community, this era of applied-driven modeling innovation peaked at the 2007 NP Bayes workshop at the Issac Newton Institute at Cambridge University. The Newton Institute is an outstanding facility and there was an energy and excited vibe permeating the workshop, with a wide variety of topics being covered, ranging from innovative modeling driven by biostatistical applications to theoretical advances on properties. One of the most exciting aspects of statistical research is the ability to fully engage in a significant applied problem, developing methods that really make a practical difference in inferences or predictions in the motivating application, as well as in other related applications. To me, it is ideal to start with an applied motivation, such as an important aspect of the data that is not captured by existing statistical approaches, and then attempt to build new models and computational algorithms that have theoretical support and make a positive difference to the bottom-line answers in the analysis. The flexibility of NP Bayes models makes this toolbox ideal for attacking challenging applied problems.

Although the expansion of the NP Bayes community and impact of the research has continued since the 2007 Newton workshop, the trajectory and flavor of the work has shifted substantially in recent years. This shift is due in part to the emergence of big data and to some important cultural hurdles, which have slowed the expansion of NP Bayes in statistics and scientific applications, while stimulating increasing growth in machine learning. Culturally, statisticians tend to be highly conservative, having a healthy skepticism of new approaches even if they seemingly improve practical performance in prediction and simulation studies. Many statisticians will not really trust an approach that lacks asymptotic justification, and there is a strong preference for simple methods that can be studied and understood more easily. This is perhaps one reason for the enormous statistical literature on minor variations of the lasso.

NP Bayes methods require more of a learning curve. Most graduate programs in statistics have perhaps one elective course on Bayesian statistics, and NP Bayes is not a simple conceptual modification of parametric Bayes. Often models are specified in terms of infinite-dimensional random probability measures and stochastic processes. On the surface, this seems daunting and the knee-jerk reaction by many statisticians is negative, mentioning unnecessary complexity, concerns about over-fitting, whether the data can really support such complexity, lack of interpretability, and limited understanding of theoretical properties such as asymptotic behavior. This reaction restricts entry into the field and makes it more difficult to get publications and grant funding.

However, these concerns are largely unfounded. In general, the perceived complexity of NP Bayes models is due to lack of familiarity. Canonical model classes, such as DPs and GPs, are really quite simple in their structure and tend to be no more difficult to implement than flexible parametric models. The intrinsic Bayesian penalty for model complexity tends to protect against over-fitting. For example, consider the DPM of Gaussians for density estimation shown in equation (25.1). The model is simple in structure, being a discrete mixture of normals, but the perceived complexity comes in through the incorporation of infinitely many components. For statisticians unfamiliar with the intricacies of such models, natural questions arise such as "how can the data inform about all these parameters" and "there certainly must be over-fitting and huge prior sensitivity." However, in practice, the prior and the penalty that comes in through integrating over the prior in deriving the marginal likelihood tends to lead to allocation of all the individuals in the sample to relatively few clusters. Hence, even though there are infinitely many components, only a few of these are used and the model behaves like a finite mixture of Gaussians, with sieve behavior in terms of using more components as the sample size increases. Contrary to the concern about over-fitting, the tendency is instead to place a high posterior weight on very few components, potentially under-fitting in small sample sizes. DPMs are a simple example but the above story applies much more broadly.

The lack of understanding in the broad statistical community of the behavior of NP Bayes procedures tempered some of the enthusiastic applications-driven modeling of the 2000s, motivating an emerging field focused on studying frequentist asymptotic properties. There is a long history of NP Bayes asymptotics, showing properties such as consistency and rates of concentration of the posterior around the true unknown distribution or function. In the past five years, this field has really taken off and there is now a rich literature showing strong properties ranging from minimax optimal adaptive rates of posterior concentration (Bhattacharya et al., 2013) to Bernstein–von Mises results characterizing the asymptotic distribution of functionals (Rivoirard and Rousseau, 2012). Such theorems can be used to justify many NP Bayes methods as also providing an optimal frequentist procedure, while allowing frequentist statisticians to exploit computational methods and probabilistic interpretations of Bayes methods. In addition, an appealing advantage of NP Bayes nonparametric methods is the allowance for uncertainty in tuning parameter choice through hyperpriors, bypassing the need for cross-validation. The 2013 NP Bayes conference in Amsterdam was notable in exhibiting a dramatic shift in topics compared with the 2007 Newton conference, away from applications-driven modeling and towards asymptotics.

The other thread that was very well represented in Amsterdam was NP Bayes machine learning, which has expanded into a dynamic and important area. The machine learning (ML) community is fundamentally different culturally from statistics, and has had a very different response to NP Bayes methods as a result. In particular, ML tends to be motivated by applications in which bottom-line performance in metrics, such as out-of-sample prediction, takes center stage. In addition, the ML community prefers peer-reviewed proceedings for conferences, such as Neural Information Processing Systems (NIPS) and the International Conference on Machine Learning Research (ICML), over journal publications. These conference proceedings are short papers, and there is an emphasis on innovative new ideas which improve bottom line performance. ML researchers tend to be aggressive and do not shy away from new approaches which can improve performance regardless of complexity. A substantial proportion of the novelty in NP Bayes modeling and computation has come out of the ML community in recent years. With the increased emphasis on big data across fields, the lines between ML and statistics have been blurring. However, publishing an initial idea in NIPS or ICML is completely different than publishing a well-developed and carefully thought out methods paper in a leading statistical theory and methods journal, such as the *Journal of the American Statistical Association*, *Biometrika* or the *Journal of the Royal Statistical Society, Series B*. My own research has greatly benefited by straddling the asymptotic, ML and applications-driven modeling threads, attempting to develop practically useful and innovative new NP Bayes statistical methods having strong asymptotic properties.

25.3 Gazing into the future

Moving into the future, NP Bayes methods have rich promise in terms of providing a framework for attacking a very broad class of 'modern' problems involving high-dimensional and complex data. In big complex data settings, it is much more challenging to do model checking and to carefully go through the traditional process of assessing the adequacy of a parametric model, making revisions to the model as appropriate. In addition, when the number of variables is really large, it becomes unlikely that a particular parametric model works well for all these variables. This is one of the reasons that ensemble approaches, which average across many models/algorithms, tend to produce state of the art performance in difficult prediction tasks. Combining many simple models, each able to express different characteristics of the data, is useful and similar conceptually to the idea of Bayesian model averaging (BMA), though BMA is typically only implemented within a narrow parametric class (e.g., normal linear regression).

In considering applications of NP Bayes in big data settings, several questions arise. The first is "Why bother?" In particular, what do we have to gain over the rich plethora of machine learning algorithms already available, and which are being refined and innovated upon daily by thousands of researchers? There are clear and compelling answers to this question. ML algorithms almost always rely on convex optimization to obtain a point estimate, and uncertainty is seldom of much interest in the ML community, given the types of applications they are faced with. In contrast, in most scientific applications, prediction is not the primary interest and one is usually focused on inferences that account for uncertainty. For example, the focus may be on assessing the conditional independence structure (graphical model) relating genetic variants, environmental exposures and cardiovascular disease outcomes (an application I'm currently working on). Obtaining a single estimate of the graph is clearly not sufficient, and would be essentially uninterpretable. Indeed, such graphs produced by ML methods such as graphical lasso have been deemed "ridiculograms." They critically depend on a tuning parameter that is difficult to choose objectively and produce a massive number of connections that cannot be effectively examined visually. Using an NP Bayes approach, we could instead make highly useful statements (at least according to my collaborators), such as (i) the posterior probability that genetic variants in a particular gene are associated with cardiovascular disease risk, adjusting for other factors, is $P\%$; or (ii) the posterior probability that air pollution exposure contributes to risk, adjusted for genetic variants and other factors, is $Q\%$. We can also obtain posterior probabilities of an edge between each variable without parametric assumptions, such as Gaussianity. This is just one example of the utility of probabilistic NP Bayes models; I could list dozens of others.

The question then is why aren't more people using and working on the development of NP Bayes methods? The answer to the first part of this question is clearly computational speed, simplicity and accessibility. As mentioned above, there is somewhat of a learning curve involved in NP Bayes, which is not covered in most graduate curriculums. In contrast, penalized optimization methods, such as the lasso, are both simple and very widely taught. In addition, convex optimization algorithms for very rapidly implementing penalized optimization, especially in big data settings, have been highly optimized and refined in countless publications by leading researchers. This has led to simple methods that are scalable to big data, and which can exploit distributed computing architectures to further scale up to enormous settings. Researchers working on these types of methods often have a computer science or engineering background, and in the applications they face, speed is everything and characterizing uncertainty in inference or testing is just not a problem they encounter. In fact, ML researchers working on NP Bayes methods seldom report inferences or use uncertainty in their analyses; they instead use NP Bayes methods combined with approximations, such as variational Bayes or expectation propagation, to improve performance on ML tasks, such as prediction. Often predictive performance can be improved, while avoiding cross-validation for tuning parameter selection, and these gains have partly led to the relative popularity of NP Bayes in machine learning.

It is amazing to me how many fascinating and important unsolved problems remain in NP Bayes, with the solutions having the potential to substantially impact practice in analyzing and interpreting data in many fields. For example, there is no work on the above nonparametric Bayes graphical modeling problem, though we have developed an initial approach we will submit for publication soon. There is very limited work on fast and scalable approximations to the posterior distribution in Bayesian nonparametric models. Markov chain Monte Carlo (MCMC) algorithms are still routinely used despite their problems with scalability due to the lack of decent alternatives. Variational Bayes and expectation propagation algorithms developed in ML lack theoretical guarantees and often perform poorly, particularly when the focus goes beyond obtaining a point estimate for prediction. Sequential Monte Carlo (SMC) algorithms face similar scalability problems to MCMC, with a daunting number of particles needed to obtain adequate approximations for high-dimensional models. There is a clear need for new models for flexible dimensionality reduction in broad settings. There is a clear lack of approaches for complex non-Euclidean data structures, such as shapes, trees, networks and other object data.

I hope that this chapter inspires at least a few young researchers to focus on improving the state of the art in NP Bayes statistics. The most effective path to success and high impact in my view is to focus on challenging real-world applications in which current methods have obvious inadequacies. Define innovative probability models for these data, develop new scalable approximations and computational algorithms, study the theoretical properties, implement the methods on real data, and provide software packages for routine use. Given how few people are working in such areas, there are many low hanging fruit and the clear possibility of major breakthroughs, which are harder to achieve when jumping on bandwagons.

References

Bhattacharya, A., Pati, D., and Dunson, D.B. (2013). Anisotropic function estimation using multi-bandwidth Gaussian processes. *The Annals of Statistics*, in press.

Blackwell, D. and MacQueen, J. (1973). Ferguson distributions via Pólya urn schemes. *The Annals of Statistics*, 1:353–355.

Chen, K., Jiang, W., and Tanner, M. (2010). A note on some algorithms for the Gibbs posterior. *Statistics & Probability Letters*, 80:1234–1241.

Chernozhukov, V. and Hong, H. (2003). An MCMC approach to classical estimation. *Journal of Econometrics*, 115:293–346.

de Jonge, R. and van Zanten, J. (2010). Adaptive nonparametric Bayesian inference using location-scale mixture priors. *The Annals of Statistics*, 38:3300–3320.

Dunson, D.B. and Park, J.-H. (2008). Kernel stick-breaking processes. *Biometrika*, 95:307–323.

Dunson, D.B. and Taylor, J. (2005). Approximate Bayesian inference for quantiles. *Journal of Nonparametric Statistics*, 17:385–400.

Ferguson, T.S. (1973). Bayesian analysis of some nonparametric problems. *The Annals of Statistics*, 1:209–230.

Ferguson, T.S. (1974). Prior distributions on spaces of probability measures. *The Annals of Statistics*, 2:615–629.

Genest, C. and Favre, A.-C. (2007). Everything you always wanted to know about copula modeling but were afraid to ask. *Journal of Hydrologic Engineering*, 12:347–368.

Genest, C. and Nešlehová, J. (2012). Copulas and copula models. *Encyclopedia of Environmetrics*, 2nd edition. Wiley, Chichester, 2:541–553.

Griffiths, T. and Ghahramani, Z. (2011). The Indian buffet process: An introduction and review. *Journal of Machine Learning Research*, 12:1185–1224.

Hoff, P. (2007). Extending the rank likelihood for semi parametric copula estimation. *The Annals of Applied Statistics*, 1:265–283.

Ishwaran, H. and James, L. (2001). Gibbs sampling methods for stick-breaking priors. *Journal of the American Statistical Association*, 96:161–173.

Jiang, W. and Tanner, M. (2008). Gibbs posterior for variable selection in high-dimensional classification and data mining. *The Annals of Statistics*, 36:2207–2231.

Lo, A. (1984). On a class of Bayesian nonparametric estimates. 1. density estimates. *The Annals of Statistics*, 12:351–357.

Murray, J.S., Dunson, D.B., Carin, L., and Lucas, J.E. (2013). Bayesian Gaussian copula factor models for mixed data. *Journal of the American Statistical Association*, 108:656–665.

Owhadi, H., Scovel, C., and Sullivan, T. (2013). Bayesian brittleness: Why no Bayesian model is "good enough." *arXiv:1304.6772* .

Rivoirard, V. and Rousseau, J. (2012). Bernstein–von Mises theorem for linear functionals of the density. *The Annals of Statistics*, 40:1489–1523.

Rodríguez, A., Dunson, D.B., and Gelfand, A. (2008). The nested Dirichlet process. *Journal of the American Statistical Association*, 103:1131–1144.

Sethuraman, J. (1994). A constructive definition of Dirichlet priors. *Statistica Sinica*, 4:639–650.

Teh, Y., Jordan, M., Beal, M., and Blei, D. (2006). Hierarchical Dirichlet processes. *Journal of the American Statistical Association*, 101:1566–1581.

26

How do we choose our default methods?

Andrew Gelman
Department of Statistics
Columbia University, New York

The field of statistics continues to be divided into competing schools of thought. In theory one might imagine choosing the uniquely best method for each problem as it arises, but in practice we choose for ourselves (and recommend to others) default principles, models, and methods to be used in a wide variety of settings. This chapter briefly considers the informal criteria we use to decide what methods to use and what principles to apply in statistics problems.

26.1 Statistics: The science of defaults

Applied statistics is sometimes concerned with one-of-a-kind problems, but statistical methods are typically intended to be used in routine practice. This is recognized in classical theory (where statistical properties are evaluated based on their long-run frequency distributions) and in Bayesian statistics (averaging over the prior distribution). In computer science, machine learning algorithms are compared using cross-validation on benchmark corpuses, which is another sort of reference distribution. With good data, a classical procedure should be robust and have good statistical properties under a wide range of frequency distributions, Bayesian inferences should be reasonable even if averaging over alternative choices of prior distribution, and the relative performance of machine learning algorithms should not depend strongly on the choice of corpus.

How do we, as statisticians, decide what default methods to use? Here I am using the term "method" broadly, to include general approaches to statistics (e.g., Bayesian, likelihood-based, or nonparametric) as well as more specific choices of models (e.g., linear regression, splines, or Gaussian processes) and options within a model or method (e.g., model averaging, L_1 regularization, or hierarchical partial pooling). There are so many choices that it is hard to imagine any statistician carefully weighing the costs and benefits of each before deciding how to solve any given problem. In addition, given the existence of multiple competing approaches to statistical inference and decision making, we can deduce that no single method dominates the others.

Sometimes the choice of statistical philosophy is decided by convention or convenience. For example, I recently worked as a consultant on a legal case involving audits of several random samples of financial records. I used the classical estimate $\hat{p} = y/n$ with standard error $\sqrt{\hat{p}(1-\hat{p})/n}$, switching to $\hat{p} = (y+2)/(n+4)$ for cases where $y = 0$ or $y = n$. This procedure is simple, gives reasonable estimates with good confidence coverage, and can be backed up by a solid reference, namely Agresti and Coull (1998), which has been cited over 1000 times according to Google Scholar. If we had been in a situation with strong prior knowledge on the probabilities p, or interest in distinguishing between $p = .99$, .999, and .9999, it would have made sense to consider something closer to a full Bayesian approach, but in this setting it was enough to know that the probabilities were high, and so the simple $(y+2)/(n+4)$ estimate (and associated standard error) was fine for our data, which included values such as $y = n = 75$.

In many settings, however, we have freedom in deciding how to attack a problem statistically. How then do we decide how to proceed?

Schools of statistical thoughts are sometimes jokingly likened to religions. This analogy is not perfect — unlike religions, statistical methods have no supernatural content and make essentially no demands on our personal lives. Looking at the comparison from the other direction, it is possible to be agnostic, atheistic, or simply live one's life without religion, but it is not really possible to do statistics without some philosophy. Even if you take a Tukeyesque stance and admit only data and data manipulations without reference to probability models, you still need some criteria to evaluate the methods that you choose.

One way in which schools of statistics *are* like religions is in how we end up affiliating with them. Based on informal observation, I would say that statisticians typically absorb the ambient philosophy of the institution where they are trained — or else, more rarely, they rebel against their training or pick up a philosophy later in their career or from some other source such as a persuasive book. Similarly, people in modern societies are free to choose their religious affiliation but it typically is the same as the religion of parents and extended family. Philosophy, like religion but not (in general) ethnicity, is something we are free to choose on our own, even if we do not usually take the opportunity to take that choice. Rather, it is common to exercise our free will in this setting by forming our own personal accommodation with the religion or philosophy bequeathed to us by our background.

For example, I affiliated as a Bayesian after studying with Don Rubin and, over the decades, have evolved my own philosophy using his as a starting point. I did not go completely willingly into the Bayesian fold — the first statistics course I took (before I came to Harvard) had a classical perspective, and in the first course I took with Don, I continued to try to frame all the inferential problems into a Neyman–Pearson framework. But it didn't take me or my fellow students long to slip into comfortable conformity.

My views of Bayesian statistics have changed over the years — in particular, I have become much more fond of informative priors than I was during the writing of the first two editions of *Bayesian Data Analysis* (published 1995 and 2004) — and I went through a period of disillusionment in 1991, when I learned to my dismay that most of the Bayesians at the fabled Valencia meeting had no interest in checking the fit of their models. In fact, it was a common view among Bayesians at the time that it was either impossible, inadvisable, or inappropriate to check the fit of a model to data. The idea was that the prior distribution and the data model were subjective and thus uncheckable. To me, this attitude seemed silly — if a model is generated subjectively, that would seem to be *more* of a reason to check it — and since then my colleagues and I have expressed this argument in a series of papers; see, e.g., Gelman et al. (1996), and Gelman and Shalizi (2012). I am happy to say that the prevailing attitude among Bayesians has changed, with some embracing posterior predictive checks and others criticizing such tests for their low power (see, e.g., Bayarri and Castellanos, 2007). I do not agree with that latter view: I think it confuses different aspects of model checking; see Gelman (2007). On the plus side, however, it represents an acceptance of the idea that Bayesian models can be checked.

But this is all a digression. The point I wanted to make here is that the division of statistics into parallel schools of thought, while unfortunate, has its self-perpetuating aspects. In particular, I can communicate with fellow Bayesians in a way that I sometimes have difficulty with others. For example, some Bayesians dislike posterior predictive checks, but non-Bayesians mostly seem to ignore the idea — even though Xiao-Li Meng, Hal Stern, and I wrote our paper in general terms and originally thought our methods might appeal more strongly to non-Bayesians. After all, those statisticians were already using p-values to check model fit, so it seemed like a small step to average over a distribution. But this was a step that, by and large, only Bayesians wanted to take. The reception of this article was what convinced me to focus on reforming Bayesianism from the inside rather than trying to develop methods one at a time that would make non-Bayesians happy.

26.2 Ways of knowing

How do we decide to believe in the effectiveness of a statistical method? Here are a few potential sources of evidence (I leave the list unnumbered so as not to imply any order of priority):

(a) mathematical theory (e.g., coherence of inference or convergence);

(b) computer simulations (e.g., demonstrating approximate coverage of interval estimates under some range of deviations from an assumed model);

(c) solutions to toy problems (e.g., comparing the partial pooling estimate for the eight schools to the no pooling or complete pooling estimates);

(d) improved performance on benchmark problems (e.g., getting better predictions for the Boston Housing Data);

(e) cross-validation and external validation of predictions;

(f) success as recognized in a field of application (e.g., our estimates of the incumbency advantage in congressional elections);

(g) success in the marketplace (under the theory that if people are willing to pay for something, it is likely to have something to offer).

None of these is enough on its own. Theory and simulations are only as good as their assumptions; results from toy problems and benchmarks don't necessarily generalize to applications of interest; cross-validation and external validation can work for some sorts of predictions but not others; and subject-matter experts and paying customers can be fooled.

The very imperfections of each of these sorts of evidence gives a clue as to why it makes sense to care about all of them. We can't know for sure so it makes sense to have many ways of knowing.

I do not delude myself that the methods I personally prefer have some absolute status. The leading statisticians of the twentieth century were Neyman, Pearson, and Fisher. None of them used partial pooling or hierarchical models (well, maybe occasionally, but not much), and they did just fine. Meanwhile, other statisticians such as myself use hierarchical models to partially pool as a compromise between complete pooling and no pooling. It is a big world, big enough for Fisher to have success with his methods, Rubin to have success with his, Efron to have success with his, and so forth. A few years ago (Gelman, 2010) I wrote of the *methodological attribution problem*:

> "The many useful contributions of a good statistical consultant, or collaborator, will often be attributed to the statistician's methods or philosophy rather than to the artful efforts of the statistician himself or herself. Don Rubin has told me that scientists are fundamentally Bayesian (even if they do not realize it), in that they interpret uncertainty intervals Bayesianly. Brad Efron has talked vividly about how his scientific collaborators find permutation tests and p-values to be the most convincing form of evidence. Judea Pearl assures me that graphical models describe how people really think about causality. And so on. I am sure that all these accomplished researchers, and many more, are describing their experiences accurately. Rubin wielding a posterior

distribution is a powerful thing, as is Efron with a permutation test or Pearl with a graphical model, and I believe that (a) all three can be helping people solve real scientific problems, and (b) it is natural for their collaborators to attribute some of these researchers' creativity to their methods.

The result is that each of us tends to come away from a collaboration or consulting experience with the warm feeling that our methods really work, and that they represent how scientists really think. In stating this, I am not trying to espouse some sort of empty pluralism — the claim that, for example, we would be doing just as well if we were all using fuzzy sets, or correspondence analysis, or some other obscure statistical method. There is certainly a reason that methodological advances are made, and this reason is typically that existing methods have their failings. Nonetheless, I think we all have to be careful about attributing too much from our collaborators' and clients' satisfaction with our methods."

26.3 The pluralist's dilemma

Consider the arguments made fifty years ago or so in favor of Bayesian inference. At that time, there were some applied successes (e.g., I.J. Good repeatedly referred to his successes using Bayesian methods to break codes in the Second World War) but most of the arguments in favor of Bayes were theoretical. To start with, it was (and remains) trivially (but not unimportantly) true that, conditional on the model, Bayesian inference gives the right answer. The whole discussion then shifts to whether the model is true, or, better, how the methods perform under the (essentially certain) condition that the model's assumptions are violated, which leads into the tangle of various theorems about robustness or lack thereof.

Forty or fifty years ago one of Bayesianism's major assets was its mathematical coherence, with various theorems demonstrating that, under the right assumptions, Bayesian inference is optimal. Bayesians also spent a lot of time writing about toy problems, for example, Basu's example of the weights of elephants (Basu, 1971). From the other direction, classical statisticians felt that Bayesians were idealistic and detached from reality.

How things have changed! To me, the key turning points occurred around 1970–80, when statisticians such as Lindley, Novick, Smith, Dempster, and Rubin applied hierarchical Bayesian modeling to solve problems in education research that could not be easily attacked otherwise. Meanwhile Box did similar work in industrial experimentation and Efron and Morris connected these approaches to non-Bayesian theoretical ideas. The key in any case was to use partial pooling to learn about groups for which there was only a small amount of local data.

Lindley, Novick, and the others came at this problem in several ways. First, there was Bayesian theory. They realized that, rather than seeing certain aspects of Bayes (for example, the need to choose priors) as limitations, they could see them as opportunities (priors can be estimated from data!) with the next step folding this approach back into the Bayesian formalism via hierarchical modeling. We (the Bayesian community) are still doing research on these ideas; see, for example, the recent paper by Polson and Scott (2012) on prior distributions for hierarchical scale parameters.

The second way that the Bayesians of the 1970s succeeded was by applying their methods on realistic problems. This is a pattern that has happened with just about every successful statistical method I can think of: an interplay between theory and practice. Theory suggests an approach which is modified in application, or practical decisions suggest a new method which is then studied mathematically, and this process goes back and forth.

To continue with the timeline: the modern success of Bayesian methods is often attributed to our ability using methods such as the Gibbs sampler and Metropolis algorithm to fit an essentially unlimited variety of models: practitioners can use programs such as Stan to fit their own models, and researchers can implement new models at the expense of some programming but without the need of continually developing new approximations and new theory for each model. I think that's right — Markov chain simulation methods indeed allow us to get out of the pick-your-model-from-the-cookbook trap — but I think the hierarchical models of the 1970s (which were fit using various approximations, not MCMC) showed the way.

Back 50 years ago, theoretical justifications were almost all that Bayesian statisticians had to offer. But now that we have decades of applied successes, that is naturally what we point to. From the perspective of Bayesians such as myself, theory is valuable (our Bayesian data analysis book is full of mathematical derivations, each of which can be viewed if you'd like as a theoretical guarantee that various procedures give correct inferences conditional on assumed models) but applications are particularly convincing. And applications can ultimately become good toy problems, once they have been smoothed down from years of teaching.

Over the years I have become pluralistic in my attitudes toward statistical methods. Partly this comes from my understanding of the history described above. Bayesian inference seemed like a theoretical toy and was considered by many leading statisticians as somewhere between a joke and a menace — see Gelman and Robert (2013) — but the hardcore Bayesians such as Lindley, Good, and Box persisted and got some useful methods out of it. To take a more recent example, the bootstrap idea of Efron (1979) is an idea that in some way is obviously wrong (as it assigns zero probability to data that did not occur, which would seem to violate the most basic ideas of statistical sampling) yet has become useful to many and has since been supported in many cases by theory.

In this discussion, I have the familiar problem that might be called the pluralist's dilemma: how to recognize that my philosophy is just one among many, that my own embrace of this philosophy is contingent on many things beyond my control, while still expressing the reasons why I believe this philosophy to be preferable to the alternatives (at least for the problems I work on).

One way out of the dilemma is to recognize that different methods are appropriate for different problems. It has been said that R.A. Fisher's methods and the associated 0.05 threshold for p-values worked particularly well for experimental studies of large effects with relatively small samples — the sorts of problems that appear over and over again in books of Fisher, Snedecor, Cochran, and their contemporaries. That approach might not work so well in settings with observational data and sample sizes that vary over several orders of magnitude. I will again quote myself (Gelman, 2010):

> "For another example of how different areas of application merit different sorts of statistical thinking, consider Rob Kass's remark: 'I tell my students in neurobiology that in claiming statistical significance I get nervous unless the p-value is much smaller than .01.' In political science, we are typically not aiming for that level of uncertainty. (Just to get a sense of the scale of things, there have been barely 100 national elections in all of US history, and political scientists studying the modern era typically start in 1946.)"

Another answer is path dependence. Once you develop facility with a statistical method, you become better at it. At least in the short term, I will be a better statistician using methods with which I am already familiar. Occasionally I will learn a new trick but only if forced to by circumstances. The same pattern can hold true with research: we are more equipped to make progress in a field along directions in which we are experienced and knowledgeable. Thus, Bayesian methods can be the most effective for me and my students, for the simple reason that we have already learned them.

26.4 Conclusions

Statistics is a young science in which progress is being made in many areas. Some methods in common use are many decades or even centuries old, but recent and current developments in nonparametric modeling, regularization, and multivariate analysis are central to state-of-the-art practice in many areas of applied statistics, ranging from psychometrics to genetics to predictive modeling in business and social science. Practitioners have a wide variety of statistical approaches to choose from, and researchers have many potential directions to study. A casual and introspective review suggests that there are many different criteria we use to decide that a statistical method is worthy of routine use. Those of us who lean on particular ways of knowing (which might include performance on benchmark problems, success in new applications, insight into toy problems, optimality as shown by simulation studies or mathematical proofs, or success in the marketplace) should remain aware of the relevance of all these dimensions in the spread of default procedures.

References

Agresti, A. and Coull, B.A. (1998). Approximate is better than exact for interval estimation of binomial proportions. *The American Statistician*, 52:119–126.

Basu, D. (1971). An essay on the logical foundations of survey sampling, part 1 (with discussion). In *Foundations of Statistical Inference* (V.P. Godambe and D.A. Sprott, Eds.). Holt, Reinhart and Winston, Toronto, pp. 203–242.

Box, G.E.P. and Tiao, G.C. (1973). *Bayesian Inference in Statistical Analysis*. Wiley, New York.

Dempster, A.P., Rubin, D.B., and Tsutakawa, R.K. (1981). Estimation in covariance components models. *Journal of the American Statistical Association*, 76:341–353.

Efron, B. (1979). Bootstrap methods: Another look at the jackknife. *The Annals of Statistics*, 7:1–26.

Efron, B. and Morris, C. (1975). Data analysis using Stein's estimator and its generalizations. *Journal of the American Statistical Association*, 70:311–319.

Gelman, A. (2010). Bayesian statistics then and now. Discussion of "The future of indirect evidence," by Bradley Efron. *Statistical Science*, 25:162–165.

Gelman, A., Meng, X.-L., and Stern, H.S. (1996). Posterior predictive assessment of model fitness via realized discrepancies (with discussion). *Statistica Sinica*, 6:733–807.

Gelman, A. and Robert, C.P. (2013). "Not only defended but also applied": The perceived absurdity of Bayesian inference (with discussion). *The American Statistician*, 67(1):1–5.

Gelman, A. and Shalizi, C. (2012). Philosophy and the practice of Bayesian statistics (with discussion). *British Journal of Mathematical and Statistical Psychology*, 66:8–38.

Lindley, D.V. and Novick, M.R. (1981). The role of exchangeability in inference. *The Annals of Statistics*, 9:45–58.

Lindley, D.V. and Smith, A.F.M. (1972). Bayes estimates for the linear model. *Journal of the Royal Statistical Society, Series B*, 34:1–41.

Polson, N.G. and Scott, J.G. (2012). On the half-Cauchy prior for a global scale parameter. *Bayesian Analysis*, 7(2):1–16.

Tukey, J.W. (1977). *Exploratory Data Analysis.* Addison-Wesley, Reading, MA.

27

Serial correlation and Durbin–Watson bounds

T.W. Anderson
Department of Economics and Department of Statistics
Stanford University, Stanford, CA

Consider the model $\boldsymbol{y} = \boldsymbol{X}\boldsymbol{\beta}+\boldsymbol{u}$, where $\boldsymbol{y}$ is an n-vector of dependent variables, $\boldsymbol{X}$ is a matrix of $n \times k$ independent variables, and $\boldsymbol{u}$ is a n-vector of unobserved disturbance. Let $\boldsymbol{z} = \boldsymbol{y} - \boldsymbol{X}\boldsymbol{b}$, where $\boldsymbol{b}$ is the least squares estimate of $\boldsymbol{\beta}$. The d-statistic tests the hypothesis that the components of $\boldsymbol{u}$ are independent versus the alternative that the components follow a Markov process. The Durbin–Watson bounds pertain to the distribution of the d-statistics.

27.1 Introduction

A time series is composed of a sequence of observations $y_1, \ldots, y_n$, where the index i of the observation y_i represents time. An important feature of a time series is the order of observations: y_i is observed after $y_1, \ldots, y_{i-1}$ are observed. The correlation of successive observations is called a serial correlation. Related to each y_i may be a vector of independent variables $(x_{1i}, \ldots, x_{ki})$. Many questions of time series analysis relate to the possible dependence of y_i on $x_{1i}, \ldots, x_{ki}$; see, e.g., Anderson (1971).

A serial correlation (first-order) of a sequence $y_1, \ldots, y_n$ is

$$\sum_{i=2}^{n} y_i y_{i-1} \Big/ \sum_{i=1}^{n} y_i^2 .$$

This coefficient measures the correlation between $y_1, \ldots, y_{n-1}$ and $y_2, \ldots, y_n$. There are various modifications of this correlation coefficient such as replacing y_i by $y_i - \bar{y}$. See below for the circular serial coefficient. The term "autocorrelation" is also used for serial correlation.

I shall discuss two papers coauthored by James Durbin and Geoffrey Watson entitled "Testing for serial correlation in least squares regression I and II," published in 1950 and 1951 respectively (Durbin and Watson, 1950, 1951). The statistical analysis developed in these papers has proved very useful in econometric research.

The Durbin–Watson papers are based on a model in which there is a set of "independent" variables $x_{1i}, \ldots, x_{ni}$ associated with each "dependent" variable y_i for $i \in \{1, \ldots, n\}$. The dependent variable of y_i is considered as the linear combination

$$y_i = \beta_1 x_{1i} + \cdots + \beta_R x_{Ri} + w_i, \quad i \in \{1, \ldots, n\},$$

where w_i is an unobservable random disturbance. The questions that Durbin and Watson address have to do with the possible dependence in a set of observations $y_1, \ldots, y_n$ beyond what is explained by the independent variables.

27.2 Circular serial correlation

R.L. Anderson (Anderson, 1942), who was Watson's thesis advisor, studied the statistic

$$\frac{\sum\limits_{i=1}^{n} (y_i - y_{i-1})^2}{\sum\limits_{i=1}^{n} y_i^2} = 2 - 2\,\frac{\sum\limits_{i=1}^{n} y_i y_{i-1}}{\sum\limits_{i=1}^{n} y_i^2},$$

where $y_0 = y_n$. The statistic

$$\sum_{i=1}^{n} y_i y_{i-1} \Big/ \sum_{i=1}^{n} y_i^2$$

is known as the "circular serial correlation coefficient." Defining $y_0 = y_n$ is a device to make the mathematics simpler. The serial correlation coefficient measures the relationship between the sequence $y_1, \ldots, y_n$ and $y_0, \ldots, y_{n-1}$.

In our exposition we make repeated use of the fact that the distribution of $\mathbf{x}^\top \mathbf{A} \mathbf{x}$ is the distribution of $\sum_{i=1}^{n} \lambda_i z_i^2$, where $\lambda_1, \ldots, \lambda_n$ are the characteristic roots (latent roots) of $\mathbf{A}$, i.e., the roots of

$$|\mathbf{A} - \lambda \mathbf{I}_n| = 0, \qquad \mathbf{A} = \mathbf{A}^\top,$$

and $\boldsymbol{x}$ and $\boldsymbol{z}$ have the density $\mathcal{N}\left(\mathbf{0}, \sigma^2 \boldsymbol{I}\right)$. The numerator of the circular serial correlation is $\mathbf{x}^\top \mathbf{A} \mathbf{x}$, where

$$\mathbf{A} = \frac{1}{2} \begin{bmatrix} 0 & 1 & 0 & \cdots & 1 \\ 1 & 0 & 1 & \cdots & 0 \\ 0 & 1 & 0 & \cdots & 0 \\ \vdots & \vdots & \vdots & & \vdots \\ 1 & 0 & 0 & \cdots & 0 \end{bmatrix}.$$

The characteristic roots are $\lambda_j = \cos 2\pi j/n$ and $\sin 2\pi j/n$, $j \in \{1, \ldots, n\}$. If n is even, the roots occur in pairs. The distribution of the circular serial correlation is the distribution of

$$\sum_{j=1}^{n} \lambda_j z_j^2 \Big/ \sum_{j=1}^{n} z_j^2 , \tag{27.1}$$

where $z_1, \ldots, z_n$ are independent standard Normal variables. Anderson studied the distribution of the circular serial correlation, its moments, and other properties.

27.3 Periodic trends

During World War II, R.L. Anderson and I were members of the Princeton Statistical Research Group. We noticed that the jth characteristic vector of $\mathbf{A}$ had the form $\cos 2\pi jh/n$ and/or $\sin 2\pi jh/n$, $h \in \{1, \ldots, n\}$. These functions are periodic and hence are suitable to represent seasonal variation. We considered the model

$$y_i = \beta_1 x_{1i} + \cdots + \beta_k x_{ki} + u_i,$$

where $x_{hi} = \cos 2\pi hi/n$ and/or $\sin 2\pi hi/n$. Then the distribution of

$$r = \frac{\sum \left(y_i - \sum \beta_h x_{hi}\right)\left(y_{i-1} - \sum \beta_h x_{h,i-1}\right)}{\sum \left(y_i - \sum \beta_h x_{hi}\right)^2}$$

is the distribution of (27.1), where the sums are over the z's corresponding to the cos and sin terms that did not occur in the trends. The distributions of the serial correlations have the same form as before.

Anderson and Anderson (1950) found distributions of r for several cyclical trends as well as moments and approximate distributions.

27.4 Uniformly most powerful tests

As described in Anderson (1948), many problems of serial correlation are included in the general model

$$K \exp\left[-\frac{\alpha}{2}\left\{(\boldsymbol{y}-\boldsymbol{\mu})^\top \boldsymbol{\Psi}(\boldsymbol{y}-\boldsymbol{\mu}) + \lambda(\boldsymbol{y}-\boldsymbol{\mu})^\top \boldsymbol{\Theta}(\boldsymbol{y}-\boldsymbol{\mu})\right\}\right],$$

where K is a constant, $\alpha > 0$, $\boldsymbol{\Psi}$ a given positive definite matrix, $\boldsymbol{\Theta}$ a given symmetric matrix, λ a parameter such that $\boldsymbol{\Psi} - \lambda\boldsymbol{\Theta}$ is positive definite, and

$\boldsymbol{\mu}$ is the expectation of $\boldsymbol{y}$,

$$\mathcal{E}\boldsymbol{y} = \boldsymbol{\mu} = \sum \beta_j \boldsymbol{\phi}_j.$$

We shall consider testing the hypothesis

$$\mathcal{H} : \lambda = 0.$$

The first theorem characterizes tests such that the probability of the acceptance region when $\lambda = 0$ does not depend on the values of $\beta_1, \ldots, \beta_k$. The second theorem gives conditions for a test being uniformly most powerful when $\lambda > 0$ is the alternative.

These theorems are applicable to the circular serial correlation when $\boldsymbol{\Psi} = \sigma^2 \boldsymbol{I}$ and $\boldsymbol{\Theta} = \sigma^2 \boldsymbol{A}$ defined above.

The equation

$$\sum (y_i - y_{i-1})^2 = \sum (y_i^2 + y_{i-1}^2) - 2 \sum y_t y_{t-1}$$

suggests that a serial correlation can be studied in terms of $\sum (y_t - y_{t-1})^2$ which may be suitable to test that $y_1, \ldots, y_n$ are independent against the alternative that $y_1, \ldots, y_n$ satisfy an autoregressive process. Durbin and Watson prefer to study

$$d = \sum (z_i - z_{i-1})^2 \Big/ \sum z_i^2,$$

where $\boldsymbol{z}$ is defined below.

27.5 Durbin–Watson

The model is

$$\underset{n\times 1}{\boldsymbol{y}} = \underset{n\times k}{\boldsymbol{X}} \underset{k\times 1}{\boldsymbol{\beta}} + \underset{n\times 1}{\boldsymbol{u}}.$$

We consider testing the null hypothesis that $\boldsymbol{u}$ has a Normal distribution with mean $\boldsymbol{0}$ and covariance $\sigma^2 \boldsymbol{I}_n$ against the alternative that $\boldsymbol{u}$ has a Normal distribution with mean $\boldsymbol{0}$ and covariance $\sigma^2 \boldsymbol{A}$, a positive definite matrix. The sample regression of $\boldsymbol{y}$ is $\boldsymbol{b} = (\boldsymbol{X}^\top \boldsymbol{X})^{-1} \boldsymbol{X}^\top \boldsymbol{y}$ and the vector of residuals is

$$\begin{aligned} \boldsymbol{z} &= \boldsymbol{y} - \boldsymbol{X}\boldsymbol{b} = \{\boldsymbol{I} - \boldsymbol{X}(\boldsymbol{X}^\top \boldsymbol{X})^{-1} \boldsymbol{X}^\top\} \boldsymbol{y} \\ &= \{\boldsymbol{I} - \boldsymbol{X}(\boldsymbol{X}^\top \boldsymbol{X})^{-1} \boldsymbol{X}^\top\}(\boldsymbol{X}\boldsymbol{\beta} + \boldsymbol{u}) \\ &= \boldsymbol{M}\boldsymbol{u}, \end{aligned}$$

where

$$\boldsymbol{M} = \boldsymbol{I} - \boldsymbol{X}(\boldsymbol{X}^\top \boldsymbol{X})^{-1} \boldsymbol{X}^\top.$$

Consider the serial correlation of the residuals

$$r = \frac{\boldsymbol{z}^\top \boldsymbol{A}\boldsymbol{z}}{\boldsymbol{z}^\top \boldsymbol{z}} = \frac{\boldsymbol{u}^\top \boldsymbol{M}^\top \boldsymbol{A}\boldsymbol{M}\boldsymbol{u}}{\boldsymbol{u}^\top \boldsymbol{M}^\top \boldsymbol{M}\boldsymbol{u}}.$$

The matrix $\boldsymbol{M}$ is idempotent, i.e., $\boldsymbol{M}^m = \boldsymbol{M}$, and symmetric. Its latent roots are 0 and 1 and it has rank $n-k$. Let the possibly nonzero roots of $\boldsymbol{M}^\top \boldsymbol{A}\boldsymbol{M}$ be $\boldsymbol{\nu}_1, \ldots, \nu_{n-k}$. There is an $n \times (n-k)$ matrix $\boldsymbol{H}$ such that $\boldsymbol{H}^\top \boldsymbol{H} = \boldsymbol{I}_{n-k}$ and

$$\boldsymbol{H}^\top \boldsymbol{M}^\top \boldsymbol{A}\boldsymbol{M}\boldsymbol{H} = \begin{bmatrix} \nu_1 & 0 & \cdots & 0 \\ 0 & \nu_2 & \cdots & 0 \\ \vdots & \vdots & \vdots & \vdots \\ 0 & 0 & \cdots & \nu_{n-k} \end{bmatrix}.$$

Let $\boldsymbol{w} = \boldsymbol{H}^\top \boldsymbol{v}$. Then

$$r = \sum_{j=1}^{n-k} \nu_j w_j^2 \Big/ \sum_{j=1}^{n-k} w_j^2.$$

Durbin and Watson prove that

$$\lambda_j \leq \nu_j \leq \lambda_{j+k}, \quad j \in \{1, \ldots, n-k\}.$$

Define

$$r_L = \sum_{j=1}^{n-k} \lambda_j w_j^2 \Big/ \sum_{j=1}^{n-k} w_j^2, \qquad r_U = \sum_{j=1}^{n-k} \lambda_{j+k} w_j^2 \Big/ \sum_{j=1}^{n-k} w_j^2.$$

Then $r_L \leq r \leq r_U$.

The "bounds procedure" is the following. If the observed serial correlation is greater than $r_U^\star$ conclude that the hypothesis of no serial correlation of the disturbances is rejected. If the observed correlation is less than $r_L^\star$, conclude that the hypothesis of no serial correlation of the disturbance is accepted. The interval $(r_L^\star, r_U^\star)$ is called "the zone of indeterminacy." If the observed correlation falls in the interval $(r_L^\star, r_U^\star)$, the data are considered as not leading to a conclusion.

References

Anderson, R.L. (1942). Distribution of the serial correlation coefficient. *The Annals of Mathematical Statistics*, 13:1–13.

Anderson, R.L. and Anderson, T.W. (1950). Distribution of the circular serial correlation coefficient for residuals from a fitted Fourier series. *The Annals of Mathematical Statistics*, 21:59–81.

Anderson, T.W. (1948). On the theory of testing serial correlation. *Skandinavisk Aktuarietidskrift*, 31:88–116.

Anderson, T.W. (1971). *The Statistical Analysis of Time Series.* Wiley, New York.

Anderson, T.W. (2003). *An Introduction to Multivariate Statistical Analysis*, 3rd edition. Wiley, New York.

Chipman, J.S. (2011). *Advanced Econometric Theory.* Routledge, London.

Durbin, J. and Watson, G.S. (1950). Testing for serial correlation in least squares regression. I. *Biometrika*, 37:409–428.

Durbin, J. and Watson, G.S. (1951). Testing for serial correlation in least squares regression. II. *Biometrika*, 38:159–178.

28

A non-asymptotic walk in probability and statistics

Pascal Massart
Département de mathématiques
Université de Paris-Sud, Orsay, France

My research is devoted to the derivation of non-asymptotic results in probability and statistics. Basically, this is a question of personal taste: I have been struggling with constants in probability bounds since the very beginning of my career. I was very lucky to learn from the elegant work of Michel Talagrand that the dream of a non-asymptotic theory of independence could actually become reality. Thanks to my long-term collaboration with my colleague and friend Lucien Birgé, I could realize the importance of a non-asymptotic approach to statistics. This led me to follow a singular path, back and forth between concentration inequalities and model selection, that I briefly describe below in this (informal) paper for the 50th anniversary of the COPSS.

28.1 Introduction

The interest in non-asymptotic tail bounds for functions of independent random variables is rather recent in probability theory. Apart from sums, which have been well understood for a long time, powerful tools for handling more general functions of independent random variables were not introduced before the 1970s. The systematic study of concentration inequalities aims at bounding the probability that such a function differs from its expectation or its median by more than a given amount. It emerged from a remarkable series of papers by Michel Talagrand in the mid-1990s.

Talagrand provided a major new insight into the problem, around the idea summarized in Talagrand (1995): "A random variable that smoothly depends on the influence of many independent random variables satisfies Chernoff type bounds." This revolutionary approach opened new directions of research and stimulated numerous applications in various fields such as discrete mathematics, statistical mechanics, random matrix theory, high-dimensional geometry, and statistics.

The study of random fluctuations of suprema of empirical processes has been crucial in the application of concentration inequalities to statistics and machine learning. It also turned out to be a driving force behind the development of the theory. This is exactly what I would like to illustrate here while focusing on the impact on my own research in the 1990s and beyond.

28.2 Model selection

Model selection is a classical topic in statistics. The idea of selecting a model by penalizing some empirical criterion goes back to the early 1970s with the pioneering work of Mallows and Akaike. The classical parametric view on model selection as exposed in Akaike's seminal paper (Akaike, 1973) on penalized log-likelihood is asymptotic in essence. More precisely Akaike's formula for the penalty depends on Wilks' theorem, i.e., on an asymptotic expansion of the log-likelihood.

Lucien Birgé and I started to work on model selection criteria based on a non-asymptotic penalized log-likelihood early in the 1990s. We had in mind that in the usual asymptotic approach to model selection, it is often unrealistic to assume that the number of observations tends to infinity while the list of models and their size are fixed. Either the number of observations is not that large (a hundred, say) and when playing with models with a moderate number of parameters (five or six) you cannot be sure that asymptotic results apply, or the number of observations is really large (as in signal de-noising, for instance) and you would like to take advantage of it by considering a potentially large list of models involving possibly large numbers of parameters.

From a non-asymptotic perspective, the number of observations and the list of models are what they are. The purpose of an ideal model selection procedure is to provide a data-driven choice of model that tends to optimize some criterion, e.g., minimum expected risk with respect to the quadratic loss or the Kullback–Leibler loss. This provides a well-defined mathematical formalization of the model selection problem but it leaves open the search for a neat generic solution.

Fortunately for me, the early 1990s turned out to be a rich period for the development of mathematical statistics, and I came across the idea that letting the size of models go to infinity with the number of observations makes it possible to build adaptive nonparametric estimators. This idea can be traced back to at least two different sources: information theory and signal analysis.

In particular, Lucien and I were very impressed by the beautiful paper of Andrew Barron and Tom Cover (Barron and Cover, 1991) on density estimation via minimum complexity model selection. The main message there (at least for discrete models) is that if you allow model complexity to grow with sample size, you can then use minimum complexity penalization to build nonparametric estimators of a density which adapts to the smoothness.

Meanwhile, David Donoho, Iain Johnstone, Gérard Kerkyacharian and Dominique Picard were developing their approach to wavelet estimation. Their striking work showed that in a variety of problems, it is possible to build adaptive estimators of a regression function or a density through a remarkably simple procedure: thresholding of the empirical wavelet coefficients. Many papers could be cited here but Donoho et al. (1995) is possibly the mostly useful review on the topic. Wavelet thresholding has an obvious model selection flavor to it, as it amounts to selecting a set of wavelet coefficients from the data.

At some point, it became clear to us that there was room for building a general theory to help reconcile Akaike's classical approach to model selection, the emerging results by Barron and Cover or Donoho et al. in which model selection is used to construct nonparametric adaptive estimators, and Vapnik's structural minimization of the risk approach for statistical learning; see Vapnik (1982).

28.2.1 The model choice paradigm

Assume that a random variable $\xi^{(n)}$ is observed which depends on a parameter n. For concreteness, you may think of $\xi^{(n)}$ as an n-sample from some unknown distribution. Consider the problem of estimating some quantity of interest, s, which is known to belong to some (large) set $\mathcal{S}$. Consider an *empirical risk criterion* γ_n based on $\xi^{(n)}$ such that the mapping

$$t \mapsto \mathrm{E}\left\{\gamma_n\left(t\right)\right\}$$

achieves a minimum at point s. One can then define a natural (non negative) loss function related to this criterion by setting, for all $t \in \mathcal{S}$,

$$\ell\left(s,t\right) = \mathrm{E}\left\{\gamma_n\left(t\right)\right\} - \mathrm{E}\left\{\gamma_n\left(s\right)\right\}.$$

When $\xi^{(n)} = (\xi_1, \ldots, \xi_n)$, the empirical risk criterion γ_n is usually defined as some empirical mean

$$\gamma_n\left(t\right) = P_n\left\{\gamma\left(t,\cdot\right)\right\} = \frac{1}{n}\sum_{i=1}^{n}\gamma\left(t,\xi_i\right) \tag{28.1}$$

of an adequate risk function γ. Two typical examples are as follows.

Example 1 (Density estimation) *Let $\xi_1, \ldots, \xi_n$ be a random sample from an unknown density s with respect to a given measure μ. Taking $\gamma(t,x) = -\ln\{t(x)\}$ in* (28.1) *leads to the log-likelihood criterion. The corresponding loss function, ℓ, is simply the Kullback–Leibler information between the probability measures $s\mu$ and $t\mu$. Indeed, $\ell(s,t) = \int s\ln(s/t)\mathrm{d}\mu$ if $s\mu$ is absolutely continuous with respect to $t\mu$ and $\ell(s,t) = \infty$ otherwise. However if*

$\gamma(t,x) = \|t\|^2 - 2t(x)$ in (28.1), where $\|\cdot\|$ denotes the norm in $L_2(\mu)$, one gets the least squares criterion and the loss function is given by $\ell(s,t) = \|s-t\|^2$ for every $t \in L_2(\mu)$.

Example 2 (Gaussian white noise) *Consider the process $\xi^{(n)}$ on $[0,1]$ defined by $\mathrm{d}\xi^{(n)}(x) = s(x) + n^{-1/2}\mathrm{d}W(x)$ with $\xi^{(n)}(0) = 0$, where W denotes the Brownian motion. The least squares criterion is defined by $\gamma_n(t) = \|t\|^2 - 2\int_0^1 t(x)\,\mathrm{d}\xi^{(n)}(x)$, and the corresponding loss function ℓ is simply the squared L_2 distance defined for all $s,t \in [0,1]$ by $\ell(s,t) = \|s-t\|^2$.*

Given a model S (which is a subset of $\mathcal{S}$), the empirical risk minimizer is simply defined as a minimizer of γ_n over S. It is a natural estimator of s whose quality is directly linked to that of the model S. The question is then: How can one choose a suitable model S? It would be tempting to choose S as large as possible. Taking S as $\mathcal{S}$ itself or as a "big" subset of $\mathcal{S}$ is known to lead either to inconsistent estimators (Bahadur, 1958) or to suboptimal estimators (Birgé and Massart, 1993). In contrast if S is a "small" model (e.g., some parametric model involving one or two parameters), the behavior of the empirical risk minimizer on S is satisfactory as long as s is close enough to S, but the model can easily end up being completely wrong.

One of the ideas suggested by Akaike is to use the risk associated to the loss function ℓ as a quality criterion for a model. To illustrate this idea, it is convenient to consider a simple example for which everything is easily computable. Consider the white noise framework. If S is a linear space with dimension D, and if $\phi_1, \ldots, \phi_D$ denotes some orthonormal basis of S, the least squares estimator is merely a projection estimator, viz.

$$\hat{s} = \sum_{j=1}^{D} \left\{ \int_0^1 \phi_j(x)\mathrm{d}\xi^{(n)}(x) \right\} \phi_j$$

and the expected quadratic risk of $\hat{s}$ is equal to

$$\mathrm{E}(\|s-\hat{s}\|^2) = d^2(s,S) + D/n.$$

This formula for the quadratic risk reflects perfectly the model choice paradigm: if the model is to be chosen in such a way that the risk of the resulting least squares estimator remains under control, a balance must be struck between the bias term $d^2(s,S)$ and the variance term D/n.

More generally, given an empirical risk criterion γ_n, each model S_m in an (at most countable and usually finite) collection $\{S_m : m \in \mathcal{M}\}$ can be represented by the corresponding empirical risk minimizer $\hat{s}_m$. One can use the minimum of $\mathrm{E}\{\ell(s,\hat{s}_m)\}$ over $\mathcal{M}$ as a benchmark for model selection. Ideally one would like to choose $m(s)$ so as to minimize the risk $\mathrm{E}\{\ell(s,\hat{s}_m)\}$ with respect to $m \in \mathcal{M}$. This is what Donoho and Johnstone called an oracle; see, e.g., Donoho and Johnstone (1994). The purpose of model selection is to design a data-driven choice $\hat{m}$ which mimics an oracle, in the sense that the risk of the selected estimator $\hat{s}_{\hat{m}}$ is not too far from the benchmark $\inf_{m\in\mathcal{M}} \mathrm{E}\{\ell(s,\hat{s}_m)\}$.

28.2.2 Non-asymptopia

The penalized empirical risk model selection procedure consists in considering an appropriate penalty function pen: $\mathcal{M} \to \mathbb{R}_+$ and choosing $\widehat{m}$ to minimize

$$\text{crit}\,(m) = \gamma_n\,(\hat{s}_m) + \text{pen}\,(m)$$

over $\mathcal{M}$. One can then define the selected model $S_{\widehat{m}}$ and the penalized empirical risk estimator $\hat{s}_{\widehat{m}}$.

Akaike's penalized log-likelihood criterion corresponds to the case where the penalty is taken as D_m/n, where D_m denotes the number of parameters defining the regular parametric model S_m. As mentioned above, Akaike's heuristics heavily relies on the assumption that the dimension and the number of models are bounded with respect to n, as $n \to \infty$. Various penalized criteria have been designed according to this asymptotic philosophy; see, e.g., Daniel and Wood (1971).

In contrast a non-asymptotic approach to model selection allows both the number of models and the number of their parameters to depend on n. One can then choose a list of models which is suitable for approximation purposes, e.g., wavelet expansions, trigonometric or piecewise polynomials, or artificial neural networks. For example, the hard thresholding procedure turns out to be a penalized empirical risk procedure if the list of models depends on n.

To be specific, consider once again the white noise framework and consider an orthonormal system $\phi_1, \ldots, \phi_n$ of $L_2[0,1]$ that depends on n. For every subset m of $\{1, \ldots, n\}$, define the model S_m as the linear span of $\{\phi_j : j \in m\}$. The complete variable selection problem requires the selection of a subset m from the collection of all subsets of $\{1, \ldots, n\}$. Taking a penalty function of the form $\text{pen}\,(m) = T^2\,|m|\,/n$ leads to an explicit solution for the minimization of crit(m) because in this case, setting $\hat{\beta}_j = \int \phi_j\,(x)\,\mathrm{d}\xi^{(n)}\,(x)$, the penalized empirical criterion can be written as

$$\text{crit}\,(m) = -\sum_{j \in m} \hat{\beta}_j^2 + \frac{T^2\,|m|}{n} = \sum_{j \in m} \left(-\hat{\beta}_j^2 + \frac{T^2}{n} \right).$$

This criterion is obviously minimized at

$$\widehat{m} = \{j \in \{1, \ldots, n\} : \sqrt{n}\,|\hat{\beta}_j| \geq T\},$$

which is precisely the hard thresholding procedure. Of course the crucial issue is to choose the level of thresholding, T.

More generally the question is: what kind of penalty should be recommended from a non-asymptotic perspective? The naive notion that Akaike's criterion could be used in this context fails in the sense that it may typically lead to under-penalization. In the preceding example, it would lead to the choice $T = \sqrt{2}$ while it stems from the work of Donoho et al. that the level of thresholding should be at least of order $\sqrt{2\ln(n)}$ as $n \to \infty$.

28.2.3 Empirical processes to the rescue

The reason for which empirical processes have something to do with the analysis of penalized model selection procedures is roughly the following; see Massart (2007) for further details. Consider the centered empirical risk process

$$\bar{\gamma}_n(t) = \gamma_n(t) - \mathrm{E}\{\gamma_n(t)\}.$$

Minimizing $\mathrm{crit}(m)$ is then equivalent to minimizing

$$\mathrm{crit}(m) - \gamma_n(s) = \ell(s, \hat{s}_m) - \{\bar{\gamma}_n(s) - \bar{\gamma}_n(\hat{s}_m)\} + \mathrm{pen}(m).$$

One can readily see from this formula that to mimic an oracle, the penalty $\mathrm{pen}(m)$ should ideally be of the same order of magnitude as $\bar{\gamma}_n(s) - \bar{\gamma}_n(\hat{s}_m)$. Guessing what is the exact order of magnitude for $\bar{\gamma}_n(s) - \bar{\gamma}_n(\hat{s}_m)$ is not an easy task in general, but one can at least try to compare the fluctuations of $\bar{\gamma}_n(s)-\bar{\gamma}_n(\hat{s}_m)$ to the quantity of interest, $\ell(s, \hat{s}_m)$. To do so, one can introduce the supremum of the weighted process

$$Z_m = \sup_{t \in S_m} \frac{\bar{\gamma}_n(s) - \bar{\gamma}_n(t)}{w\{\ell(s,t)\}},$$

where w is a conveniently chosen non-decreasing weight function. For instance if $w\{\ell(s,t)\} = 2\sqrt{\ell(s,t)}$ then, for every $\theta > 0$,

$$\ell(s, \hat{s}_m) - \{\bar{\gamma}_n(s) - \bar{\gamma}_n(\hat{s}_m)\} \geq (1-\theta)\,\ell(s, \hat{s}_m) - Z_m^2/\theta.$$

Thus by choosing $\mathrm{pen}(m)$ in such a way that $Z_m^2 \leq \theta\,\mathrm{pen}(m)$ (with high probability), one can hope to compare the model selection procedure with the oracle.

We are at the very point where the theory of empirical processes comes in, because the problem is now to control the quantity Z_m, which is indeed the supremum of an empirical process — at least when the empirical risk is defined through (28.1). Lucien Birgé and I first used this idea in 1994, while preparing our contribution to the Festschrift for Lucien Le Cam to mark his 70th birthday. The corresponding paper, Birgé and Massart (1997), was published later and we generalized it in Barron et al. (1999).

In the context of least squares density estimation that we were investigating, the weight function w to be considered is precisely of the form $w(x) = 2\sqrt{x}$. Thus if the model S_m happens to be an finite-dimensional subspace of $L_2(\mu)$ generated by some orthonormal basis $\{\phi_\lambda : \lambda \in \Lambda_m\}$, the quantity of interest

$$Z_m = \sup_{t \in S_m} \frac{\bar{\gamma}_n(s) - \bar{\gamma}_n(t)}{2\,\|s-t\|} \tag{28.2}$$

can easily be made explicit. Indeed, assuming that s belongs to S_m (this assumption is not really needed but makes the analysis much more illuminating),

$$Z_m = \sqrt{\sum_{\lambda \in \Lambda_m} (P_n - P)^2(\phi_\lambda)}, \tag{28.3}$$

where $P = s\mu$. In other words, Z_m is simply the square root of a χ^2-type statistic. Deriving exponential bounds for Z_m from (28.3) does not look especially easier than starting from (28.2). However, it is clear from (28.3) that $\mathrm{E}(Z_m^2)$ can be computed explicitly. As a result one can easily bound $\mathrm{E}(Z_m)$ using Jensen's inequality, viz.

$$\mathrm{E}(Z_m) \leq \sqrt{\mathrm{E}(Z_m^2)}.$$

By shifting to concentration inequalities, we thus hoped to escape the heavy-duty "chaining" machinery, which was the main tool available at that time to control suprema of empirical processes.

It is important to understand that this is not merely a question of taste or elegance. The disadvantage with chaining inequalities is that even if you optimize them, at the end of the day the best you can hope for is to derive a bound with the right order of magnitude; the associated numerical constants are typically ridiculously large. When the goal is to validate (or invalidate) penalized criteria such as Mallows' or Akaike's criterion from a non-asymptotic perspective, constants do matter. This motivated my investigation of the fascinating topic of concentration inequalities.

28.3 Welcome to Talagrand's wonderland

Motivated by the need to understand whether one can derive concentration inequalities for suprema of empirical processes, I intensified my readings on the concentration of measures. By suprema of empirical processes, I mean

$$Z = \sup_{t \in \mathcal{T}} \sum_{i=1}^{n} X_{i,t},$$

where $\mathcal{T}$ is a set and $X_{1,t}, \ldots, X_{n,t}$ are mutually independent random vectors taking values in $\mathbb{R}^{\mathcal{T}}$.

For applications such as that which was described above, it is important to cover the case where $\mathcal{T}$ is infinite, but for the purpose of building structural inequalities like concentration inequalities, $\mathcal{T}$ finite is in fact the only case that matters because one can recover the general case from the finite case by applying monotone limit procedures, i.e., letting the size of the index set grow to infinity. Henceforth I will thus assume the set $\mathcal{T}$ to be finite.

When I started investigating the issue in 1994, the literature was dominated by the Gaussian Concentration Theorem for Lipschitz functions of independent standard Gaussian random variables. This result was proved independently by Borell (1975) and by Cirel'son and Sudakov (1974). As a side remark, note that these authors actually established the concentration of Lipschitz functions around the median; the analogous result for the mean is due to Cirel'son et al. (1976). At any rate, it seemed to me to be somewhat of a beautiful but isolated mountain, given the abundance of results by Michel Talagrand on the concentration of product measures. In the context of empirical processes, the Gaussian Concentration Theorem implies the following spectacular result.

Assume that $X_{1,t}, \ldots, X_{n,t}$ are Gaussian random vectors centered at their expectation. Let v be the maximal variance of $X_{1,t} + \cdots + X_{n,t}$ when t varies, and use M to denote either the median or the mean of Z. Then

$$\Pr(Z \geq M + z) \leq \exp\left(-\frac{z^2}{2v}\right).$$

In the non-Gaussian case, the problem becomes much more complex. One of Talagrand's major achievements on the topic of concentration inequalities for functions on a product space $\mathcal{X} = \mathcal{X}_1 \times \cdots \times \mathcal{X}_n$ is his celebrated convex distance inequality. Given any vector $\alpha = (\alpha_1, \ldots, \alpha_n)$ of non-negative real numbers and any $(x, y) \in \mathcal{X} \times \mathcal{X}$, the weighted Hamming distance d_α is defined by

$$d_\alpha(x, y) = \sum_{i=1}^{n} \alpha_i \mathbf{1}(x_i \neq y_i),$$

where $\mathbf{1}(A)$ denotes the indicator of the set A. Talagrand's convex distance from a point x to some measurable subset A of $\mathcal{X}$ is then defined by

$$d_T(x, A) = \sup_{|\alpha|_2^2 \leq 1} d_\alpha(x, A),$$

where $|\alpha|_2^2 = \alpha_1^2 + \cdots + \alpha_n^2$.

If P denotes some product probability measure $P = \mu_1 \otimes \cdots \otimes \mu_n$ on $\mathcal{X}$, the concentration of P with respect to d_T is specified by Talagrand's convex distance inequality, which ensures that for any measurable set A, one has

$$P\{d_T(\cdot, A) \geq z\} \leq P(A) \exp\left(-\frac{z^2}{4}\right). \tag{28.4}$$

Typically, it allows the analysis of functions that satisfy the regularity condition

$$f(x) - f(y) \leq \sum_{i=1}^{n} \alpha_i(x) \mathbf{1}(x_i \neq y_i). \tag{28.5}$$

One can then play the following simple but subtle game. Choose $A = \{f \leq M\}$ and observe that in view of condition (28.5), one has $f(x) < M + z$ for all x such that $d_{\alpha(x)}(x, A) < z$. Thus if $v \geq \sup_x \{\alpha_1^2(x) + \cdots + \alpha_n^2(x)\}$, one finds

$$\{f \geq M + z\} \subseteq \{d_{\alpha(\cdot)}(\cdot, A) \geq z\} \subseteq \left\{ \sup_{|\alpha|_2^2 \leq v} d_\alpha(\cdot, A) \geq z \right\}.$$

Hence by Talagrand's convex distance inequality (28.4), one gets

$$P(f \geq M + z) \leq 2 \exp\left(-\frac{z^2}{4v}\right) \tag{28.6}$$

whenever M is a median of f under P. The preceding inequality applies to a Rademacher process, which is a special case of an empirical process. Indeed, setting $\mathcal{X} = \{-1, 1\}^n$ and defining

$$f(x) = \sup_{t \in \mathcal{T}} \sum_{i=1}^{n} \alpha_{i,t} x_i = \sum_{i=1}^{n} \alpha_{i,t^*(x)} x_i$$

in terms of real numbers $\alpha_{i,t}$, one can see that, for every x and y,

$$f(x) - f(y) \leq \sum_{i=1}^{n} \alpha_{i,t^*(x)} (x_i - y_i) \leq 2 \sum_{i=1}^{n} \left|\alpha_{i,t^*(x)}\right| \mathbf{1}(x_i \neq y_i).$$

This means that the function f satisfies the regularity condition (28.5) with $\alpha_i(x) = 2\left|\alpha_{i,t^*(x)}\right|$. Thus if $X = (X_{1,t}, \ldots, X_{n,t})$ is uniformly distributed on the hypercube $\{-1, 1\}^n$, it follows from (28.6) that the supremum of the Rademacher process

$$Z = \sup_{t \in \mathcal{T}} \sum_{i=1}^{n} \alpha_{i,t} X_{i,t} = f(X)$$

satisfies the sub-Gaussian tail inequality

$$\Pr(Z \geq M + z) \leq 2 \exp\left(-\frac{z^2}{4v}\right),$$

where the variance factor v can be taken as $v = 4 \sup_{t \in \mathcal{T}} (\alpha_{1,t}^2 + \cdots + \alpha_{n,t}^2)$.

This illustrates the power of Talagrand's convex distance inequality. Alas, while condition (28.5) is perfectly suited for the analysis of Rademacher processes, it does not carry over to more general empirical processes.

At first, I found it a bit frustrating that there was no analogue of the Gaussian concentration inequality for more general empirical processes and that Talagrand's beautiful results were seemingly of no use for dealing with suprema of empirical processes like (28.2). Upon reading Talagrand (1994) carefully, however, I realized that it contained at least one encouraging result. Namely, Talagrand (1994) proved a sub-Gaussian Bernstein type inequality for $Z - C\,\mathrm{E}(Z)$, where C is a universal constant. Of course in Talagrand's version, C is not necessarily equal to 1 but it was reasonable to expect that this should be the case. This is exactly what Lucien Birgé and I were able to show. We presented our result at the 1994 workshop organized at Yale in honor of Lucien Le Cam. A year later or so, I was pleased to hear from Michel Talagrand that, motivated in part by the statistical issues described above, and at the price of some substantial deepening of his approach to concentration of product measures, he could solve the problem and obtain his now famous concentration inequality for the supremum of a bounded empirical process; see Talagrand (1996).

28.4 Beyond Talagrand's inequality

Talagrand's new result for empirical processes stimulated intense research, part of which was aimed at deriving alternatives to Talagrand's original approach. The interested reader will find in Boucheron et al. (2013) an account of the transportation method and of the so-called entropy method that we developed in a series of papers (Boucheron et al., 2000, 2003, 2005; Massart, 2000) in the footsteps of Michel Ledoux (1996). In particular, using the entropy method established by Olivier Bousquet (2002), we derived a version of Talagrand's inequality for empirical processes with optimal numerical constants in the exponential bound.

Model selection issues are still posing interesting challenges for empirical process theory. In particular, the implementation of non-asymptotic penalization methods requires data-driven penalty choice strategies. One possibility is to use the concept of "minimal penalty" that Lucien Birgé and I introduced in Birgé and Massart (2007) in the context of Gaussian model selection and, more generally, the "slope heuristics" (Arlot and Massart, 2009), which basically relies on the idea that the empirical loss

$$\gamma_n(s) - \gamma_n(\hat{s}_m) = \sup_{t \in S_m} \{\gamma_n(s) - \gamma_n(t)\}$$

has a typical behavior for large dimensional models. A complete theoretical validation of these heuristics is yet to be developed but partial results are available; see, e.g., Arlot and Massart (2009), Birgé and Massart (2007), and Saumard (2013).

A fairly general concentration inequality providing a non-asymptotic analogue to Wilks' Theorem is also established in Boucheron and Massart (2011) and used in Arlot and Massart (2009). This result stems from the entropy method, which is flexible enough to capture the following rather subtle self-localization effect. The variance of $\sup_{t \in S_m} \{\gamma_n(s) - \gamma_n(t)\}$ can be proved to be of the order of the variance of $\gamma_n(s) - \gamma_n(t)$ at $t = \hat{s}_m$, which may be much smaller than the maximal variance. This is typically the quantity that would emerge from a direct application of Talagrand's inequality for empirical processes.

The issue of calibrating model selection criteria from data is of great importance. In the context where the list of models itself is data dependent (think, e.g., of models generated by variables selected from an algorithm such as LARS), the problem is related to the equally important issue of choosing regularization parameters; see Meynet (2012) for more details. This is a new field of investigation which is interesting both from a theoretical and a practical point of view.

References

Akaike, H. (1973). Information theory and an extension of the maximum likelihood principle. *Proceedings of the Second International Symposium on Information Theory*. Akademia Kiado, Budapest, pp. 267–281.

Arlot, S. and Massart, P. (2009). Data driven calibration of penalties for least squares regression. *Journal of Machine Learning Research*, 10:245–279.

Bahadur, R.R. (1958). Examples of inconsistency of maximum likelihood estimates. *Sankhyā, Series A*, 20:207–210.

Barron, A.R., Birgé, L., and Massart, P. (1999). Risk bounds for model selection via penalization. *Probability Theory and Related Fields*, 113:301–413.

Barron, A.R. and Cover, T.M. (1991). Minimum complexity density estimation. *IEEE Transactions on Information Theory*, 37:1034–1054.

Birgé, L. and Massart, P. (1993). Rates of convergence for minimum contrast estimators. *Probability Theory and Related Fields*, 97:113–150.

Birgé, L. and Massart, P. (1997). From model selection to adaptive estimation. In *Festschrift for Lucien Le Cam: Research Papers in Probability and Statistics* (D. Pollard, E. Torgersen, and G. Yang, Eds.). Springer, New York, pp. 55–87.

Birgé, L. and Massart, P. (2007). Minimal penalties for Gaussian model selection. *Probability Theory and Related Fields*, 138:33–73.

Borell, C. (1975). The Brunn–Minkowski inequality in Gauss space. *Inventiones Mathematicae*, 30:207–216.

Boucheron, S., Bousquet, O., Lugosi, G., and Massart, P. (2005). Moment inequalities for functions of independent random variables. *The Annals of Probability*, 33:514–560.

Boucheron, S., Lugosi, G., and Massart, P. (2000). A sharp concentration inequality with applications. *Random Structures and Algorithms*, 16:277–292.

Boucheron, S., Lugosi, G., and Massart, P. (2003). Concentration inequalities using the entropy method. *The Annals of Probability*, 31:1583–1614.

Boucheron, S., Lugosi, G., and Massart, P. (2013). *Concentration Inequalities: A Nonasymptotic Theory of Independence*. Oxford University Press, Oxford, UK.

Boucheron, S. and Massart, P. (2011). A high-dimensional Wilks phenomenon. *Probability Theory and Related Fields*, 150:405–433.

Bousquet, O. (2002). A Bennett concentration inequality and its application to suprema of empirical processes. *Comptes rendus mathématiques de l'Académie des sciences de Paris*, 334:495–500.

Cirel'son, B.S., Ibragimov, I.A., and Sudakov, V.N. (1976). Norm of Gaussian sample function. In *Proceedings of the 3rd Japan–USSR Symposium on Probability Theory*, Lecture Notes in Mathematics 550. Springer, Berlin, pp. 20–41.

Cirel'son, B.S. and Sudakov, V.N. (1974). Extremal properties of half spaces for spherically invariant measures. *Journal of Soviet Mathematics* 9, 9–18 (1978) [Translated from *Zapiski Nauchnykh Seminarov Leningradskogo Otdeleniya Matematicheskogo Instituta im. V.A. Steklova AN SSSR*, 41:14–24 (1974)].

Daniel, C. and Wood, F.S. (1971). *Fitting Equations to Data.* Wiley, New York.

Donoho, D.L. and Johnstone, I.M. (1994). Ideal spatial adaptation by wavelet shrinkage. *Biometrika*, 81:425–455.

Donoho, D.L., Johnstone, I.M., Kerkyacharian, G., and Picard, D. (1995). Wavelet shrinkage: Asymptopia? (with discussion). *Journal of the Royal Statistical Society, Series B*, 57:301–369.

Ledoux, M. (1996). On Talagrand deviation inequalities for product measures. *ESAIM: Probability and Statistics*, 1:63–87.

Massart, P. (2000). About the constants in Talagrand's concentration inequalities for empirical processes. *The Annals of Probability*, 28:863–884.

Massart, P. (2007). *Concentration Inequalities and Model Selection.* École d'été de probabilités de Saint-Flour 2003. Lecture Notes in Mathematics 1896, Springer, Berlin.

Meynet, C. (2012). *Sélection de variables pour la classification non supervisée en grande dimension.* Doctoral dissertation, Université Paris-Sud XI, Orsay, France.

Saumard, A. (2013). Optimal model selection in heteroscedastic regression using piecewise polynomial functions. *Electronic Journal of Statistics*, 7:1184–1223.

Talagrand, M. (1994). Sharper bounds for empirical processes. *The Annals of Probability*, 22:28–76.

Talagrand, M. (1995). Concentration of measure and isoperimetric inequalities in product spaces. *Publications mathématiques de l'Institut des hautes études supérieures*, 81:73–205.

Talagrand, M. (1996). New concentration inequalities in product spaces. *Inventiones Mathematicae*, 126:505–563.

Vapnik, V.N. (1982). *Estimation of Dependencies Based on Empirical Data.* Springer, New York.

29

The past's future is now: What will the present's future bring?

Lynne Billard
Department of Statistics
University of Georgia, Athens, GA

Articles published in the early years of the *Journal of the American Statistical Association*, i.e., 1888–1910s, posited new theories mostly by using arithmetical arguments. Starting around the mid-1910s the arguments became algebraic in nature and by the 1920s this trend was well established. Today, a century later, in addition to cogent mathematical arguments, new statistical developments are becoming computational, such is the power and influence of the modern computer (a device un-dreamed of in those earlier days). Likewise, we see enormous changes in the size and nature of assembled data sets for our study. Therefore, entirely new paradigms are entering our discipline, radically changing the way we go about our art. This chapter focuses on one such method wherein the data are symbolically valued, i.e., hypercubes in p-dimensional space $\mathbb{R}^p$, instead of the classically valued points in $\mathbb{R}^p$.

29.1 Introduction

The advent of modern computer capabilities has a consequence that entirely new paradigms are entering our discipline radically changing the way we go about our art. One hundred years ago, researchers were transitioning from using arithmetical arguments when developing their new mathematically-based ideas to using algebraic arguments (i.e., mathematical tools, algebra, calculus, and the like). Today's transition lies more along the lines of computational mathematical/statistical developments as we struggle with the massively huge data sets at hand. This chapter focuses on one such method — symbolic data — projected by Goodman (2011) as one of the two most important new developments on the horizon wherein the data are symbolically valued, i.e., hypercubes in p-dimensional space $\mathbb{R}^p$, instead of points in $\mathbb{R}^p$ as for classical data. In Section 29.2, we describe briefly what symbolic data are and how they might arise. Then, in Section 29.3, we illustrate some symbolic methodological analyses and compare the results with those obtained when using classical surrogates. Some concluding remarks about the future of such data are presented in Section 29.4.

29.2 Symbolic data

Symbolic data consist of lists, intervals, histograms and the like, and arise in two broadly defined ways. One avenue is when data sets of classical point observations are aggregated into smaller data sets. For example, consider a large medical data set of millions of individual observations with information such as demographic (e.g., age, gender, etc.), geographical (e.g., town of residence, country, region, ...), basic medical diagnostics (pulse rate, blood pressure, weight, height, previous maladies and when, etc.), current ailments (e.g., cancer type such as liver, bone, etc.; heart condition, etc.), and so on. It is unlikely the medical insurer (or medical researcher, or...) is interested in the details of your specific visit to a care provider on a particular occasion; indeed, the insurer may not even be interested in your aggregated visits over a given period of time. Rather, interest may focus on all individuals (and their accumulated history) who have a particular condition (such as heart valve failure), or, maybe interest centers on the collection of individuals of a particular gender-age group with that condition. Thus, values are aggregated across all individuals in the specific categories of interest. It is extremely unlikely that all such individuals will have the same pulse rate, the same weight, and so forth. Instead, the aggregated values can take values across an interval, as a histogram, as a list of possible values, etc. That is, the data set now consists of so-called symbolic data.

Automobile insurers may be interested in accident rates of categories such as 26-year-old male drivers of red convertibles, and so on. Census data are frequently in the form of symbolic data; e.g., housing characteristics for regions may be described as {owner occupied, .60; renter occupied, .35; vacant, .05} where 60% of the homes are owner occupied, etc.

There are countless examples. The prevailing thread is that large data sets of single classical observations are aggregated in some way with the result that symbolic data perforce emerge. There are a myriad of ways these original data sets can be aggregated, with the actual form being driven by the scientific question/s of interest.

On the other hand, some data are naturally symbolic in nature. For example, species are typically described by symbolic values; e.g., the mushroom species *bernardi* has a pileus cap width of $[6,7]$ cm. However, the particular mushroom in your hand may have a cap width of 6.2 cm, say. Pulse rates bounce around, so that an apparent rate of 64 (say) may really be 64 ± 2, i.e., the interval $[62, 66]$. There are numerous examples.

Detailed descriptions and examples can be found in Bock and Diday (2000) and Billard and Diday (2006). A recent review of current methodologies is available in Noirhomme-Fraiture and Brito (2011), with a non-technical introduction in Billard (2011). The original concept of symbolic data was introduced in Diday (1987). Note that symbolic data are not the same as fuzzy data; however, while they are generally different from the coarsening and grouping concepts of, e.g., Heitjan and Rubin (1991), there are some similarities.

The major issue then is how do we analyse these intervals (or, histograms, ...)? Taking classical surrogates, such as the sample mean of aggregated values for each category and variable, results in a loss of information. For example, the intervals $[10, 20]$ and $[14, 16]$ both have the same midpoint; yet they are clearly differently valued observations. Therefore, it is important that analytical methods be developed to analyse symbolic data directly so as to capture these differences. There are other underlying issues that pertain such as the need to develop associated logical dependency rules to maintain the integrity of the overall data structure; we will not consider this aspect herein however.

29.3 Illustrations

Example 29.1 Table 29.1 displays (in two parts) a data set of histogram valued observations, extracted from Falduti and Taibaly (2004), obtained by aggregating by airline approximately 50,000 classical observations for flights arriving at and departing from a major airport. For illustrative simplicity, we take three random variables Y_1 = flight time, Y_2 = arrival-delay-time, and Y_3 = departure-delay-time for 10 airlines only. Thus, for airline $u = 1, \ldots, 10$ and variable $j = 1, 2, 3$, we denote the histogram valued observation by $Y_{uj} = \{[a_{ujk}, b_{ujk}), p_{ujk} : k = 1, \ldots, s_{uj}\}$ where the histogram sub-interval $[a_{ujk}, b_{ujk})$ has relative frequency p_{ujk} with $\sum_k p_{ujk} = 1$. The number of subintervals s_{uj} can vary across observations (u) and across variables (j).

Figure 29.1 shows the tree that results when clustering the data by a Ward's method agglomerative hierarchy algorithm applied to these histogram data when the Euclidean extended Ichino–Yaguchi distance measure is used; see Ichino and Yaguchi (1994) and Kim and Billard (2011, 2013), for details. Since there are too many classical observations to be able to build an equivalent tree on the original observations themselves, we resort to using classical surrogates. In particular, we calculate the sample means for each variable and airline. The resulting Ward's method agglomerative tree using Euclidean distances between the means is shown in Figure 29.2.

TABLE 29.1
Airline data.

Y_1	$[0,70)$	$[70,110)$	$[110,150)$	$[150,190)$	$[190,230)$	$[230,270)$
1	.00017	.10568	.33511	.20430	.12823	.045267
2	.13464	.10799	.01823	.37728	.35063	.01122
3	.70026	.22415	.07264	.00229	.00065	.00000
4	.26064	.21519	.34916	.06427	.02798	.01848
5	.17867	.41499	.40634	.00000	.00000	.00000
6	.28907	.41882	.28452	.00683	.00076	.00000
7	.00000	.00000	.00000	.00000	.03811	.30793
8	.39219	.31956	.19201	.09442	.00182	.00000
9	.00000	.61672	.36585	.00348	.00174	.00000
10	.76391	.20936	.01719	.00645	.00263	.00048
Y_2	$[-40,-20)$	$[-20,0)$	$[0,20)$	$[20,40)$	$[40,60)$	$[60,80)$
1	.09260	.38520	.28589	.09725	.04854	.03046
2	.09537	.45863	.30014	.07433	.03226	.01683
3	.12958	.41361	.21008	.09097	.04450	.02716
4	.06054	.44362	.33475	.08648	.03510	.01865
5	.08934	.44957	.29683	.07493	.01729	.03746
6	.07967	.36646	.28376	.10698	.06070	.03794
7	.14024	.30030	.29573	.18293	.03659	.01067
8	.03949	.40899	.33727	.12483	.04585	.02224
9	.07840	.44599	.21603	.10627	.04530	.03310
10	.10551	.55693	.22989	.06493	.02363	.01074
Y_3	$[-15,5)$	$[5,25)$	$[25,45)$	$[45,65)$	$[65,85)$	$[85,105)$
1	.67762	.16988	.05714	.03219	.01893	.01463
2	.84993	.07293	.03086	.01964	.01683	.00421
3	.65249	.14071	.06872	.04025	.02749	.01669
4	.77650	.14516	.04036	.01611	.01051	.00526
5	.63112	.24784	.04323	.02017	.02882	.00288
6	.70030	.12064	.06297	.04628	.02049	.01290
7	.73323	.16463	.04726	.01677	.01220	.00305
8	.78711	.12165	.05311	.01816	.00772	.00635
9	.71080	.12369	.05749	.03310	.01916	.00523
10	.83600	.10862	.03032	.01408	.00573	.00286

It is immediately apparent that the trees differ, even though both have the same "determinant" — agglomerative, Ward's method, and Euclidean distances. However, one tree is based on the means only while the other is based on the histograms; i.e., the histogram tree of Figure 29.1, in addition to the information in the means, also uses information in the internal variances of the observed values. Although the details are omitted, it is easy to show that, e.g., airlines $(1,2,4)$ have similar means and similar variances overall; however, by omitting the information contained in the variances (as in Figure 29.2),

TABLE 29.1
Airline data (continued).

Y_1	$[270, 310)$	$[310, 350)$	$[350, 390)$	$[390, 430)$	$[430, 470)$	$[470, 540]$
1	.07831	.07556	.02685	.00034	.00000	.00000
2	.00000	.00000	.00000	.00000	.00000	.00000
3	.00000	.00000	.00000	.00000	.00000	.00000
4	.03425	.02272	.00729	.00000	.00000	.00000
5	.00000	.00000	.00000	.00000	.00000	.00000
6	.00000	.00000	.00000	.00000	.00000	.00000
7	.34299	.21494	.08384	.01220	.00000	.00000
8	.00000	.00000	.00000	.00000	.00000	.00000
9	.00523	.00174	.00348	.00000	.00000	.00174
10	.00000	.00000	.00000	.00000	.00000	.00000
Y_2	$[80, 100)$	$[100, 120)$	$[120, 140)$	$[140, 160)$	$[160, 200)$	$[200, 240]$
1	.01773	.01411	.00637	.00654	.01532	.00000
2	.01403	.00281	.00281	.00000	.00281	.00000
3	.02094	.01440	.01276	.00884	.02716	.00000
4	.00797	.00661	.00356	.00051	.00220	.00000
5	.00865	.00576	.00576	.00576	.00865	.00000
6	.02883	.00835	.01366	.00835	.00531	.00000
7	.00762	.00305	.00152	.00762	.01372	.00000
8	.00817	.00635	.00227	.00136	.00318	.00000
9	.01916	.01394	.00871	.01220	.02091	.00000
10	.00286	.00143	.00167	.00095	.00143	.00000
Y_3	$[105, 125)$	$[125, 145)$	$[145, 165)$	$[165, 185)$	$[185, 225)$	$[225, 265]$
1	.00878	.00000	.00361	.00947	.00775	.00000
2	.00281	.00000	.00000	.00140	.00140	.00000
3	.01407	.00000	.01014	.01407	.01538	.00000
4	.00305	.00000	.00085	.00068	.00153	.00000
5	.00865	.00000	.00865	.00000	.00865	.00000
6	.01897	.00000	.00986	.00607	.00152	.00000
7	.00457	.00000	.00152	.00762	.00915	.00000
8	.00227	.00000	.00136	.00045	.00182	.00000
9	.01045	.00000	.01742	.01394	.00871	.00000
10	.00095	.00000	.00072	.00048	.00024	.00000

while airlines $(1, 2)$ have comparable means, they differ from those for airline 4. That is, the classical surrogate analysis is based on the means only.

A polythetic divisive tree built on the Euclidean extended Ichino–Yaguchi distances for the histograms is shown in Figure 29.3; see Kim and Billard (2011) for this algorithm. The corresponding monothetic divisive tree is comparable. This tree is different again from those of Figures 29.1 and 29.2; these differences reflect the fact that different clustering algorithms, along with different distance matrices and different methods, can construct quite different trees.

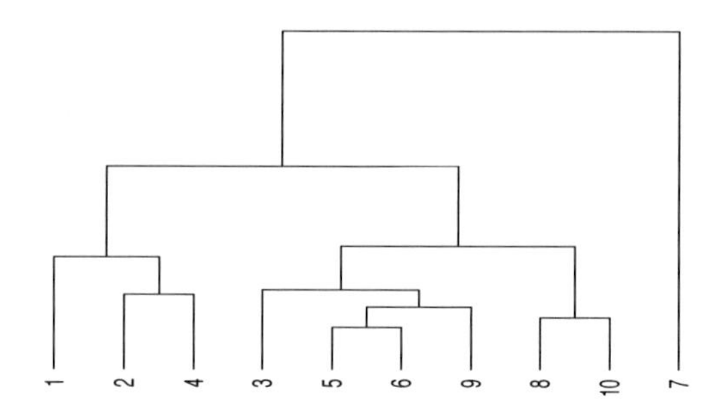

FIGURE 29.1
Ward's method agglomerative tree based on histograms.

Example 29.2 Figure 29.4 displays simulated individual classical observations (Y_1, Y_2) drawn from bivariate normal distributions $\mathcal{N}_2(\boldsymbol{\mu}, \boldsymbol{\Sigma})$. There are five samples each with $n = 100$ observations. Sample $S = 1$ has mean $\boldsymbol{\mu} = (5, 0)$, standard deviations $\sigma_1 = \sigma_2 = .25$ and correlation coefficient $\rho = 0$; samples S=2,3 have $\boldsymbol{\mu} = (1, 1)$, $\sigma_1 = \sigma_2 = .25$ and $\rho = 0$; and samples $S = 4, 5$ have $\boldsymbol{\mu} = (1, 1)$, $\sigma_1 = \sigma_2 = 1$ and $\rho = .8$. Each of the samples can be aggregated to produce a bivariate histogram observation $\mathbf{Y}_s$, $s = 1, \ldots, 5$.

When a divisive algorithm for histogram data is applied to these data, three clusters emerge containing the observations $C_1 = \{Y_1\}$, $C_2 = \{Y_2, Y_3\}$, and $C_3 = \{Y_4, Y_5\}$, respectively. In contrast, applying algorithms, e.g., a K-means

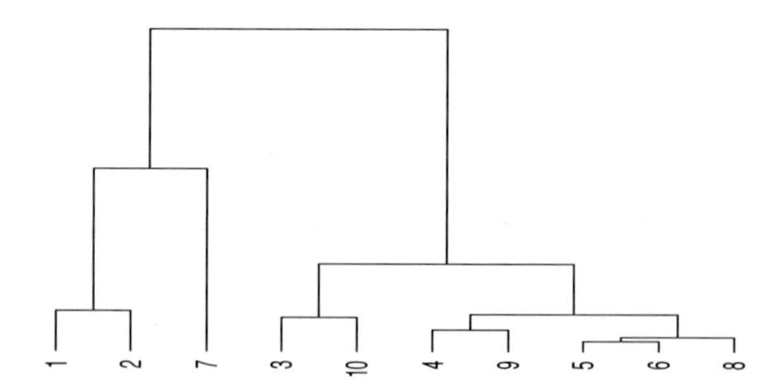

FIGURE 29.2
Ward's method agglomerative tree based on means.

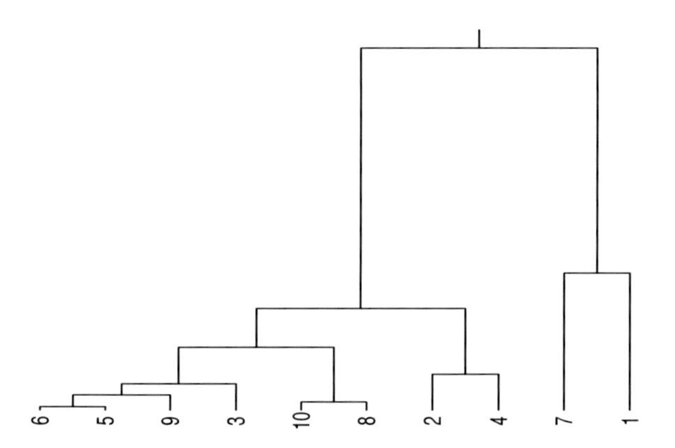

FIGURE 29.3
Polythetic divisive tree based on histograms.

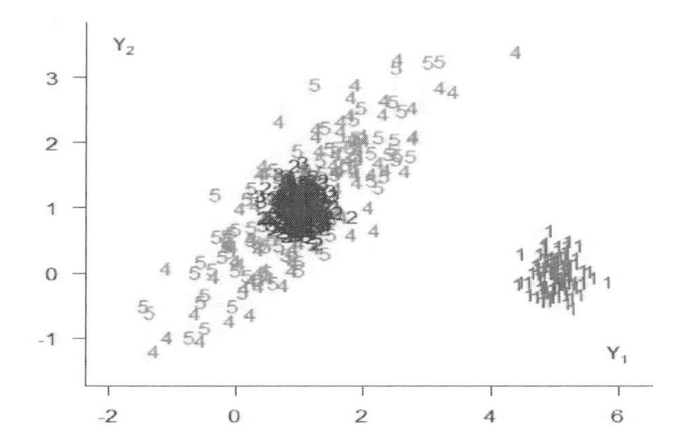

FIGURE 29.4
Simulated data — How many clusters?

method, to the classical observations (or, to classical surrogates such as the means), identifies only two clusters, viz., $C_1 = \{Y_1\}$ and $C_2 = \{Y_2, Y_3, Y_4, Y_5\}$. That is, since information such as the internal variations is not part of a classical analysis, the classical clustering analyses are unable to identify observations Y_2 and Y_3 as being different from observations Y_4 and Y_5.

Example 29.3 Consider the data of Table 29.2 where for simplicity, we restrict attention to minimum and maximum monthly temperatures for four months only, January, April, July, and October, $Y_1 - Y_4$, respectively, in 1988 for $n = 10$ weather stations. The interval values for station $u = 1, \ldots, 10$ and variable $j = 1, \ldots, 4$ are denoted by $Y_{uj} = [a_{uj}, b_{uj}]$. Elevation is also included as Y_5; note Y_5 is a classical value and so is a special case of an interval value with $a_{u5} \equiv [a_{u5}, a_{u5}]$. The data are extracted from `http://dss.ucar.edu/datasets/ds578.5` which contain annual monthly weather values for several variables for many stations in China over many years.

TABLE 29.2
Temperature intervals and elevation.

Station u	January $[a_{u1}, b_{u1}]$	April $[a_{u2}, b_{u2}]$	July $[a_{u3}, b_{u3}]$	October $[a_{u4}, b_{u4}]$	Elevation a_{u5}
1	$[-18.4, -7.5]$	$[-0.1, 13.2]$	$[17.0, 26.5]$	$[0.6, 13.1]$	4.82
2	$[-20.0, -9.6]$	$[0.2, 11.9]$	$[17.8, 27.2]$	$[-0.2, 12.5]$	3.44
3	$[-23.4, -15.5]$	$[-4.5, 9.5]$	$[12.9, 23.0]$	$[-4.0, 8.9]$	14.78
4	$[-27.9, -16.0]$	$[-1.5, 12.0]$	$[16.1, 25.0]$	$[-2.6, 10.9]$	4.84
5	$[-8.4, 9.0]$	$[1.7, 16.4]$	$[10.8, 23.2]$	$[1.4, 18.7]$	73.16
6	$[2.3, 16.9]$	$[9.9, 24.3]$	$[17.4, 22.8]$	$[14.5, 23.5]$	32.96
7	$[2.8, 16.6]$	$[10.4, 23.4]$	$[16.9, 24.4]$	$[12.4, 19.7]$	37.82
8	$[10.0, 17.7]$	$[15.8, 23.9]$	$[24.2, 33.8]$	$[19.2, 27.6]$	2.38
9	$[11.5, 17.7]$	$[17.8, 24.2]$	$[25.8, 33.5]$	$[20.3, 26.9]$	1.44
10	$[11.8, 19.2]$	$[16.4, 22.7]$	$[25.6, 32.6]$	$[20.4, 27.3]$	0.02

A principal component analysis on these interval-valued data produces the projections onto the $PC_1 \times PC_2$ space shown in Figure 29.5. Details of the methodology and the visualization construction can be found in Le-Rademacher and Billard (2012), and further details of this particular data set are in Billard and Le-Rademacher (2012).

Notice in particular that since the original observations are hypercubes in $\mathbb{R}^p$ space, so we observe that the corresponding principal components are hypercubes in PC-space. The relative sizes of these PC hypercubes reflect the relative sizes of the data hypercubes. For example, if we compare the observed values of stations $u = 5$ and $u = 10$ in Table 29.2, it is clear that the intervals across the variables for $u = 5$ are on balance wider than are those for

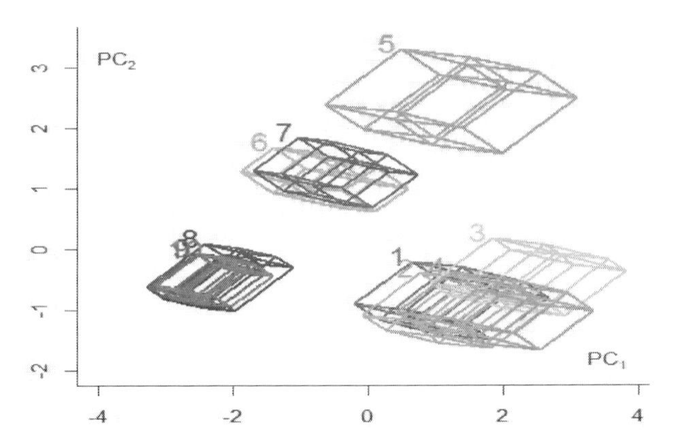

FIGURE 29.5
PCA based on intervals.

$u = 10$; thus, the principal component hypercube is larger for $u = 5$ than for $u = 10$. That is, the observation $u = 5$ has a larger internal variation. These internal variations are a component of the covariance terms in the covariance (and correlation) matrix. This feature is not possible in a classical analysis, with the point observation in $\mathbb{R}^p$ being transformed into but a point value in PC-space, as shown in Figure 29.6 for the classical principal component analysis performed on the interval means. While both the symbolic and classical analyses showed the temperatures as being of comparable importance to PC_1 with elevation being important only for PC_2, the visualizations through the PC hypercubes of Figure 29.5 are more informative than are the PC points of Figure 29.6.

29.4 Conclusion

By the time that Eddy (1986) considered the future of computers in statistical research, it was already clear that a computer revolution was raising its head over the horizon. This revolution was not simply focussed on bigger and better computers to do traditional calculations on a larger scale, though that too was a component, then and now. Rather, more expansively, entirely new ways of approaching our art were to be the new currency of the looming 21st century. Early signs included the emergence of new methodologies such as the bootstrap (Efron, 1979) and Gibbs sampler (Geman and Geman, 1984), though both owed their roots to earlier researchers. While clearly these and similar computational methodologies had not been feasible in earlier days thereby being a product of computer advances, they are still classical approaches per se.

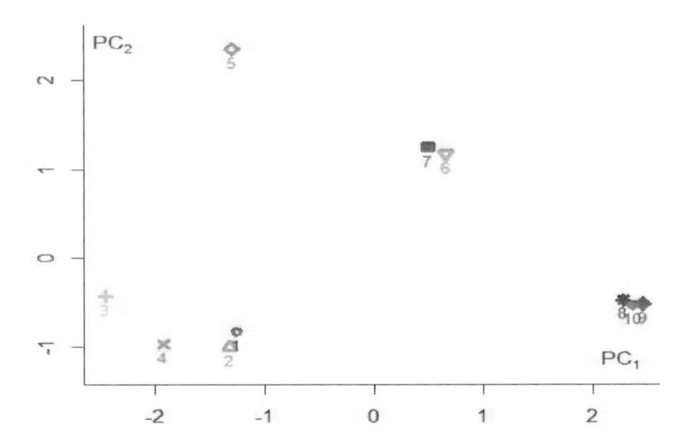

FIGURE 29.6
PCA based on means.

By the 1990s, COPSS Presidential Addresses referred to the upcoming information and technological revolution, its waves already heading for the unwary; see, e.g., Kettenring (1997) and Billard (1995, 1997).

However, the real advances will take quite different formats to those predicted in the 1990s. In a very prescient comment, Schweizer (1984) declared that "distributions are the numbers of the future." The present is that future. Furthermore, today's future consists of a new paradigm whereby new methodologies, and new theories to support those methodologies, must be developed if we are to remain viable players as data analysts. These new methods must also be such that the classical models of the still-present and past come out correctly as special cases of the new approaches.

In this chapter, one such approach, viz., symbolic data, has been described albeit ever so briefly. While a study of the literature may at first suggest there are many symbolic techniques currently available, in reality there are very few and even then those few handle relatively narrowly defined situations.

There are two major directions for future work: one is to develop the new methodologies for new data structures and to extend the plethora of situations that a century or more of research in so-called classical statistics produced, while the other is to establish mathematical underpinning to support those new methods (somewhat akin to the theoretical foundations provided initially by Bickel and Freedman (1981), and Singh (1981), which validated the early bootstrap work). One certainty is sure — the present-future demands that we engage our energies in addressing the myriad of issues surrounding large and complex data sets. It is an exciting time to be a part of this undertaking.

References

Bickel, P.J. and Freedman, D.A. (1981). Some asymptotic theory for the bootstrap. *The Annals of Statistics*, 9:1196–1217.

Billard, L. (1995). The roads travelled. 1995. *Biometrics*, 51:1–11.

Billard, L. (1997). A voyage of discovery. *Journal of American Statistical Association*, 92:1–12.

Billard, L. (2011). Brief overview of symbolic data and analytic issues. *Statistical Analysis and Data Mining*, 4:149–156.

Billard, L., and Diday, E. (2006). *Symbolic Data Analysis: Conceptual Statistics and Data Mining*. Wiley, Chichester.

Billard, L., and Le-Rademacher, J. (2012). Principal component analysis for interval data. *Wiley Interdisciplinary Reviews: Computational Statistics*, 4:535–540.

Bock, H.-H. and Diday, E. (2000). *Analysis of Symbolic Data: Exploratory Methods for Extracting Statistical Information from Complex Data.* Springer, Berlin.

Diday, E. (1987). Introduction à l'approche symbolique en analyse des données. *Premières Journées Symbolique-Numérique*, CEREMADE, Université Paris–Dauphine, Paris, France, pp. 21–56.

Eddy, W.F. (1986). Computers in statistical research. *Statistical Science*, 1:419–437.

Efron, B. (1979). Bootstrap methods: Another look at the jackknife. *The Annals of Statistics*, 7:1–26.

Falduti, N. and Taibaly, H. (2004). *Étude des retards sur les vols des compagnies aériennes.* Report CEREMADE, Université Paris–Dauphine, Paris, France.

Geman, S. and Geman, D. (1984). Stochastic relaxation, Gibbs distributions, and the Bayesian restoration of images. *IEEE Transactions on Pattern Analysis and Machine Intelligence*, 6:721–744.

Goodman, A. (2011). Emerging topics and challenges for statistical analysis and data mining. *Statistical Analysis and Data Mining*, 4:3–8.

Heitjan, D.F. and Rubin, D.B. (1991). Ignorability and coarse data. *The Annals of Statistics*, 19:2244–2253.

Ichino, M. and Yaguchi, H. (1994). Generalized Minkowski metrics for mixed feature type data analysis. *IEEE Transactions on Systems, Man and Cybernetics*, 24:698–708.

Kettenring, J.R. (1997). Shaping statistics for success in the 21st century. *Journal of the American Statistical Association*, 92:1229–1234.

Kim, J. and Billard, L. (2011). A polythetic clustering process for symbolic observations and cluster validity indexes. *Computational Statistics and Data Analysis*, 55:2250–2262.

Kim, J. and Billard, L. (2013). Dissimilarity measures for histogram-valued observations. *Communications in Statistics: Theory and Methods*, 42:283–303.

Le-Rademacher, J. and Billard, L. (2012). Symbolic-covariance principal component analysis and visualization for interval-valued data. *Journal of Computational and Graphical Statistics*, 21:413–432.

Noirhomme-Fraiture, M. and Brito, M.P. (2011). Far beyond the classical data models: Symbolic data analysis. *Statistical Analysis and Data Mining*, 4:157–170.

Schweizer, B. (1984). Distributions are the numbers of the future. In *Proceedings of the Mathematics of Fuzzy Systems Meeting* (A. di Nola and A. Ventes, Eds.). Università di Napoli, Napoli, Italy, pp. 137–149.

Singh, K. (1981). On the asymptotic accuracy of Efron's bootstrap. *The Annals of Statistics*, 9:1187–1195.

30

Lessons in biostatistics

Norman E. Breslow
Department of Biostatistics
University of Washington, Seattle, WA

Today's medical journals are full of factual errors and false conclusions arising from lack of statistical common sense. Reflecting on personal experiences, I argue that statisticians can substantially improve medical science by informed application of standard statistical principles. Two specific areas are identified where lack of such input regularly produces faulty research. Statisticians are needed more than ever to bring rigor to clinical research.

30.1 Introduction

Biostatisticians develop and apply statistical concepts and methods to clinical medicine, to laboratory medicine and to population medicine or public health. During the fifty years since COPSS was organized, their work has become increasingly important. Major medical journals often insist on biostatistical review of submitted articles. Biostatistics graduates are in high demand for work in industry, government and academia. They occupy prominent positions as heads of corporations and universities, deans of schools of public health and directors of major research programs.

In spite of the heightened visibility of the profession, much of today's medical research is conducted without adequate biostatistical input. The result is not infrequently a waste of public resources, the promulgation of false conclusions and the exposure of patients to possible mistreatment. I describe a few of the more common episodes of flawed research with which I have come in contact, which involve "immortal time" in follow-up studies and lack of proper validation of discriminant rules. I discuss the lessons learned both from these episodes and more generally from my decades of work in childhood cancer. The primary focus of the chapter is on biostatistics in clinical medicine. Other chapters in this volume discuss the role of statistics in laboratory medicine, especially genetics, and in public health.

30.2 It's the science that counts

My introduction to biostatistics was in graduate school. During the school year a small group from the Stanford statistics department made the trek to the medical school for a weekly seminar. There we learned from medical faculty and our professors about the research problems on which they were collaborating. During the summer we took jobs with local research organizations. At weekly meetings back on campus, we presented the problems stemming from our work and got advice from each other and the professors on how to approach them.

One summer I worked at the state health department. There was considerable interest in the possibility of an infectious origin for leukemia and speculation that transmission of the putative infectious agent might occur between animals and humans. The health department was conducting a census of cancer occurrence in dogs and cats in Alameda county, and the epidemiologists wanted to evaluate possible space-time clustering of leukemia cases in people and in cats. The maps at their disposal, however, were inaccurate. Ascertainment of the geographic coordinates needed for quantitative analysis was subject to substantial error. My assignment was to read up on spatial statistical distributions and develop a measurement error model. I was having considerable difficulty.

I will never forget the stern advice I received from Professor Lincoln Moses following my presentation at the weekly meeting back on campus. "What you need is a good set of maps," he said. "Try the water company!" Obviously, in his mind, as later in mine, the best method of dealing with measurement error was to avoid it! Bradford Hill gave similar advice:

> "One must go and seek more facts, paying less attention to the techniques of handling the data and far more to the development and perfection of the methods of obtaining them." (Hill, 1953)

As it turned out, the East Bay Municipal Utilities District (EBMUD) had just completed a very extensive and costly mapping program. The maps were so accurate that you had to decide where in the residence to plot the case to determine the coordinates. Executives in charge of the program were delighted to learn that their maps would serve not only corporate interests but also those of public health. Instead of working on a statistical methods problem, I spent my remaining time that summer on administrative issues related to the use of the maps by the health department. A photograph of me with health department and EBMUD officials poring over the maps was published in the corporate magazine and hangs in my office today. The lesson learned was invaluable.

I had a similar experience shortly after my arrival at the University of Washington in 1968. Having applied for a position in the Mathematics Department, not realizing it was in the process of downsizing and discharging most of its statisticians, I wound up as a biostatistician in the Medical School. My support came mainly from service as statistician to the Children's Cancer Study Group. In those days the MDs who chaired the protocol study committees sometimes compiled the data themselves (one dedicated researcher meticulously arranged the flow sheets on her living room floor) and sent me simple data summaries with a request for calculation of some standard statistic. I was appalled by the routine exclusion from randomized treatment comparisons of patients who had "inadequate trials" of chemotherapy due to early discontinuation of the assigned treatment regimen or early death. It was clear that a more systematic approach to data collection and analysis was needed.

My colleague Dick Kronmal, fortunately, had just developed a computer system to store and summarize data from longitudinal studies that generated multiple records per patient (Kronmal et al., 1970). This system was perfect for the needs of the children's cancer group. It allowed me to quickly establish a Data and Statistical Center both for the group and for the National Wilms Tumor Study (NWTS), whose steering committee I joined as a founding member in 1969. (Wilms is an embryonal tumor of the kidney that occurs almost exclusively in children.) Once again the lesson learned was that "development and perfection of the methods of obtaining the data" were at least as important to the overall scientific enterprise as were the statistical methods I subsequently helped develop to "handle" right censored survival data. Having me, as statistician, take control of data collection and processing, while sharing responsibility for data quality with the clinicians, made it easier for me to then also exercise some degree of control over which patients were included in any given analysis.

My experience was not unusual. The role of biostatisticians in cooperative clinical research was rapidly evolving as the importance of their contributions became more widely appreciated. It soon became commonplace for them to occupy leadership positions within the cooperative group structure, for example, as heads of statistics departments or as directors of independently funded coordinating centers.

A steady diet of service to clinical trial groups, however, can with time become tedious. It also interferes with production of the first-authored papers needed for promotion in academia. One way to relieve the tedium, and to generate the publications, is to get more involved in the science. For example, the biostatistician can propose and conduct ancillary studies that utilize the valuable data collected through the clinical trials mechanism. The first childhood leukemia study in which I was involved was not atypical in demonstrating that treatment outcomes varied much more with baseline host and disease characteristics, in this case age and the peripheral white blood cell count, than with the treatments the study was designed to assess (Miller et al., 1974). This result was apparently a revelation to the clinicians. They jumped on it to propose treatment stratification based on prognostic factor groups in subsequent trials, so that the most toxic and experimental treatments were reserved for those who actually needed them. Subsequently, I initiated several studies of prognosis in Wilms tumor that resulted in greater appreciation for the adverse outcomes associated with tumor invasion of regional lymph nodes and ultimately to changes in the staging system. Fascinated by how well Knudson's (Knudson, Jr., 1971) 2-hit mutational model explained the genetic epidemiology of retinoblastoma, another embryonal tumor of a paired organ (in this case the eye rather than the kidney), I conducted studies of the epidemiology of Wilms tumor that provided strong evidence for genetic heterogeneity, an explanation for its lower incidence and younger ages-at-onset in Asians and a hypothesis regarding which survivors were at especially high risk of end stage renal disease in young adulthood (Breslow and Beckwith, 1982; Breslow et al., 2006; Lange et al., 2011). Since 1991, I have served as Principal Investigator on the NIH grant that funds the continued follow-up of NWTS survivors for "late effects" associated with Wilms tumor and its treatment. This study has occupied an increasingly important place in my research repertoire.

30.3 Immortal time

In my opening lecture to a class designed primarily for second year doctoral students in epidemiology, I state the First Rule of survival analysis: *Selection into the study cohort, or into subgroups to be compared in the analysis, must not depend on events that occur after the start of follow-up.* While this point may be obvious to a statistician, certainly one trained to use martingale arguments to justify inferences about how past history influences rates of future events, it was not obvious to many of the epidemiologists. The "immortal time" bias that results from failure to follow the rule has resulted, and continues regularly to result, in grossly fraudulent claims in papers published in the most prestigious medical journals.

My first exposure to the issue came soon after I started work with the children's cancer group. The group chair was puzzled by a recently published article that called into question the standard criteria for evaluation of treatment response in acute leukemia. These included the stipulation that patients with a high bone marrow lymphocyte count (BMLC) be excluded from the excellent response category. Indeed, a high BMLC often presaged relapse, defined as 5% or higher blast cells in the marrow. The article in question, however, reported that patients whose lymphocytes remained below the threshold level of 20% of marrow cells throughout the period of remission tended to have shorter remissions than patients whose BMLC exceeded 20% on at least one occasion. Although I knew little about survival analysis, and had not yet articulated the First Rule, I was familiar with random variation and the tendency of maxima to increase with the length of the series. Intuition suggested that the article's conclusion, that there was "no justification for excluding a patient from complete remission status because of bone marrow lymphocytosis," was erroneous.

Accordingly, using new data from the children's cancer group, I attempted to convince my clinical colleagues that the reasoning was fallacious (Breslow and Zandstra, 1970). I first replicated the earlier findings by demonstrating that, when patients were classified into three categories according to the BMLC values observed during remission, the "remission duration" (progression-free survival) curve for the group with highest *maximum* BMLC was on top and that for the group with lowest maximum BMLC was on the bottom. When patients were classified according to the *average* of their BMLC values during remission, however, the ordering was reversed. Both comparisons were highly statistically significant. Of course, even the analysis based on average counts violated the First Rule. Nowadays one would employ time-dependent covariates or stratification to evaluate how the history of BMLC affected future relapse rates. The experience was a valuable lesson about the importance of "statistical thinking" in clinical research.

Many biostatisticians were sensitized to the issue of immortal time by Mitch Gail's critique of early claims of the efficacy of heart transplantation (Gail, 1972). To illustrate the problems with the statistical approach taken by cardiac surgeons in those days, he compared survival curves from time of admission as a transplant candidate according to whether or not the patient had subsequently received a transplant. He pointed out that patients who died early had less opportunity to receive a transplant, whereas those who did receive one were guaranteed, by definition, to have survived long enough for a suitable donor to be found. In effect, person-months of observation prior to transplant were unfairly subtracted from the total person-months for the control group, biasing their survival rate downwards, and added to the person-months for the transplant group, biasing their survival rate upwards. Correct accounting for the timing of transplant in the statistical comparison was subsequently undertaken by several statistical teams, for example, by use of time-dependent covariates in the Cox model (Crowley and Hu, 1977). When the data were properly analyzed, transplant as performed at the time was found to have little benefit.

Nick Day and I, in the section of our second IARC monograph (Breslow and Day, 1987) on allocation of person-time to time-dependent exposure categories, called attention to a fallacious claim of decreasing death rates with increasing duration of work in the polyvinyl-chloride industry. Here the investigators had contrasted standardized mortality ratios (of numbers of deaths observed to those expected from age-specific population rates) among workers employed for 0–14 versus 15+ years in the industry. Not only all deaths occurring beyond 15 years, however, but also all person-time accumulated by persons employed for 15+ years, had been allocated to the latter group. Day and I stated: "The correct assignment of each increment in person-years of follow-up is to that same exposure category to which a death would be assigned should it occur at that time." In other words, the first 15 years of employment time for the vinyl-chloride workers whose employment continued beyond that point should have been assigned to the 0–14 group. When this correction was made, the 15+ year exposure group had a slightly higher mortality ratio than did the 0–14 year group.

Faculty at McGill University in Montréal, Canada, have repeatedly called attention to erroneous conclusions in the medical literature stemming from immortal time bias. One recent article takes issue with the finding that actors who won an Oscar lived on average nearly four years longer than those in a matched control group (Sylvestre et al., 2006). The authors pointed out that, as long ago as 1843, William Farr warned against the hazards of "classifying persons by their status at the end of follow-up and analyzing them as if they had been in these categories from the outset" (Farr, 1975). Farr continued

> "... certain professions, stations and ranks are only attained by persons advanced in years; and some occupations are followed only in youth; hence it requires no great amount of sagacity to perceive that 'the mean age at death' [· · ·] cannot be depended on in investigating the influence of occupation, rank and profession upon health and longevity."

Noting the relatively early ages at death of Cornets, Curates and Juvenile Barristers, he concluded wryly: "It would be almost necessary to make them Generals, Bishops and Judges — for the sake of their health."

Mistakes are made even when investigators are seemingly aware of the problem. A 2004 report in *The New England Journal of Medicine* examined the effect on survival of a delay in kidney transplantation among children with end stage renal disease. The authors stated:

> "Delay in kidney transplantation as a potential risk factor for early death was analyzed by comparing mortality among groups with different lengths of time until transplantation. To account for survival bias, delay as a predictor of death was analyzed beginning 2 years after the initiation of renal replacement therapy. There was no significant difference in mortality observed among groups with different lengths of time until transplantation (Fig 3)" (McDonald and Craig, 2004).

Close examination of their Figure 3, however, leads to a different conclusion. Survival curves from two years after onset of renal replacement therapy (dialysis or transplant) were shown separately for those with preemptive transplant (no delay), less than one-year delay and 1–2 years delay, categories based on information available at the start of follow-up at two years. They are in the anticipated order, with the survival outcomes best for those having had an immediate transplant followed in succession by those having had a 0–1 or 1–2 year delay. Had the authors simply added a fourth curve for those not yet transplanted by year 2, they would have found that it lay below the others. This would have confirmed the anticipated rank order in survival outcomes under the hypothesis that longer delay increased subsequent mortality. However, they mistakenly split the fourth group into those who never received a transplant and those who did so at some point after two years. The survival curve for the "no transplant" group was far below all the others, with many deaths having occurred early on prior to a suitable donor becoming available, while the curve for the "$\geq$ 2 years" group was second from highest due to immortal time. The clear message in the data was lost. I have used this widely cited paper as the basis for several exam and homework questions. Students often find the lessons about immortal time to be the most important they learned from the class.

I mentioned earlier my dissatisfaction with the exclusion of patients with "inadequate trials" from randomized treatment comparisons, a policy that was widely followed by the children's cancer group when I joined it. Such "per protocol" analyses constitute another common violation of the First Rule. Exclusion of patients based on events that occur after the start of follow-up, in particular, the failure to receive protocol treatment, invites bias that is avoided by keeping all eligible patients in the study from the moment they are randomized. Analyses using all the eligible patients generate results that apply to a real population and that are readily compared with results from like studies. Attempts to clearly describe the fictitious populations to which the per protocol analyses apply are fraught with difficulty. My colleague Tom Fleming has thoughtfully discussed the fundamental principle that all patients be kept in the analysis following randomization, its rationale and its ramifications (Fleming, 2011).

30.4 Multiplicity

Whether from cowardice or good sense, I consciously strived throughout my career to avoid problems involving vast amounts of data collected on individual subjects. There seemed to be enough good clinical science to do with the limited number of treatment and prognostic variables we could afford to collect for the childhood cancer patients. The forays into the epidemiology of Wilms tumor similarly used limited amounts of information on gender, ages at onset, birth weights, histologic subtypes, precursor lesions, congenital malformations and the like. This allowed me to structure analyses using a small number of variables selected a priori to answer specific questions based on clearly stated hypotheses.

My successors do not have this luxury. Faced with the revolution in molecular biology, they must cope with increasingly high dimensional data in an attempt to assist clinicians deliver "personalized medicine" based on individual "omics" (genomics, epigenomics, proteomics, transcriptomics, metabolomics, etc.) profiles. I hope that widespread enthusiasm for the new technologies does not result in a tremendous expenditure of resources that does little to advance public health. This can be avoided if statisticians demand, and are given, a meaningful role in the process. I am impressed by how eagerly my younger colleagues, as well as some of my peers, have responded to the challenge.

The problems of multiplicity were brought home to me in a forceful way when I read an article based on data from the 3rd and 4th NWTS trials supplied by our pathologist to a group of urologists and pathologists at the prestigious Brady Institute at Johns Hopkins Hospital (JHH); see Partin et al. (1994). Regrettably, they had not solicited my input. I was embarrassed that a publication based on NWTS data contained such blatant errors. For one thing, although our pathologist had supplied them with a case-control sample that was overweighted with children who had relapsed or had advanced disease at onset, they ignored the design and analysed the data as a simple random sample. Consequently their Kaplan–Meier estimates of progression-free survival were seriously in error, suggesting that nearly half the patients with "favorable histology" relapsed or died within five years of diagnosis, whereas the actual fraction who did so was about 11%.

A more grievous error, however, was using the same data both to construct and to validate a predictive model based on a new technology that produced moderately high dimensional quantitative data. Determined to improve on the subjectivity of the pathologist, the JHH team had developed a technique they called nuclear morphometry to quantify the malignancy grading of Wilms and other urologic tumors, including prostate. From the archived tumor slide submitted by our pathologist for each patient, they selected 150 blastemal nuclei for digitizing. The digitized images were then processed using a commercial software package known as Dyna CELL. This produced for each nucleus a series of 16 shape descriptors including, for example, area, perimeter, two measures of roundness and two of ellipticity. For each such measure 17 descriptive statistics were calculated from the distribution of 150 values: Mean, variance, skewness, kurtosis, means of five highest and five lowest values, etc. This yielded $16 \times 17 = 242$ nuclear morphometric observations per patient. Among these, the *skewness* of the nuclear roundness factor (SNRF) and the average of the lowest five values for ellipticity as measured by the feret diameter (distance between two tangents on opposite sides of a planar figure) method (LEFD) were found to best separate cases from controls, each yielding $p = .01$ by univariate logistic regression. SNRF, LEFD and age, a variable I had previously identified as an important prognostic factor, were confirmed by stepwise regression analysis as the best three of the available univariate predictors. They were combined into a discriminant function that, needless to say, did separate the cases from the controls used in its development, although only with moderate success.

TABLE 30.1
Regression coefficients (± SEs) in multivariate nuclear morphometric discriminant functions fitted to three data sets[†].

Risk Factor	Case-Control Sample NWTS + JHH ($n = 108$)*	Case-Control Sample NWTS Alone ($n = 95$)	Prospective Sample NWTS ($n = 218$)
Age (yr)	.02	$.013 \pm .008$	$.017 \pm .005$
SNRF	1.17	$1.23 \pm .52$	$-.02 \pm .26$
LEFD	90.6	121.6 ± 48.4	$.05 \pm 47.5$

[†]From Breslow et al. (1999). Reproduced with permission.
*From Partin et al. (1994)

I was convinced that most of this apparent success was due to the failure to account for the multiplicity of comparisons inherent in the selection of the best 2 out of 242 measurements for the discriminant function. With good cooperation from JHH, I designed a prospective study to validate the ability of their nuclear morphometric score to predict relapse in Wilms tumor (Breslow et al., 1999). I identified 218 NWTS-4 patients who had not been included in the case-control study, each of whom had an archived slide showing a diagnosis by our pathologist of a Wilms tumor having the same "favorable" histologic subtype as considered earlier. The slides were sent to the JHH investigators, who had no knowledge of the treatment outcomes, and were processed in the same manner as for the earlier case-control study. We then contrasted results obtained by re-analysis of data for the 95 NWTS patients in the case-control study, excluding 13 patients from JHH who also had figured in the earlier report, with those obtained by analysis of data for the 218 patients in the prospective study.

The results, reproduced in Table 30.1, were instructive. Regression coefficients obtained using a Cox regression model fitted to data for the 95 NWTS patients in the original study are shown in the third column. They were comparable to those reported by the JHH group based on logistic regression analysis of data for the 95 NWTS plus 13 JHH patients. These latter coefficients, shown in the second column of the table, were used to construct the nuclear morphometric score. Results obtained using Cox regression fitted to the 218 patients in the prospective study, of whom 21 had relapsed and one had died of toxicity, were very different. As I had anticipated, the only variable that was statistically significant was the known prognostic factor age. Coefficients for the two nuclear morphometric variables were near zero. When the original nuclear morphometric score was applied to the prospective data, using the same cutoff value as in the original report, the sensitivity was reduced from

75% to 71% and the specificity from 69% to 56%. Only the inclusion of age in the score gave it predictive value when applied to the new data.

No further attempts to utilize nuclear morphometry to predict outcomes in patients with Wilms tumor have been reported. Neither the original paper from JHH nor my attempt to correct its conclusions have received more than a handful of citations. Somewhat more interest was generated by use of the same technique to grade the malignancy of prostate cancer, for which the JHH investigators identified the *variance* of the nuclear roundness factor as the variable most predictive of disease progression and disease related death. While their initial studies on prostate cancer suffered from the same failure to separate test and validation data that compromised the Wilms tumor case-control study, variance of the nuclear roundness factor did apparently predict adverse outcomes in a later prospective study.

Today the public is anxiously awaiting the anticipated payoff from their investment in omics research so that optimal medical treatments may be selected based on each patient's genomic or epigenomic make-up. Problems of multiplicity inherent in nuclear morphometry pale in comparison to those posed by development of personalized medicine based on omics data. A recent report from the Institute of Medicine (IOM) highlights the important role that statisticians and statistical thinking will play in this development (IOM, 2012). This was commissioned following the exposure of serious flaws in studies at Duke University that had proposed tests based on gene expression (microarray) profiles to identify cancer patients who were sensitive or resistant to specific chemotherapeutic agents (Baggerly and Coombes, 2009). Sloppy data management led to major data errors including off-by-one errors in gene lists and reversal of some of the sensitive/resistant labels. The corrupted data, coupled with inadequate information regarding details of computational procedures, made it impossible for other researchers to replicate the published findings. Questions also were raised regarding the integrity of the validation process. Ultimately, dozens of papers were retracted from major journals, three clinical trials were suspended and an investigation was launched into financial and intellectual/professional conflicts of interest.

The IOM report recommendations are designed to prevent a recurrence of this saga. They emphasize the need for evaluation of a completely "locked down" computational procedure using, preferably, an independent validation sample. Three options are proposed for determining when a fully validated test procedure is ready for clinical trials that use the test to direct patient management. To ensure that personalized treatment decisions based on omics tests truly do advance the practice of medicine, I hope eventually to see randomized clinical trials where test-based patient management is compared directly with current standard care.

30.5 Conclusion

The past 50 years have witnessed many important developments in statistical theory and methodology, a few of which are mentioned in other chapters of this COPSS anniversary volume. I have focussed on the place of statistics in clinical medicine. While this sometimes requires the creation of new statistical methods, more often it entails the application of standard statistical principles and techniques. Major contributions are made simply by exercising the rigorous thinking that comes from training in mathematics and statistics. Having statisticians take primary responsibility for data collection and management often improves the quality and integrity of the entire scientific enterprise.

The common sense notion that definition of comparison groups in survival analyses should be based on information available at the beginning of follow-up, rather than at its end, has been around for over 150 years. When dealing with high-dimensional biomarkers, testing of a well defined discriminant rule on a completely new set of subjects is obviously the best way to evaluate its predictive capacity. Related cross-validation concepts and methods have been known for decades. As patient profiles become more complex, and biology more quantitative, biostatisticians will have an increasingly important role to play in advancing modern medicine.

References

Baggerly, K.A. and Coombes, K.R. (2009). Deriving chemosensitivity from cell lines: Forensic bioinformatics and reproducible research in high-throughput biology. *The Annals of Applied Statistics*, 3:1309–1334.

Breslow, N.E. and Beckwith, J.B. (1982). Epidemiological features of Wilms tumor — Results of the National Wilms Tumor Study. *Journal of the National Cancer Institute*, 68:429–436.

Breslow, N.E., Beckwith, J.B., Perlman, E.J., and Reeve, A.E. (2006). Age distributions, birth weights, nephrogenic rests, and heterogeneity in the pathogenesis of Wilms tumor. *Pediatric Blood Cancer*, 47:260–267.

Breslow, N.E. and Day, N.E. (1987). *Statistical Methods in Cancer Research II: The Design and Analysis of Cohort Studies.* IARC Scientific Publications. International Agency for Research on Cancer, Lyon, France.

Breslow, N.E., Partin, A.W., Lee, B.R., Guthrie, K.A., Beckwith, J.B., and Green, D.M. (1999). Nuclear morphometry and prognosis in favorable histology Wilms tumor: A prospective reevaluation. *Journal of Clinical Oncology*, 17:2123–2126.

Breslow, N.E. and Zandstra, R. (1970). A note on the relationship between bone marrow lymphocytosis and remission duration in acute leukemia. *Blood*, 36:246–249.

Crowley, J. and Hu, M. (1977). Covariance analysis of heart transplant survival data. *Journal of the American Statistical Association*, 72:27–36.

Farr, W. (1975). *Vital Statistics: A Memorial Volume of Selections from the Writings of William Farr*. Scarecrow Press, Metuchen, NJ.

Fleming, T.R. (2011). Addressing missing data in clinical trials. *Annals of Internal Medicine*, 154:113–117.

Gail, M.H. (1972). Does cardiac transplantation prolong life? A reassessment. *Annals of Internal Medicine*, 76:815–817.

Hill, A.B. (1953). Observation and experiment. *New England Journal of Medicine*, 248:995–1001.

Institute of Medicine (2012). *Evolution of Translational Omics: Lessons Learned and the Path Forward*. The National Acadamies Press, Washington, DC.

Knudson, A.G. Jr. (1971). Mutation and cancer: Statistical study of retinoblastoma. *Proceedings of the National Academy of Sciences*, 68:820–823.

Kronmal, R.A., Bender, L., and Mortense, J. (1970). A conversational statistical system for medical records. *Journal of the Royal Statistical Society, Series C*, 19:82–92.

Lange, J., Peterson, S.M., Takashima, J.R., Grigoriev, Y., Ritchey, M.L., Shamberger, R.C., Beckwith, J.B., Perlman, E., Green, D.M., and Breslow, N.E. (2011). Risk factors for end stage renal disease in non-WT1-syndromic Wilms tumor. *Journal of Urology*, 186:378–386.

McDonald, S.P. and Craig, J.C. (2004). Long-term survival of children with end-stage renal disease. *New England Journal of Medicine*, 350:2654–2662.

Miller, D.R., Sonley, M., Karon, M., Breslow, N.E., and Hammond, D. (1974). Additive therapy in maintenance of remission in acute lymphoblastic leukemia of childhood — Effect of initial leukocyte count. *Cancer*, 34:508–517.

Partin, A.W., Yoo, J.K., Crooks, D., Epstein, J.I., Beckwith, J.B., and Gearhart, J.P. (1994). Prediction of disease-free survival after therapy in Wilms tumor using nuclear morphometric techniques. *Journal of Pediatric Surgery*, 29:456–460.

Sylvestre, M.P., Huszti, E., and Hanley, J.A. (2006). Do Oscar winners live longer than less successful peers? A reanalysis of the evidence. *The Annals of Internal Medicine*, 145:361–363.

31

A vignette of discovery

Nancy Flournoy

Department of Statistics
University of Missouri, Columbia, MO

This story illustrates the power of statistics as a learning tool. Through an interplay of exploration and carefully designed experiments, each with their specific findings, an important discovery is made. Set in the 1970s and 80s, procedures implemented with the best intentions were found to be deadly. Before these studies, only hepatitis was known to be transmitted through contaminated blood products. We discovered that the cytomegalovirus could be transferred through contaminated blood products and developed novel blood screening techniques to detect this virus just before it become well known for its lethality among persons with AIDS. We conclude with some comments regarding the design of experiments in clinical trials.

31.1 Introduction

Today blood banks have institutionalized sophisticated procedures for protecting the purity of blood products. The need for viral screening procedures are now taken for granted, but 40 years ago transmission of viral infections through the blood was understood only for hepatitis. Here I review the genesis of a hypothesis that highly lethal cytomegalovirus (CMV) pneumonia could result from contaminated blood products and the experiments that were conducted to test this hypothesis. The question of cytomegalovirus infection resulting from contaminated blood products arose in the early days of bone marrow transplantation. So I begin by describing this environment and how the question came to be asked.

E. Donnell Thomas began to transplant bone marrow into patients from donors who were not their identical twins in 1969. By 1975, his Seattle transplant team had transplanted 100 patients with acute leukemia (Thomas et al., 1977). Bone marrow transplantation is now a common treatment for childhood leukemia with a good success rate for young people with a well-matched donor.

Early attempts to transplant organs, such as kidneys, livers and hearts, were not very successful until it was determined that matching patient and donor at a few key genetic loci would substantially reduce the risk of rejection. Drugs to suppress the natural tendency of the patient's immune system to attack a foreign object further reduced the risk of rejection. In bone marrow transplantation, a good genetic match was also needed to prevent rejection. However, because a new and foreign immune system was being transplanted with the bone marrow, drugs were used not only to reduce the risk of rejection, but to keep the transplanted marrow from deciding that the whole patient was a foreign object and mounting an auto-immune-like attack called graph versus host disease (GVHD). Furthermore, in order both to destroy diseases of the blood and to prevent rejection of the new bone marrow, high doses of irradiation and drugs were given prior to the transplant. Eradicating as completely as possible all the patient's bone marrow destroys the bulk of the patient's existing immune system.

Since typically two to three weeks are required before the transplanted bone marrow's production of blood cells resumes, the Seattle team tried to anticipate problems that could result from a patient having an extended period of severely compromised blood production, and hence extremely poor immune function. It was well known that granulocytes (the white blood cells) fight infection. In order to protect the patient from bacterial infection, an elaborate and expensive system was devised to assure that the patient would be supported with plenty of donated granulocytes. When the transplant team moved into the newly built Fred Hutchinson Cancer Research Center in 1975, a large portion of one floor was dedicated to this task. On rows of beds, the bone marrow donors lay for hours each day with needles in both arms. Blood was taken from one arm, and passed through a machine that filtered off the granulocytes and returned the rest of the blood to the donor through the other arm.

Typically, the same person donated both the bone marrow and the granulocytes, with the hope that the genetic match would prevent the patient from becoming allergic to the granulocyte transfusions. The bone marrow donor was expected to stay in town for at least six weeks and lie quietly with needles in both arms every day so that the transplant patient could fight off threats of infection. Being a marrow donor required a huge time commitment.

31.2 CMV infection and clinical pneumonia

Early in the development of the bone marrow transplant procedure, it was clear that patients with sibling donors who were not identical twins were at high risk of death caused by CMV pneumonia (Neiman et al., 1977). Figure 31.1 — displaying ten years of data from Meyers et al. (1982) — shows

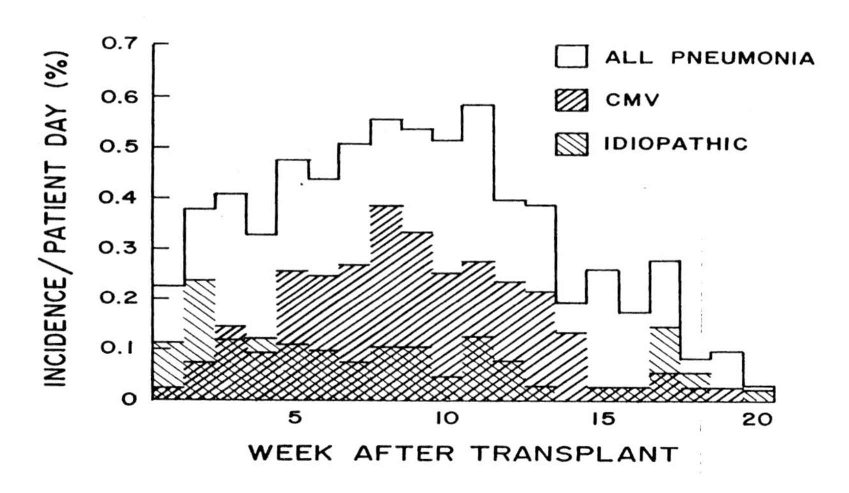

FIGURE 31.1
Incidence of CMV and all nonbacterial pneumonias expressed as percentage per patient-day for each week after allogeneic marrow transplant.

the incidence of pneumonia as a function of time following transplantation from genetically matched siblings who were not identical twins. The incidence distribution, expressed as the percentage per patient-day each week, is slightly skewed toward the time of transplant with a mode at about seven weeks. Of 525 patients, 215 (38%) had nonbacterial pneumonia, with CMV isolated in 40% of the cases and other viruses identified in 29% of the cases.

Eighty-four percent of the 215 pneumonia cases were fatal. In contrast, CMV pneumonia was not a cause of death among those patients whose bone marrow donors were identical twins (Appelbaum et al., 1982). At the time, we erroneously speculated that this difference was due to the fact that patients without identical twin donors received more drugs to suppress the immune system than did patients with identical twin donors with no risk of graft rejection. However, the identical twins did not receive their transplant care at the Fred Hutchinson Cancer Research Center, but at University of Washington Hospital, where there was no provision for providing prophylactic granulocytes. We failed to recognize that the twins' reduced drug therapy was confounded with their not getting prophylactic granulocyte transfusions.

CMV is a common virus in the environment and by the time people reach adulthood, it can be found in about half the population. CMV does not cause problems in healthy individuals, but because it was diagnosed most frequently in pneumonia cases and these were the most fatal, the push was on to characterize the course of the illness, identify prognostic factors and find therapies. In order to standardize diagnostic procedures so that the course of the risk period could be established and to identify cases of viral infection early so that intervention trials might be feasible, we instituted a standard schedule of testing blood and urine samples for the presence of CMV virus and measuring anti-CMV antibody titers. CMV infection was defined to be present if the CMV virus was isolated in the routine blood and urine tests, or if it was found in the course of diagnosing pneumonia, or if antibody titers rose four fold (seroconverted). Between October 1977 and August 1979, surveillance testing of blood and urine samples and antibody to CMV was measured in 158 patients and their donors prior to transplantation and periodically following transplant. The incidence of CMV infection in urine samples was approximately the same, regardless of the presence or absence of antibody to CMV before transplant, in either the donor or the recipient (Meyers et al., 1980). However, antibody titers increased (as measured by a summary statistic, the mean response stimulation index) after 41–60 days following transplant among patients who contracted CMV pneumonia (Figure 31.2). Note how the mode of the simulation index among patients with CMV pneumonia coincides with the time of peak incidence shown in Figure 31.1. Among patients whose CMV titers were positive pretransplant (seropositive), average titers remained high (see Figure 31.3). But among patients whose pretransplant titers were negative (seronegative), the stimulation index remained low until about 60 days after transplant and then began to rise without regard to the marrow donor's pretransplant titer. So although we dismissed the idea that CMV infection was being transferred through the donor's bone marrow, this study suggested prognostic factors that might be manipulated in an intervention.

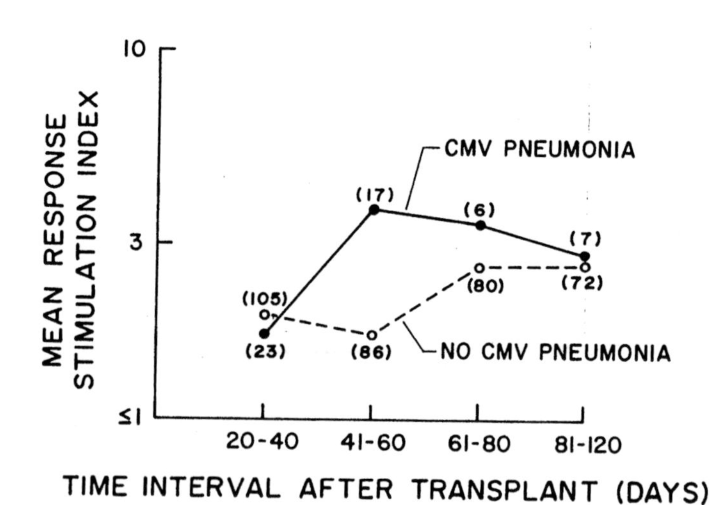

FIGURE 31.2
Mean response of lymphocytes to cytomegalovirus antigen. Numbers in parentheses represent the sample size in each group.

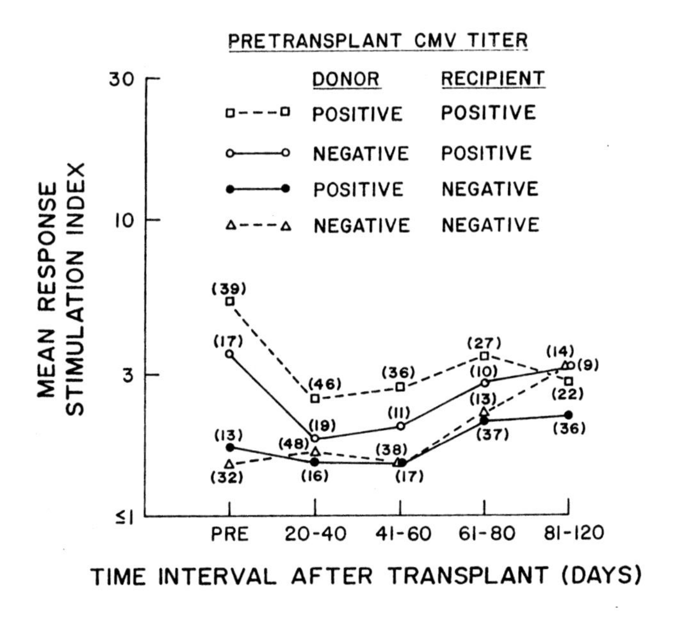

FIGURE 31.3
Mean response of lymphocytes to cytomegalovirus. Numbers in parentheses represent the sample size in each group.

Having designed and implemented a multidisciplinary research information system (Flournoy and Hearne, 1990) before the advent of commercial systems, we began mining the data for prognostic factors for CMV infection among those patients transplanted between 1979 and 1982 who had at least four surveillance cultures (Meyers et al., 1986). The surveillance data showed that just over half (51.5%) of the 545 recipients of marrow transplants without an identical twin donor became infected with CMV. CMV was cultured from 280 (51.4%) of the patients; 168 (30.8%) had at least a four-fold rise in titers (seroconverted). Much attention in this study focused on the relationship between the surveillance test results and the subsequent development of pneumonia. Also, the relationship between surveillance results, pneumonia and the complication of the transplant procedure GVHD were investigated. An association between GVHD and CMV clearly existed, suggesting that fatalities due to CMV would be reduced by eliminating GVHD. This was a false lead down another blind alley.

Among patients who had CMV in their blood prior to transplant, 69% subsequently became infected (i.e., they either seroconverted and/or began to excrete CMV in their urine). Among patients without CMV in their blood prior to transplant, 57% of those whose donors did and 28% of those whose donors did not have CMV in their blood subsequently became infected. These observations suggested that patients having CMV in their blood prior to transplant were at high risk of infections; and that in patients without CMV in their blood prior to transplant, the donor might be passing the infection to the patient, either through the marrow transplant itself or through the blood transfusions given after transplant. This was our first clue that the granulocyte transfusions might be transmitting CMV; but it was impossible to fathom that our large effort dedicated to providing prophylactic granulocyte transfusions was so harmful. We believed that a randomized clinical trial would confirm that there was some unknown confounding variable that would explain away this association.

A proportional hazards regression analysis was performed separately for patients with and without CMV in their blood prior to transplant. Among seropositive patients, all the significant covariates were demographic variables, disease characteristics or treatment complications for which no known control was possible. Thus the models did not suggest possible interventions. However, among seronegative patients, the relative rate of CMV infection was 2.3 times greater ($p = .0006$) if the granulocyte transfusions were also found to be positive for CMV. This was the second clue.

31.3 Interventions

To run a clinical trial of subjects receiving and not receiving prophylactic granulocytes required developing a higher throughput procedure for identifying CMV in blood products. While exploratory discussions began with the King County Blood Bank about how to develop the needed screening procedures, we began an alternative clinical trial that did not require novel blood analytic methods be developed.

In light of the data mining results, we focused on the patients whose pretransplant titers were negative. Thinking that prophylactic anti-CMV immune globulin might prevent CMV infection from developing, eligible patients were randomized to receive globulin or nothing, with stratifications for the use of prophylactic granulocyte transfusions and the donor's titer to CMV. At the onset of the CMV immune globulin study, we took the association between CMV infection and granulocyte transfusions seriously enough to stratify for it, but not so seriously as to study it directly. To be eligible for this study (Meyers et al., 1983), a patient had to be seronegative for antibody to CMV prior to transplant and to not excrete the virus into the urine for the first two weeks after transplantation. Patients excreting virus during this period were presumed to have been infected with CMV before transplantation and were excluded from final analysis.

Figure 31.4 compares Kaplan–Meier estimates of the probability of CMV infection as a function of week after transplant for globulin recipients and

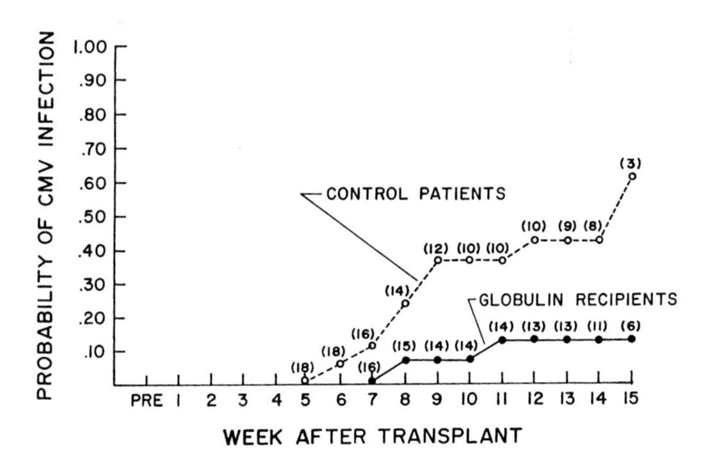

FIGURE 31.4
Kaplan–Meier probabilities of acquiring cytomegalovirus infection. The numbers in parentheses indicate the sample size of patients still at risk of infection at the beginning of each interval. The risk is different for globulin recipients and controls at $p = .03$ by the Mantel–Cox test.

control patients. The overall difference in infection rates between globulin recipients and controls was not significant. CMV infection rates by strata are shown in Table 31.1. Seeking any lead to act upon, the difference observed among patients receiving no granulocytes provided hope that globulin might be effective in a larger study of some subset of subjects. The difference in rates depending upon whether or not the granulocyte donor was seronegative or seropositive finally led us to question seriously the role of granulocyte transfusions in CMV infection.

We were thunderstruck by the possibility that we were transmitting CMV through the blood. The impact should this observation be confirmed in a controlled randomized study is described by Meyers et al. (1983):

> "Screening blood products for antibody to cytomegalovirus, or more appropriately for virus or viral antigens (techniques that are not yet available), increases the burden on blood-banking facilities, decreases the pool of blood donors, and, most importantly, decreases the rapid availability of fresh blood products such as platelets. The use of an immune globulin is therefore an attractive practical alternative among patients who need large amounts of fresh blood products such as platelets and for whom screening of blood products is less practical."

The New England Journal of Medicine rejected our paper (Meyers et al., 1983) because our observations concerning granulocyte transfusions were external to the hypotheses postulated in the initial experimental design. But these

TABLE 31.1
Incidence of CMV infection.

Patient Treatment	Globulin		No Globulin	
Granulocytes from seropositive donors	7/8	(88.5%)	6/7	(85.7%)
Granulocytes from seronegative donors	1/5	(20.0%)	0/6	(00.0%)
No granulocytes	2/17	(11.8%)	8/19	(42.1%)

observations led to new hypotheses and the next randomized study. While it is extremely important to distinguish between observations obtained by controlled randomized intervention and those obtained otherwise, hypothesis generation is an essential task.

We spent a year working with King County Blood Bank to develop screening procedures, set up laboratory equipment and train technicians in order to conduct a randomized clinical trial. Although we restricted the study to patients who were seronegative for CMV in two consecutive tests and who had not received any unscreened blood recently, more patients were available for study than the blood bank could handle. Therefore, we studied the prophylactic capability of immune globulin at the same time in a randomized 2×2 factorial design. CMV immune globulin had no effect [data not shown] on the rate of CMV infection (Bowden et al., 1986).

The effect of only giving CMV seronegative blood transfusions, controlling for the marrow donor's CMV status, is summarized in Table 31.2.

Among patients whose marrow donors were seronegative, those randomized to receive seronegative granulocyte transfusions had a 4.5% infection rate, whereas those randomized to receive unscreened transfusions had a 32% infection rate. What is more, the one patient with a seronegative donor who was assigned to receive seronegative blood products and subsequently became infected with CMV actually mistakenly received several seropositive transfusions.

TABLE 31.2
Incidence of CMV infection among 85 patients studied for at least 62 days after transplantation.

Marrow Donor's CMV Status	Randomized to Granulocytes: Seronegative	Randomized to Granulocytes: Unscreened
Seronegative	1/22 (04.5%)	8/25 (32.0%)
Seropositive	3/12 (25.0%)	5/16 (31.3%)

As the study proceeded, blood bank personnel became increasingly agitated as they considered the ramifications of a significant finding. Blood banks all over the country would have to set up screening programs; the cost would be enormous, they warned. The study results went out to blood banks, however, and viral screening procedures were put into place. The timing was fortuitous because the AIDS crisis was building. Today the idea that viral infections can be transmitted through the blood is taken for granted.

31.4 Conclusions

Our experience suggests three key steps to successful experimental designs. First, ask an important, well defined question. Too often this step receives insufficient attention and resources, and it may be the most difficult. Determining a well structured and important question can involve considerable data collection, exploratory analysis and data mining. Data mining without data collection targeted to a specific question may yield valuable findings, but many important questions now go begging in the push to devote time and resources to analyzing existing databases. (When medical fellows began rotating through the Fred Hutchinson Cancer Center wards, they had to conceive of a study, execute it and draft a paper. The questions they raised were inventive and some resulted in life-saving findings. When I left in 1986, having created a shared interdisciplinary research information system, the fellows typically looked at what data was available and studied a question that could be answered with the available data. I found the relative lack of imagination in the questions being asked extremely distressing, and feel responsible for enabling it.)

The problem associated with moving too quickly to confirmatory studies has fairly recently been acknowledged by the pharmaceutical industry. But this acknowledgement seems slow to translate into substantially increased resources being devoted to early phase studies.

The second key step is to develop interventions that focus sharply on the question at hand and to randomize subjects to the interventions and to the standard of care or a placebo. This step is operationally well developed. The third step is to replicate the experiment and encourage others to do likewise.

While this series of studies resulted in important discoveries, other series of studies were not so fruitful. In particular, two-arm randomized studies of chemotherapy and radiation schedules were largely uninformative. These variables are virtually continuous, whereas the important variables in the studies of CMV were mostly inherently discrete. With such multidimensional continuous random variables in an environment in which ethics preclude utilizing a large sample space covering unknown territory, our unbridled enthusiasm for the two-arm clinical trial as a learning tool was misplaced. Coming to appreciate the necessity of exploratory analysis in such settings led to my current work in adaptive designs and their analysis.

References

Appelbaum, F., Meyers, J., Fefer, A., Flournoy, N., Cheever, N., Greenberg, M., Greenberg, P., Hackman, R., and Thomas, E. (1982). Nonbacterial nonfungal pneumonia following marrow transplantation in 100 identical twins. *Transplantation*, 33:265–268.

Bowden, R., Sayers, M., Flournoy, N., Newton, B., Banaji, M., Thomas, E., and Meyers, J. (1986). Cytomegalovirus immune globulin and seronegative blood products to prevent primary cytomegalovirus infection after marrow transplant. *New England Journal of Medicine*, 314:1006–1010.

Flournoy, N. and Hearne, L. (1990). *Quality Control for a Shared Multidisciplinary Database*. Marcel Dekker, New York, pp. 43–56.

Meyers, J.D., Flournoy, N., and Thomas, E.D. (1980). Cytomegalovirus infection and specific cell-mediated immunity after marrow transplant. *The Journal of Infectious Diseases*, 142:816–824.

Meyers, J.D., Flournoy, N., and Thomas, E.D. (1982). Nonbacterial pneumonia after allogeneic marrow transplantation. *Reviews of Infectious Diseases*, 4:1119–1132.

Meyers, J.D., Flournoy, N., and Thomas, E.D. (1986). Risk factors for cytomegalovirus infection after human marrow transplant. *The Journal of Infectious Diseases*, 153:478–488.

Meyers, J.D., Leszczynski, J., Zaia, J.A., Flournoy, N., Newton, B., Snydman, D.R., Wright, G.G., Levin, M.L., and Thomas, E.D. (1983). Prevention of cytomegalovirus infection by cytomegalovirus immune globulin after marrow transplantation. *Annals of Internal Medicine*, 98:442–446.

Neiman, P.E., Reeves, W., Ray, G., Flournoy, N., Lerner, K.G., Sale, G.E., and Thomas, E.D. (1977). A prospective analysis of interstitial pneumonia and opportunistic viral infection among recipients of allogeneic bone marrow graphs. *The Journal of Infectious Diseases*, 136:754–767.

Thomas, E.D., Buckner, C.D., Banaji, M., Clift, R.A., Fefer, A., Flournoy, N., Goodell, B.W., Hickman, R.O., Lerner, K.G., Neiman, P.E., Sale, G.E., Sanders, J.E., Singer, J., Stevens, M., Storb, R., and Weiden, P.L. (1977). One hundred patients with acute leukemia treated by chemotherapy, total body irradiation, and allogeneic marrow transplantation. *Blood*, 49:511–533.

32

Statistics and public health research

Ross L. Prentice
Division of Public Health Sciences
Fred Hutchinson Cancer Research Center, Seattle, WA

Statistical thinking and methods have much to contribute in the area of multidisciplinary public health research. Often barriers to research progress can only be addressed using innovative statistical methods. Furthermore, statisticians have an important role to play in helping to shape the population science research agenda. These points will be illustrated using topics in nutritional epidemiology, preventive intervention development, and randomized trial design and analysis.

32.1 Introduction

It is a pleasure to join the celebration of the 50th anniversary of COPSS, which continues to fulfill a stimulatory and valuable coordinating role among the participating statistical societies. This anniversary provides a reminder of the impressive advances in statistical theory and application over the past 50 years, as is certainly the case in the biomedical research context, and in the public health research area more specifically.

Much of biomedical research involves the follow-up of cohorts of individuals to observe health-related outcomes. Most frequently this work involves human studies, but the statistical methods employed may also apply to studies in model systems. A typical study may involve relating some treatment, or some set of study subject characteristics or exposures, to the time until a disease event occurs. Early statistical proposals for the analysis of such "failure time" data in therapeutic trial settings involved the use of linear models, usually for the logarithm of failure time. Because of the usual presence of right censorship, error distributions having closed-form right tails were often employed, rather than traditional Normal models. At the same time, methodology for epidemiological applications were developing a focus on relative disease rates, or closely related odds ratios with the Mantel and Haenszel (1959) summary odds ratio estimator standing out as a key statistical and epidemiological contribution. Also nonparametric methods, and hazard rate estimators, entered through the Kaplan and Meier (1958) survivor function estimator.

These modeling approaches came together in the Cox (1972) regression model, one of the most influential and highly cited statistical papers of all time. The semiparametric Cox model extended the ratio modeling into a full hazard ratio regression approach, while also incorporating a nonparametric baseline hazard rate that valuably relaxed parametric models, such as the Weibull model, that had previously been used. Furthermore, the parametric hazard ratio component of this semiparametric model could be relaxed in important ways by including, for example, stratification on key confounding factors, treatment by time interactions to relax proportionality assumptions, and stochastic time-dependent covariates to examine associations for covariates collected during study follow-up. For relatively rare outcomes, the Cox model proportional hazards special case is well approximated by a corresponding odds ratio regression model, and logistic regression soon became the mainstay approach to the analysis of case-control epidemiological data (Prentice and Pyke, 1979).

Over the past 30 years, valuable statistical methods have been developed for data structures that are more complex than a simple cohort follow-up with a univariate failure time outcome. Many such developments were motivated by substantive challenges in biomedicine. These include nested case-control and case-cohort sampling procedures to enhance estimation efficiency with rare disease outcomes; methods for the joint analysis of longitudinal and failure time data; sequential data analysis methods; missing and mismeasured data methods; multivariate failure time methods, including recurrent event and correlated/clustered failure time methods; and event history models and methods more generally. Many of these developments along with corresponding statistical theory have been summarized in book form where pertinent references may be found; see, e.g., Andersen et al. (1993) and Kalbfleisch and Prentice (2002).

In the last decade, foci for the development of statistical methods in biomedical applications have included the incorporation of high-dimensional genomic data, with regularization approaches to deal with dimensionality and data sparsity as in, e.g., Tibshirani (1994); methods for the development, evaluation and utilization of biomarkers for many purposes, including early disease detection, disease recurrence detection, and objective exposure assessment; and methods for disease risk prediction that integrate with concepts from the diagnostic testing literature. Relative disease rate modeling, and the Cox model in particular, provided a foundation for many of these developments.

32.2 Public health research

While many of the developments just listed were motivated by clinical research applications, the public health and disease prevention areas also provided considerable stimulation. In fact, the public health research arena is one that merits consideration as a career emphasis for many more statistically trained investigators: If we look back over this same 50-year period, we will see that 50 years ago papers were just starting to emerge on cigarette smoking and lung cancer, with tobacco companies advertising the many virtues of smoking, including the ability to "soothe the throat." Currently, with the smoking patterns in place since women adopted patterns of smoking behavior similar to those for men over the past 20–30 years, the estimated lung cancer mortality rate among current smokers, whether male or female, is a whopping 25 times that in never smokers, with substantial elevations also in mortality rates for chronic obstructive pulmonary disease, multiple other cancers, and coronary heart disease as well; see, e.g., Thun et al. (2013). Vigorous and organized research programs are needed for exposures having these types of horrendous health consequences to be identified early, and for the responsible exposures to be eliminated or reduced.

An example of the potential of public health research when supported by regulatory action is provided by the use of postmenopausal hormones. Postmenopausal estrogens came on the market about 50 years ago, and the use of estrogens throughout the lifespan was promoted as the way for women to retain youth and vitality, while avoiding the vasomotor symptoms associated with menopause. By the mid-1970s, it was apparent that the widely used estrogens that derived from the urine of pregnant mares led to a 5–10 fold increase in the risk of uterine cancer, so a progestin was added to protect the uterus. Observational epidemiological data mostly collected over the subsequent 20 years seemed supportive of the utility of those preparations (conjugated equine estrogens for women who were post-hysterectomy; estrogens plus medroxyprogesterone acetate for women with a uterus), with reported benefits for heart disease, fracture and dementia prevention, among other health benefits.

However, a different picture emerged when these regimens were put to the test in randomized controlled trials (Writing Group for the Women's Health Initiative Investigators, 2002; Anderson et al., 2004). This was especially the case for combined estrogens plus progestin, where health benefits were exceeded by health risks, including an early elevation in heart disease, sustained elevations in stroke, a major elevation in breast cancer risk, and an increase in probable dementia. These trial results were instrumental in leading to suitable package insert warnings by the US Food and Drug Administration, and to a major change in the use of these regimens, with about 70% of women taking estrogens plus progestin stopping abruptly in 2002, along with about 40% of women taking estrogens alone. One can project that this and subsequent changes in hormone therapy practices have led, for example, to about 15,000 to 20,000 fewer women developing breast cancer each year since 2003 in the United States alone, along with tens of thousands of additional such women elsewhere in the world.

Moving to another public health topic, obesity is the epidemic of our time. It is clear that overweight and obesity arise from a sustained imbalance over time in energy consumed in the diet compared to energy expended at rest and through physical activity. Obesity is an established risk factor for many of the chronic diseases that are experienced in great excess in Western societies, including vascular diseases and several major cancers, and diabetes. However, specific knowledge from nutritional or physical activity epidemiology as to which dietary and activity patterns can be recommended are substantially lacking and, at any rate, are not sufficiently compelling to stimulate the societal changes that may be needed to begin to slow and reverse the obesity epidemic. For example, needed changes may involve personal choices in food selection, preparation and consumption patterns; choices away from a sedentary lifestyle; food supply and distribution changes; changes in city design; restrictions in advertising; and taxation changes, to cite but a few. Of course, favorable dietary and physical activity patterns may have health benefits that go well beyond normal weight maintenance.

The remainder of this short contribution will elaborate some of the research barriers to public health research progress in some areas just mentioned, with focus on statistical issues.

32.3 Biomarkers and nutritional epidemiology

While other application areas also grapple with exposure assessment issues, these problems appear to dominate in the nutritional epidemiology research area. For example, an international review (World Health Organization, 2003) of nutritional epidemiology research identified few dietary exposures that are associated with vascular diseases or cancer, with most reports based on self-reported diet, typically using a food frequency questionnaire (FFQ) approach where the study subject reports consumption frequency and serving size over the preceding few months for a list of foods.

A lack of consistency among epidemiological reports on specific dietary associations has stimulated a modest focus on dietary assessment measurement error over the past 25–30 years. Much of this work has involved comparisons of FFQ data to corresponding data using other self-report tools, such as 24-hour dietary recalls (24-HRs), or several days of food records, to examine FFQ measurement error properties. However, for a few important dietary factors, including total energy consumption and protein consumption, one can obtain objective dietary assessments, at least for relativity short periods of time, from urinary excretion markers. When those biomarker measures are compared to self-report data, one sees strong positive measurement error correlations among FFQs, 24-HRs and four-day food records (4-DFRs); see Prentice et al. (2011).

The implication is that biomarker data, but not data using another self-report, need to be used to assess self-report measurement error, and to calibrate the self-report data for use in nutritional epidemiology association studies. Studies to date using this type of regression calibration approach tend to give quite different results from traditional analyses based on self-report data alone, for example, with strong positive associations between total energy consumption with heart disease and breast, colorectal and total cancer incidence; see, e.g., Prentice and Huang (2011).

From a statistical modeling perspective, calibrated dietary exposure estimates typically arise from linear regression of (log-transformed) biomarker values on corresponding self-report estimates and on such study subject characteristics as body mass index, age, and ethnicity. These latter variables are quite influential in explaining biomarker variation, as may in part reflect systematic biases in dietary reporting. For example, while persons of normal weight tend to show little energy under-reporting, obese persons underestimate substantially, in the 30–50% range on average (Heitmann and Lissner, 1995). These types of systematic biases can play havoc with disease association analyses if not properly addressed.

Measurement error correction methods are not easy for nutritional epidemiologists to grasp, and are not so easy even for nutritionally-oriented statisticians. A logical extension of the biomarker calibration work conducted to date is a major research emphasis on nutritional biomarker development, to produce measurement error–corrected consumption estimates for many more nutrients and foods. Statisticians, in conjunction with nutritional and epidemiological colleagues, can play a major role in establishing the rationale for, and the design of, such a nutritional biomarker development enterprise, which may entail the conduct of sizeable human feeding studies. For example, such a feeding study among 150 free-living Seattle participants in the Women's Health Initiative is currently nearing completion, and will examine candidate biomarkers and higher dimensional metabolic profiles for novel nutritional biomarker development.

32.4 Preventive intervention development and testing

Closely related to the development of biomarkers of exposure, is the use of biomarkers for preventive intervention development. While there is a rather large enterprise for the development of therapeutic interventions, the development of innovative disease prevention interventions is less impressive. One reason for this discrepancy is that prevention may not share the same fiscal incentives as treatment, since persons whose disease has been delayed or averted are usually not individually identifiable. Furthermore, the types of specimens needed for relevant biological measures (e.g., gene expression profiles), are frequently not available in the context of studies of large cohorts of healthy persons. As a result, preventive interventions that have been studied in randomized controlled trials have mostly involved pill taking approaches, with rationale derived from observational epidemiology or borrowed from preceding therapeutic trials.

Specifically, there have been few trials of behavioral interventions with chronic disease outcomes. As an exception, the Diabetes Prevention Program Research Group (Diabetes Prevention Program Research Group, 2002) randomized trial among 3234 persons having impaired glucose tolerance demonstrated a major benefit for Type 2 diabetes incidence with a combined dietary and physical activity intervention. Also, the Women's Health Initiative low-fat dietary modification trial (Prentice et al., 2006) among 48,835 ostensibly healthy postmenopausal women demonstrated a modest reduction in its breast cancer primary outcome, but the reduction didn't meet the usual requirements for statistical significance (log-rank significance level of .07). There has never been a full-scale physical activity intervention trial with chronic disease outcomes.

Statistical methods for high-dimensional data analysis and biological network development may be able to help fill the preventive intervention development gap. For example, changes in proteomic or metabolomic profiles may be able to combine with changes in conventional risk factors for targeted diseases in intermediate outcome intervention trials of practical size and expense to select among, and provide the initial evaluation of, potential preventive interventions in a manner that considers both efficacy and safety.

Also, because of cost and logistics, few full-scale disease prevention trials can be conducted, regardless of the nature of the intervention. Innovative hybrid designs that combine the rather comprehensive profiling of the previous paragraph with case-control data for targeted outcomes that also include the same types of high-dimensional biologic data may be able to produce tests of intervention effects on chronic disease of acceptable reliability for most purposes, at costs that are not extreme. Interventions meeting criteria in such hybrid designs, that also have large public health potential, could then be put forward with a strong rationale for the few full-scale randomized trials with disease outcomes that can be afforded.

32.5 Clinical trial data analysis methods

As noted in the Introduction, statistical methods are rather well developed for the comparison of failure times between randomized groups in clinical trials. However, methods for understanding the key biological pathways leading to an observed treatment effect are less well developed. Efforts to explain treatment differences in terms of post-randomization biomarker changes may be limited by biomarker sample timing issues and temporal aspects of treatment effects. Furthermore, such efforts may be thwarted by measurement error issues in biomarker assessment. Biomarker change from baseline may be highly correlated with treatment assignment, implying likely sensitivity of mediation analysis to even moderate error in intermediate variable assessment.

Another area in need of statistical methodology development is that of multivariate failure time data analysis. While Kaplan–Meier curves, censored data rank tests, and Cox regression provide well-developed tools for the analysis of univariate failure time data, corresponding established tools have not stabilized for characterizing dependencies among failure times, and for examining treatment effects jointly with a set of failure time outcomes. For example, in the context of the postmenopausal hormone therapy trials mentioned earlier (Anderson et al., 2004; Writing Group for the Women's Health Initiative Investigators, 2002), one could ask whether data on stroke occurrence can be used to strengthen the estimation of treatment effects on coronary heart disease, and vice versa, in a nonparametric manner. The lack of standardized approaches to addressing this type of question can be traced to the lack of a suitable nonparametric maximum likelihood estimation of the multivariate survivor function, which could point the way to nonparametric and semiparametric likelihood approaches to the analysis of more complex multivariate failure time data structures.

32.6 Summary and conclusion

Statistical thinking and innovation have come to play a major role throughout biomedical research during the 50 years of COPSS' existence. Public health aspects of these developments have lagged somewhat due to the need to rely substantially on purely observational data for most purposes, for practical reasons. Such observational data are valuable and adequate for many purposes, but they may require innovative biomarker supplementation for exposures that are difficult to assess, as in nutritional and physical activity epidemiology. This could include supplementation by intermediate outcome, or full-scale, randomized prevention trials for topics of great public health importance, such as postmenopausal hormone therapy; and supplementation by mechanistic and biological network data for timely identification of the health effects of exposures, such as cigarette smoking, that are not amenable to human experimentation.

These are among the most important research needs for the health of the populations we serve. These populations are keenly interested in, and highly supportive of, public health and disease prevention research. Statisticians have as crucial a role as any other disciplinary group in responding to this interest and trust, and statistical training is highly valuable for participation and leadership roles in shaping and carrying out the needed research.

Acknowledgements

This work was supported by grants CA53996 and CA119171, and contract HHSN268720110046C from the National Institutes of Health.

References

Andersen, P.K., Borgan, Ø., Gill, R.D., and Keiding, N. (1993). *Statistical Models Based on Counting Processes*. Springer, New York.

Anderson, G.L., Limacher, M.C., Assaf, A.R., Bassford, T., Beresford, S.A., Black, H.R., Bonds, D.E., Brunner, R.L., Brzyski, R.G., Caan, B. et al. (2004). Effects of conjugated equine estrogen in postmenopausal women with hysterectomy: The Women's Health Initiative randomized controlled trial. *Journal of the American Medical Association*, 291:1701–1712.

Cox, D.R. (1972). Regression models and life-tables (with discussion). *Journal of the Royal Statistical Society, Series B*, 34:187–220.

Diabetes Prevention Program Research Group (2002). Reduction in the incidence of Type 2 diabetes with lifestyle intervention or metformin. *New England Journal of Medicine*, 346:393–403.

Heitmann, B.L. and Lissner, L. (1995). Dietary underreporting by obese individuals—Is it specific or non-specific? *British Medical Journal*, 311:986–989.

Kalbfleisch, J.D. and Prentice, R.L. (2002). *The Statistical Analysis of Failure Time Data*, 2nd edition. Wiley, New York.

Kaplan, E.L. and Meier, P. (1958). Nonparametric estimation from incomplete observations. *Journal of the American Statistical Association*, 53:457–481.

Mantel, N. and Haenszel, W. (1959). Statistical aspects of the analysis of data from retrospective studies of disease. *Journal of the National Cancer Institute*, 22:719–748.

Prentice, R.L., Caan, B., Chlebowski, R.T., Patterson, R., Kuller, L.H., Ockene, J.K., Margolis, K.L., Limacher, M.C., Manson, J.E., Parker, L.M. et al. (2006). Low-fat dietary pattern and risk of invasive breast cancer. *Journal of the American Medical Association*, 295:629–642.

Prentice, R.L. and Huang, Y. (2011). Measurement error modeling and nutritional epidemiology association analyses. *The Canadian Journal of Statistics*, 39:498–509.

Prentice, R.L., Mossavar-Rahmani, Y., Huang, Y., Van Horn, L., Beresford, S.A., Caan, B., Tinker, L., Schoeller, D., Bingham, S., Eaton, C.B. et al. (2011). Evaluation and comparison of food records, recalls, and frequencies for energy and protein assessment by using recovery biomarkers. *American Journal of Epidemiology*, 174:591–603.

Prentice, R.L. and Pyke, R. (1979). Logistic disease incidence models and case-control studies. *Biometrika*, 66:403–411.

Thun, M.J., Carter, B.D., Feskanich, D., Freedman, N.D., Prentice, R.L., Lopez, A.D., Hartge, P., and Gapstur, S.M. (2013). 50-year trends in smoking-related mortality in the United States. *New England Journal of Medicine*, 368:351–364.

Tibshirani, R.J. (1994). Regression shrinkage and selection via the LASSO. *Journal of the Royal Statistical Society, Series B*, 58:267–288.

World Health Organization (2003). *Diet, Nutrition and the Prevention of Chronic Diseases*. Report of a Joint WHO/FAO Expert Consultation, vol. 916. World Health Organization, Genève, Switzerland.

Writing Group for the Women's Health Initiative Investigators (2002). Risks and benefits of estrogen plus progestin in healthy postmenopausal women: Principal results from the Women's Health Initiative randomized controlled trial. *Journal of the American Medical Association*, 288:321–33.

33

Statistics in a new era for finance and health care

Tze Leung Lai
Department of Statistics
Stanford University, Stanford, CA

We are entering a new era in finance in the wake of the recent financial crisis and financial reform, and in health care as the provisions of the Affordable Care Act are being implemented from 2010 to 2020. We discuss opportunities and challenges for the field of statistics in this new era.

33.1 Introduction

The past few years witnessed the beginning of a new era in financial markets and in the US health care system. In March 2010, landmark health care reform was passed through two federal statutes: the Patient Protection and Affordable Care Act and subsequently amended by the Health Care and Education Reconciliation Act. A few months later, the Dodd–Frank Wall Street Reform and Consumer Protection Act was signed into federal law on July 21, 2010, in response to widespread calls for regulatory reform following the 2007–08 financial crisis. Since this year marks the 50th anniversary of the COPSS, it seems timely to discuss how statistical science can help address the challenges of this new era for finance and health care and to suggest some outreach opportunities for the field of statistics.

We begin with health care in Section 33.2. One of the provisions of the Patient Protection and Affordable Care Act is the establishment of a non-profit Patient-Centered Outcomes Research Institute to undertake comparative effectiveness research (CER), examining the "relative health outcomes, clinical effectiveness, and appropriateness" of different medical treatments. This involves the design of comparative studies of the treatments and their statistical analysis, and Section 33.2 discusses the limitations of standard study designs when applied to CER and describes some innovative designs for comparative effectiveness trials. Section 33.3 proceeds further to discuss innovative designs to improve the efficiency of clinical trials in the development of new treatments.

In Section 33.4, after reviewing the flaws in credit risk modeling and management that led to the 2007–08 financial crisis, we discuss how these flaws can be addressed by using better statistical methods. Section 33.5 continues this discussion on the role of statistics in financial and risk modeling in the new era after the Dodd–Frank Act. Some concluding remarks are given in Section 33.6.

33.2 Comparative effectiveness research clinical studies

One approach to CER is to use observational studies, including analysis of claims or registry data; see, e.g., Stukel et al. (2007). As pointed out by Shih and Lavori (2013), such an approach involves "confounding by indication," the tendency for clinicians and patients to choose treatments with their anticipated effects in mind. This leads to bias in estimating the effectiveness, which has to be handled by statistical adjustments and modeling techniques, or instrumental variables methods, or some combination. An obvious way to remove confounding is a randomized trial. However, conventional randomized trial designs are not only too costly but also ineffective in changing medical practice. An example is the Antihypertensive and Lipid Lowering Treatment to Prevent Heart Attack Trial (ALLHAT), which was a randomized, double-blind, multi-center clinical trial designed to recruit 40,000 hypertensive patients to be randomized to a diuretic treatment (chlorthalidone) and three alternative antihypertensive pharmacologic treatments. Patients were followed every three months for the first year and every four months thereafter for an average of six years of follow-up. This landmark CER trial cost over $100 million. The results showed no difference in the prevention of heart attack and the superiority of chlorthalidone in preventing one or more forms of cardiovascular disease (ALLHAT Collaborative Research Group, 2002). Yet, a few years later, the impact of the trial was found to be disappointing because of difficulty in pursuading doctors to change, scientific disagreement about the interpretation of the results, and heavy marketing by the pharmaceutical companies of their own drugs; see Lai and Lavori (2011).

Section 4 of Lai and Lavori (2011) describes some innovative approaches that are promising to meet the challenges of designing CER clinical trials. One is sequential multiple-assignment randomization for dynamic treatment strategies in the management of patients with chronic diseases, as in Thall et al. (2007) who describe a two-stage randomized trial of twelve different strategies of first-line and second-line treatments of androgen-independent prostate cancer. Another is equipoise-stratified randomization used by the STAR*D trial

(Rush et al., 2004) that compares seven treatment options in patients who did not attain a satisfactory response with citalopram, an inhibitor antidepressant. After receiving citalopram, participants without sufficient symptomatic benefit were eligible for randomization among these options. A novel feature of the study design is that it ascertains before randomization the set of options that the patient-clinician dyad consider to be equally reasonable, given the patient's preferences, and his or her state after a trial of citalopram. This set of options characterizes the patient's Equipoise Stratum (ES). A total of 1429 patients were randomized under this scheme. The largest ES were the "Medication Switch Only" group, allowing randomization among the three medications (40%) and the "Medication Augmentation Only," allowing randomization between two options (29%). The "Any Augmentation" (10%) and "Any Switch" (7%) were the next largest, and only 5% of patients were randomized among options that contrasted a switch and augment condition. In retrospect, it became clear that patients (and their clinicians) were roughly divided into two groups, those who obtained partial benefit from citalopram and therefore were interested in augmentation, and those who obtained no benefit and were interested only in switching. Thus, the ES design allowed the study to self-design in assigning patients to the parts of the experiment that were relevant to current practice and to patient preferences.

A third approach is to design point-of-care (POC) clinical trials which can be regarded as experiments embedded into clinical care; see Fiore et al. (2011) on a VA-sponsored trial that compares the effectiveness of two insulin dosing regimens for hospitalized diabetic patients. In POC trials, subjects are randomized at the health care encounter, clinician equipoise defines the reference population, and baseline and/or outcome data are captured through electronic medical records. By using outcome-adaptive randomization, POC trials integrate experimentation into implementation and learn sequentially the superior treatment(s). This is similar to the classical multi-arm bandit problem (Lai, 2001) except that POC uses adaptive randomization to implement it in a clincial setting with clinician equipoise and patient's informed consent. Group sequential generalized likelihood ratio tests with efficient outcome-adaptive randomization to multiple arms have recently been developed and can be used for POC trials (Lai and Liao, 2012; Shih and Lavori, 2013).

33.3 Innovative clinical trial designs in translational medicine

Besides CER studies, Lai and Lavori (2011) describe novel design methods for clinical studies of personalized treatments and targeted cancer therapies in translational medicine. "From bench to bedside" — in which "bench" refers to laboratory experiments to study new biochemical principles and novel molecular compounds and "bedside" refers to new treatments developed after preclinical animal studies and Phase I, II, and III trials involving human subjects — is a maxim of evidence-based translational medical research.

The development of imatinib, the first drug to target the genetic defects of a particular cancer while leaving healthy cells unharmed, exemplifies this maxim and has revolutionized the treatment of cancer. A Phase I clinical trial treating CML (chronic myeloid leukemia) patients with the drug began in June 1998, and within six months remissions had occurred in all patients as determined by their white blood cell counts returning to normal. In a subsequent five-year study on survival, which followed 553 CML patients who had received imatinib as their primary therapy, only 5% of the patients died from CML and 11% died from all causes during the five-year period. Moreover, there were few significant side effects; see Druker et al. (2006). Such remarkable success of targeted therapies has led to hundreds of kinase inhibitors and other targeted drugs that are in various stages of development in the present anticancer drug pipeline.

Most new targeted treatments, however, have resulted in only modest clinical benefit, with less than 50% remission rates and less than one year of progression-free survival, unlike a few cases such as trastuzumab in HER2-positive breast cancer, imatinib in CML and GIST, and gefitinib and erlotinib in non-small cell lung cancer. While the targeted treatments are devised to attack specific targets, the "one size fits all" treatment regimens commonly used may have diminished their effectiveness, and genomic-guided and risk-adapted personalized therapies that are tailored for individual patients are expected to substantially improve the effectiveness of these treatments. To achieve this potential for personalized therapies, the first step is to identify and measure the relevant biomarkers. The markers can be individual genes or proteins or gene expression signatures. The next step is to select drugs (standard cytotoxins, monoclonal antibodies, kinase inhibitors and other targeted drugs) based on the genetics of the disease in individual patients and biomarkers of drug sensitivity and resistance. The third step is to design clinical trials to provide data for the development and verification of personalized therapies. This is an active area of research and several important developments are reviewed in Lai and Lavori (2011) and Lai et al. (2012). It is an ongoing project of ours at Stanford's Center for Innovative Study Design.

Despite the sequential nature of Phase I–III trials, in which Phase I studies are used to determine a safe dose or dosage regimen, Phase II trials are used to evaluate the efficacy of the drug for particular indications (endpoints) in patients with the disease, and Phase III trials aim to demonstrate the effectiveness of the drug for its approval by the regulatory agency, the trials are often planned separately, treating each trial as an independent study whose design depends on studies in previous phases. Since the sample sizes of the trials are often inadequate because of separate planning, an alternative strategy is to expand a trial seamlessly from one phase into the next phase, as in the Phase II–III cancer trial designs of Inoue et al. (2002) and Lai et al.

(2012). The monograph by Bartroff et al. (2012) gives an overview of recent developments in sequential and adaptive designs of Phase I, II, and III clinical trials, and statistical analysis and inference following these trials.

33.4 Credit portfolios and dynamic empirical Bayes in finance

The 2007–08 financial crisis began with unpreparedly high default rates of subprime mortgage loans in 2007 and culminated in the collapse of large financial institutions such as Bear Stearns and Lehman Brothers in 2008. Parallel to the increasing volume of subprime mortgage loans whose value was estimated to be $1.3 trillion by March 2007, an important development in financial markets from 2000 to 2007 was the rapid growth of credit derivatives, culminating in $32 trillion worth of notional principal for outstanding credit derivatives by December 2009. These derivative contracts are used to hedge against credit loss of either a single corporate bond, as in a credit default swap (CDS), or a portfolio of corporate bonds, as in a cash CDO (collateralized debt obligation), or variant thereof called synthetic CDO. In July 2007, Bear Stearns disclosed that two of its subprime hedge funds which were invested in CDOs had lost nearly all their value following a rapid decline in the subprime mortgage market, and Standard & Poor's (S&P) downgraded the company's credit rating. In March 2008, the Federal Reserve Bank of New York initially agreed to provide a $25 billion collateralized 28-day loan to Bear Stearns, but subsequently changed the deal to make a $30 billion loan to JPMorgan Chase to purchase Bear Stearns. Lehman Brothers also suffered unprecedented losses for its large positions in subprime and other lower-rated mortgage-backed securities in 2008. After attempts to sell it to Korea Development Bank and then to Bank of America and to Barclays failed, it filed for Chapter 11 bankruptcy protection on September 15, 2008, making the largest bankruptcy filing, with over $600 billion in assets, in US history. A day after Lehman's collapse, American International Group (AIG) needed bailout by the Federal Reserve Bank, which gave the insurance company a secured credit facility of up to $85 billion to enable it to meet collateral obligations after its credit ratings were downgraded below AA, in exchange for a stock warrant for 79.9% of its equity. AIG's London unit had sold credit protection in the form of CDS and CDO to insure $44 billion worth of securities originally rated AAA. As Lehman's stock price was plummeting, investors found that AIG had valued its subprime mortgage-backed securities at 1.7 to 2 times the values used by Lehman and lost confidence in AIG. Its share prices had fallen over 95% by September 16, 2008. The "contagion" phenomenon, from increased default probabilities of subprime mortgages to those of counterparties in credit derivative contracts whose values vary with credit ratings, was mostly neglected in the models of joint default intensities that were used to price CDOs and mortgage-backed securities. These models also failed to predict well the "frailty" traits of latent macroeconomic variables that underlie mortgages and mortgage-backed securities.

For a multiname credit derivative such as CDO involving k firms, it is important to model not only the individual default intensity processes but also the joint distribution of these processes. Finding tractable models that can capture the key features of the interrelationships of the firms' default intensities has been an active area of research since intensity-based (also called reduced-form) models have become a standard approach to pricing the default risk of a corporate bond; see Duffie and Singleton (2003) and Lando (2004). Let Φ denote the standard Normal distribution function, and let G_i be the distribution function of the default time τ_i for the ith firm, where $i \in \{1, \ldots, M\}$. Then $Z_i = \Phi^{-1}\{G_i(\tau_i)\}$ is standard Normal. Li (2000) went on to assume that $(Z_1, \ldots, Z_M)$ is multivariate Normal and specifies its correlation matrix Γ by using the correlations of the stock returns of the M firms. This is an example of a copula model; see, e.g., Genest and Favre (2007) or Genest and Nešlehová (2012).

Because it provides a simple way to model default correlations, the Gaussian copula model quickly became a widely used tool to price CDOs and other multi-name credit derivatives that were previously too complex to price, despite the lack of convincing argument to connect the stock return correlations to the correlations of the Normally distributed transformed default times. In a commentary on "the biggest financial meltdown since the Great Depression," Salmon (2012) mentioned that the Gaussian copula approach, which "looked like an unambiguously positive breakthrough," was used uncritically by "everybody from bond investors and Wall Street banks to rating agencies and regulators" and "became so deeply entrenched — and was making people so much money — that warnings about its limitations were largely ignored." In the wake of the financial crisis, it was recognized that better albeit less tractable models of correlated default intensities are needed for pricing CDOs and risk management of credit portfolios. It was also recognized that such models should include relevant firm-level and macroeconomic variables for default prediction and also incorporate frailty and contagion.

The monograph by Lai and Xing (2013) reviews recent works on dynamic frailty and contagion models in the finance literature and describes a new approach involving dynamic empirical Bayes and generalized linear mixed models (GLMM), which have been shown to compare favorably with the considerably more complicated hidden Markov models for the latent frailty processes or the additive intensity models for contagion. The empirical Bayes (EB) methodology, introduced by Robbins (1956) and Stein (1956), considers n independent and structurally similar problems of inference on the parameters θ_i from observed data $Y_1, \ldots, Y_n$, where Y_i has probability density $f(y|\theta_i)$. The θ_i are assumed to have a common prior distribution G that has unspecified hyperparameters. Letting $d_G(y)$ be the Bayes decision rule (with respect to some loss function and assuming known hyperparameters) when $Y_i = y$ is observed, the basic principle underlying EB is that a parametric form of G (as in Stein, 1956) or even G itself (as in Robbins, 1956) can be consistently estimated from $Y_1, \ldots, Y_n$, leading to the EB rule $d_{\hat{G}}$. Dynamic EB extends this idea to longitudinal data Y_{it}; see Lai et al. (2013). In the context of insurance claims over time for n contracts belonging to the same risk class, the conventional approach to insurance rate-making (called "evolutionary credibility" in actuarial science) assumes a linear state-space for the longitudinal claims data so that the Kalman filter can be used to estimate the claims' expected values, which are assumed to form an autoregressive time series. Applying the EB principle to the longitudinal claims from the n insurance contracts, Lai and Sun (2012) have developed a class of linear mixed models as an alternative to linear state-space models for evolutionary credibility and have shown that the predictive performance is comparable to that of the Kalman filter when the claims are generated by a linear state-space model. This approach can be readily extended to GLMMs not only for longitudinal claims data but also for default probabilities of n firms, incorporating frailty, contagion, and regime switching. Details are given in Lai and Xing (2013).

33.5 Statistics in the new era of finance

Statistics has been assuming an increasingly important role in quantitative finance and risk management after the financial crisis, which exposed the weakness and limitations of traditional financial models, pricing and hedging theories, risk measures and management of derivative securities and structured products. Better models and paradigms, and improvements in risk management systems are called for. Statistics can help meet these challenges, which in turn may lead to new methodological advances for the field.

The Dodd–Frank Act and recent financial reforms in the European Union and other countries have led to new financial regulations that enforce transparency and accountability and enhance consumer financial protection. The need for good and timely data for risk management and regulatory supervision is well recognized, but how to analyze these massive datasets and use them to give early warning and develop adaptive risk control strategies is a challenging statistical problem that requires domain knowledge and interdisciplinary collaboration. The monograph by Lai and Xing (2013) describes some recent research in sequential surveillance and early warning, particularly for systemic risk which is the risk of a broad-based breakdown in the financial system as experienced in the recent financial crisis. It reviews the critical financial market infrastructure and core-periphery network models for mathematical representation of the infrastructure; such networks incorporate the transmission of risk and liquidity to and from the core and periphery nodes of the network. After an overview of the extensive literature in statistics and engineering on sequential change-point detection and estimation, statistical process control, and stochastic adaptive control, it discusses how these methods can be modified and further developed for network models to come up with early warning indicators for financial instability and systemic failures. This is an ongoing research project with colleagues at the Financial and Risk Modeling Institute at Stanford, which is an interdisciplinary research center involving different schools and departments.

Besides risk management, the field of statistics has played an important role in algorithmic trading and quantitative investment strategies, which have gained popularity after the financial crisis when hedge funds employing these strategies outperformed many equity indices and other risky investment options. Statistical modeling of market microstructure and limit-order book dynamics is an active area of research; see for example, Ait-Sahalia et al. (2005) and Barndorff-Nielsen et al. (2008). Even the foundational theory of mean-variance portfolio optimization has received a new boost from contemporaneous developments in statistics during the past decade. A review of these developments is given by Lai et al. (2011) who also propose a new statistical approach that combines solution of a basic stochastic optimization problem with flexible modeling to incorporate time series features in the analysis of the training sample of historical data.

33.6 Conclusion

There are some common threads linking statistical modeling in finance and health care for the new era. One is related to "Big Data" for regulatory supervision, risk management and algorithmic trading, and for emerging health care systems that involve electronic medical records, genomic and proteomic biomarkers, and computer-assisted support for patient care. Another is related to the need for collaborative research that can integrate statistics with domain knowledge and subject-matter issues. A third thread is related to dynamic panel data and empirical Bayes modeling in finance and insurance. Health insurance reform is a major feature of the 2010 Affordable Care Act, and has led to a surge of interest in the new direction of health insurance in actuarial science. In the US, insurance contracts are predominantly fee-for-service (FFS). In such arrangements the FFS contracts offer perverse incentives for providers, typically resulting in over-utilization of medical care. Issues with FFS would be mitigated with access to more reliable estimates of patient heterogeneity based on administrative claims information. If the health insurer can better predict future claims, it can better compensate providers for caring for patients with complicated conditions. This is an area where the field of statistics can have a major impact.

References

Ait-Sahalia, Y., Mykland, P., and Zhang, L. (2005). How often to sample a continuous-time process in the presence of market microstructure noise. *Review of Financial Studies*, 18:351–416.

ALLHAT Collaborative Research Group (2002). Major outcomes in high-risk hypertensive patients randomized to angiotensin-converting enzyme inhibitor or calcium channel blocker vs diuretic: The Antihypertensive and Lipid-Lowering Treatment to Prevent Heart Attack Trial (ALLHAT). *Journal of the American Medical Association*, 288:2981–2997.

Barndorff-Nielsen, O.E., Hansen, P.R., Lunde, A., and Shephard, N. (2008). Designing realized kernels to measure the ex post variation of equity prices in the presence of noise. *Econometrica*, 76:1481–1536.

Bartroff, J., Lai, T.L., and Shih, M.-C. (2012). *Sequential Experimentation in Clinical Trials: Design and Analysis*. Springer, New York.

Druker, B.J., Guilhot, F., O'Brien, S.G., Gathmann, I., Kantarjian, H., Gattermann, N., Deininger, M.W.N., Silver, R.T., Goldman, J.M., Stone, R.M., Cervantes, F., Hochhaus, A., Powell, B.L., Gabrilove, J.L., Rousselot, P., Reiffers, J., Cornelissen, J.J., Hughes, T., Agis, H., Fischer, T., Verhoef, G., Shepherd, J., Saglio, G., Gratwohl, A., Nielsen, J.L., Radich, J.P., Simonsson, B., Taylor, K., Baccarani, M., So, C., Letvak, L., Larson, R.A., and IRIS Investigators (2006). Five-year follow-up of patients receiving imatinib for chronic myeloid leukemia. *New England Journal of Medicine*, 355:2408–2417.

Duffie, D. and Singleton, K.J. (2003). *Credit Risk: Pricing, Measurement, and Management*. Princeton University Press, Princeton, NJ.

Fiore, L., Brophy, M., D'Avolio, L., Conrad, C., O'Neil, G., Sabin, T., Kaufman, J., Hermos, J., Swartz, S., Liang, M., Gaziano, M., Lawler, E., Ferguson, R., Lew, R., Doras, G., and Lavori, P. (2011). A point-of-care clinical trial comparing insulin administered using a sliding scale versus a weight-based regimen. *Clinical Trials*, 8:183–195.

Genest, C. and Favre, A.-C. (2007). Everything you always wanted to know about copula modeling but were afraid to ask. *Journal of Hydrologic Engineering*, 12:347–368.

Genest, C. and Nešlehová, J. (2012). Copulas and copula models. *Encyclopedia of Environmetrics*, 2nd edition. Wiley, Chichester, 2:541–553.

Inoue, L.Y., Thall, P.F., and Berry, D.A. (2002). Seamlessly expanding a randomized phase II trial to phase III. *Biometrics*, 58:823–831.

Lai, T.L. (2001). Sequential analysis: Some classical problems and new challenges (with discussion). *Statistica Sinica*, 11:303–408.

Lai, T.L. and Lavori, P.W. (2011). Innovative clinical trial designs: Toward a 21st-century health care system. *Statistics in Biosciences*, 3:145–168.

Lai, T.L., Lavori, P.W., and Shih, M.-C. (2012). Sequential design of Phase II–III cancer trials. *Statistics in Medicine*, 31:1944–1960.

Lai, T.L., Lavori, P.W., Shih, M.-C., and Sikic, B.I. (2012). Clinical trial designs for testing biomarker-based personalized therapies. *Clinical Trials*, 9:141–154.

Lai, T.L. and Liao, O.Y.-W. (2012). Efficient adaptive randomization and stopping rules in multi-arm clinical trials for testing a new treatment. *Sequential Analysis*, 31:441–457.

Lai, T.L., Su, Y., and Sun, K. (2013). Dynamic empirical Bayes models and their applications to longitudinal data analysis and prediction. *Statistica Sinica*. To appear.

Lai, T.L. and Sun, K.H. (2012). Evolutionary credibility theory: A generalized linear mixed modeling approach. *North American Actuarial Journal*, 16:273–284.

Lai, T.L. and Xing, H. (2013). *Active Risk Management: Financial Models and Statistical Methods*. Chapman & Hall, London.

Lai, T.L., Xing, H., and Chen, Z. (2011). Mean-variance portfolio optimization when means and covariances are unknown. *The Annals of Applied Statistics*, 5:798–823.

Lando, D. (2004). *Credit Risk Modeling: Theory and Applications*. Princeton University Press, Princeton, NJ.

Li, D.X. (2000). On default correlation: A copula function approach. *Journal of Fixed Income*, 9:43–54.

Robbins, H. (1956). An empirical Bayes approach to statistics. In *Proceedings of the Third Berkeley Symposium on Mathematical Statistics and Probability, 1954–1955*, vol. 1. University of California Press, Berkeley, CA, pp. 157–163.

Rush, A., Fava, M., Wisniewski, S., Lavori, P., Trivedi, M., Sackeim, H., Thase, M., Nierenberg, A., Quitkin, F., Kashner, T., Kupfer, D., Rosenbaum, J., Alpert, J., Stewart, J., McGrath, P., Biggs, M., Shores-Wilson, K., Lebowitz, B., Ritz, L., Niederehe, G. and STAR D Investigators Group (2004). Sequenced treatment alternatives to relieve depression (STAR*D): Rationale and design. *Controlled Clinical Trials*, 25:119–142.

Salmon, F. (2012). The formula that killed Wall Street. *Significance*, 9(1):16–20.

Shih, M.-C. and Lavori, P.W. (2013). Sequential methods for comparative effectiveness experiments: Point of care clinical trials. *Statistica Sinica*, 23:1775–1791.

Stein, C. (1956). Inadmissibility of the usual estimator for the mean of a multivariate normal distribution. In *Proceedings of the Third Berkeley Symposium on Mathematical Statistics and Probability, 1954–1955*, vol. 1. University of California Press, Berkeley, CA, pp. 197–206.

Stukel, T.A., Fisher, E.S., Wennberg, D.E., Alter, D.A., Gottlieb, D.J., and Vermeulen, M.J. (2007). Analysis of observational studies in the presence of treatment selection bias: Effects of invasive cardiac management on AMI survival using propensity score and instrumental variable methods. *Journal of the American Medical Association*, 297:278–285.

Thall, P.F., Logothetis, C., Pagliaro, L.C., Wen, S., Brown, M.A., Williams, D., and Millikan, R.E. (2007). Adaptive therapy for androgen-independent prostate cancer: A randomized selection trial of four regimens. *Journal of the National Cancer Institute*, 99:1613–1622.

34

Meta-analyses: Heterogeneity can be a good thing

Nan M. Laird
Department of Biostatistics
Harvard School of Public Health, Boston, MA

Meta-analysis seeks to summarize the results of a number of different studies on a common topic. It is widely used to address important and dispirit problems in public health and medicine. Heterogeneity in the results of different studies is common. Sometimes perceived heterogeneity is a motivation for the use of meta-analysis in order to understand and reconcile differences. In other cases the presence of heterogeneity is regarded as a reason not to summarize results. An important role for meta-analysis is the determination of design and analysis factors that influence the outcome of studies. Here I review some of the controversies surrounding the use of meta-analysis in public health and my own experience in the field.

34.1 Introduction

Meta-analysis has become a household word in many scientific disciplines. The uses of meta-analysis vary considerably. It can be used to increase power, especially for secondary endpoints or when dealing with small effects, to reconcile differences in multiple studies, to make inferences about a very particular treatment or intervention, to address more general issues, such as what is the magnitude of the placebo effect or to ask what design factors influence the outcome of research? In some cases, a meta-analysis indicates substantial heterogeneity in the outcomes of different studies.

With my colleague, Rebecca DerSimonian, I wrote several articles on meta-analysis in the early 1980s, presenting a method for dealing with heterogeneity. In this paper, I provide the motivation for this work, advantages and difficulties with the method, and discuss current trends in handling heterogeneity.

34.2 Early years of random effects for meta-analysis

I first learned about meta-analysis in the mid-1970s while I was still a graduate student. Meta-analysis was being used then and even earlier in the social sciences to summarize the effectiveness of treatments in psychotherapy, the effects of class size on educational achievement, experimenter expectancy effects in behavioral research and results of other compelling social science research questions.

In the early 1980s, Fred Mosteller introduced me to Warner Slack and Doug Porter at Harvard Medical School who had done a meta-analysis on the effectiveness of coaching students for the Scholastic Aptitude Tests (SAT). Fred was very active in promoting the use of meta-analysis in the social sciences, and later the health and medical sciences. Using data that they collected on sixteen studies evaluating coaching for verbal aptitude, and thirteen on math aptitude, Slack and Porter concluded that coaching is effective on raising aptitude scores, contradicting the principle that the SATs measure "innate" ability (Slack and Porter, 1980).

What was interesting about Slack and Porter's data was a striking relationship between the magnitude of the coaching effect and the degree of control for the coached group. Many studies evaluated only coached students and compared their before and after coaching scores with national norms provided by the ETS on the average gains achieved by repeat test takers. These studies tended to show large gains for coaching. Other studies used convenience samples as comparison groups, and some studies employed either matching or randomization. This last group of studies showed much smaller gains. Fred's private comment on their analysis was "Of course coaching is effective; otherwise, we would all be out of business. The issue is what kind of evidence is there about the effect of coaching?"

The science (or art) of meta-analysis was then in its infancy. Eugene Glass coined the phrase meta-analysis in 1976 to mean the statistical analysis of the findings of a collection of individual studies (Glass, 1976). Early papers in the field stressed the need to systematically report relevant details on study characteristics, not only about design of the studies, but characteristics of subjects, investigators, interventions, measures, study follow-up, etc. However a formal statistical framework for creating summaries that incorporated heterogeneity was lacking. I was working with Rebecca DerSimonian on random effects models at the time, and it seemed like a natural approach that could be used to examine heterogeneity in a meta-analysis. We published a follow-up article to Slack and Porter in the *Harvard Education Review* that introduced a random effects approach for meta-analysis (DerSimonian and Laird, 1983). The approach followed that of Cochran who wrote about combining the effects of different experiments with measured outcomes (Cochran, 1954). Cochran introduced the idea that the observed effect of each study could be partitioned into the sum of a "true" effect plus a sampling (or within-study) error. An estimate of the within-study error should be available from each individual study, and can be used to get at the distribution of "true" effects. Many of the social science meta-analyses neglected to identify within-study error, and in some cases, within-study error variances were not reported. Following Cochran, we proposed a method for estimating the mean of the "true" effects, as well as the variation in "true" effects across studies. We assumed that the observed study effect was the difference in two means on a measured scale and also assumed normality for the distribution of effects and error terms. However normality is not necessary for validity of the estimates.

The random effects method for meta-analysis is now widely used, but introduces stumbling blocks for some researchers who find the concept of a distribution of effects for treatments or interventions unpalatable. A major conceptual problem is imagining the studies in the analysis as a sample from a recognizable population. As discussed in Laird and Mosteller (1990), absence of a sampling frame to draw a random sample is a ubiquitous problem in scientific research in most fields, and so should not be considered as a special problem unique to meta-analysis. For example, most investigators treat patients enrolled in a study as a random sample from some population of patients, or clinics in a study as a random sample from a population of clinics and they want to make inferences about the population and not the particular set of patients or clinics. This criticism does not detract from the utility of the random effects method. If the results of different research programs all yield similar results, there would not be great interest in a meta-analysis. The principle behind a random effects approach is that a major purpose of meta-analysis is to quantify the variation in the results, as well as provide an overall mean summary.

Using our methods to re-analyze Slack and Porter's results, we concluded that any effects of coaching were too small to be of practical importance (DerSimonian and Laird, 1983). Although the paper attracted considerable media attention (articles about the paper were published in hundreds of US news papers), the number of citations in the scientific literature is comparatively modest.

34.3 Random effects and clinical trials

In contrast to our paper on coaching, a later paper by DerSimonian and myself has been very highly cited, and led to the moniker "DerSimonian and Laird method" when referring to a random-effects meta-analysis (DerSimonian and Laird, 1986). This paper adapted the random effects model for meta-analysis of clinical trials; the basic idea of the approach is the same, but here the treatment effect was assumed to be the difference in Binomial cure rates between a treated and control group. Taking the observed outcome to be a difference in Binomial cure rates raised various additional complexities. The difference in cure rates is more relevant and interpretable in clinical trials, but statistical methods for combining a series of 2×2 tables usually focus on the odds ratio. A second issue is that the Binomial mean and the variance are functionally related. As a result, the estimate of the within-study variance (which was used to determine the weight assigned to each study) is correlated with the estimated study effect size. We ignored this problem, with the result that the method can be biased, especially with smaller samples, and better approaches are available (Emerson et al., 1996; Wang et al., 2010). The final issue is estimating and testing for heterogeneity among the results, and how choice of effect measure (rate difference, risk ratio, or odds ratio) can affect the results. Studies have shown that choice of effect measure has relatively little effect on assessment of heterogeneity (Berlin et al., 1989).

The clinical trials setting can be very different from data synthesis in the social sciences. The endpoint of interest may not correspond to the primary endpoint of the trial and the number of studies can be much smaller. For example, many clinical trials may be designed to detect short term surrogate endpoints, but are under-powered to detect long term benefits or important side effects (Hine et al., 1989). In this setting a meta-analysis can be the best solution for inferences about long term or untoward side effects. Thus the primary purpose of the meta-analysis may be to look at secondary endpoints when the individual studies do not have sufficient power to detect the effects of secondary endpoints. Secondly, it is common to restrict meta-analyses to randomized controlled trials (RTCs) and possibly also to trials using the same treatment. This is in direct contrast to meta-analyses that seek to answer very broad questions; such analyses can include many types of primary studies, studies with different outcome measures, different treatments, etc. In one of the earliest meta-analyses, Beecher (1955) sought to measure the "placebo" effect by combining data from 15 clinical studies of pain for a variety of different indications treated by different placebo techniques.

Yusuf et al. (1985) introduced a "fixed effects" method for combining the results of a series of controlled clinical trials. They proposed a fixed-effect approach to the analysis of all trials ever done, published or not, that presumes we are only interested in the particular set of studies we have found in our search (which is in principle all studies ever done). In practice, statisticians are rarely interested in only the particular participants in our data collection efforts, but want findings that can be generalized to similar participants, whether they be clinics, hospitals, patients, investigators, etc. In fact, the term "fixed" effects is sometimes confused with the equal effects setting, where the statistical methods used implicitly assume that the "true" effects for each study are the same. Yusuf et al. (1985) may have partly contributed to this confusion by stating that their proposed estimate and standard error of the overall effect do *not* require equality of the effects, but then cautioning that the interpretation of the results is restricted to the case where the effects are "approximately similar." As noted by Cochran, ignoring variation in the effects of different studies generally gives smaller standard errors that do not account for this variance.

34.4 Meta-analysis in genetic epidemiology

In recent years, meta-analysis is increasingly used in a very different type of study: large scale genetic epidemiology studies. Literally thousands of reports are published each year on associations between genetic variants and diseases. These reports may look at only a few genetic variants in specified genes, or they may test hundreds of thousands of variants in all the chromosomes in Genome Wide Association Scans (GWAS). These GWAS reports are less than ten years old, but now are the standard approach for gene "discovery" in complex diseases. Because there are so many variants being tested, and because the effect sizes tend to be quite small, replication of positive findings in independent samples has been considered a requirement for publication right from the beginning. But gradually there has been a shift from reporting the results of the primary study and a small but reasonable number of replications, to pooling the new study and the replication studies, and some or all available GWAS studies using the same endpoint.

In contrast to how the term meta-analysis is used elsewhere, the term meta-analysis is often used in the genetic epidemiology literature to describe what is typically called "pooling" in the statistical literature, that is, analyzing individual level data together as a single study, stratifying or adjusting by source. The pooling approach is popular and usually viewed as more desirable, despite the fact that studies (Lin and Zeng, 2010) have shown that combining the summary statistics in a meta-analysis is basically equivalent to pooling under standard assumptions. I personally prefer meta-analysis because it enables us to better account for heterogeneity and inflated variance estimates.

Like all epidemiological studies, the GWAS is influenced by many design factors which will affect the results, and they do have a few special features which impact on the use of pooling or meta-analysis, especially in the context of heterogeneity. First, the cost of genotyping is great, and to make these studies affordable, samples have been largely opportunistic. Especially early on, most GWAS used pre-existing samples where sufficient biological material was available for genotyping, and the traits of interest were already available. This can cause considerable heterogeneity. For example, I was involved in the first GWAS to find association between a genetic variant and obesity as measured by body mass index (BMI); it illustrates the importance of study design.

The original GWAS (with only 100,000 genetic markers) was carried out in the Framingham Heart Study (Herbert et al., 2006). We used a novel approach to the analysis and had data from five other cohorts for replication. All but one of the five cohorts reproduced the result. This was an easy study to replicate, because virtually every epidemiological study of disease has also measured height and weight, so that BMI is available, even though the study was not necessarily designed to look at factors influencing BMI. In addition, replication required genotyping only one genetic marker. In short order, hundreds of reports appeared in the literature, many of them non-replications. A few of us undertook a meta-analysis of these reported replications or non-replications with the explicit hypothesis that the study population influences the results; over 76,000 subjects were included in this analysis. We considered three broad types of study populations. The first is general population cohort where subjects were drawn from general population samples without restricting participants on the basis of health characteristics. Examples of this are the Framingham Heart Study and a German population based sample (KORA). The second is healthy population samples where subjects are drawn from populations known to be healthier than those in the general population, typically from some specific work force. The third category of studies included those specifically designed to study obesity; those studies used subjects chosen on the basis of obesity, including case-control samples where obese and non-obese subjects were selected to participate and family-controlled studies that used only obese subjects and their relatives.

In agreement with our hypothesis, the strongest result we found was that the effect varied by study population. The general population samples and the selected samples replicated the original study in finding a significant association, but the healthy population studies showed no evidence of an effect. This is a critically important finding. Many fields have shown that randomized versus non-randomized, blinded versus unblinded, etc., can have major effects, but this finding is a bit different. Using healthy subjects is not intrinsically a poor design choice, but may be so for many common, complex disorders. One obvious reason is that genetic studies of healthy subjects may lack sufficient variation in outcome to have much power. More subtle factors might include environmental or other genetic characteristics which interact to modify the gene effect being investigated. In any event, it underscores the desirability of assessing and accounting for heterogeneity in meta-analyses of genetic associations.

A second issue is related to the fact that there are still relatively few GWAS of specific diseases, so many meta-analyses involve only a handful of studies. The random effects approach of DerSimonian and Laird does not work well with only a handful of studies because it estimates a variance, where the sample size for the variance estimate is the number of studies. Finally for a GWAS, where hundreds of thousands of genetic markers are tested, often on thousands of subjects, any meta-analysis method needs to be easily implemented in order to be practically useful. Software for meta-analysis is included in the major genetic analysis statistical packages (Evangelou and Ioannidis, 2013), but most software packages only implement a fixed effects approach. As a result, the standard meta-analyses of GWAS use the fixed effects approach and potentially overstate the precision in the presence of heterogeneity.

34.5 Conclusions

The DerSimonian and Laird method has weathered a great deal of criticism, and undoubtedly we need better methods for random effects analyses, especially when the endpoints of interest are proportions and when the number of studies being combined is small and or the sample sizes within each study are small. Most meta-analyses involving clinical trials acknowledge the importance of assessing variation in study effects, and new methods for quantifying this variation are widely used (Higgins and Thompson, 2002). In addition, meta-regression methods for identifying factors influencing heterogeneity are available (Berkey et al., 1995; Thompson and Higgins, 2002); these can be used to form subsets of studies which are more homogeneous. There is an extensive literature emphasizing the necessity and desirability of assessing heterogeneity, and many of these reinforce the role of study design in connection with heterogeneity. The use of meta-analysis in genetic epidemiology to find disease genes is still relatively new, but the benefits are widely recognized (Ioannidis et al., 2007). Better methods for implementing random effects methods with a small number of studies will be especially useful here.

References

Beecher, H.K. (1955). The powerful placebo. *Journal of the American Medical Association*, 159:1602–1606.

Berkey, C.S., Hoaglin, D.C., Mosteller, F., and Colditz, G.A. (1995). A random-effects regression model for meta-analysis. *Statistics in Medicine*, 14:395–411.

Berlin, J.A., Laird, N.M., Sacks, H.S., and Chalmers, T.C. (1989). A comparison of statistical methods for combining event rates from clinical trials. *Statistics in Medicine*, 8:141–151.

Cochran, W.G. (1954). The combination of estimates from different experiments. *Biometrics*, 10:101–129.

DerSimonian, R. and Laird, N.M. (1983). Evaluating the effect of coaching on SAT scores: A meta-analysis. *Harvard Educational Review*, 53:1–15.

DerSimonian, R. and Laird, N.M. (1986). Meta-analysis in clinical trials. *Controlled Clinical Trials*, 7:177–188.

Emerson, J.D., Hoaglin, D.C., and Mosteller, F. (1996). Simple robust procedures for combining risk differences in sets of 2×2 tables. *Statistics in Medicine*, 15:1465–1488.

Evangelou, E. and Ioannidis, J.P.A. (2013). Meta-analysis methods for genome-wide association studies and beyond. *Nature Reviews Genetics*, 14:379–389.

Glass, G.V. (1976). Primary, secondary, and meta-analysis of research. *Educational Researcher*, 5:3–8.

Herbert, A., Gerry, N.P., McQueen, M.B., Heid, I.M., Pfeufer, A., Illig, T., Wichmann, H.-E., Meitinger, T., Hunter, D., Hu, F.B. et al. (2006). A common genetic variant is associated with adult and childhood obesity. *Science*, 312:279–283.

Higgins, J. and Thompson, S.G. (2002). Quantifying heterogeneity in a meta-analysis. *Statistics in Medicine*, 21:1539–1558.

Hine, L., Laird, N.M., Hewitt, P., and Chalmers, T. (1989). Meta-analytic evidence against prophylactic use of lidocaine in acute myocardial infarction. *Archives of Internal Medicine*, 149:2694–2698.

Ioannidis, J.P.A., Patsopoulos, N.A., and Evangelou, E. (2007). Heterogeneity in meta-analyses of genome-wide association investigations. *PloS one*, 2:e841.

Laird, N.M. and Mosteller, F. (1990). Some statistical methods for combining experimental results. *International Journal of Technological Assessment of Health Care*, 6:5–30.

Lin, D. and Zeng, D. (2010). Meta-analysis of genome-wide association studies: No efficiency gain in using individual participant data. *Genetic Epidemiology*, 34:60–66.

Slack, W. and Porter, D. (1980). The scholastic aptitude test: A critical appraisal. *Harvard Educational Review*, 66:1–27.

Thompson, S.G. and Higgins, J. (2002). How should meta-regression analyses be undertaken and interpreted? *Statistics in Medicine*, 21:1559–1573.

Wang, R., Tian, L., Cai, T., and Wei, L. (2010). Nonparametric inference procedure for percentiles of the random effects distribution in meta-analysis. *The Annals of Applied Statistics*, 4:520–532.

Yusuf, S., Peto, R., Lewis, J., Collins, R., and Sleight, P. (1985). Beta blockade during and after myocardial infarction: an overview of the randomized trials. *Progress in Cardiovascular Diseases*, 27:335–371.

35

Good health: Statistical challenges in personalizing disease prevention

Alice S. Whittemore
Department of Health Research and Policy
Stanford University, Stanford, CA

Increasingly, patients and clinicians are basing healthcare decisions on statistical models that use a person's covariates to assign him/her a probability of developing a disease in a given future time period. In this chapter, I describe some of the statistical problems that arise when evaluating the accuracy and utility of these models.

35.1 Introduction

Rising health care costs underscore the need for cost-effective disease prevention and control. To achieve cost-efficiency, preventive strategies must focus on individuals whose genetic and lifestyle characteristics put them at highest risk. To identify these individuals, statisticians and public health professionals are developing personalized risk models for many diseases and other adverse outcomes. The task of checking the accuracy and utility of these models requires new statistical methods and new applications for existing methods.

35.2 How do we personalize disease risks?

We do this using a personalized risk model, which is an algorithm that assigns a person a probability of developing an adverse outcome in a given future time period (say, five, ten or twenty years). The algorithm combines his/her values for a set of risk-associated covariates with regional incidence and mortality data and quantitative evidence of the covariates' effects on risk of the outcome.

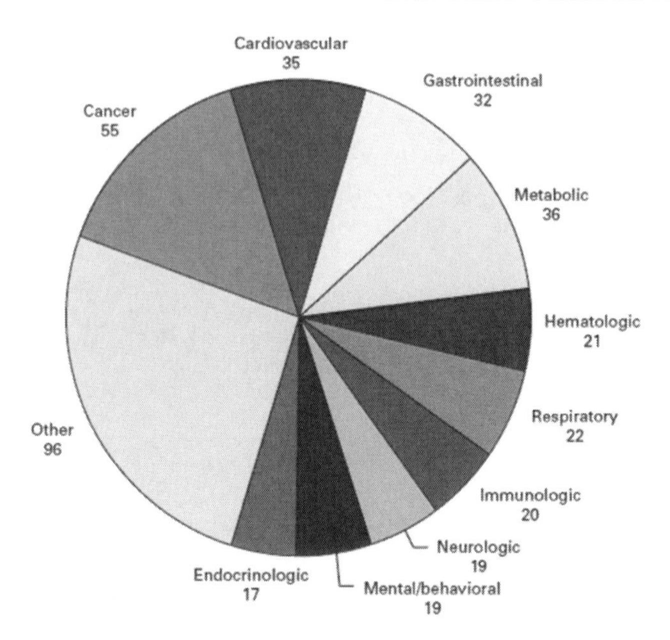

FIGURE 35.1
Genetic tests sold directly to consumers for medical conditions.

For example, the Breast Cancer Risk Assessment Tool (BCRAT) assigns a woman a probability of developing breast cancer in the next five years, using her self-reported risk factors, relative risk estimates obtained from randomized trials or observational studies, and national rates for breast cancer incidence and death from all causes (Costantino et al., 1999; Gail et al., 1989).

Personal risk models play important roles in the practice of medicine, as standards for clinical care become increasingly tailored to patients' individual characteristics and preferences. The need to evaluate risk models is increasing as personalized medicine evolves. A PUBMED search using the key words validating risk models produced nearly 4000 hits, indicating substantial interest in this topic in current medical practice. Moreover there are now more than 370 online direct-to-consumer genetic tests with risk assessments for adverse health outcomes (Figure 35.1). The Food and Drug Administration is considering regulating these assessments, so we need reliable methods to validate them.

Determining the probability of a future adverse outcome for a particular person is not unlike determining the chance of a hurricane or earthquake in a particular area during a given time period. Not surprisingly therefore, many of the statistical problems involved in developing and evaluating risk models have intrigued meteorologists and seismologists for decades, and their findings form a useful foundation for this work; see, e.g., Brier (1950), Hsu and Murphy (1986), Murphy (1973), and Wilks (1995).

35.3 How do we evaluate a personal risk model?

Risk models for long-term future outcomes are commonly assessed with respect to two attributes. Their calibration reflects how well their assigned risks agree with observed outcome occurrence within subgroups of the population. Their discrimination (also called precision or resolution) reflects how well they distinguish those who ultimately do and do not develop the outcome. Good calibration does not imply good discrimination. For example, if the actual disease risks of a population show little inter-personal variation, discrimination will be poor even for a perfectly calibrated risk model. Conversely, good discrimination does not imply good calibration. Discrimination depends only on the ranks of a model's assigned risks, so any rank-invariant transformation of a model's risks will affect its calibration but not its discrimination.

An important task is to quantify how much a model's calibration and discrimination can be improved by expanding it with additional covariates, such as newly discovered genetic markers. However, the discrimination of a risk model depends on the distribution of risk-associated covariates in the population of interest. As noted in the previous paragraph, no model can discriminate well in a population with a homogeneous covariate distribution. Thus while large discrimination gains from adding covariates to a model are informative (indicating substantial additional risk variation detected by the expanded model), a small precision gain is less so, as it may merely reflect underlying risk homogeneity in the population.

Several metrics have been proposed to assess and compare models with respect to their calibration and discrimination. Their usefulness depends on how they will be used, as shown by the following examples.

Example 1. Risk models are used to determine eligibility for randomized clinical trials involving treatments with serious potential side effects. For instance, the BCRAT model was used to determine eligibility for a randomized trial to determine if tamoxifen can prevent breast cancer (Fisher et al., 1998, 2005). Because tamoxifen increases the risks of stroke, endometrial cancer and deep-vein thrombosis, eligibility was restricted to women whose breast cancer risks were deemed high enough to warrant exposure to these side effects. Thus eligible women were those whose BCRAT-assigned five-year breast cancer risk exceeded 1.67%. For this type of application, a good risk model should yield a decision rule with few false positives, i.e., one that excludes women who truly are at low breast cancer risk. A model without this attribute could inflict tamoxifen's side effects on women with little chance of gaining from the experience.

Example 2. Risk models are used to improve the cost-efficiency of preventive interventions. For instance, screening with breast magnetic resonance imaging (MRI) detects more breast cancers but costs more and produces more false positive scans, compared to mammography. Costs and false positives can be reduced by restricting MRI to women whose breast cancer risk exceeds some threshold (Plevritis et al., 2006). For this type of application, a good risk model should give a classification rule that assigns mammography to those truly at low risk (i.e., has a low false positive rate), but also assigns MRI to those truly at high risk (i.e., has a high true positive rate).

Example 3. Risk models are used to facilitate personal health care decisions. Consider, for instance, a postmenopausal woman with osteoporosis who must choose between two drugs, raloxifene and alendronate, to prevent hip fracture. Because she has a family history of breast cancer, raloxifene would seem a good choice, since it also reduces breast cancer risk. However she also has a family history of stroke, and raloxifene is associated with increased stroke risk. To make a rational decision, she needs a risk model that provides accurate information about her own risks of developing three adverse outcomes (breast cancer, stroke, hip fracture), and the effects of the two drugs on these risks.

The first two examples involve classifying people into "high" and "low" categories; thus they require risk models with low false positive and/or false negative rates. In contrast, the third example involves balancing one person's risks for several different outcomes, and thus it requires risk models whose assigned risks are accurate enough at the individual level to facilitate rational healthcare decisions. It is common practice to summarize a model's calibration and discrimination with a single statistic, such as a chi-squared goodness-of-fit test. However, such summary measures do not reveal subgroups whose risks are accurately or inaccurately pegged by a model. This limitation can be addressed by focusing on subgroup-specific performance measures. Evaluating performance in subgroups also helps assess a model's value for facilitating personal health decisions. For example, a woman who needs to know her breast cancer risk is not interested in how a model performs for others in the population; yet summary performance measures involve the distribution of covariates in the entire population to which she belongs.

35.4 How do we estimate model performance measures?

Longitudinal cohort studies allow comparison of actual outcomes to model-assigned risks. At entry to a cohort, subjects report their current and past covariate values. A risk model then uses these baseline covariates to assign each subject a risk of developing the outcome of interest during a specified subsequent time period. For example, the Breast Cancer Family Registry (BCFR), a consortium of six institutions in the United States, Canada and Australia, has been monitoring the vital statuses and cancer occurrences of registry participants for more than ten years (John et al., 2004). The New York site of the BCFR has used the baseline covariates of some 1900 female registry participants to assign each of them a ten-year probability of breast cancer development according to one of several risk models (Quante et al., 2012). These assigned risks are then compared to actual outcomes during follow-up. Subjects who die before outcome occurrence are classified as negative for the outcome, so those with life-threatening co-morbidities may have low outcome risk because they are likely to die before outcome development.

Using cohort data to estimate outcome probabilities presents statistical challenges. For example, some subjects may not be followed for the full risk period; instead they are last observed alive and outcome-free after only a fraction of the period. An analysis that excludes these subjects may yield biased estimates. Instead, censored time-to-failure analysis is needed, and the analysis must accommodate the competing risk of death (Kalbfleisch and Lawless, 1998; Kalbfleisch and Prentice, 2002; Putter et al., 2007). Another challenge arises when evaluating risk models that include biomarkers obtained from blood collected at cohort entry. Budgetary constraints may prohibit costly biomarker assessment for the entire cohort, and cost-efficient sampling designs are needed, such as a nested case-control design (Ernster, 1994), a case-cohort design (Prentice, 1986), or a two-stage sampling design (Whittemore and Halpern, 2013).

Model calibration is often assessed by grouping subjects into quantiles of assigned risk, and comparing estimated outcome probability to mean assigned risk within each quantile. Results are plotted on a graph called an attribution diagram (AD) (Hsu and Murphy, 1986). For example, the top two panels of Figure 35.2 show ADs for subjects from the NY-BCFR cohort who have been grouped in quartiles of breast cancer risk as assigned by two breast cancer risk models, the BCRAT model and the International Breast Cancer Intervention Study (IBIS) model (Tyrer et al., 2004). The null hypothesis of equality between quantile-specific mean outcome probabilities and mean assigned risks is commonly tested by classifying subjects into risk groups and applying a chi-squared goodness-of-fit statistic. This approach has limitations: 1) the quantile grouping is arbitrary and varies across cohorts sampled from the same population; 2) the averaging of risks over subjects in a quantile can obscure subsets of subjects with poor model fit; 3) when confidence intervals for estimated outcome probabilities exclude the diagonal line it is difficult to trouble-shoot the risk model; 4) assuming a chi-squared asymptotic distribution for the goodness-of-fit statistic ignores the heterogeneity of risks within quantiles.

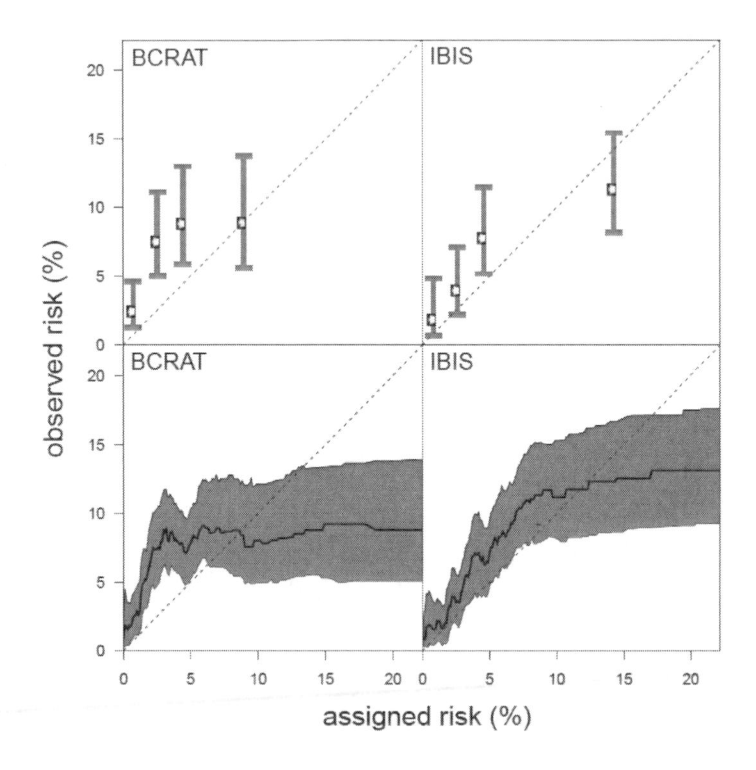

FIGURE 35.2
Grouped and individualized goodness-of-fit of BCRAT- and IBIS-assigned breast cancer risks in NY BCFR data.

Some of these limitations are addressed by an alternative approach that uses nearest-neighbor (NN) methods to estimate an outcome probability for each set of subjects with a given assigned risk (Akritas, 1994; Heagerty et al., 2000). The NN estimate of actual risk for a subject with assigned risk r is based on the set of all individuals with risks r' such that $|G(r) - G(r')| < \epsilon$, where $G(r)$ the empirical cumulative distribution function of assigned risks and ϵ (which determines the maximum neighborhood size) is a small positive number that decreases with total cohort sample size. The lower two panels of Figure 35.2 show the NN ADs for BCRAT and IBIS risk models as applied to the NY BCFR data. Notice that these ADs are similar to, but more informative than, the discretized ADs in the upper panels. The NN ADs show for which assigned risks (and thus which individuals) the 95% CIs exclude the diagonal line, suggesting significant poor fit.

The NN estimates are not without problems. The lower panels of Figure 35.2 suggest that both models fit poorly in the right tails of their assigned risk distributions. However, unpublished simulations using a perfectly-calibrated model reveal that this conclusion is likely to be erroneous, and that the observed flattening of the estimated outcome probability curves in the right tails is an artifact of the NN method. Such flattening reflects the clumping of sparse subjects in the tails into the same neighborhood to estimate a single common outcome probability. New methods are needed to address these issues.

Model discrimination is commonly assessed using the concordance statistic or C-statistic, also called the area under the receiver-operating-characteristic curve (Hanley and McNeil, 1982; Pepe, 2003). This statistic estimates the probability that the risk assigned to a randomly sampled individual who develops the outcome exceeds that of a randomly sampled individual who does not. The C-statistic has several limitations. Like all summary statistics, it fails to indicate subgroups for whom a model discriminates poorly, or subgroups for which one model discriminates better than another. In addition, patients and health professionals have difficulty interpreting it. A more informative measure is the Case Risk Percentile (CRP), defined for each outcome-positive subject (case) as the percentile of his/her assigned risk in the distribution of assigned risks of all outcome-negative subjects. The CRP equals 1 SPV, where SPV denotes her standardized placement value (Pepe and Cai, 2004; Pepe and Longton, 2005). The CRP can be useful for comparing the discrimination of two risk models.

For example, Figure 35.3 shows the distribution of CRPs for 81 breast cancer cases in the NY-BCFR data, based on the BCRAT & IBIS models. Each point in the figure corresponds to a subject who developed breast cancer within 10 years of baseline. Each of the 49 points above the diagonal represents a case whose IBIS CRP exceeds her BCRAT CRP (i.e., IBIS better discriminates her risk from that of non-cases than does BCRAT), and the 32 points below the line represent cases for whom BCRAT discriminates better than IBIS. (Note that CRPs can be computed for any assigned risk, not just those of cases.) A model's C-statistic is just the mean of its CRPs, averaged over all cases. Importantly, covariates associated with having a CRP above or below the diagonal line can indicate which subgroups are better served by one model than the other. The CRPs are individualized measures of model sensitivity.

Research is needed to develop alternatives to the C-statistic that are more useful for evaluating model discrimination. Further discussion of this issue can be found in Pepe et al. (2010) and Pepe and Janes (2008).

35.5 Can we improve how we use epidemiological data for risk model assessment?

We need better methods to accommodate the inherent limitations of epidemiological data for assessing risk model performance. For example, the subjects in large longitudinal cohort studies are highly selected, so that findings may

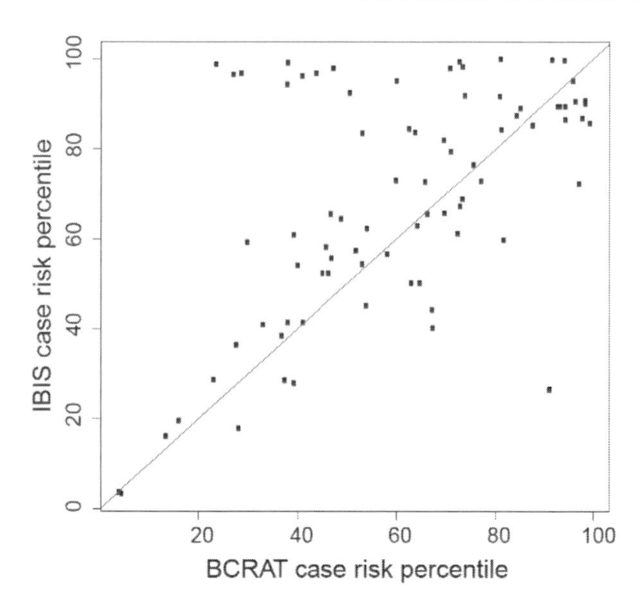

FIGURE 35.3
Scatterplot of BCRAT and IBIS case risk percentiles for NY BCFR data.

not be generalizable. Thus we need methods for accommodating bias in estimated performance measures due to cohort selection. Also, because large cohort studies are costly, we need ways to evaluate model discrimination using case-control data that is not nested within a cohort. Finally, we need methods for developing, applying and evaluating multi-state models for multiple adverse events. The following is a brief description of these problem areas.

35.5.1 Cohort selection bias

The covariate distributions of individuals in the general population are not well represented by those of the highly selected participants in large, long-term cohort studies. For example, we found that a published ovarian cancer risk model developed using postmenopausal women in the Nurses' Health Study (Rosner et al., 2005) was well-calibrated to postmenopausal subjects in the California Teachers Study (CTS) (Bernstein et al., 2002) but poorly calibrated to those in the Women's Health Initiative (WHI) (Luo et al., 2011). We found that although covariate-specific hazard-ratios are similar in the two cohorts, their covariate distributions are very different: e.g., parity is much higher in WHI than CTS. Moreover the distributions of covariates like education and parity among cohort subjects tend to be more homogeneous than those of the general population. Work is needed to compare the distributions of covariates among subjects in cohort studies with those of the general US population, as represented, for example, by participants in one of the cross-sectional studies conducted by the National Health Interview Survey (NHANES). Methods are needed to use these distributions to estimate the model performance measures we would see if the model were applied to subjects whose covariates reflect those of the general population.

35.5.2 Evaluating risk models with case-control data

Data from case-control studies nested within a cohort are not useful for evaluating model calibration, which concerns the agreement between a model's assigned risks and the actual probabilities of adverse outcome occurrence within a future risk period. Sampling only the outcome-positive and outcome-negative subjects (ignoring the time at risk contributed by censored subjects) can lead to severe bias in calibration measures, due to overestimation of outcome probabilities (Whittemore and Halpern, 2013). However under certain assumptions, unbiased (though inefficient) estimates of discrimination measures can be obtained from nested case-control studies. The critical assumption is that the censoring be uninformative; i.e., that subjects censored at a given follow-up time are a random sample of all cohort members alive and outcome-free at that time (Heagerty et al., 2000). This assumption is reasonable for the type of censoring encountered in most cohort studies. There is a need to evaluate the efficiency loss in estimated discrimination measures associated with excluding censored subjects.

However when interest centers on special populations, such as those at high risk of the outcome, it may not be feasible to find case-control data nested within a cohort to evaluate model discrimination. For example, we may want to use breast cancer cases and cancer-free control women ascertained in a high-risk cancer clinic to determine and compare discrimination of several models for ten-year breast cancer risk. Care is needed in applying the risk models to non-nested case-control data such as these, and interpreting the results. To mimic the models' prospective setting, two steps are needed: 1) the models must assign outcome risks conditional on the absence of death during the risk period; and 2) subjects' covariates must be assessed at a date ten years before outcome assessment (diagnosis date for cases, date of interview for controls). In principle, the data can then be used to estimate ten-year breast cancer probabilities ignoring the competing risk of death. In practice, the rules for ascertaining cases and controls need careful consideration to avoid potential selection bias (Wacholder et al., 1992).

35.5.3 Designing and evaluating models for multiple outcomes

Validating risk models that focus on a single adverse outcome (such as developing breast cancer within ten years) involves estimating a woman's ten-year breast cancer probability in the presence of co-morbidities causing her death

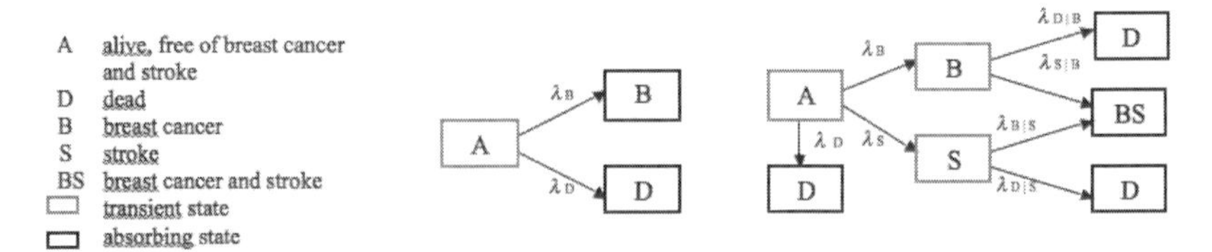

FIGURE 35.4
Graphs showing transition probabilities for transient and absorbing states for breast cancer (single outcome; left graph), and for breast cancer and stroke (two outcomes; right graph). In both graphs death before outcome occurrence is a competing risk.

before breast cancer occurrence. This competing risk of death is illustrated in the left graph of Figure 35.4. Based on her follow-up during the risk period she is classified as: a) outcome-positive (develops breast cancer); b) outcome-negative (dies before breast cancer or is alive and breast-cancer-free at end of period); or c) censored (last observed alive and free of breast cancer before end of period). Competing risk theory is needed to estimate her breast cancer probability in these circumstances. Most risk models assume that mortality rates depend only on age at risk, sex and race/ethnicity. However covariates for co-morbidities are likely to be available and could be important in risk model performance and validation among older cohort subjects. Thus we need to expand existing risk models to include covariates associated with mortality risk, and to examine the effect of this expansion on risk model performance.

There also is need to examine the feasibility of expanding existing risk models to include multiple outcomes of interest. For example, an osteoporotic woman might need to weigh the risks and benefits of several fracture-preventive options (e.g., tamoxifen, a bisphosphonate, or no drug). If she has a strong family history of certain chronic diseases (e.g., breast cancer, stroke) she needs a model that provides accurate estimates of her risks of these outcomes under each of the options she is considering. Her marginal outcome risks may be estimable from existing single-outcome risk models, but these models do not accommodate correlated risks for different outcomes. Also they were calibrated to cohorts with different selection factors and different covariate distributions, so their estimates may not be comparable. The graph on the right of Figure 35.4 indicates that the complexity of multi-state models for stochastic processes increases exponentially with the number of outcomes considered. (Here breast cancer (B) and stroke (S) are transient states, since subjects in these states are at risk for the other outcome, while death (D) and development of both breast cancer and stroke (BS) are absorbing states.)

Work is needed to determine whether the rich body of work on multi-state stochastic processes can be applied to cohort data to provide more realistic risk estimates for multiple, competing and noncompeting outcomes.

Consider, for example, the simple problem of combining existing models for breast cancer and stroke to assign to each cohort subject a vector of the form $(P(B), P(S), P(D), P(BS))$, where, for example, $P(B)$ is her assigned risk of developing breast cancer, and $P(BS)$ is her assigned risk of developing both breast cancer and stroke, during the risk period. The resulting multistate model would allow the possibility of within-person correlation in outcome-specific risks.

35.6 Concluding remarks

This chapter has outlined some statistical problems that arise when assigning people individualized probabilities of adverse health outcomes, and when evaluating the utility of these assignments. Like other areas in statistics, this work rests on foundations that raise philosophical issues. For example, the discussion of risk model calibration assumes that each individual has an unknown probability of developing the outcome of interest in the given time period. This presumed probability depends on his/her values for a set of risk factors, only some of which are known. But, unlike a parameter that governs a statistical distribution, one person's "true risk" does not lend itself to straightforward definition. Yet, much of the previous discussion requires this assumption.

Even when the assumption is accepted, fundamental issues arise. How can we estimate one woman's breast cancer probability without aggregating her survival data with those of others who may have different risks? And how much of a woman's breast cancer risk is due purely to chance? If we knew the combined effects of all measurable breast cancer risk factors, and if we could apply this knowledge to assign risks to disease-free women, how much residual variation in subsequent outcomes might we see?

These issues notwithstanding, it seems clear that the need for cost-efficient, high quality health care will mandate individualized strategies for prevention and treatment. Difficult cost-benefit tradeoffs will become increasingly common as we discover new drugs and therapies with adverse side effects. Patients and their clinical caregivers need rigorous, evidence-based guidance in making the choices confronting them.

References

Akritas, M.G. (1994). Nearest neighbor estimation of a bivariate distribution under random censoring. *The Annals of Statistics*, 22:1299–1327.

Bernstein, L., Allen, M., Anton-Culver, H., Deapen, D., Horn-Ross, P.L., Peel, D., Pinder, R., Reynolds, P., Sullivan-Halley, J., West, D., Wright, W., Ziogas, A., and Ross, R.K. (2002). High breast cancer incidence rates among California teachers: Results from the California Teachers Study (United States). *Cancer Causes Control*, 13:625–635.

Brier, G.W. (1950). Verification of forecasts expressed in terms of probability. *Monthly Weather Review*, 78:1–3.

Costantino, J.P., Gail, M.H., Pee, D., Anderson, S., Redmond, C.K., Benichou, J., and Wieand, H.S. (1999). Validation studies for models projecting the risk of invasive and total breast cancer incidence. *Journal of the National Cancer Institute*, 91:1541–1548.

Ernster, V.L. (1994). Nested case-control studies. *Preventive Medicine*, 23:587–590.

Fisher, B., Costantino, J.P., Wickerham, D.L., Cecchini, R.S., Cronin, W.M., and Robidoux, A. (2005). Tamoxifen for the prevention of breast cancer: current status of the National Surgical Adjuvant Breast and Bowel Project P–1 study. *Journal of the National Cancer Institute*, 97:1652–1662.

Fisher, B., Costantino, J.P., Wickerham, D.L., Redmond, C.K., Kavanah, M., Cronin, W.M., Vogel, V., Robidoux, A., Dimitrov, N., Atkins, J., Daly, M., Wieand, S., Tan-Chiu, E., Ford, L., and Wolmark, N. (1998). Tamoxifen for prevention of breast cancer: Report of the National Surgical Adjuvant Breast and Bowel Project P–1 Study. *Journal of the National Cancer Institute*, 90:1371–1388.

Gail, M.H., Brinton, L.A., Byar, D.P., Corle, D.K., Green, S.B., Schairer, C., and Mulvihill, J.J. (1989). Projecting individualized probabilities of developing breast cancer for white females who are being examined annually. *Journal of the National Cancer Institute*, 81:1879–1886.

Hanley, J.A. and McNeil, B.J. (1982). The meaning and use of the area under a receiver operating characteristic (ROC) curve. *Radiology*, 143:29–36.

Heagerty, P.J., Lumley, T., and Pepe, M.S. (2000). Time-dependent ROC curves for censored survival data and a diagnostic marker. *Biometrics*, 56:337–344.

Hsu, W.R. and Murphy, A.H. (1986). The attributes diagram a geometrical framework for assessing the quality of probability forecasts. *International Journal of Forecasting*, 2:285–293.

John, E.M., Hopper, J.L., Beck, J.C., Knight, J.A., Neuhausen, S.L., Senie, R.T., Ziogas, A., Andrulis, I.L., Anton-Culver, H., Boyd, N., Buys, S.S., Daly, M.B., O'Malley, F.P., Santella, R.M., Southey, M.C., Venne, V.L.,

Venter, D.J., West, D., Whittemore, A.S., Seminara, D. and the Breast Cancer Family Registry (2004). The Breast Cancer Family Registry: an infrastructure for cooperative multinational, interdisciplinary and translational studies of the genetic epidemiology of breast cancer. *Breast Cancer Research*, 6:R375–89.

Kalbfleisch, J.D. and Lawless, J.F. (1998). Likelihood analysis of multi-state models for disease incidence and mortality. *Statistics in Medicine*, 7:140–160.

Kalbfleisch, J.D. and Prentice, R.L. (2002). *The Statistical Analysis of Failure Time Data*, 2nd edition. Wiley, New York.

Luo, J., Horn, K., Ockene, J.K., Simon, M.S., Stefanick, M.L., Tong, E., and Margolis, K.L. (2011). Interaction between smoking and obesity and the risk of developing breast cancer among postmenopausal women: The Women's Health Initiative Observational Study. *American Journal of Epidemiology*, 174:919–928.

Murphy, A.H. (1973). A new vector partition of the probability score. *Journal of Applied Meteorology*, 12:595–600.

Pepe, M.S. (2003). *The Statistical Evaluation of Medical Tests for Classification and Prediction.* Oxford University Press, Oxford.

Pepe, M.S. and Cai, T. (2004).The analysis of placement values for evaluating discriminatory measures. *Biometrics*, 60:528–535.

Pepe, M.S., Gu, J.W., and Morris, D.E. (2010). The potential of genes and other markers to inform about risk. *Cancer Epidemiology, Biomarkers & Prevention*, 19:655–665.

Pepe, M.S. and Janes, H.E. (2008). Gauging the performance of SNPs, biomarkers, and clinical factors for predicting risk of breast cancer. *Journal of the National Cancer Institute*, 100:978–979.

Pepe, M.S. and Longton, G. (2005). Standardizing diagnostic markers to evaluate and compare their performance. *Epidemiology*, 16:598–603.

Plevritis, S.K., Kurian, A.W., Sigal, B.M., Daniel, B.L., Ikeda, D.M., Stockdale, F.E., and Garber, A.M. (2006). Cost-effectiveness of screening BRCA1/2 mutation carriers with breast magnetic resonance imaging. *Journal of the American Medical Association*, 295:2374–2384.

Prentice, R.L. (1986). A case-cohort design for epidemiologic cohort studies and disease prevention trials. *Biometrika*, 73:1–11.

Putter, H., Fiocco, M., and Geskus, R.B. (2007). Tutorial in biostatistics: Competing risks and multistate models. *Statistics in Medicine*, 26:2389–2430.

Quante, A.S., Whittemore, A.S., Shriver, T., Strauch, K., and Terry, M.B. (2012). Breast cancer risk assessment across the risk continuum: Genetic and nongenetic risk factors contributing to differential model performance. *Breast Cancer Research*, 14:R144.

Rosner, B.A., Colditz, G.A., Webb, P.M., and Hankinson, S.E. (2005). Mathematical models of ovarian cancer incidence. *Epidemiology*, 16:508–515.

Tyrer, J., Duffy, S.W., and Cuzick, J. (2004). A breast cancer prediction model incorporating familial and personal risk factors. *Statistics in Medicine*, 23:1111–1130.

Wacholder, S., McLaughlin, J.K., Silverman, D.T., and Mandel, J.S. (1992). Selection of controls in case-control studies. I. Principles. *American Journal of Epidemiology*, 135:1019–1049.

Whittemore, A.S. and Halpern, J. (2013). Two-stage sampling designs for validating personal risk models. *Statistical Methods in Medical Research*, in press.

Wilks, D.S. (1995). *Statistical Methods in the Atmospheric Sciences*. Academic Press, London.

36

Buried treasures

Michael A. Newton
Departments of Statistics and of Biostatistics and Medical Informatics
University of Wisconsin, Madison, WI

Keeping pace with the highly diversified research frontier of statistics is hard enough, but I suggest that we also pay ever closer attention to great works of the past. I offer no prescription for how to do this, but reflect instead on three cases from my own research where my solution involved realizing a new interpretation of an old, interesting but possibly uncelebrated result which had been developed in a different context.

36.1 Three short stories

36.1.1 Genomics meets sample surveys

Assessing differential expression patterns between cancer subtypes provides some insight into their biology and may direct further experimentation. On similar tissues cancer may follow distinct developmental pathways and thus produce distinct expression profiles. These differences may be captured by the sample variance statistic, which would be large when some members of a gene set (functional category) have high expression in one subtype compared to the other, and other members go the opposite way. A case in point is a collection of cell-cycle regulatory genes and their expression pattern in tumors related to human papilloma virus (HPV) infection. Pyeon et al. (2007) studied the transcriptional response in $n = 62$ head, neck and cervical cancer samples, some of which were positive for virus (HPV+) and some of which were not (HPV−). Gene-level analysis showed significant differential expression in both directions. Set-level analysis showed that one functional category stood out from the several thousands of known categories in having an especially large value of between-gene/within-set sample variance. This category was detected using a standardized sample variance statistic. The detection launched a series of experiments on the involved genes, both in the same tissues under alternative measurement technology and on different tissues. The findings lead to a new hypothesis about how HPV+/− tumors differentially deregulate the cell-cycle processes during tumorigenesis as well as to biomarkers for HPV−associated cancers (Pyeon et al., 2011). Figure 36.1 shows a summary of gene-level differential expression scores between HPV+ and HPV− cancers (so-called log fold changes), for all genes in the genome (left), as well as for $m = 99$ genes from a cell-cycle regulatory pathway.

A key statistical issue in this case was how to standardize a sample variance statistic. The gene-level data were first reduced to the log-scale fold change between HPV+ and HPV− cell types; these x_g, for genes g, were then considered fixed in subsequent calculations. For a known functional category $c \subseteq \{1, \ldots, G\}$ of size m, the statistic $u(x, c)$ measured the sample variance of the x_g's within c. This statistic was standardized by imagining the distribution of $u(x, C)$, for random sets C, considered to be drawn uniformly from among all $\binom{G}{m}$ possible size-m subsets of the genome. Well forgetting about all the genomics, the statistical question concerned the distribution of the sample variance in without-replacement finite-population sampling; in particular, I needed an expected value and variance of $u(x, C)$ under this sampling. Not being especially well versed in the findings of finite-population sampling, I approached these moment questions from first principles and with a novice's vigor, figuring that something simple was bound to emerge. I did not make much progress on the variance of $u(x, C)$, but was delighted to discover a beautiful solution in Tukey (1950, p. 517), which had been developed far from the context of genomics and which was not widely cited. Tukey's buried treasure used so-called K functions, which are set-level statistics whose expected value equals the same statistic computed on the whole population. Subsequently I learned that earlier R.A. Fisher had also derived this variance; see also Cho et al. (2005). In any case, I was glad to have gained some insight from Tukey's general framework.

36.1.2 Bootstrapping and rank statistics

Researchers were actively probing the limits of bootstrap theory when I began my statistics career. A case of interest concerned generalized bootstrap means. From a real-valued random sample $X_1, \ldots, X_n$, one studied the conditional distribution of the randomized statistic

$$\bar{X}_n^W = \frac{1}{n} \sum_{i=1}^{n} W_{n,i} X_i,$$

conditional on the data X_i, and where the random weights $W_{n,i}$ were generated by the statistician to enable the conditional distribution of $\bar{X}_n^W$ to approximate the marginal sampling distribution of $\bar{X}_n$. Efron's bootstrap corresponds to weights having a certain multinomial distribution, but indications were that useful approximations were available for beyond the multinomial.

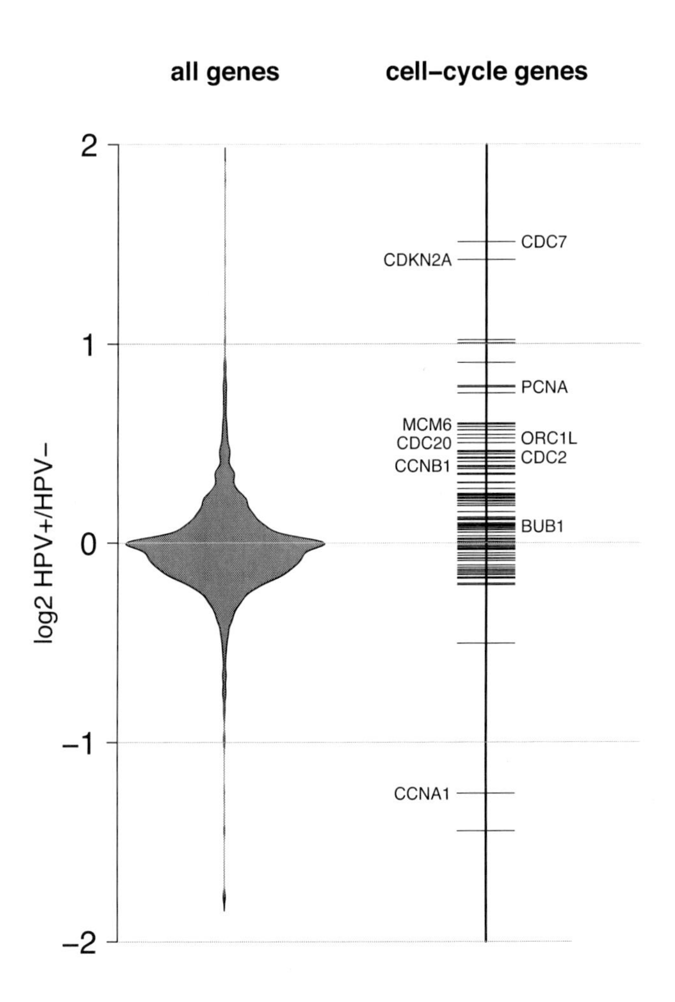

FIGURE 36.1
The relative positions of $m = 99$ cell-cycle genes (KEGG 04110) (right) are shown in the context of all measured genes (left) when genes are sorted by log fold change between HPV+ and HPV− tumors (vertical axis). Widths in the red violin plot indicate the empirical density. KEGG 04110 had higher standardized sample variance than any functional category in GO or KEGG. Based on this high variance, further experiments were performed on the 10 named genes (right) leading to a new hypothesis about how the HPV virus deregulates the control of cell cycle, and to biomarkers for HPV-associated cancer.

In a most rewarding collaboration, David Mason and I tackled the case where the $W_{n,i}$ were exchangeable, making the seemingly superfluous observation that $\bar{X}_n^W$ must have the same conditional distribution, given data X_i, as the additionally randomized

$$T_n = \frac{1}{n} \sum_{i=1}^{n} W_{n,\pi_{n,i}} X_i,$$

where, for each n, $\pi_{n,i}$ is a uniform random permutation of the integers $1, \ldots, n$. While the usual bootstrap statistic has two sources of randomness (one from the data and from the bootstrap weights), this T_n had yet a third source, neither generated by nature or the statistician, but just imagined owing to the exchangeability of the weights. Having all three sources allowed us to condition on both the data X_i and the statistician-generated weights $W_{n,i}$, and still have some randomness in T_n.

A quite unconnected and somewhat amazing treasure from the theory of linear rank statistics now became relevant. Given two triangular arrays of constants, $a_{n,i}$ and $b_{n,i}$, the randomized mean

$$S_n = \sum_{i=1}^{n} a_{n,\pi_{n,i}} b_{n,i}$$

had been studied extensively in nonparametric testing, because this is the form of the linear rank statistic. Hájek (1961) presented weak conditions on the triangular arrays such that S_n is asymptotically normal, owing to the random shuffling caused by the $\pi_{n,i}$. Thus, reconsidering Hájek's result in the new bootstrap context was the key to making progress on the weighted bootstrap problem (Mason and Newton, 1992).

36.1.3 Cancer genetics and stochastic geometry

A tumor is monoclonal in origin if all its cells trace by descent to a single initiated cell that is aberrant relative to the surrounding normal tissue (e.g., incurs some critical genetic mutation). Tumors are well known to exhibit internal heterogeneity, but this does not preclude monoclonal origin, since mutation, clonal expansion, and selection are dynamic evolutionary processes occurring within a tumor that move the single initiated cell to a heterogeneous collection of descendants. Monoclonal origin is the accepted hypothesis for most cancers, but evidence is mounting that tumors may initiate through some form of molecular interaction between distinct clones. As advanced as biotechnology has become, the cellular events at the point of tumor initiation remain beyond our ability to observe directly, and so the question of monoclonal versus polyclonal origin has been difficult to resolve. I have been fortunate to work on the question in the context of intestinal cancer, in series of projects with W.F. Dove, A. Thliveris, and R. Halberg.

When measured at several months of age, intestinal tracts from mice used in the experiments were dotted with tumors. By some rather elaborate experimental techniques cell lineages could be marked by one of two colors: some tumors were pure in color, as one would expect under monoclonal origin, yet some contained cells of both colors, and were thus overtly polyclonal. The presence of such polyclonal tumors did not raise alarm bells, since it was possible that separate tumors were forming in close proximity, and that they had merged into a single tumor mass by the time of observation. If so, the polyclonality was merely a consequence of random collision of independently initiated clones, and did not represent a mechanistically important phenomenon. The investigators suspected, however, that the frequency of these overtly polyclonal (heterotypic) tumors was too high to be explained by random collision, especially considering the tumor size, the overall tumor frequency, and the lineage marker patterns. It may have been, and subsequent evidence has confirmed, that cellular interactions are critical in the initial stages of tumor development. The statistical task at hand was to assess available data in terms of evidence against the random collision hypothesis.

In modeling data on frequencies of various tumor types, it became necessary to calculate the expected number of monoclonal tumors, biclonal tumors, and triclonal tumors when initiation events occur randomly on the intestinal surface. This is a problem in stochastic geometry, as clones will collide if they are sufficiently close. Like in the gene-set-variance problem, I tackled the expected value using first principles and with hopes that a simple approximation might emerge. The monoclonal and biclonal expectations were not so hard, but the triclonal calculation gave me fits. And then I found Armitage (1949). In a problem on the overlap of dust particles on a sampling plate, Armitage had faced the same expected value calculation and had provided a rather thorough solution, with error bounds. If N particles land at random in a region of area A, and if they clump when they lie within δ units, then the expected numbers of singletons, clumps-of-two, and clumps-of-three particles are approximately

$$\mu_1 = Ne^{-4\psi}, \quad \mu_2 = 2N\left(\psi - \frac{4\pi + 3\sqrt{3}}{\pi}\psi^2\right), \quad \mu_3 = N\left\{\frac{4(2\pi + 3\sqrt{3})}{3\pi}\psi^2\right\},$$

where $\psi = N\pi\delta^2/(4A)$. Fortunately I could use the framework of stochastic geometry to link the quite different contexts (particle counting and tumor formation) and identify a path to testing the random collision hypothesis (Newton et al., 2006). The biological consequences continue to be investigated.

36.2 Concluding remarks

I have found great utility in beautiful statistical findings that have been relatively uncelebrated by the field and that were developed in response to problems different than I was facing. I expect there are many such buried treasures, and I encourage statisticians to seek them out even as they push forward addressing all kinds of new statistical problems. Perhaps there is very little to what I'm saying. Had I been more prepared when launching into any of the three cases above I might have known right away how to use the available statistical results. But this seems like a lot to ask; our training programs are bursting with course work and cannot be expected to explain all of the discipline's treasures. You might also argue that the great thing about statistics and mathematics is that a single formalism works equally in all kinds of different contexts; my case studies do no more than express how the formalism is not dependent upon context. Perhaps my point is more that we must continue to exercise this formalism, continue to find analogies between distinct problems, and continue to support and develop tools that make these connections easier to identify.

Thank goodness for archiving efforts like `JSTOR` and the modern search engines that help us find these treasures. All of us can help by continuing to support efforts, like open access, aiming to minimize barriers to information flow. Authors and journals can help by making a greater effort to cite key background references and suggest links to related problems. Instructors, especially of courses in mathematical statistics, can help by emphasizing the distinct contexts that enliven each statistical fact. Grant reviewers and tenure committees can help by recognizing that innovation comes not only in conjuring up new theory and methodology but also by the thoughtful development of existing statistical ideas in new and important contexts. Finally, thanks to John Tukey, Peter Armitage, and Jaroslav Hájek and others for the wonderful results they've left for us to find.

> "There is more treasure in books than in all the pirate's loot on Treasure Island and best of all, you can enjoy these riches every day of your life."
>
> –Walt Disney

References

Armitage, P. (1949). An overlap problem arising in particle counting. *Biometrika*, 45:501–519.

Cho, E., Cho, M.J., and Eltinge, J. (2005). The variance of the sample variance from a finite population. *International Journal of Pure and Applied Mathematics*, 21:389–396.

Hájek, J. (1961). Some extensions of the Wald–Wolfowitz–Noether theorem. *The Annals of Mathematical Statistics*, 32:506–523.

Mason, D.M. and Newton, M.A. (1992). A rank statistics approach to the consistency of a general bootstrap. *The Annals of Statistics*, 20:1611–1624.

Newton, M.A., Clipson, L., Thliveris, A.T., and Halberg, R.B. (2006). A statistical test of the hypothesis that polyclonal intestinal tumors arise by random collision of initiated clones. *Biometrics*, 62:721–727.

Pyeon, D., Lambert, P.F., Newton, M.A., and Ahlquist, P.G. (2011). Biomarkers for human papilloma virus-associated cancer. US Patent No. 8,012,678 B2.

Pyeon, D., Newton, M.A., Lambert, P.F., den Boon, J.A., Sengupta, S., Marsit, C.J., Woodworth, C.D., Connor, J.P., Haugen, T.H., Smith, E.M., Kelsey, K.T., Turek, L.P., and Ahlquist, P. (2007). Fundamental differences in cell cycle deregulation in human papillomavirus-positive and human papillomavirus-negative head/neck and cervical cancers. *Cancer Research*, 67:4605–4619.

Tukey, J.W. (1950). Some sampling simplified. *Journal of the American Statistical Association*, 45:501–519.

37

Survey sampling: Past controversies, current orthodoxy, and future paradigms

Roderick J.A. Little
Department of Biostatistics
University of Michigan, Ann Arbor, MI

37.1 Introduction

My contribution to this historic celebration of the COPSS concerns the field of survey sampling, its history and development since the seminal paper by Neyman (1934), current orthodoxy, and a possible direction for the future. Many encounter survey sampling through the dull prism of moment calculations, but I have always found the subject fascinating. In my first sampling course, I remember being puzzled by the different forms of weighting in regression — by the inverse of the probability of selection, or by the inverse of the residual variance (Brewer and Mellor, 1973). If they were different, which was right? My early practical exposure was at the World Fertility Survey, where I learnt some real-world statistics, and where the sampling guru was one of the giants in the field, Leslie Kish (Kish et al., 1976). Kish was proud that the developing countries in the project were more advanced than developed countries in publishing appropriate estimates of standard error that incorporated the sample design. Always engaging, he shared my love of western classical music and tolerated my model-based views. More recently, I spent time helping to set up a research directorate at the US Census Bureau, an agency that was at the forefront of advances in applied sampling under the leadership of Maurice Hansen.

What distinguishes survey sampling from other branches of statistics? The genesis of the subject is a simple and remarkable idea — by taking a simple random sample from a population, reasonably reliable estimates of population quantities can be obtained with quantifiable accuracy by sampling around a thousand units, whether the population size is ten thousand or twenty million. Simple random sampling is neither optimal or even practical in many real-world settings, and the main developments in the field concerned complex sample designs, which include features like stratification, weighting and clustering. Another important aspect is its primary focus on finite population quantities rather than parameters of models. The practical concerns of how to do probability sampling in the real world, such as the availability of sampling frames, how to exploit administrative data, and alternative modes of survey administration, are an important part of the field; valuable, since currently statistical training tends to focus on estimation and inference, neglecting designs for collecting data.

Survey sampling is notable as the one field of statistics where the prevailing philosophy is design-based inference, with models playing a supporting role. The debates leading up to this current status quo were heated and fascinating, and I offer one view of them here. I also present my interpretation of the current status quo in survey sampling, what I see as its strengths and drawbacks, and an alternative compromise between design-based and model-based inference, Calibrated Bayes, which I find more satisfying.

The winds of change can be felt in this field right now. Robert Groves, a recent Director of the US Census Bureau, wrote:

> "For decades, the Census Bureau has created 'designed data' in contrast to 'organic data' [$\cdots$] What has changed is that the volume of organic data produced as auxiliary to the Internet and other systems now swamps the volume of designed data. In 2004 the monthly traffic on the internet exceeded 1 exabyte or 1 billion gigabytes. The risk of confusing data with information has grown exponentially... The challenge to the Census Bureau is to discover how to combine designed data with organic data, to produce resources with the most efficient information-to-data ratio. This means we need to learn how surveys and censuses can be designed to incorporate transaction data continuously produced by the internet and other systems in useful ways. Combining data sources to produce new information not contained in any single source is the future. I suspect that the biggest payoff will lie in new combinations of designed data and organic data, not in one type alone." (Groves, 2011)

I believe that the standard design-based statistical approach of taking a random sample of the target population and weighting the results up to the population is not adequate for this task. Tying together information from traditional surveys, administrative records, and other information gleaned from cyberspace to yield cost-effective and reliable estimates requires statistical modeling. However, robust models are needed that have good repeated sampling properties.

I now discuss two major controversies in survey sampling that shaped the current state of the field.

37.2 Probability or purposive sampling?

The first controversy concerns the utility of probability sampling itself. A probability sample is a sample where the selection probability of each of the samples that could be drawn is known, and each unit in the population has a non-zero chance of being selected. The basic form of probability sample is the simple random sample, where every possible sample of the chosen size n has the same chance of being selected.

When the distribution of some characteristics is known for the population, a measure of representativeness of a sample is how close the sample distribution of these characteristics matches the population distribution. With simple random sampling, the match may not be very good, because of chance fluctuations. Thus, samplers favored methods of purposive selection where samples were chosen to match distributions of population characteristics. The precise nature of purposive selection is often unclear; one form is quota sampling, where interviewers are given a quota for each category of a characteristic (such as age group) and told to sample until that quota is met.

In a landmark early paper on sampling, Neyman (1934) addressed the question of whether the method of probability sampling or purposive selection was better. His resolution was to advocate a method that gets the best of both worlds, stratified sampling. The population is classified into strata based on values of known characteristics, and then a random sample of size n_j is taken from stratum j, of size N_j. If $f_j = n_j/N_j$, the sampling fraction in stratum j, is a constant, an equal probability sample is obtained where the distribution of the characteristics in the sample matches the distribution of the population.

Stratified sampling was not new; see, e.g., Kaier (1897); but Neyman expanded its practical utility by allowing f_j to vary across strata, and weighting sampled cases by $1/f_j$. He proposed what is now known as Neyman allocation, which optimizes the allocations for given variances and costs of sampling within each strata. Neyman's paper expanded the practical utility of probability sampling, and spurred the development of other complex sample designs by Mahalanobis, Hansen, Cochran, Kish and others, greatly extending the practical feasibility and utility of probability sampling in practice. For example, a simple random sampling of people in a country is not feasible since a complete list of everyone in the population from which to sample is not available. Multistage sampling is needed to implement probability sampling in this setting.

There were dissenting views — simple random sampling (or equal probability sampling in general) is an all-purpose strategy for selecting units to achieve representativeness "on average" — it can be compared with randomized treatment allocation in clinical trials. However, statisticians seek optimal properties, and random sampling is very suboptimal for some specific purposes. For example, if the distribution of X is known in population, and the objective is the slope of the linear regression of Y on X, it's obviously much more efficient to locate half the sample at each of the extreme values of X — this minimizes the variance of the least squares slope, achieving large gains of efficiency over equal probability sampling (Royall, 1970). But this is not a probability sample — units with intermediate values of X have zero chance of selection. Sampling the extremes of X does not allow checks of linearity, and lacks robustness. Royall argues that if this is a concern, choose sample sizes at intermediate values of X, rather than letting these sizes be determined by chance. The concept of *balanced sampling* due to Royall and Herson (1973) achieves robustness by matching moments of X in the sample and population. Even if sampling is random within categories of X, this is not probability sampling since there is no requirement that all values of X are included. Royall's work is persuasive, but random sampling has advantages in multipurpose surveys, since optimizing for one objective often comes at the expense of others.

Arguments over the utility of probability sampling continue to this day. A recent example concerns the design of the National Children's Study (Michael, 2008; Little, 2010), planned as the largest long-term study of children's health and development ever to be conducted in the US. The study plans to follow 100,000 children from before birth to early adulthood, together with their families and environment, defined broadly to include chemical, physical, behavioral, social, and cultural influences. Lively debates were waged over the relative merits of a national probability sample over a purposive sample from custom-chosen medical centers. In discussions, some still confused "probability sample" with "simple random sample." Probability sampling ideas won out, but pilot work on a probability sample of households did not produce enough births. The latest plan is a form of national probability sample based on hospitals and prenatal clinics.

An equal probability design is indicated by the all-purpose nature of the National Children's Study. However, a sample that includes high pollution sites has the potential to increase the variability of exposures, yielding more precise estimates of health effects of contaminants. A compromise with attractions is to do a combination — say choose 80% of the sample by equal probability methods, but retain 20% of the sample to ensure coverage of areas with high contaminant exposures.

37.3 Design-based or model-based inference?

The role of probability sampling relates to ideas about estimation and inference — how we analyze the data once we have it. Neyman (1934) is widely celebrated for introducing confidence intervals as an alternative to "inverse probability" for inference from a probability sample. This laid the foundation for the "design-based approach" to survey inference, where population values are fixed and inferences are based on the randomization distribution in the selection of units... although Neyman never clearly states that he regards population values as fixed, and his references to Student's t distribution suggest that he had a distribution in mind. This leads me to the other topic of controversy, concerning design-based vs model-based inference; see, e.g., Smith (1976, 1994), Kish and Frankel (1974), Hansen et al. (1983), Kish (1995), and Chambers and Skinner (2003).

In design-based inference, population values are fixed, and inference is based on the probability distribution of sample selection. Obviously, this assumes that we have a probability sample (or "quasi-randomization," where we pretend that we have one). In model-based inference, survey variables are assumed to come from a statistical model. Probability sampling is not the basis for inference, but is valuable for making the sample selection ignorable; see Rubin (1976), Sugden and Smith (1984), and Gelman et al. (1995). There are two main variants of model-based inference: Superpopulation modeling, where frequentist inference is based on repeated samples from a "superpopulation" model; and Bayesian modeling, where fixed parameters in the superpopulation model are assigned a prior distribution, and inferences about finite population quantities or parameters are based on their posterior distributions. The argument about design-based or model-based inference is a fascinating component of the broader debate about frequentist versus Bayesian inference in general: Design-based inference is inherently frequentist, and the purest form of model-based inference is Bayes.

37.3.1 Design-based inference

More formally, for $i \in \{1, \ldots, N\}$, let y_i be the survey (or outcome) variable of the ith unit, where $N < \infty$ is the number of units in the population, and let I_i be the inclusion indicator variable of the ith unit. Let Z represent design information, such as stratum or cluster indicators. We consider inference about a finite population quantity $Q(Y, Z)$, for example the population total $Q(Y, Z) = y_1 + \cdots + y_N$, where $Y = (y_1, \ldots, y_N)$.

In the design-based or randomization approach as described, e.g., by Cochran (1977), inferences are based on the distribution of $I = (I_1, \ldots, I_N)$, and the outcome variables $y_1, \ldots, y_N$ are treated as fixed quantities. Inference involves (a) the choice of an estimator for Q, $\widehat{q} = \widehat{q}(Y_{\text{inc}}, I, Z)$, where Y_{inc} is the included part of Y; and (b) the choice of a variance estimator $\widehat{\nu} = \widehat{\nu}(Y_{\text{inc}}, I, Z)$ that is unbiased or approximately unbiased for the variance of $\widehat{q}$ with respect to the distribution of I. Inferences are then generally based on normal large-sample approximations. For example, a 95% confidence interval for Q is $\widehat{q} \pm 1.96\sqrt{\widehat{\nu}}$.

Estimators $\widehat{q}$ are chosen to have good design-based properties, such as

(a) *Design unbiasedness*: $\mathrm{E}(\widehat{q}|Y) = Q$, or

(b) *Design consistency*: $\widehat{q} \to Q$ as the sample size gets large (Brewer, 1979; Isaki and Fuller, 1982).

It is natural to seek an estimate that is design-efficient, in the sense of having minimal variance. However, it became clear that that kind of optimality is not possible without an assumed model (Horvitz and Thompson, 1952; Godambe, 1955). Design-unbiasedness tends to be too stringent, and design-consistency is a weak requirement (Firth and Bennett, 1998), leading to many choices of estimates; in practice, choices are motivated by implicit models, as discussed further below. I now give some basic examples of the design-based approach.

Example 1 (Estimate of a population mean from a simple random sample): Suppose the target of inference is the population mean $Q = \overline{Y} = (y_1 + \cdots + y_N)/N$ and we have a simple random sample of size n, $(y_1, \ldots, y_n)$. The usual unbiased estimator is the sample mean $\widehat{q} = \overline{y} = (y_1 + \cdots + y_n)/n$, which has sampling variance $V = (1 - n/N)S_y^2$, where S_y^2 is the population variance of Y. The estimated variance $\widehat{v}$ is obtained by replacing S_y^2 in V by its sample estimate s_y^2. A 95% confidence interval for $\overline{Y}$ is $\overline{y} \pm 1.96\sqrt{\widehat{v}}$.

Example 2 (Design weighting): Suppose the target of inference is the population total $T = (y_1 + \cdots + y_N)$, and we have a sample $(y_1, \ldots, y_n)$ where the ith unit is selected with probability π_i, $i \in \{1, \ldots, n\}$. Following Horvitz and Thompson (1952), an unbiased estimate of T is given by

$$\widehat{t}_{\mathrm{HT}} = \sum_{i=1}^{N} w_i y_i I_i,$$

where $w_i = 1/\pi_i$ is the sampling weight for unit i, namely the inverse of the probability of selection. Estimates of variance depend on the specifics of the design.

Example 3 (Estimating a population mean from a stratified random sample: For a stratified random sample with selection probability $\pi_j = n_j/N_j$ in stratum j, the Horvitz–Thompson estimator of the population mean $Q = \overline{Y} = (y_1 + \cdots + y_N)/N$ is the stratified mean, viz.

$$\overline{y}_{\mathrm{HT}} = \frac{1}{N}\sum_{j=1}^{J}\sum_{i=1}^{n_j}\frac{N_j}{n_j}y_{ij} = \overline{y}_{\mathrm{st}} = \sum_{j=1}^{J} P_j \overline{y}_j,$$

where $P_j = N_j/N$ and $\overline{y}_j$ is the sample mean in stratum j. The corresponding estimate of variance is

$$\widehat{v}_{\text{st}} = \sum_{j=1}^{J} \left(1 - \frac{n_j}{N_j}\right) \frac{s_j^2}{n_j},$$

where s_j^2 is the sample variance of Y in stratum j. A corresponding 95% confidence interval for $\overline{Y}$ is $\overline{y}_{\text{st}} \pm 1.96\sqrt{\widehat{v}_{\text{st}}}$.

Example 4 (Estimating a population mean from a PPS sample): In applications such as establishment surveys or auditing, it is common to have measure of size X available for all units in the population. Since large units often contribute more to summaries of interest, it is efficient to sample them with higher probability. In particular, for probability proportional to size (PPS) sampling, unit i with size $X = x_i$ is sampled with probability cx_i, where c is chosen to yield the desired sample size; units that come in with certainty are sampled and removed from the pool. Simple methods of implementation are available from lists of population units, with cumulated ranges of size. The Horvitz–Thompson estimator

$$\widehat{t}_{\text{HT}} = c \sum_{i=1}^{N} \frac{y_i}{x_i} I_i$$

is the standard estimator of the population total in this setting.

The Horvitz–Thompson estimator often works well in the context of PPS sampling, but it is dangerous to apply it to all situations. A useful guide is to ask when it yields sensible predictions of nonsampled values from a modeling perspective. A model corresponding to the HT estimator is the HT model

$$y_i \overset{\text{iid}}{\sim} \mathcal{N}(\beta x_i, \sigma^2 x_i^2), \tag{37.1}$$

where $\mathcal{N}(\mu, \tau^2)$ denotes the Normal distribution with mean μ and variance τ^2. This leads to predictions $\widehat{\beta} x_i$, where

$$\widehat{\beta} = n^{-1} \sum_{i=1}^{N} \frac{y_i}{x_i} I_i,$$

so $\widehat{t}_{\text{HT}} = \widehat{\beta}(x_1 + \cdots + x_N)$ is the result of using this model to predict the sampled and nonsampled values. If the HT model makes very little sense, the HT estimator and associated estimates of variance can perform poorly. The famous elephant example of Basu (1971) provides an extreme and comic illustration.

Models like the HT estimator often motivate the choice of estimator in the design-based approach. Another, more modern use of models is in model-assisted inference, where predictions from a model are adjusted to protect against model misspecification. A common choice is the generalized regression (GREG) estimator, which for a total takes the form:

$$\widehat{t}_{\text{GREG}} = \sum_{i=1}^{N} \widehat{y}_i + \sum_{i=1}^{N} \frac{y_i - \widehat{y}_i}{\pi_i},$$

where $\widehat{y}_i$ are predictions from a model; see, e.g., Särndal et al. (1992). This estimator is design-consistent whether or not the model is correctly specified, and foreshadows "doubly-robust" estimators in the mainline statistics literature.

37.3.2 Model-based inference

The model-based approach treats both $I = (I_1, \ldots, I_N)$ and $Y = (y_1, \ldots, y_N)$ as random variables. A model is assumed for the survey outcomes Y with underlying parameters θ, and this model is used to predict the nonsampled values in the population, and hence the finite population total. Inferences are based on the joint distribution of Y and I. Rubin (1976) and Sugden and Smith (1984) show that under probability sampling, inferences can be based on the distribution of Y alone, provided the design variables Z are conditioned in the model, and the distribution of I given Y is independent of the distribution of Y conditional on the survey design variables. In frequentist superpopulation modeling, the parameters θ are treated as fixed; see, e.g., Valliant et al. (2000). In Bayesian survey modeling, the parameters are assigned a prior distribution, and inferences for $Q(Y)$ are based on its posterior predictive distribution, given the sampled values; see, e.g., Ericson (1969), Binder (1982), Rubin (1987), Ghosh and Meeden (1997), Little (2004), Sedransk (2008), Fienberg (2011), and Little (2012). I now outline some Bayesian models for the examples discussed above.

Example 1 continued (Bayes inference for a population mean from a simple random sample): A basic model for simple random sampling is

$$y_i | \mu, \sigma^2 \overset{\text{iid}}{\sim} \mathcal{N}(\mu, \sigma^2),$$

with a Jeffreys' prior on the mean and variance $p(\mu, \log \sigma^2) =$ a constant. A routine application of Bayes theorem yields a t distribution for the posterior distribution of $\overline{Y}$, with mean $\overline{y}$, scale $s\sqrt{1 - n/N}$ and degrees of freedom $n-1$. The 95% credibility interval is the same as the frequentist confidence interval above, except that the normal percentile, 1.96, is replaced by the t percentile, as is appropriate since the variance is estimated. Arguably this interval is superior to the normal interval even if the data is not normal, although better models might be developed for that situation.

Example 2 continued (Bayesian approaches to design weighting): Weighting of cases by the inverse of the probability of selection is not really a model-based tool, although (as in the next example) model-based estimates correspond to design-weighted estimators for some problems. Design weights are conceived more as covariates in a prediction model, as illustrated in Example 4 below.

Example 3 continued (Estimating a population mean from a stratified random sample): For a stratified random sample, the design variables Z consist of the stratum indicators, and conditioning on Z suggests that models need to have distinct stratum parameters. Adding a subscript j for stratum to the normal model for Example 1 leads to

$$y_i|\mu_j, \sigma_j^2 \overset{\text{ind}}{\sim} \mathcal{N}(\mu_j, \sigma_j^2),$$

with prior $p(\mu_j, \log \sigma_j^2) =$ a constant. The resulting posterior distribution recovers the stratified mean as the posterior mean and the stratified variance for the posterior variance, when the variances σ_j^2 are assumed known. Estimating the variances leads to the posterior distribution as a mixture of t distributions. Many variants of this basic normal model are possible.

Example 4 continued (Estimating a population mean from a PPS sample): The posterior mean from the HT model (37.1) is equivalent to the HT estimator, aside from finite population. Zhen and Little (2003) relax the linearity assumption of the mean structure, modeling the mean of Y given size X as a penalized spline; see also and Zheng and Little (2005). Simulations suggest that this model yields estimates of the total that have superior mean squared error than the HT estimator when the HT model is misspecified. Further, posterior credible intervals from the expanded model have better confidence coverage.

37.3.3 Strengths and weakness

A simplified overview of the two schools of inference is that weighting is a fundamental feature of design-based methods, with models playing a secondary role in guiding the choice of estimates and providing adjustments to increase precision. Model-based inference is much more focused on predicting non-sampled (or nonresponding) units with estimates of uncertainty. The model needs to reflect features of the design like stratification and clustering to limit the effects of model misspecification, as discussed further below. Here is my personal assessment of the strengths and weaknesses of the approaches.

The attraction of the design-based perspective is that it avoids direct dependence on a model for the population values. Models can help the choice of estimator, but the inference remains design-based, and hence somewhat nonparametric. Models introduce elements of subjectivity — all models are wrong, so can we trust results? Design-based properties like design consistency are desirable since they apply regardless of the validity of a model. Computationally, weighting-based methods have attractions in that they can be applied uniformly to a set of outcomes, and to domain and cross-class means, whereas modeling needs more tailoring to these features.

A limitation of the design-based perspective is that inference is based on probability sampling, but true probability samples are harder and harder to come by. In the household sample setting, contact is harder — there are fewer telephone land-lines, and more barriers to telephonic contact; nonresponse is increasing, and face-to-face interviews are increasingly expensive. As Groves noted in the above-cited quote, a high proportion of available information is now not based on probability samples, but on ill-defined population frames.

Another limitation of design-based inference is that it is basically asymptotic, and provides limited tools for small samples, such as for small area estimation. The asymptotic nature leads to (in my opinion) too much emphasis on estimates and estimated standard errors, rather than obtaining intervals with good confidence coverage. This is reflected by the absence of t corrections for estimating the variances in Examples 1 and 3 above.

On a more theoretical level, design-based inference leads to ambiguities concerning what to condition on in the "reference set" for repeated sampling. The basic issue is whether to condition on ancillary statistics — if conditioning on ancillaries is taken seriously, it leads to the likelihood principle (Birnbaum, 1962), which design-based inference violates. Without a model for predicting non-sampled cases, the likelihood is basically uninformative, so approaches that follow the likelihood principle are doomed to failure.

As noted above, design-based inference is not explicitly model-based, but attempting design-based inference without any reference to implicit models is unwise. Models are needed in design-based approach, as in the "model-assisted" GREG estimator given above.

The strength of the model-based perspective is that it provides a flexible, unified approach for all survey problems — models can be developed for surveys that deal with frame, nonresponse and response errors, outliers, small area models, and combining information from diverse data sources. Adopting a modeling perspective moves survey sample inference closer to mainstream statistics, since other disciplines like econometrics, demography, public health, rely on statistical modeling. The Bayesian modeling requires specifying priors, but has that benefit that it is not asymptotic, and can provide better small-sample inferences. Probability sampling justified as making sampling mechanism ignorable, improving robustness.

The disadvantage of the model-based approach is more explicit dependence on the choice of model, which has subjective elements. Survey statisticians are generally conservative, and unwilling to trust modeling assumptions, given the consequences of lack of robustness to model misspecification. Developing good models requires thought and an understanding of the data, and models have the potential for more complex computations.

37.3.4 The design-model compromise

Emerging from the debate over design-based and model-based inference is the current consensus, which I have called the Design-Model Compromise (DMC); see Little (2012). Inference is design-based for aspects of surveys that are amenable to that approach, mainly inferences about descriptive statistics in large probability samples. These design-based approaches are often model assisted, using methods such as regression calibration to protect against model misspecification; see, e.g., Särndal et al. (1992). For problems where the design-based approach is infeasible or yields estimates with insufficient precision, such as small area estimation or survey nonresponse, a model-based approach is adopted. The DMC approach is pragmatic, and attempts to exploit the strengths of both inferential philosophies. However, it lacks a cohesive over-arching philosophy, involving a degree of "inferential schizophrenia" (Little, 2012).

I give two examples of "inferential schizophrenia." More discussion and other examples are given in Little (2012). Statistical agencies like the US Census Bureau have statistical standards that are generally written from a design-based viewpoint, but researchers from social science disciplines like economics are trained to build models. This dichotomy leads to friction when social scientists are asked to conform to a philosophy they view as alien. Social science models need to incorporate aspects like clustering and stratification to yield robust inferences, and addressing this seems more likely to be successful from a shared modeling perspective.

Another example is that the current paradigm generally employs direct design-based estimates in large samples, and model-based estimates in small samples. Presumably there is some threshold sample size where one is design based for larger samples and model based for smaller samples. This leads to inconsistency, and ad-hoc methods are needed to match direct and model estimates at different levels of aggregation. Estimates of precision are less easily reconciled, since confidence intervals from the model tend to be smaller than direct estimates because the estimates "borrow strength." Thus, it is quite possible for a confidence interval for a direct estimate to be wider than a confidence interval for a model estimate based on a smaller sample size, contradicting the notion that uncertainty decreases as information increases.

37.4 A unified framework: Calibrated Bayes

Since a comprehensive approach to survey inference requires models, a unified theory has to be model-based. I have argued (Little, 2012) that the appropriate framework is calibrated Bayes inference (Box, 1980; Rubin, 1984; Little, 2006), where inferences are Bayesian, but under models that yield inferences with good design-based properties; in other words, Bayesian credibility intervals when assessed as confidence intervals in repeated sampling should have close to nominal coverage. For surveys, good calibration requires that Bayes models should incorporate sample design features such as weighting, stratification and clustering. Weighting and stratification is captured by included weights and stratifying variables as covariates in the prediction model; see, e.g., Gelman (2007). Clustering is captured by Bayesian hierarchical models, with clusters as random effects. Prior distributions are generally weakly informative, so that the likelihood dominates the posterior distribution.

Why do I favor Bayes over frequentist superpopulation modeling? Theoretically, Bayes has attractive properties if the model is well specified, and putting weakly informative prior distributions over parameters tends to propagate uncertainty in estimating these parameters, yielding better frequentist confidence coverage than procedures that fix parameters at their estimates. The penalized spline model in Example 4 above is one example of a calibrated Bayes approach, and others are given in Little (2012). Here is one more concluding example.

Example 5 (Calibrated Bayes modeling for stratified sampling with a size covariate): A common model for estimating a population mean of a variable Y from a simple random sample $(y_1, \ldots, y_n)$, with a size variable X measured for all units in the population, is the simple ratio model

$$y_i | x_i, \mu, \sigma^2 \overset{\text{ind}}{\sim} \mathcal{N}(\beta x_i, \sigma^2 x_i),$$

for which predictions yield the ratio estimator $\overline{y}_{\text{rat}} = \overline{X} \times \overline{y}/\overline{x}$, where $\overline{y}$ and $\overline{x}$ are sample means of Y and X and $\overline{X}$ is the population mean of X. Hansen et al. (1983) suggest that this model is deficient when the sample is selected by disproportionate stratified sampling, yielding biased inferences under relatively minor deviations from the model. From a calibrated Bayes perspective, the simple ratio model does not appropriately reflect the sample design. An alternative model that does this is the separate ratio model

$$y_i | x_i, z_i = j, \mu_j, \sigma_j^2 \overset{\text{ind}}{\sim} \mathcal{N}(\beta_j x_i, \sigma_j^2 x_i),$$

where $z_i = j$ indicates stratum j. Predictions from this model lead to the separate ratio estimator

$$\overline{y}_{\text{sep}} = \sum_{j=1}^{J} \frac{\overline{y}_j}{\overline{x}_j} P_j \overline{X}_j,$$

where P_j is the proportion of the population in stratum j. This estimator can be unstable if sample sizes in one or more strata are small. A Bayesian modification is to treat the slopes β_j as $\mathcal{N}(\beta, \tau^2)$, which smooths the estimate towards something close to the simple ratio estimate. Adding prior distributions for the variance components provides Bayesian inferences that incorporate errors for estimating the variances, and also allows smoothing of the stratum-specific variances.

37.5 Conclusions

I am a strong advocate of probability sampling, which has evolved into a flexible and objective design tool. However, probability samples are increasingly hard to achieve, and the strict design-based view of survey inference is too restrictive to handle all situations. Modeling is much more flexible, but models need to be carefully considered, since poorly chosen models lead to poor inferences. The current design-model compromise is pragmatic, but lacks a coherent unifying principle. Calibrated Bayes provides a unified perspective that blends design-based and model-based ideas. I look forward to further development of this approach, leading to more general acceptance among survey practitioners. More readily-accessible and general software is one area of need.

Hopefully this brief traverse of survey sampling in the last eighty years has piqued your interest. It will be interesting to see how the field of survey sampling evolves in the next eighty years of the existence of COPSS.

Acknowledgements

This work was supported as part of an Interagency Personnel Agreement with the US Census Bureau. The views expressed on statistical, methodological, technical, or operational issues are those of the author and not necessarily those of the US Census Bureau.

References

Basu, D. (1971). An essay on the logical foundations of survey sampling, part I (with discussion). In *Foundations of Statistical Inference* (V.P. Godambe and D.A. Sprott, Eds.). Holt, Rinehart and Winston, Toronto, pp. 203–242.

Binder, D.A. (1982). Non-parametric Bayesian models for samples from finite populations. *Journal of the Royal Statistical Society, Series B*, 44:388–393.

Birnbaum, A. (1962). On the foundations of statistical inference (with discussion). *Journal of the American Statistical Association*, 57:269–326.

Box, G.E.P. (1980). Sampling and Bayes inference in scientific modelling and robustness (with discussion). *Journal of the Royal Statistical Society, Series A*, 143:383–430.

Brewer, K.R.W. (1979). A class of robust sampling designs for large-scale surveys. *Journal of the American Statistical Association*, 74:911–915.

Brewer, K.R.W. and Mellor, R.W. (1973). The effect of sample structure on analytical surveys. *Australian Journal of Statistics*, 15:145–152.

Chambers, R.L. and Skinner, C.J. (2003). *Analysis of Survey Data.* Wiley, New York.

Cochran, W.G. (1977). *Sampling Techniques*, 3rd edition. Wiley, New York.

Ericson, W.A. (1969). Subjective Bayesian models in sampling finite populations (with discussion). *Journal of the Royal Statistical Society, Series B*, 31:195–233.

Fienberg, S.E. (2011). Bayesian models and methods in public policy and government settings. *Statistical Science*, 26:212–226.

Firth, D. and Bennett, K.E. (1998). Robust models in probability sampling. *Journal of the Royal Statistical Society, Series B*, 60:3–21.

Gelman, A. (2007). Struggles with survey weighting and regression modeling (with discussion). *Statistical Science*, 22:153–164.

Gelman, A., Carlin, J.B., Stern, H.S., and Rubin, D.B. (1995). *Bayesian Data Analysis.* Chapman & Hall, London.

Ghosh, M. and Meeden, G. (1997). *Bayesian Methods for Finite Population Sampling.* Chapman & Hall, London.

Godambe, V.P. (1955). A unified theory of sampling from finite populations. *Journal of the Royal Statistical Society, Series B*, 17:269–278.

Groves, R.M. (2011) The future of producing social and economic statistical information, Part I. *Director's Blog*, `www.census.gov`, September 8, 2011. US Census Bureau, Department of Commerce, Washington DC.

Hansen, M.H., Madow, W.G., and Tepping, B.J. (1983). An evaluation of model-dependent and probability-sampling inferences in sample surveys (with discussion). *Journal of the American Statistical Association*, 78:776–793.

Horvitz, D.G. and Thompson, D.J. (1952). A generalization of sampling without replacement from a finite universe. *Journal of the American Statistical Association*, 47:663–685.

Isaki, C.T. and Fuller, W.A. (1982). Survey design under the regression superpopulation model. *Journal of the American Statistical Association*, 77:89–96.

Kaier, A.N. (1897). *The Representative Method of Statistical Surveys* [1976, English translation of the original Norwegian]. Statistics Norway, Oslo, Norway.

Kish, L. (1995). The hundred years' wars of survey sampling. *Statistics in Transition*, 2:813–830. [Reproduced as Chapter 1 of *Leslie Kish: Selected Papers* (G. Kalton and S. Heeringa, Eds.). Wiley, New York, 2003].

Kish, L. and Frankel, M.R. (1974). Inferences from complex samples (with discussion). *Journal of the Royal Statistical Society, Series B*, 36:1–37.

Kish, L., Groves, L.R., and Krotki, K.P. (1976). Standard errors from fertility surveys. *World Fertility Survey Occasional Paper 17*, International Statistical Institute, The Hague, Netherlands.

Little, R.J.A. (2004). To model or not to model? Competing modes of inference for finite population sampling. *Journal of the American Statistical Association*, 99:546–556.

Little, R.J.A. (2006). Calibrated Bayes: A Bayes/frequentist roadmap. *The American Statistician*, 60:213–223.

Little, R.J.A. (2010). Discussion of articles on the design of the National Children's Study. *Statistics in Medicine*, 29:1388–1390.

Little, R.J.A. (2012). Calibrated Bayes: An alternative inferential paradigm for official statistics (with discussion). *Journal of Official Statistics*, 28:309–372.

Michael, R.T. and O'Muircheartaigh, C.A. (2008). Design priorities and disciplinary perspectives: the case of the US National Children's Study. *Journal of the Royal Statistical Society, Series A*, 171:465–480.

Neyman, J. (1934). On the two different aspects of the representative method: The method of stratified sampling and the method of purposive selection. *Journal of the Royal Statistical Society*, 97:558–606.

Rao, J.N.K. (2011). Impact of frequentist and Bayesian methods on survey sampling practice: A selective appraisal. *Statistical Science*, 26:240–256.

Royall, R.M. (1970). On finite population sampling theory under certain linear regression models. *Biometrika*, 57:377–387.

Royall, R.M. and Herson, J.H. (1973). Robust estimation in finite populations, I and II. *Journal of the American Statistical Association*, 68:880–893.

Rubin, D.B. (1976). Inference and missing data. *Biometrika*, 53:581–592.

Rubin, D.B. (1984). Bayesianly justifiable and relevant frequency calculations for the applied statistician. *The Annals of Statistics*, 12:1151–1172.

Rubin, D.B. (1987). *Multiple Imputation for Nonresponse in Surveys.* Wiley, New York.

Särndal, C.-E., Swensson, B., and Wretman, J.H. (1992). *Model Assisted Survey Sampling.* Springer, New York.

Sedransk, J. (2008). Assessing the value of Bayesian methods for inference about finite population quantities. *Journal of Official Statistics*, 24:495–506.

Smith, T.M.F. (1976). The foundations of survey sampling: A review (with discussion). *Journal of the Royal Statistical Society, Series A*, 139:183–204.

Smith, T.M.F. (1994). Sample surveys 1975–1990: An age of reconciliation? (with discussion). *International Statistical Review*, 62:5–34.

Sugden, R.A. and Smith, T.M.F. (1984). Ignorable and informative designs in survey sampling inference. *Biometrika*, 71:495–506.

Valliant, R., Dorfman, A.H., and Royall, R.M. (2000). *Finite Population Sampling and Inference: A Prediction Approach.* Wiley, New York.

Zheng, H. and Little, R.J.A. (2003). Penalized spline model-based estimation of the finite population total from probability-proportional-to-size samples. *Journal of Official Statistics*, 19:99–117.

Zheng, H. and Little, R.J.A. (2005). Inference for the population total from probability-proportional-to-size samples based on predictions from a penalized spline nonparametric model. *Journal of Official Statistics*, 21:1–20.

38

Environmental informatics: Uncertainty quantification in the environmental sciences

Noel Cressie
NIASRA, School of Mathematics and Applied Statistics
University of Wollongong, Wollongong, NSW, Australia

38.1 Introduction

This exposition of environmental informatics is an attempt to bring current thinking about uncertainty quantification to the environmental sciences. Environmental informatics is a term that I first heard being used by Bronwyn Harch of Australia's Commonwealth Scientific and Industrial Research Organisation to describe a research theme within her organization. Just as bioinformatics has grown and includes biostatistics as a sub-discipline, environmental informatics, or EI, has the potential to be much broader than classical environmental statistics; see, e.g., Barnett (2004).

Which came first, the hypothesis or the data? In EI, we start with environmental data, but we use them to reveal, quantify, and validate scientific hypotheses with a panoply of tools from statistics, mathematics, computing, and visualization.

There is a realization now in science, including the environmental sciences, that there is uncertainty in the data, the scientific models, and the parameters that govern these models. Quantifying that uncertainty can be approached in a number of ways. To some, it means smoothing the data to reveal interpretable patterns; to data miners, it often means looking for unusual data points in a sea of "big data"; and to statisticians, it means all of the above, using statistical modeling to address questions like, "Are the patterns real?" and "Unusual in relation to what?"

In the rest of this chapter, I shall develop a vision for EI around the belief that Statistics is the science of uncertainty, and that behind every good data-mining or machine-learning technique is an implied statistical model. Computing even something as simple as a sample mean and a sample variance can be linked back to the very simplest of statistical models with a location parameter and additive homoscedastic errors. The superb book by

Hastie, Tibshirani, and Friedman (Hastie et al., 2009) shows the fecundity of establishing and developing such links. EI is a young discipline, and I would like to see it develop in this modern and powerful way, with uncertainty quantification through Statistics at its core.

In what follows, I shall develop a framework that is fundamentally about environmental data and the processes that produced them. I shall be particularly concerned with big, incomplete, noisy datasets generated by processes that may be some combination of non-linear, multi-scale, non-Gaussian, multivariate, and spatio-temporal. I shall account for all the known uncertainties coherently using hierarchical statistical modeling, or HM (Berliner, 1996), which is based on a series of conditional-probability models. Finally, through loss functions that assign penalties as a function of how far away an estimate is from its estimand, I shall use a decision-theoretic framework (Berger, 1985) to give environmental policy-makers a way to make rational decisions in the presence of uncertainty, based on competing risks (i.e., probabilities).

38.2 Hierarchical statistical modeling

The building blocks of HM are the data model, the (scientific) process model, and the parameter model. If Z represents the data, Y represents the process, and θ represents the parameters (e.g., measurement-error variance and reaction-diffusion coefficients), then the data model is

$$[Z|Y,\theta], \tag{38.1}$$

the process model is

$$[Y|\theta], \tag{38.2}$$

and the parameter model is

$$[\theta], \tag{38.3}$$

where $[A|B,C]$ is generic notation for the conditional-probability distribution of the random quantity A given B and C.

A statistical approach represents the uncertainties coherently through the joint-probability distribution, $[Z,Y,\theta]$. Using the building blocks (38.1)–(38.3), we can write

$$[Z,Y,\theta] = [Z|Y,\theta] \times [Y|\theta] \times [\theta]. \tag{38.4}$$

The definition of entropy of a random quantity A is $\mathrm{E}(\ln[A])$ By re-writing (38.4) as

$$\mathrm{E}(\ln[Z,Y,\theta]) = \mathrm{E}(\ln[Z|Y,\theta]) + \mathrm{E}(\ln[Y|\theta]) + \mathrm{E}(\ln[\theta]),$$

we can see that the joint entropy can be partitioned into data-model entropy, process-model entropy, and parameter-model entropy. This results in a "divide and conquer" strategy that emphasizes where scientists can put effort into understanding the sources of uncertainty and into designing scientific studies that control (and perhaps minimize some of) the entropy components.

The process Y and the parameters θ are unknown, but the data Z are known. (Nevertheless, the observed Z is still thought of as one of many possible that could have been observed, with a distribution $[Z]$.) At the beginning of all statistical inference is a step that declares what to condition on, and I propose that EI follow the path of conditioning on what is known, namely Z. Then the conditional probability distribution of all the unknowns given Z is

$$[Y, \theta | Z] = [Z, Y, \theta]/[Z] = [Z|Y, \theta] \times [Y|\theta] \times [\theta]/[Z], \tag{38.5}$$

where the first equality is known as Bayes' Theorem (Bayes, 1763); (38.5) is called the posterior distribution, and we call (38.1)–(38.3) a Bayesian hierarchical model (BHM). Notice that $[Z]$ on the right-hand side of (38.5) is a normalizing term that ensures that the posterior distribution integrates (or sums) to 1.

There is an asymmetry associated with the role of Y and θ, since (38.2) very clearly emphasises that $[Y|\theta]$ is where the "science" resides. It is equally true that $[Y, \theta] = [\theta|Y] \times [Y]$. However, probability models for $[\theta|Y]$ and $[Y]$ do not follow naturally from the way that uncertainties are manifested. The asymmetry emphasizes that Y is often the first priority for inference. As a consequence, we define the predictive distribution, $[Y|Z]$, which can be obtained from (38.5) by marginalization:

$$[Y|Z] = \int [Z|Y, \theta] \times [Y|\theta] \times [\theta] \, \mathrm{d}\theta/[Z]. \tag{38.6}$$

Then inference on Y is obtained from (38.6). While (38.5) and (38.6) are conceptually straightforward, in EI we may be trying to evaluate them in global spatial or spatio-temporal settings where Z might be on the order of Gb or Tb, and Y might be of a similar order. Thus, HM requires innovative conditional-probability modeling in (38.1)–(38.3), followed by innovative statistical computing in (38.5) and (38.6). Leading cases involve spatial data (Cressie, 1993; Banerjee et al., 2004) and spatio-temporal data (Cressie and Wikle, 2011). Examples of dynamical spatio-temporal HM are given in Chapter 9 of Cressie and Wikle (2011), and we also connect the literature in data assimilation, ensemble forecasting, blind-source separation, and so forth to the HM paradigm.

38.3 Decision-making in the presence of uncertainty

Let $\widehat{Y}(Z)$ be one of many decisions about Y based on Z. Some decisions are better than others, which can be quantified through a (non-negative) loss function, $L\{Y, \widehat{Y}(Z)\}$. The Bayes expected loss is $\mathrm{E}\{L(Y, \widehat{Y})\}$, and we minimize this with respect to $\widehat{Y}$. Then it is a consequence of decision theory (Berger, 1985) that the optimal decision is

$$Y^*(Z) = \arg\inf_{\widehat{Y}} \mathrm{E}\{L(Y, \widehat{Y})|Z\}, \tag{38.7}$$

where for some generic function g, the notation $\mathrm{E}\{g(Y)|Z\}$ is used to represent the conditional expectation of $g(Y)$ given Z.

Sometimes $\mathrm{E}\{L(Y, \widehat{Y})|\theta\}$ is called the risk, but I shall call it the expected loss; sometimes $\mathrm{E}\{L(Y, \widehat{Y})\}$ is called the Bayes risk, but see above where I have called it the Bayes expected loss. In what follows, I shall reserve the word risk to be synonymous with probability.

Now, if θ were known, only Y remains unknown, and HM involves just (38.1)–(38.2). Then Bayes' Theorem yields

$$[Y|Z, \theta] = [Z|Y, \theta] \times [Y|\theta]/[Z|\theta]. \tag{38.8}$$

In this circumstance, (38.8) is both the posterior distribution and the predictive distribution; because of the special role of Y, I prefer to call it the predictive distribution. The analogue to (38.7) when θ is known is, straightforwardly,

$$Y^*(Z) = \arg\inf_{\widehat{Y}} \mathrm{E}\{L(Y, \widehat{Y})|Z, \theta\}. \tag{38.9}$$

Clearly, $Y^*(Z)$ in (38.9) also depends on θ.

Using the terminology of Cressie and Wikle (2011), an empirical hierarchical model (EHM) results if an estimate $\hat{\theta}(Z)$, or $\hat{\theta}$ for short, is used in place of θ in (38.8): Inference on Y is then based on the empirical predictive distribution,

$$[Y|Z, \hat{\theta}] = [Z, Y, \hat{\theta}] \times [Y|\hat{\theta}]/[Z|\hat{\theta}], \tag{38.10}$$

which means that $\hat{\theta}$ is also used in place of θ in (38.9).

BHM inference from (38.5) and (38.6) is coherent in the sense that it emanates from the well defined joint-probability distribution (38.4). However, the BHM requires specification of the prior $[\theta]$, and one often consumes large computing resources to obtain (38.5) and (38.6). The EHM's inference from (38.10) can be much more computationally efficient, albeit with an empirical predictive distribution that has smaller variability than the BHM's predictive distribution (Sengupta and Cressie, 2013). Bayes' Theorem applied to BHM or EHM for spatio-temporal data results in a typically very-high-dimensional predictive distribution, given by (38.6) or (38.10), respectively, whose computation requires dimension reduction and statistical-computing algorithms such as EM (McLachlan and Krishnan, 2008), MCMC (Robert and Casella, 2004), and INLA (Rue et al., 2009). For additional information on dimension reduction, see, e.g., Wikle and Cressie (1999), Wikle et al. (2001), Cressie and Johannesson (2006), Banerjee et al. (2008), Cressie and Johannesson (2008), Kang and Cressie (2011), Katzfuss and Cressie (2011), Lindgren et al. (2011), and Nguyen et al. (2012).

In the last 20 years, methodological research in Statistics has seen a shift from mathematical statistics towards statistical computing. Deriving an analytical form for (38.6) or (38.10) is almost never possible, but being able to sample realizations from them often is. This shift in emphasis has enormous potential for EI.

For economy of exposition, I feature the BHM in the following discussion. First, if I can sample from the posterior distribution, $[Y, \theta|Z]$, I can automatically sample from the predictive distribution, $[Y|Z]$, by simply ignoring the θ's in the posterior sample of (Y, θ). This is called a marginalization property of sampling. Now suppose there is scientific interest in a summary $g(Y)$ of Y (e.g., regional averages, or regional extremes). Then an equivariance property of sampling implies that samples from $[g(Y)|Z]$ are obtained by sampling from $[Y|Z]$ and simple evaluating each member of the sample at g. This equivariance property is enormously powerful, even more so when the sampling does not require knowledge of the normalizing term $[Z]$ in (38.5). The best known statistical computing algorithm that samples from the posterior and predictive distributions is MCMC; see, e.g., Robert and Casella (2004).

Which summary of the predictive distribution $[g(Y)|Z]$ will be used to estimate the scientifically interesting quantity $g(Y)$? Too often, the posterior mean,

$$\mathrm{E}\{g(Y)|Z\} = \int g(Y)[Y|Z]\,\mathrm{d}Y,$$

is chosen as a "convenient" estimator of $g(Y)$. This is an optimal estimator when the loss function is squared-error: $L\{g(Y), \widehat{g}\} = \{\widehat{g} - g(Y)\}^2$; see, e.g., Berger (1985). However, squared-error loss assumes equal consequences (i.e., loss) for under-estimation as for over-estimation. When a science or policy question is about extreme events, the squared-error loss function is strikingly inadequate, yet scientific inference based on the posterior mean is ubiquitous.

Even if squared-error loss were appropriate, it would be incorrect to compute $\mathrm{E}(Y|Z)$ and produce $g\{\mathrm{E}(Y|Z)\}$ as an optimal estimate, unless g is a linear functional of Y. However, this is also common in the scientific literature. Under squared-error loss, the optimal estimate is $\mathrm{E}\{g(Y)|Z\}$, which is defined above. Notice that aggregating over parts of Y defines a linear functional g, but that taking extrema over parts of Y results in a highly non-linear functional g. Consequently, the supremum/infimum of the optimal estimate of Y (i.e., $g\{\mathrm{E}(Y|Z)\}$) is a severe under-estimate/over-estimate of the supremum/infimum of Y, i.e., $g(Y)$.

38.4 Smoothing the data

EI is fundamentally linked to environmental data and the questions that resulted in their collection. Questions are asked of the scientific process Y, and the data Z paint an imperfect and incomplete picture of Y. Often, the first tool that comes to a scientist's hand is a "data smoother," which here I shall call f. Suppose one defines

$$\widetilde{Y} \equiv f(Z); \tag{38.11}$$

notice that f "de-noises" (i.e., filters out highly variable components) and "fills in" where there are missing data. The scientist might be tempted to think of $\widetilde{Y}$ as data coming directly from the process model, $[Y|\theta]$, and use classical statistical likelihoods based on $[Y = \widetilde{Y}|\theta]$ to fit θ and hence the model $[Y|\theta]$. But this paradigm is fundamentally incorrect; science should incorporate uncertainty using a different paradigm. Instead of (38.11), suppose I write

$$\widetilde{Z} \equiv f(Z). \tag{38.12}$$

While the difference between (38.11) and (38.12) seems simply notational, conceptually it is huge.

The smoothed data $\widetilde{Z}$ should be modelled according to $[\widetilde{Z}|Y, \theta]$, and the process Y can be incorporated into an HM through $[Y|\theta]$. Scientific inference then proceeds from $[Y|\widetilde{Z}]$ in a BHM according to (38.6) or from $[Y|\widetilde{Z}, \hat{\theta}]$ in an EHM according to (38.10). The definition given by (38.12) concentrates our attention on the role of data, processes, and parameters in an HM paradigm and, as a consequence, it puts uncertainty quantification on firm inferential foundations (Cressie and Wikle, 2011, Chapter 2).

Classical frequentist inference could also be implemented through a marginal model (i.e., the likelihood),

$$[\widetilde{Z}|\theta] = \int [\widetilde{Z}|Y, \theta] \times [Y|\theta] \, \mathrm{d}Y,$$

although this fact is often forgotten when likelihoods are formulated. As a consequence, these marginal models can be poorly formulated or unnecessarily complicated when they do not recognize the role of Y in the probability modelling.

38.5 EI for spatio-temporal data

This section of the chapter gives two examples from the environmental sciences to demonstrate the power of the statistical-modelling approach to uncertainty quantification in EI.

38.5.1 Satellite remote sensing

Satellite remote sensing instruments are remarkable in terms of their optical precision and their ability to deliver measurements under extreme conditions.

Once the satellite has reached orbit, the instrument must function in a near vacuum with low-power requirements, sensing reflected light (in the case of a passive sensor) through a highly variable atmospheric column.

The specific example I shall discuss here is that of remote sensing of atmospheric CO_2, a greenhouse gas whose increase is having, and will have, a large effect on climate change. The global carbon cycle describes where carbon is stored and the movement of carbon between these reservoirs. The oceans and vegetation/soil are examples of CO_2 sinks, and fires and anthropogenic emissions are examples of CO_2 sources; of the approximately 8 Gt per year that enters the atmosphere, about half is anthropogenic. About 4 Gt stays in the atmosphere and the other 4 Gt is absorbed roughly equally by the oceans and terrestrial processes. This global increase of approximately 4 Gt of atmospheric CO_2 per year is unsustainable in the long term.

It is of paramount importance to be able to characterize precisely where and when sinks (and sources) occur. Because of a lack of globally extensive, extremely precise, and very densely sampled CO_2 data, these are largely unknown. Once the spatial and temporal variability of the carbon cycle is understood, regional climate projections can be made, and rational mitigation/adaptation policies can be implemented.

Although the atmosphere mixes rapidly (compared to the oceans), there is a lot of spatial variability as a function of both surface location and (geopotential) height. There is also a lot of temporal variability at any given location, as is clear from the US National Oceanic and Atmospheric Administration's CO_2 daily measurements from their Mauna Loa (Hawaii) observatory. Hence, we define atmospheric CO_2 as a spatio-temporal process, $Y(\mathbf{s};t)$, at spatial co-ordinates $\mathbf{s}$ and time t. Here, $\mathbf{s}$ consists of $(\text{lon}, \text{lat}) = (x,y)$ and geo-potential height h, that belongs to the spatial domain D_s, the extent of the atmosphere around Earth; and t belongs to a temporal domain D_t (e.g., t might index days in a given month).

There are several remote sensing instruments that measure atmospheric CO_2 (e.g., NASA's AIRS instrument and the Japanese space agency's GOSAT instrument); to improve sensitivity to near-surface CO_2, NASA built the OCO-2 instrument. (The original OCO satellite failed to reach orbit in 2009.) It allows almost pinpoint spatial-locational accuracy (the instrument's footprint is 1.1×2.25 km), resulting in high global data densities during any given month. However, its small footprint results in quite a long repeat-cycle of 16 days, making it harder to capture daily temporal variability at high spatial resolution. I am a member of NASA's OCO-2 Science Team that is concerned with all components of the data-information-knowledge pyramid referred to below in Section 38.6.

The physics behind the CO_2 retrieval requires measurements of CO_2 in the so-called strong CO_2 and weak CO_2 bands of the spectrum, and of O_2 in the oxygen A-band (Crisp et al., 2004). The result is a data vector of radiances $\mathbf{Z}(x,y;t)$, where $(x,y) = (\text{lon}, \text{lat})$ is the spatial location on the geoid, $D_g \equiv (-180°, +180°) \times (-90°, +90°)$, of the atmospheric column from footprint to satellite; and t is the time interval (e.g., a day or a week) during which the measurements (i.e., radiances) for that column were taken, where t ranges over the period of interest D_t. This vector is several-thousand dimensional, and there are potentially many thousands of such vectors per time interval. Hence, datasets can be very large.

The data are "noisy" due to small imperfections in the instrument, ubiquitous detector noise, and the presence of aerosols and clouds in the column. After applying quality-control flags based on aerosol, cloud, albedo conditions, some data are declared unreliable and hence "missing." The ideal is to estimate the (dry air mole fraction) CO_2 amount, $Y(x, y, h; t)$ in units of ppm, as h varies down the atmospheric column centred at (x, y), at time t. When the column is divided up into layers centred at geo-potential heights $h_1, \ldots, h_K$, we may write

$$\mathbf{Y}_0(x, y; t) \equiv (Y(x, y, h_1; t), \ldots, Y(x, y, h_K; t))^\top, \tag{38.13}$$

as the scientific process (i.e., state) of interest. The dimension of the state vector (38.13) is 20 for OCO-2, although 40 or so additional state variables, $\mathbf{Y}_1(x, y; t)$, are incorporated into $\mathbf{Y}(x, y; t) \equiv (\mathbf{Y}_0(x, y; t)^\top, \mathbf{Y}_1(x, y; t)^\top)^\top$, from which the radiative-transfer relation can be modeled as

$$\mathbf{Z}(x, y; t) = \mathbf{F}_\theta\{\mathbf{Y}(x, y; t)\} + \varepsilon(x, y; t). \tag{38.14}$$

In (38.14), the functional form of $\mathbf{F}_\theta$ is known (approximately) from the physics, but typically it requires specification of parameters θ. If θ were known, (38.14) is simply the data model, $[\mathbf{Z}(x, y; t)|Y, \theta]$ on the right-hand side of (38.4). The process model, $[Y|\theta]$, on the right-hand side of (38.4) is the joint distribution of $Y \equiv \{\mathbf{Y}(x, y; t) : (x, y) \in D_g, t \in D_t\}$, whose individual multivariate distributions are specified by OCO-2 ATB Document (2010) to be multivariate Gaussian with mean vectors and covariance matrices calculated from forecast fields produced by the European Centre for Medium-Range Weather Forecasting (ECMWF). However, this specification of the multivariate marginal distributions does not specify spatial dependencies in the joint distribution, $[Y|\theta]$. Furthermore, informed guesses are made for the parameters in θ. The predictive distribution is given by (38.8), but this is not computed; a summary is typically used (e.g., the predictive mode). For more details, see Crisp et al. (2012) and O'Dell et al. (2012). Validation of the estimated CO_2 values is achieved through TCCON data from a globally sparse but carefully calibrated network of land-based, upward-looking CO_2 monitoring sites; see, e.g., Wunch et al. (2011).

Ubiquitously in the literature on remote sensing retrievals (Rodgers, 2000), it is the predictive mode of $[\mathbf{Y}(x, y; t)|\mathbf{Z}(x, y; t), \theta]$ that is chosen as the optimal estimator of $\mathbf{Y}(x, y; t)$. The subsequent error analysis in that literature is then concerned with deriving the mean vector and the covariance matrix of this estimator assuming that $\mathbf{F}_\theta$ is a linear function of its state variables (Connor et al., 2008). However, the atmosphere involves highly complex interactions, and the radiative transfer function is known to be highly non-linear.

FIGURE 38.1
Locations of six GOSAT soundings where retrievals of XCO2 were obtained (between June 5, 2009 and July 26, 2009).

In Cressie and Wang (2013), we enhanced the linear approximation by including the quadratic term of a second-order Taylor-series approximation, and we calculated the non-linearity biases of retrievals of CO_2 that were obtained from data collected by the GOSAT satellite at the locations in Australia shown in Figure 38.1. For the six retrievals (i.e., predictive modes), we calculated the following biases of column-averaged CO_2, or XCO2 (in units of ppm): $.86, 1.15, .19, 1.15, -.78$, and 1.40. Biases of this order of magnitude are considered to be important, and hence a systematic error analysis of remote sensing retrievals should recognize the non-linearity in $\mathbf{F}_\theta$.

It is important to note here that the predictive distribution, $[\mathbf{Y}(x,y;t)| \mathbf{Z}(x,y;t),\theta]$, is different from the predictive distribution, $[\mathbf{Y}(x,y;t)|Z,\theta]$, and I propose that it is the latter that we should use when computing the optimal estimate of $\mathbf{Y}(x,y;t)$ from (38.9). This is based on the left-hand side of (38.8), which represents the "gold standard" to which all approximations should be compared. In practice, it would be difficult to obtain the predictive distribution, $[\mathbf{Y}(x,y;t)|Z,\theta]$, for every retrieval, so it makes sense to summarize it with its first two moments. In future research, I shall compare the linear approximations of Connor et al. (2008) to the quadratic approximations of Cressie and Wang (2013) by comparing them to the gold standard.

The mode should be considered to be just one possible summary of the predictive distribution; its corresponding loss function is

$$L(Y, \widehat{Y}) = \begin{cases} 0 & \text{if } Y = \widehat{Y}, \\ 1 & \text{if } Y \neq \widehat{Y}; \end{cases}$$

see, e.g., Berger (1985). I shall refer to this as the 0–1 loss function. That is, should even one element of the approximately 60-dimensional estimated state vector miss its target, a fixed loss is declared, no matter how close it is to the missed target. And the same fixed loss is declared when all or some of the elements miss their targets, by a little or a lot. From this decision-theoretic point of view, the predictive mode looks to be an estimate that in this context is difficult to justify.

The next phase of the analysis considers the dry air mole fraction (in ppm) of CO_2 averaged through the column from Earth's surface to the satellite, which recall is called XCO2. Let $\mathbf{Y}_0^*(x, y; t)$ denote the predictive mode obtained from (38.8), which is the optimal estimate given by (38.9) with the 0-1 loss function. Then $\text{XCO2}(x, y; t)$ is estimated by

$$\widehat{\text{X}}\text{CO2}(x, y; t) \equiv \mathbf{Y}_0^*(x, y; t)^\top \mathbf{w}, \tag{38.15}$$

where the weights $\mathbf{w}$ are given in OCO-2 ATB Document (2010). From this point of view, $\widehat{\text{X}}\text{CO2}(x, y; t)$ is the result of applying a smoother f to the raw radiances $\mathbf{Z}(x, y; t)$. The set of "retrieval data" over the time period D_t are $\{\widehat{\text{X}}\text{CO2}(x_i, y_i; t_i) : i = 1, \ldots, n\}$ given by (38.15), which we saw from (38.12) can be written as $\widetilde{Z}$; and Y is the multivariate spatio-temporal field $\{\mathbf{Y}(x, y; t) : (x, y) \in D_g, t \in D_t\}$, where recall that D_g is the geoid and the period of interest D_t might be a month, say.

The true column-averaged CO_2 field over the globe is a function of Y, viz.

$$g_V(Y) \equiv \{\text{XCO2}(x, y; t) : (x, y) \in D_g, t \in D_t\}, \tag{38.16}$$

where the subscript V signifies vertical averaging of Y through the column of atmosphere from the satellite's footprint on the Earth's surface to the satellite. Then applying the principles set out in the previous sections, we need to construct spatio-temporal probability models for $[\widetilde{Z}|Y, \theta]$ and $[Y|\theta]$, and either a prior $[\theta]$ or an estimate $\hat{\theta}$ of θ. This will yield the predictive distribution of Y and hence that of $g_V(Y)$. Katzfuss and Cressie (2011, 2012) have implemented both the EHM where θ is estimated and the BHM where θ has a prior distribution, to obtain respectively, the empirical predictive distribution and the predictive distribution of $g_V(Y)$ based on $\widetilde{Z}$. The necessary computational efficiency is achieved by dimension reduction using the Spatio-Temporal Random Effects (STRE) model; see, e.g., Katzfuss and Cressie (2011). Animated global maps of the predictive mean of $g_V(Y)$ using both approaches, based on AIRS CO_2 column averages, are shown in the SSES Web-Project, "Global Mapping of CO_2" (see Figure 2 at `www.stat.osu.edu/~sses/collab_co2.html`). The regional and seasonal nature of CO_2 becomes obvious by looking at these maps. Uncertainty is quantified by the predictive standard deviations, and their heterogeneity (due to different atmospheric conditions and different sampling rates in different regions) is also apparent from the animated maps.

It is worth pointing out that the "smoothed" data, $\widetilde{Z} \equiv \{\widehat{\text{X}}\text{CO2}(x_i, y_i; t_i) : i = 1, \ldots, n\}$, are different from the original radiances, $Z \equiv \{\mathbf{Z}(x_i, y_i; t_i) : i = 1, \ldots, n\}$. Thus, $[Y|\widetilde{Z}, \theta]$ is different from $[Y|Z, \theta]$. Basing scientific inference on the latter, which contains all the data, is to be preferred, but practical considerations and tradition mean that the information-reduced, $\widetilde{Z} = f(Z)$, is used for problems such as flux estimation.

Since there is strong interest from the carbon-cycle-science community in regional surface fluxes, horizontal averaging should be a more interpretable summary of Y than vertical averaging. Let $g_1\{\mathbf{Y}(x, y; t)\}$ denote the surface CO_2 concentration with units of mass/area. For example, this could be obtained by extrapolating the near-surface CO_2 information in $\mathbf{Y}_0(x, y; t)$. Then define

$$\overline{Y}(x, y; t) \equiv \int_{R(x,y)} g_1\{Y(u, v; t)\} \, dudv \Big/ \int_{R(x,y)} dudv$$

and

$$g_H(Y) \equiv \{\overline{Y}(x, y; t) : (x, y) \in D_g, t \in D_t\}, \tag{38.17}$$

where the subscript H signifies horizontal averaging, and where $R(x, y)$ is a pre-specified spatial process of areal regions on the geoid that defines the horizontal averaging. (It should be noted that R could also be made a function of t, and indeed it probably should change with season.) For a pre-specified time increment τ, define

$$\Delta(x, y; t) \equiv \frac{\overline{Y}(x, y; t + \tau) - \overline{Y}(x, y; t)}{\tau},$$

with units of mass/(area $\times$ time). Then the flux field is

$$g_F(Y) \equiv \{\Delta(x, y; t) : (x, y) \in D_g, t \in D_t\}. \tag{38.18}$$

At this juncture, it is critical that the vector of estimated CO_2 in the column, namely, $\mathbf{Y}_0^*(x_i, y_i; t_i)$, replaces $\widehat{\text{X}}\text{CO2}(x_i, y_i; t_i)$ to define the smoothed data, $\widetilde{Z}$. Then the data model $[\widetilde{Z}|Y, \theta]$ changes, but critically the spatio-temporal statistical model for $[Y|\theta]$ is the same as that used for vertical averaging. Recall the equivariance property that if Y is sampled from the predictive distribution (38.6) or (38.8), the corresponding samples from $g_H(Y)$ and $g_F(Y)$ yield their corresponding predictive distributions. The HM paradigm allows other data sources (e.g., *in situ* TCCON measurements, data from other remote sensing instruments) to be incorporated into $\widetilde{Z}$ seamlessly; see, e.g., Nguyen et al. (2012).

The choice of τ is at the granularity of D_t, and the choice of R depends on the question being asked and the roughness of Earth's surface relative to the question. In a classical bias-variance trade-off, one wants $R(x, y)$ to be large enough for $g_F(x, y; t)$ to capture the dominant variability and small enough that the flux in $R(x, y)$ is homogeneous.

Carbon-cycle science has accounted for much of the dynamics of CO_2, but the carbon budget has consistently shown there to be a missing sink (or sinks). The OCO-2 instrument, with its almost pinpoint accuracy and high sensitivity near Earth's surface, offers an unprecedented opportunity to accurately estimate the sinks. From that point of view, the parts of Y that are of interest are lower quantiles of $g_F(Y)$, along with the (lon, lat)-regions where those quantiles occur. In Section 38.6, I argue that these queries of the process $g_F(Y)$ can be formalized in terms of loss functions; Zhang et al. (2008) give an illustration of this for decadal temperature changes over the Americas.

This different approach to flux estimation is centrally statistical, and it is based on a spatio-temporal model for $[Y|\theta]$. There is another approach, one that bases $[Y|\theta]$ on an atmospheric transport model to incorporate the physical movement of voxels in the atmosphere and, consequently, the physical movement of CO_2; see, e.g., Houweling et al. (2004), Chevallier et al. (2007), Gourdji et al. (2008), and Lauvaux et al. (2012). Motivated by articles such as Gourdji et al. (2008), I expect that the two approaches could be combined, creating a physical-statistical model.

When $[Y|\theta]$ is different, the predictive distribution given by (38.8) is different, and clearly when L in (38.9) is different, the optimal estimate given by (38.9) is different. This opens up a whole new way of thinking about flux estimation and quantifying its uncertainty, which is something I am actively pursuing as part of the OCO-2 Science Team.

38.5.2 Regional climate change projections

Climate is not weather, the latter being something that interests us on daily basis. Generally speaking, climate is the empirical distribution of temperature, rainfall, air pressure, and other quantities over long time scales (30 years, say). The empirical mean (i.e., average) of the distribution is one possible summary, often used for monitoring trends, although empirical quantiles and extrema may often be more relevant summaries for natural-resource management. Regional climate models (RCMs) at fine scales of resolution (20–50 km) produce these empirical distributions over 30-year time periods and can allow decision-makers to project what environmental conditions will be like 50–60 years in the future.

Output from an RCM is obtained by discretizing a series of differential equations, coding them efficiently, and running the programs on a fast computer. From that point of view, an RCM is deterministic, and there is nothing stochastic or uncertain about it. However, uncertainties in initial and boundary conditions, in forcing parameters, and in the approximate physics associated with the spatial and temporal discretizations (Fennessy and Shukla, 2000; Xue et al., 2007; Evans and Westra, 2012), allow us to introduce a probability model for the output, from which we can address competing risks (i.e., probabilities) of different projected climate scenarios. The RCM output can certainly be summarised statistically; in particular, it can be mapped. There is a small literature on spatial statistical analyses of RCMs, particularly of output from the North American Regional Climate Change Assessment Program (NARCCAP), administered by NCAR in Boulder, Colorado; see Kaufman and Sain (2010), Salazar et al. (2011), Kang et al. (2012), and Kang and Cressie (2013). My work in this area has involved collaboration with NCAR scientists.

Kang and Cressie (2013) give a comprehensive statistical analysis of the 11,760 regions (50×50 km pixels) in North America, for projected average temperature change, projected out to the 30-year-averaging period, 2041–70. The technical features of our article are: it is fully Bayesian; data dimension is reduced from the order of 100,000 down to the order of 100 through a Spatial Random Effects, or SRE, model (Cressie and Johannesson, 2008); seasonal variability is featured; and consensus climate-change projections are based on more than one RCM. Suppose that the quantity of scientific interest, Y, is temperature change in degrees Celsius by the year 2070, which is modeled statistically as

$$Y(\mathbf{s}) = \mu(\mathbf{s}) + \mathbf{S}(\mathbf{s})^\top \boldsymbol{\eta} + \xi(\mathbf{s}), \quad \mathbf{s} \in \text{North America} \tag{38.19}$$

where μ captures large-scale trends, and the other two terms on the right-hand side of (38.19) are Gaussian processes that represent, respectively, small-scale spatially dependent random effects and fine-scale spatially independent variability. The basis functions in $\mathbf{S}$ include 80 multi-resolutional bisquare functions and five indicator functions that capture physical features such as elevation and proximity to water bodies. This defines $[Y|\theta]$.

Importantly, the 30-year-average temperature change obtained from the NARCCAP output (i.e., the data, Z) is modeled as the sum of a spatial process Y and a spatial error term that in fact captures spatio-temporal interaction, viz.

$$Z(\mathbf{s}) = Y(\mathbf{s}) + \varepsilon(\mathbf{s}), \quad \mathbf{s} \in \text{North America} \tag{38.20}$$

where ε is a Gaussian white-noise process with a variance parameter σ_ε^2. This defines $[Z|Y, \theta]$. The target for inference is the spatial climate-change process Y, which is "hidden" behind the spatio-temporal "noisy" process Z. A prior distribution, $[\theta]$, is put on θ, and (38.6) defines the predictive distribution.

Here, θ is made up of the vector of spatial-mean effects $\boldsymbol{\mu}$, $\text{cov}(\boldsymbol{\eta})$, $\text{var}(\xi)$, and σ_ε^2, and the prior $[\theta]$ is specified in the Appendix of Kang and Cressie (2013). From (38.6), we can deduce the shaded zone of North America in Figure 38.2. There, with 97.5% probability calculated pixel-wise, any 50×50 km pixel's $Y(\mathbf{s})$ that is above a 2°C sustainability threshold is shaded. Here, Y and θ together are over $100,000$ dimensional, but the computational algorithms based on dimension reduction in the SRE model do not "break."

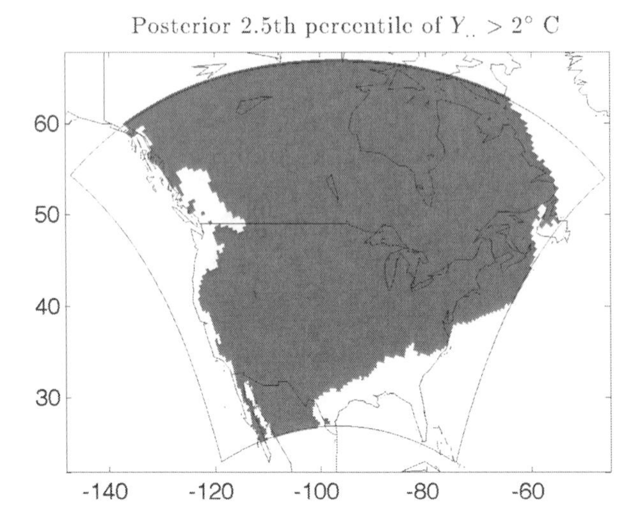

FIGURE 38.2
Regions of unsustainable ($> 2°$C with predictive probability .975) temperature increase obtained from pixel-wise predictive distributions, $[Y(\mathbf{s})|Z]$, where $\mathbf{s} \in$ North America.

Since MCMC samples are taken from a (more than) 100,000-dimensional posterior distribution, many such probability maps like Figure 38.2 can be produced. For example, there is great interest in extreme temperature changes, so let k denote a temperature-change threshold; define the spatial probability field,

$$\Pr(Y > k|Z), \quad k \geq 0. \tag{38.21}$$

As k increases in (38.21), the regions of North America that are particularly vulnerable to climate change stand out. Decision-makers can query the BHM where, for NARCCAP, the query might involve the projected temperature increase in the 50×50 km pixel containing Columbus, Ohio. Or, it might involve the projected temperature increase over the largely agricultural Olentangy River watershed (which contains Columbus). From (38.6), one can obtain the probabilities (i.e., risks) of various projected climate-change scenarios, which represent real knowledge when weighing up mitigation and adaptation strategies at the regional scale.

This HM approach to uncertainty quantification opens up many possibilities: Notice that the occurrence-or-not of the events referred to above can be written as

$$\mathbf{1}\{Y(\mathbf{s}_C) > k\} \quad \text{and} \quad \mathbf{1}\left\{\int_O Y(\mathbf{s})\,\mathrm{d}\mathbf{s} \Big/ \int_O \mathrm{d}\mathbf{s} > k\right\},$$

where $\mathbf{1}$ is an indicator function, $\mathbf{s}_C$ is the pixel containing Columbus, and O

is the set of all pixels in the Olentangy River watershed. Then squared-error loss implies that $\text{E}[\mathbf{1}\{Y(\mathbf{s}_C) > k\}|Z] = \Pr\{Y(\mathbf{s}_C) > k|Z\}$ given by (38.21) is optimal for estimating $\mathbf{1}\{Y(\mathbf{s}_C) > k\}$. Critically, other loss functions would yield different optimal estimates, since (38.7) depends on the loss function L. A policy-maker's question translated into a loss function yields a tailored answer to that question. Quite naturally in statistical decision theory, different questions are treated differently and result in different answers.

More critically, the average (climate change over 30 years) Y can be replaced with an extreme quantile, say the .95 quantile, which I denote here as $g^{(.95)}(Y)$; this hidden process corresponds to temperature change that could cause extreme stress to agricultural production in, for example, the Hunter Valley, NSW, Australia. Such projections for farmers in the "stressed" regions would be invaluable for planning crop varieties that are more conducive to higher temperature/lower rainfall conditions. That is, I propose making inference directly on extremal processes, and decisions should be made with loss functions that are tailor-made to the typical "what if" queries made by decision-makers.

Furthermore, output could have non-Gaussian distributions; for example, quantiles of temperature or rainfall would be skewed, for which spatial generalised linear models (Diggle et al., 1998; Sengupta and Cressie, 2013) would be well suited: In this framework, (38.20) is replaced with the data model,

$$[Z(\mathbf{s}) = z|Y, \theta] = \text{EF}[z; \text{E}\{Z(\mathbf{s})|Y(\mathbf{s}), \theta\}], \tag{38.22}$$

which are conditionally independent for pixels $\mathbf{s} \in$ North America. In (38.22), EF denotes the one-parameter exponential family of probability distributions. Now consider a link function ℓ that satisfies, $\ell[\text{E}\{Z(\mathbf{s})|Y(\mathbf{s}), \theta\}] = Y(\mathbf{s})$; on this transformed scale, climate change Y is modelled as the spatial process given by (38.19). In Sengupta and Cressie (2013) and Sengupta et al. (2012), we have developed spatial-statistical methodology for very large remote sensing datasets based on (38.22), that could be adapted to RCM projections. That methodology gives the predictive distribution, (38.10), which is summarised by mapping the predictive means, the predictive standard deviations (a measure of uncertainty), and the predictive extreme quantiles. Other data models could also be used in place of (38.20), such as the extreme-value distributions.

Increases in temperature generally lead to decreases in water availability, due to an increase in evaporation. By developing conditional-probability distributions of [Temperature] and [Rainfall | Temperature], we can infer the joint behaviour of [Rainfall, Temperature]. This is in contrast to the bivariate analysis in Sain et al. (2011), and it is a further example of the utility of a conditional-probability modelling approach, here embedded in a multivariate hierarchical statistical model.

38.6 The knowledge pyramid

The knowledge pyramid has data at its base, information at its next tier, knowledge at its third tier, and decision-making at its apex. In the presence of uncertainty, I propose that EI have at its core the following steps: convert data into information by exploring the data for structure; convert information into knowledge by modeling the variability and inferring the etiology; and prepare the knowledge for decision-makers by translating queries into loss functions. These may not be the usual squared-error and 0–1 loss functions, which are often chosen for convenience. They may be asymmetric and multivariable, to reflect society's interest in extreme environmental events. Associated with each loss function (i.e., a query) is an optimal estimator (i.e., a wise answer) based on minimising the predictive expected loss; see (38.7) where the predictive risks (i.e., probabilities) and the loss interact to yield an optimal estimator.

The societal consequences of environmental change, mitigation, and adaptation will lead to modeling of complex, multivariate processes in the social and environmental sciences. Difficult decisions by governments will involve choices between various mitigation and adaptation scenarios, and these choices can be made, based on the risks together with the losses that are built into EI's uncertainty quantification.

38.7 Conclusions

Environmental informatics has an important role to play in quantifying uncertainty in the environmental sciences and giving policy-makers tools to make societal decisions. It uses data on the world around us to answer questions about how environmental processes interact and ultimately how they affect Earth's organisms (including *Homo sapiens*). As is the case for bioinformatics, environmental informatics not only requires tools from statistics and mathematics, but also from computing and visualisation. Although uncertainty in measurements and scientific theories mean that scientific conclusions are uncertain, a hierarchical statistical modelling approach gives a probability distribution on the set of all possibilities. Uncertainty is no reason for lack of action: Competing actions can be compared through competing Bayes expected losses.

The knowledge pyramid is a useful concept that data analysis, HM, optimal estimation, and decision theory can make concrete. Some science and policy questions are very complex, so I am advocating an HM framework to capture the uncertainties and a series of queries (i.e., loss functions) about the scientific process to determine an appropriate course of action. Thus, a major challenge is to develop rich classes of loss functions that result in wise answers to important questions.

Acknowledgements

I would like to thank Eddy Campbell for his comments on an earlier draft, Rui Wang for his help in preparing Figure 38.1, Emily Kang for her help in preparing Figure 38.2, and Andrew Holder for his help in preparing the manuscript. This research was partially supported by the NASA Program, NNH11ZDA001N–OCO2 (Science Team for the OCO-2 Mission).

References

Banerjee, S., Carlin, B.P., and Gelfand, A.E. (2004). *Hierarchical Modeling and Analysis for Spatial Data.* Chapman and Hall/CRC, Boca Raton, FL.

Banerjee, S., Gelfand, A.E., Finley, A.O., and Sang, H. (2008). Gaussian predictive process models for large spatial data sets. *Journal of the Royal Statistical Society, Series B*, 70:825–848.

Barnett, V.D. (2004). *Environmental Statistics: Methods and Applications.* Wiley, New York.

Bayes, T. (1763). An essay towards solving a problem in the doctrine of chances. *Philosophical Transactions of the Royal Society of London*, 53:370–418.

Berger, J.O. (1985). *Statistical Decision Theory and Bayesian Analysis.* Springer, New York.

Berliner, L.M. (1996). Hierarchical Bayesian time-series models. In *Maximum Entropy and Bayesian Methods* (K. Hanson and R. Silver, Eds.). Kluwer, Dordrecht, pp. 15–22.

Chevallier, F., Bréon, F.-M., and Rayner, P.J. (2007). Contribution of the Orbiting Carbon Observatory to the estimation of CO2 sources and sinks: Theoretical study in a variational data assimilation framework. *Journal of Geophysical Research*, 112, doi:10.1029/2006JD007375.

Connor, B.J., Boesch, H., Toon, G., Sen, B., Miller, C., and Crisp, D. (2008). Orbiting Carbon Observatory: Inverse method and prospective error analysis. *Journal of Geophysical Research: Atmospheres*, 113, doi: 10.1029/2006JD008336.

Cressie, N. (1993). *Statistics for Spatial Data.* Wiley, New York.

Cressie, N. and Johannesson, G. (2006). Spatial prediction for massive data sets. In *Australian Academy of Science Elizabeth and Frederick White Conference.* Australian Academy of Science, Canberra, Australia, pp. 1–11.

Cressie, N. and Johannesson, G. (2008). Fixed rank kriging for very large spatial data sets. *Journal of the Royal Statistical Society, Series B*, 70:209–226.

Cressie, N. and Wang, R. (2013). Statistical properties of the state obtained by solving a nonlinear multivariate inverse problem. *Applied Stochastic Models in Business and Industry*, 29:424–438.

Cressie, N. and Wikle, C.K. (2011). *Statistics for Spatio-Temporal Data.* Wiley, Hoboken, NJ.

Crisp, D., Deutscher, N.M., Eldering, A., Griffith, D., Gunson, M., Kuze, A., Mandrake, L., McDuffie, J., Messerschmidt, J., Miller, C.E., Morino, I., Fisher, B.M., Natraj, V., Notholt, J., O'Brien, D.M., Oyafuso, F., Polonsky, I., Robinson, J., Salawitch, R., Sherlock, V., Smyth, M., Suto, H., O'Dell, C., Taylor, T.E., Thompson, D.R., Wennberg, P.O., Wunch, D., Yung, Y.L., Frankenberg, C., Basilio, R., Bosch, H., Brown, L.R., Castano, R. and Connor, B. (2012). The ACOS XCO2 retrieval algorithm — Part II: Global XCO2 data characterization. *Atmospheric Measurement Techniques*, 5:687–707.

Crisp, D., Jacob, D.J., Miller, C.E., O'Brien, D., Pawson, S., Randerson, J.T., Rayner, P., Salawitch, R.J., Sander, S.P., Sen, B., Stephens, G.L., Atlas, R.M., Tans, P.P., Toon, G.C., Wennberg, P.O., Wofsy, S.C., Yung, Y.L., Kuang, Z., Chudasama, B., Sprague, G., Weiss, B., Pollock, R., Bréon, F.-M., Kenyon, D., Schroll, S., Brown, L.R., Burrows, J.P., Ciais, P., Connor, B.J., Doney, S.C., and Fung, I.Y. (2004). The Orbiting Carbon Observatory (OCO) mission. *Advances in Space Research*, 34:700–709.

Diggle, P.J., Tawn, J.A., and Moyeed, R.A. (1998). Model-based geostatistics (with discussion). *Journal of the Royal Statistical Society, Series C*, 47:299–350.

Evans, J.P. and Westra, S. (2012). Investigating the mechanisms of diurnal rainfall variability using a regional climate model. *Journal of Climate*, 25:7232–7247.

Fennessy, M.J. and Shukla, J. (2000). Seasonal prediction over North America with a regional model nested in a global model. *Journal of Climate*, 13:2605–2627.

Gourdji, S.M., Mueller, K.L., Schaefer, K., and Michalak, A.M. (2008). Global monthly averaged CO_2 fluxes recovered using a geostatistical inverse modeling approach: 2. Results including auxiliary environmental data. *Journal of Geophysical Research: Atmospheres*, 113, doi:10.1029/2007JD009733.

Hastie, T., Tibshirani, R.J., and Friedman, J.H. (2009). *The Elements of Statistical Learning: Data Mining, Inference, and Prediction*, 2nd edition. Springer, New York.

Houweling, S., Bréon, F.-M., Aben, I., Rödenbeck, C., Gloor, M., Heimann, M., and Ciasis, P. (2004). Inverse modeling of CO_2 sources and sinks using satellite data: A synthetic inter-comparison of measurement techniques and their performance as a function of space and time. *Atmospheric Chemistry and Physics*, 4:523–538.

Kang, E.L. and Cressie, N. (2011). Bayesian inference for the spatial random effects model. *Journal of the American Statistical Association*, 106:972–983.

Kang, E.L. and Cressie, N. (2013). Bayesian hierarchical ANOVA of regional climate-change projections from NARCCAP Phase II. *International Journal of Applied Earth Observation and Geoinformation*, 22:3–15.

Kang, E.L., Cressie, N., and Sain, S.R. (2012). Combining outputs from the NARCCAP regional climate models using a Bayesian hierarchical model. *Journal of the Royal Statistical Society, Series C*, 61:291–313.

Katzfuss, M. and Cressie, N. (2011). Spatio-temporal smoothing and EM estimation for massive remote-sensing data sets. *Journal of Time Series Analysis*, 32:430–446.

Katzfuss, M. and Cressie, N. (2012). Bayesian hierarchical spatio-temporal smoothing for very large datasets. *Environmetrics*, 23:94–107.

Kaufman, C.G. and Sain, S.R. (2010). Bayesian ANOVA modeling using Gaussian process prior distributions. *Bayesian Analysis*, 5:123–150.

Lauvaux, T., Schuh, A.E., Uliasz, M., Richardson, S., Miles, N., Andrews, A.E., Sweeney, C., Diaz, L.I., Martins, D., Shepson, P.B., and Davis, K. (2012). Constraining the CO_2 budget of the corn belt: Exploring uncertainties from the assumptions in a mesoscale inverse system. *Atmospheric Chemistry and Physics*, 12:337–354.

Lindgren, F., Rue, H., and Lindström, J. (2011). An explicit link between Gaussian fields and Gaussian Markov random fields: The stochastic partial differential equation approach (with discussion). *Journal of the Royal Statistical Society, Series B*, 73:423–498.

McLachlan, G.J. and Krishnan, T. (2008). *The EM Algorithm and Extensions*, 2nd edition. Wiley, New York.

Nguyen, H., Cressie, N., and Braverman, A. (2012). Spatial statistical data fusion for remote-sensing applications. *Journal of the American Statistical Association*, 107:1004–1018.

OCO-2 ATB Document (2010). OCO-2 level 2 full physics retrieval algorithm theoretical basis. `http://disc.sci.gsfc.nasa.gov/acdisc/documentation/OCO-2_L2_FP_ATBD_v1_rev4_Nov10.pdf`.

O'Dell, C.W., Fisher, B., Gunson, M., McDuffie, J., Miller, C.E., Natraj, V., Oyafuso, F., Polonsky, I., Smyth, M., Taylor, T., Toon, G., Connor, B., Wennberg, P.O., Wunch, D., Bosch, H., O'Brien, D., Frankenberg, C., Castano, R., Christi, M., Crisp, D., and Eldering, A. (2012). The ACOS CO2 retrieval algorithm — Part I: Description and validation against synthetic observations. *Atmospheric Measurement Techniques*, 5:99–121.

Robert, C.P. and Casella, G. (2004). *Monte Carlo Statistical Methods*, 2nd edition. Springer, New York.

Rodgers, C.D. (2000). *Inverse Methods for Atmospheric Sounding*. World Scientific Publishing, Singapore.

Rue, H., Martino, S., and Chopin, N. (2009). Approximate Bayesian inference for latent Gaussian models by using integrated nested Laplace approximations (with discussion). *Journal of the Royal Statistical Society, Series B*, 71:319–392.

Sain, S.R., Furrer, R., and Cressie, N. (2011). Combining ensembles of regional climate model output via a multivariate Markov random field model. *The Annals of Applied Statistics*, 5:150–175.

Salazar, E.S., Finley, A., Hammerling, D., Steinsland, I., Wang, X., and Delamater, P. (2011). Comparing and blending regional climate model predictions for the American southwest. *Journal of Agricultural, Biological, and Environmental Statistics*, 16:586–605.

Sengupta, A. and Cressie, N. (2013). Hierarchical statistical modeling of big spatial datasets using the exponential family of distributions. *Spatial Statistics*, 4:14–44.

Sengupta, A., Cressie, N., Frey, R., and Kahn, B. (2012). Statistical modeling of MODIS cloud data using the spatial random effects model. In *2012 Proceedings of the Joint Statistical Meetings*, American Statistical Association, Alexandria, VA, pp. 3111–3123.

Wikle, C.K. and Cressie, N. (1999). A dimension-reduced approach to space-time Kalman filtering. *Biometrika*, 86:815–829.

Wikle, C.K., Milliff, R.F., Nychka, D. and Berliner, L.M. (2001). Spatiotemporal hierarchical Bayesian modeling: Tropical ocean surface winds. *Journal of the American Statistical Association*, 96:382–397.

Wunch, D., Ahonen, P., Biraud, S.C., Castano, R., Cressie, N., Crisp, D., Deutscher, N.M., Eldering, A., Fisher, M.L., Griffith, D. W.T., Gunson, M., Wennberg, P.O., Heikkinen, P., Keppel-Aleks, G., Kyro, E., Lindenmaier, R., Macatangay, R., Mendonca, J., Messerschmidt, J., Miller, C.E., Morino, I., Notholt, J., Toon, G.C., Oyafuso, F.A., Rettinger, M., Robinson, J., Roehl, C.M., Salawitch, R.J., Sherlock, V., Strong, K., Sussmann, R., Tanaka, T., Thompson, D.R., Connor, B.J., Uchino, O., Warneke, T., Wofsy, S.C., Fisher, B., Osterman, G.B., Frankenberg, C., Mandrake, L., and O'Dell, C. (2011). A method for evaluating bias in global measurements of CO2 total columns from space. *Atmospheric Chemistry and Physics*, 11:12317–12337.

Xue, Y., Vasic, R., Janjic, Z., Mesinger, F., and Mitchell, K.E. (2007). Assessment of dynamic downscaling of the continental US regional climate using the Eta/SSiB regional climate model. *Journal of Climate*, 20:4172–4193.

Zhang, J., Craigmile, P.F., and Cressie, N. (2008). Loss function approaches to predict a spatial quantile and its exceedance region. *Technometrics*, 50:216–227.

39

A journey with statistical genetics

Elizabeth A. Thompson
Department of Statistics
University of Washington, Seattle, WA

With the work of R.A. Fisher and Sewall Wright, the early years of the development of methods for analysis of genetic data were closely paralleled by the broader development of methods for statistical inference. In many ways, the parallel over the last 40 years is no less striking, with genetic and genomic data providing an impetus for development of broad new areas of statistical methodology. While molecular and computational technologies have changed out of all recognition over the last 40 years, the basic questions remain the same: Where are the genes? What do they do? How do they do it? These questions continue to provide new challenges for statistical science.

39.1 Introduction

"Plus ça change, plus c'est la même chose."

– Alphonse Karr (1849)

No doubt when things work out well, past events seem opportune, but I never cease to marvel how incredibly lucky I was to enter the field of statistical genetics in 1970. Two foundational books on the theory (Crow and Kimura, 1970) and application (Cavalli-Sforza and Bodmer, 1971) of population genetics were newly published. Together with the new edition of Stern (1973), these were the bibles of my early graduate-student years. While the available data seem primitive by today's standards, the extensive updates in 1976 of the earlier work of Mourant (1954) and colleagues provided a wider view of the genetic variation among human populations than had been previously available. Computing power for academic research was also fast expanding, with the new IBM 370 series in 1970, and virtual memory capabilities soon after.

TABLE 39.1
The changing study designs and data structures and relationship to developing statistical approaches.

Date	Data Design/Structure	Statistical Approach
1970s	pedigrees; evolutionary trees	Latent variables and EM
1980s	genetic maps	HMM methods
	linkage analysis	Graphical models
1990s	more complex traits	Monte Carlo likelihood; MCMC
	more complex pedigrees	Complex stochastic systems
2000s	large scale association mapping	FDR, $p \gg n$
2010s	descent of genomes; IBD	MC realization of latent structure

It is often commented, how, in an earlier era, genome science and statistical science developed in parallel. In 1922, Fisher started to develop likelihood inference (Fisher, 1922a), while in the same year his first example of maximum likelihood estimation was of estimating genetic recombination frequencies in *drosophila* (Fisher, 1922b). His first use of the term *variance* was in developing the theory of genetic correlations among relatives (Fisher, 1918), while analysis of variance was established in Fisher (1925). In parallel, Wright (1922) was also developing the theory of the dependence structure of quantitative genetic variation among related individuals, leading to the theory of path coefficients (Wright, 1921) and structural equation modeling (Pearl, 2000). The changes in genetics and genomics, statistical science, and both molecular and computational technologies over the last 40 years (1970–2010) are arguably many times greater than over the preceding 40 (1930–1970), but the same complementary developments of statistics and genetics are as clear as those of Fisher and Wright; see Table 39.1. Moreover the basic scientific questions remain the same: Where are the genes? What do they do? How do they do it?

39.2 The 1970s: Likelihood inference and the EM algorithm

The basic models of genetics are fundamentally parametric. The dependence structure of data on a pedigree dates to Elston and Stewart (1971) and is shown in Figure 39.1(a). First, population-level parameters provide the probabilities of genotypes $G^{(F)}$ of founder members, across a small set of marker (M) or trait (T) loci. Parameters of the process of transmission of DNA from parents to offspring then determine the probabilities of the genotypes (G)

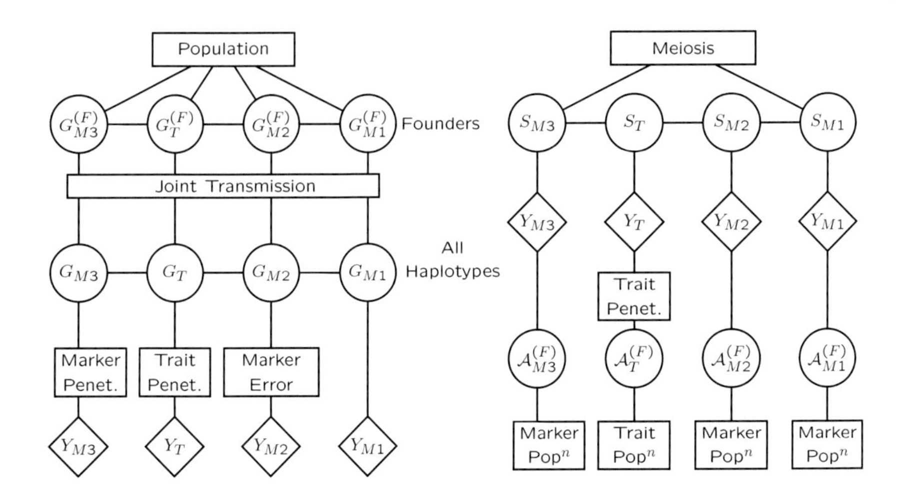

FIGURE 39.1
The two orthogonal conditional independence structures of genetic data on related individuals. (a) Left: The conditional independence of haplotypes among individuals. (b) Right: The conditional independence of inheritance among genetic loci. Figures from Thompson (2011), reproduced with permission of S. Karger AG, Basel.

of descendant individuals. Finally, penetrance parameters specify the probabilistic relationship between these genotypes and observable genetic data (Y) on individuals, again at each trait (T) or marker (M) locus. These penetrance models can incorporate typing error in marker genotypes, as well as more complex relationships between phenotype and genotype. The structured parametric models of statistical genetics lead naturally to likelihood inference (Edwards, 1972), and it is no accident that from Fisher (1922b) onwards, a major focus has been the computation of likelihoods and of maximum likelihood estimators.

Methods for the computation of likelihoods on pedigree structures make use of the conditional independence structure of genetic data. Under the laws of Mendelian genetics, conditional on the genotypes of parents, the genotypes of offspring are independent of each other, and of those of any ancestral and lateral relatives of the parents. Data on individuals depends only on their genotypes; see Figure 39.1(a). Methods for computation of probabilities of observed data on more general graphical structures are now widely known (Lauritzen and Spiegelhalter, 1988), but these methods were already standard in pedigree analysis in the 1970s. In fact the first use of this conditional independence in computing probabilities of observed data on three-generation pedigrees dates to Haldane and Smith (1947), while the Elston–Stewart algorithm (Elston and

Stewart, 1971) provided a general approach for simple pedigrees. Extension of this approach to complex pedigree structures (Cannings et al., 1978) and to more general trait models (Cannings et al., 1980) was a major advance in statistical genetics of the 1970s, enabling inference of gene ancestry in human populations (Thompson, 1978), the computation of risk probabilities in the complex pedigrees of genetic isolates (Thompson, 1981) and the analysis of gene extinction in the pedigrees of endangered species (Thompson, 1983).

Statistical genetics is fundamentally a latent variable problem. The underlying processes of descent of DNA cannot be directly observed. The observed data on individuals result from the types of the DNA they carry at certain genome locations, but these locations are often unknown. In 1977, the EM algorithm (Dempster et al., 1977) was born. In particular cases it had existed much earlier, in the gene-counting methods of Ceppelini et al. (1955), in the variance component methods of quantitative genetics (Patterson and Thompson, 1971) and in the reconstruction of human evolutionary trees (Thompson, 1975), but the EM algorithm provided a framework unifying these examples, and suggesting approaches to maximum likelihood estimation across a broad range of statistical genetics models.

39.3 The 1980s: Genetic maps and hidden Markov models

The statistical methodology of human genetic linkage analysis dates back to the 1930s work of J.B.S. Haldane (1934) and R.A. Fisher (1934), and the likelihood framework for inferring and estimating linkage from human family data was established in the 1950s by work of C.A.B. Smith (1953) and N.E. Morton (1955). However, genetic linkage findings were limited: there were no genetic marker maps.

That suddenly changed in 1980 (Botstein et al., 1980), with the arrival of the first DNA markers, the *restriction fragment polymorphisms* or RFLPs. For the first time, there was the vision we now take for granted, of genetic markers available at will throughout the genome, providing the framework against which traits could be mapped. This raised new statistical questions in the measurement of linkage information (Thompson et al., 1978). The development of DNA markers progressed from RFLP to (briefly) multilocus *variable number of tandem repeat* or VNTR loci used primarily for relationship inference (Jeffreys et al., 1991; Geyer et al., 1993) and then to STR (*short tandem repeat* or *microsatellite*) loci (Murray et al., 1994); see Table 39.2. These DNA markers, mapped across the genome, brought a whole new framework of conditional independence to the computation of linkage likelihoods (Lander and Green, 1987; Abecasis et al., 2002). Rather than the conditional independence in the transmission of DNA from parents to offspring, the rel-

TABLE 39.2
The changing pattern of genetic data (1970–2010).

Date	Marker Type	Data Structure	Trait Type
1970	Blood types	Nuclear families	Mendelian
1980	RFLPs	Large pedigrees	Simple traits
1990	STRs (Microsatellites)	Small pedigrees	Quantitative traits
2000	SNPs and mRNA expression data	Case/Control ("unrelated")	Complex traits
2010	RNAseq and Sequence data	Relatives in populations	Complex quantitative traits

evant structure became the Markov dependence of inheritance (S) of DNA at successive marker (M) or hypthesized trait (T) locations across a chromosome, as shown in Figure 39.1(b). As before the population model provides probabilities of the allelic types ($\mathcal{A}$) of founders (F), at trait (T) or marker (M) loci. At a locus, the observable data (Y) is determined by the founder allelic types ($\mathcal{A}$) and the inheritance (S) at that locus, possibly through a penetrance model in the case of trait loci.

39.4 The 1990s: MCMC and complex stochastic systems

The earlier methods (Figure 39.1(a)) are computationally exponential in the number of genetic loci analyzed jointly, the *hidden Markov model* (HMM) methods (Figure 39.1(b)) are exponential in the number of meioses in a pedigree. Neither could address large numbers of loci, observed on large numbers of related individuals. However, the same conditional independence structures that make possible the computation of linkage likelihoods for few markers or for small pedigrees, lend themselves to *Markov chain Monte Carlo* (MCMC). Genetic examples, as well as other scientific areas, gave impetus to the huge burst of MCMC in the early 1990s. However, unlike other areas, where MCMC was seen as a tool for Bayesian computation (Gelfand and Smith, 1990; Besag and Green, 1993) in statistical genetics the focus on likelihood inference led rather to Monte Carlo likelihood (Penttinen, 1984; Geyer and Thompson, 1992).

The discreteness and the constraints of genetic models provided challenges for MCMC algorithms. Earlier methods (Lange and Sobel, 1991) used the genotypes of individuals as latent variables (Figure 39.1(a)) and encoun-

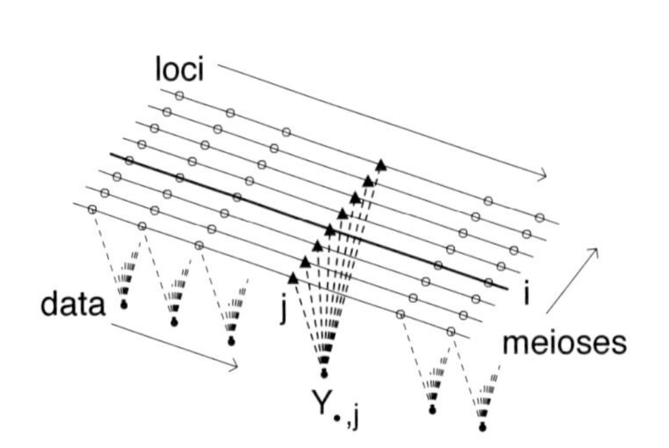

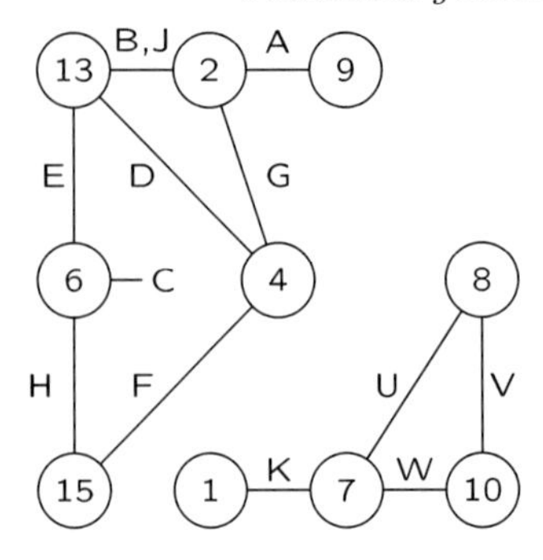

FIGURE 39.2
Imputation and specification of inheritance of DNA. (a) Left: The MCMC structure of meiosis and linkage (Thompson, 2000). (b) Right: The IBD graph specifying DNA shared by descent among observed individuals (Thompson, 2003).

tered problems of non-irreducibility of samplers (Sheehan and Thomas, 1993) and other mixing problems (Lin et al., 1993). Significant improvements were obtained by instead using the inheritance patterns of Figure 39.1(b) as latent variables (Thompson, 1994). The *a priori* independence of meioses, the Markov dependence of inheritance at successive loci, and the dependence of observable data on the inheritance pattern at a locus (Figure 39.2(a)) lead to a variety of block-Gibbs samplers of increasing computational efficiency (Heath, 1997; Thompson and Heath, 1999; Tong and Thompson, 2008).

An additional aspect of these later developments is the change from use of the conditional independence structure of a pedigree to that of an IBD-graph (Figure 39.2(b)). This graph specifies the individuals who share genome *identical by descent* (IBD) at a locus; that is, DNA that has descended to current individuals from a single copy of the DNA in a recent common ancestor. The observed trait phenotypes of individuals are represented by the edges of the IBD graph. The nodes of the graph represent the DNA carried by these individuals; the DNA types of the nodes are independent. Each individual's edge joins the two DNA nodes which he/she caries at the locus, and his/her trait phenotype is determined probabilistically by the latent allelic types of these two nodes. In the example shown in Figure 39.2(b), individuals D, G, and F all share IBD DNA at this locus, as represented by the node labeled 4. Also, individuals B and J share both their DNA nodes, while C carries two copies of a single node. Computation of probabilities of observed data on a graph such as that of Figure 39.2(b) is described by Thompson (2003). This approach gives a much closer parallel to graphical models in other areas of statistical science; see, e.g., Lauritzen (1996) and Pearl (2000).

39.5 The 2000s: Association studies and gene expression

STR markers are highly variable, but expensive to type, and occur relatively sparsely in the genome. The advent of *single nucleotide polymorphism* (SNP) markers in essentially unlimited numbers (International HapMap Consortium, 2005) brought a new dimension and new issues to the genetic mapping of complex traits. Genome-wide association studies (GWAS) became highly popular due to their expected ability to locate causal genes without the need for pedigree data. However, early GWAS were underpowered. Only with large-scale studies (Wellcome Trust, 2007) and better methods to control for population structure and heterogeneity (Price et al., 2006) did association methods start to have success. Modern GWAS typically consider a few thousand individuals, each typed for up to one million SNPs. New molecular technologies also provided new measures of gene expression variation based on the abundance of mRNA transcripts (Schena et al., 1995). Again the statistical question is one of association of a trait or sample phenotype with the expression of some small subset of many thousands of genes.

The need to make valid statistical inferences from both GWAS and from gene expression studies prompted the development of new general statistical approaches. Intrinsic to these problems is that the truth may violate the null hypothesis in many (albeit a small fraction) of the tests of significance made. This leads to a focus on false discovery rates rather than p-values (Storey, 2002, 2003). Both GWAS and gene expression studies also exhibit the modern phenomenon of high-dimensional data ($p \gg n$) or very large numbers of observations on relatively few subjects (Cai and Shen, 2010), giving scope for new methods for dimension reduction and inducing sparsity (Tibshirani et al., 2005).

Genomic technologies continue to develop, with cDNA sequence data replacing SNPs (Mardis, 2008) and RNAseq data (Shendure, 2008) replacing the more traditional microarray expression data. Both raise new statistical challenges. The opportunities for new statistical modeling and inference are immense:

> "... next-generation [sequencing] platforms are helping to open entirely new areas of biological inquiry, including the investigation of ancient genomes, the characterization of ecological diversity, and the identification of unknown etiologic agents." (Mardis, 2008)

But so also are the challenges:

> "Although these new [RNAseq] technologies may improve the quality of transcriptome profiling, we will continue to face what has probably been the larger challenge with microarrays — how best to generate biologically meaningful interpretations of complex datasets that are

sufficiently interesting to drive follow-up experimentation." (Shendure, 2008)

39.6 The 2010s: From association to relatedness

For several reasons there is a move back from population studies to a consideration of related individuals. As the sizes of case-control GWAS grow, problems of population structure increase (Price et al., 2006). Further, these samples often contain related individuals. Closely related individuals may be discarded, but large numbers of distant relatives also impact results. In addition to other heterogeneity between cases and controls, the relationship structure of the case sample may differ from that of the controls. Secondly, GWAS are predicated on the *"common disease, common variant"* model, but there is growing recognition of the role of rare variants in many diseases (Cohen et al., 2004). There are many different mutations that can affect the function of any given gene, and many different genes that function jointly in gene networks. While association tests for rare variant effects in GWAS designs have been developed (Madsen and Browning, 2009), the use of inferred shared descent can provide a more powerful approach (Browning and Thompson, 2012).

Not only does using family information in conjunction with association testing provide more power (Thornton and McPeek, 2007, 2010), but, using modern SNP data, genome shared IBD (Section 39.4) can be detected among individuals not known to be related (Brown et al., 2012; Browning and Browning, 2012). Once IBD in a given region of the genome is inferred from genetic marker data, whether using a known pedigree or from population data, its source is irrelevant. The IBD graph (Figure 39.2(b)) summarizes all the relevant information for the analysis of trait data on the observed individuals. The use of inferred IBD, or more generally estimated relatedness (Lee et al., 2011), is becoming the approach of choice in many areas of genetic analysis.

39.7 To the future

Computational and molecular technologies change ever faster, and the relevant probability models and statistical methodologies will likewise change. For the researchers of the future, more important than any specific knowledge is the approach to research. As has been said by statistical philosopher Ian Hacking:

> "Statisticians change the world not by new methods and techniques but by ways of thinking and reasoning under uncertainty."

I have given many lectures in diverse academic settings, and received many generous and kind introductions, but one which I treasure was given at a recent seminar visit to a Department of Statistics. I was introduced by one of my former PhD students, who said that I had taught him three things:

Think science: Think positive: Think why.

Think science: For me, the scientific questions motivate the statistical thinking. Although, as a student taking exams, I did better with the clearly defined world of mathematical proofs than with the uncertainties of statistical thinking, I could never have become a research mathematician. Answering exam questions was easy; knowing what questions to ask was for me impossible. Fortunately, genetic science came to my rescue: there the questions are endless and fascinating.

Think positive: Another of my former students has said that he got through his (excellent) PhD work, because whenever he came to me in despair that his results were not working out, my response was always *"But that's really interesting"*. Indeed, many things in research do not work out the way we expect, and often we learn far more from what does not work than from what does.

Think why: And when it does not work (or even when it does) the first and most important question is *"Why?"* (Thompson, 2004). If there is anything that distinguishes the human species from other organisms it is not directly in our DNA, but in our capacity to ask *"Why?"*. In research, at least, this is the all-important question.

Think science: Think positive: Think why.

If my students have learned this from me, this is far more important to their futures and to their students' futures than any technical knowledge I could have provided.

References

Abecasis, G.R., Cherny, S.S., Cookson, W.O., and Cardon, L.R. (2002). Merlin — rapid analysis of dense genetic maps using sparse gene flow trees. *Nature Genetics*, 30:97–101.

Besag, J.E. and Green, P.J. (1993). Spatial statistics and Bayesian computation. *Journal of the Royal Statistical Society, Series B*, 55:25–37.

Botstein, D., White, R.L., Skolnick, M.H., and Davis, R.W. (1980). Construction of a linkage map in man using restriction fragment polymorphism. *American Journal of Human Genetics*, 32:314–331.

Brown, M.D., Glazner, C.G., Zheng, C., and Thompson, E.A. (2012). Inferring coancestry in population samples in the presence of linkage disequilibrium. *Genetics*, 190:1447–1460.

Browning, S.R. and Browning, B.L. (2012). Identity by descent between distant relatives: Detection and applications. *Annual Review of Genetics*, 46:617–633.

Browning, S.R. and Thompson, E.A. (2012). Detecting rare variant associations by identity by descent mapping in case-control studies. *Genetics*, 190:1521–1531.

Cai, T. and Shen, X., Eds. (2010). *High-Dimensional Data Analysis*, vol. 2 of *Frontiers of Science*. World Scientific, Singapore.

Cannings, C., Thompson, E.A., and Skolnick, M.H. (1978). Probability functions on complex pedigrees. *Advances in Applied Probability*, 10:26–61.

Cannings, C., Thompson, E.A., and Skolnick, M.H. (1980). Pedigree analysis of complex models. In *Current Developments in Anthropological Genetics*, J. Mielke and M. Crawford, Eds. Plenum Press, New York, pp. 251–298.

Cavalli-Sforza, L.L. and Bodmer, W.F. (1971). *The Genetics of Human Populations*. W.H. Freeman, San Francisco, CA.

Ceppelini, R., Siniscalco, M., and Smith, C.A.B. (1955). The estimation of gene frequencies in a random mating population. *Annals of Human Genetics*, 20:97–115.

Cohen, J.C., Kiss, R.S., Pertsemlidis, A., Marcel, Y.L., McPherson, R., and Hobbs, H.H. (2004). Multiple rare alleles contribute to low plasma levels of HDL cholesterol. *Science*, 305:869–872.

Crow, J. and Kimura, M. (1970). *An Introduction to Population Genetics Theory*. Harper and Row, New York.

Dempster, A.P., Laird, N.M., and Rubin, D.B. (1977). Maximum likelihood from incomplete data via the EM algorithm (with discussion). *Journal of the Royal Statistical Society, Series B*, 39:1–37.

Edwards, A.W.F. (1972). *Likelihood*. Cambridge University Press, Cambridge, UK.

Elston, R.C. and Stewart, J. (1971). A general model for the analysis of pedigree data. *Human Heredity*, 21:523–542.

Fisher, R.A. (1918). The correlation between relatives on the supposition of Mendelian inheritance. *Transactions of the Royal Society of Edinburgh*, 52:399–433.

Fisher, R.A. (1922). On the mathematical foundations of theoretical statistics. *Philosophical Transactions of the Royal Society of London, Series A*, 222:309–368.

Fisher, R.A. (1922). The systematic location of genes by means of crossover observations. *The American Naturalist*, 56:406–411.

Fisher, R.A. (1925). *Statistical Methods for Research Workers.* Oliver & Boyd, Edinburgh, UK.

Fisher, R.A. (1934). The amount of information supplied by records of families as a function of the linkage in the population sampled. *Annals of Eugenics*, 6:66–70.

Gelfand, A.E. and Smith, A.F.M. (1990). Sampling based approaches to calculating marginal densities. *Journal of the American Statistical Association*, 46:193–227.

Geyer, C.J., Ryder, O.A., Chemnick, L.G., and Thompson, E.A. (1993). Analysis of relatedness in the California condors from DNA fingerprints. *Molecular Biology and Evolution*, 10:571–589.

Geyer, C.J. and Thompson, E.A. (1992). Constrained Monte Carlo maximum likelihood for dependent data (with discussion). *Journal of the Royal Statistical Society, Series B*, 54:657–699.

Haldane, J.B.S. (1934). Methods for the detection of autosomal linkage in man. *Annals of Eugenics*, 6:26–65.

Haldane, J.B.S. and Smith, C.A.B. (1947). A new estimate of the linkage between the genes for colour-blindness and haemophilia in man. *Annals of Eugenics*, 14:10–31.

Heath, S.C. (1997). Markov chain Monte Carlo segregation and linkage analysis for oligogenic models. *American Journal of Human Genetics*, 61:748–760.

International HapMap Consortium (2005). A haplotype map of the human genome. *Nature*, 237:1299–1319.

Jeffreys, A.J., Turner, M., and Debenham, P. (1991). The efficiency of multilocus dna fingerprint probes for individualization and establishment of family relationships, determined from extensive casework. *American Journal of Human Genetics*, 48:824–840.

Karr, J.-B.A. (1849). *Les guêpes.* Michel Lévy Frères, Paris.

Lander, E.S. and Green, P. (1987). Construction of multilocus genetic linkage maps in humans. *Proceedings of the National Academy of Sciences (USA)*, 84:2363–2367.

Lange, K. and Sobel, E. (1991). A random walk method for computing genetic location scores. *American Journal of Human Genetics*, 49:1320–1334.

Lauritzen, S.L. (1996). *Graphical Models.* Oxford University Press, Oxford, UK.

Lauritzen, S.L. and Spiegelhalter, D.J. (1988). Local computations with probabilities on graphical structures and their application to expert systems. *Journal of the Royal Statistical Society, Series B*, 50:157–224.

Lee, S.H., Wray, N.R., Goddard, M.E., and Visscher, P.M. (2011). Estimating missing heritability for disease from genome-wide association studies. *American Journal of Human Genetics*, 88:294–305.

Lin, S., Thompson, E.A., and Wijsman, E.M. (1993). Achieving irreducibility of the Markov chain Monte Carlo method applied to pedigree data. *IMA Journal of Mathematics Applied in Medicine and Biology*, 10:1–17.

Madsen, B.E. and Browning, S.R. (2009). A groupwise association test for rare mutations using a weighted sum statistic. *PLOS Genetics*, 5:e1000384.

Mardis, E.R. (2008). Next-generation DNA sequencing methods. *Annual Review of Genomics and Human Genetics*, 9:387–402.

Morton, N.E. (1955). Sequential tests for the detection of linkage. *American Journal of Human Genetics*, 7:277–318.

Mourant, A.E. (1954). *The Distribution of the Human Blood Groups.* Blackwell, Oxford, UK.

Murray, J.C., Buetow, K.H., Weber, J.L., Ludwigsen, S., Heddema, S.T., Manion, F., Quillen, J., Sheffield, V.C., Sunden, S., Duyk, G.M. et al. (1994). A comprehensive human linkage map with centimorgan density. *Science*, 265:2049–2054.

Patterson, H.D. and Thompson, R. (1971). Recovery of inter-block information when blocks are unequal. *Biometrika*, 58:545–554.

Pearl, J. (2000). *Causality: Models, Reasoning, and Inference.* Cambridge University Press, Cambridge, UK.

Penttinen, A. (1984). Modelling interaction in spatial point patterns: Parameter estimation by the maximum likelihood method. *Jyväskylä Studies in Computer Science, Economics, and Statistics*, vol. 7.

Price, A.L., Patterson, N.J., Plenge, R.M., Weinblatt, M.E., Shadick, N.A., and Reich, D. (2006). Principal components analysis corrects for stratification in genome-wide association studies. *Nature Genetics*, 38:904–909.

Schena, M., Shalon, D., Davis, R.W., and Brown, P.O. (1995). Quantitative monitoring of gene expression patterns with a complementary DNA microarray. *Science*, 270:467–470.

Sheehan, N.A. and Thomas, A.W. (1993). On the irreducibility of a Markov chain defined on a space of genotype configurations by a sampling scheme. *Biometrics*, 49:163–175.

Shendure, J. (2008). The beginning of the end for microarrays? *Nature Methods*, 5:585–587.

Smith, C.A.B. (1953). Detection of linkage in human genetics. *Journal of the Royal Statistical Society, Series B*, 15:153–192.

Stern, C. (1973). *Principles of Human Genetics*, 3rd edition. W.H. Freeman, San Francisco, CA.

Storey, J.D. (2002). A direct approach to false discovery rates. *Journal of the Royal Statistical Society, Series B*, 64:479–498.

Storey, J.D. (2003). The positive false discovery rate: A Bayesian interpretation and the q-value. *The Annals of Statistics*, 31:2013–2035.

Thompson, E.A. (1975). *Human Evolutionary Trees.* Cambridge University Press, Cambridge, UK.

Thompson, E.A. (1978). Ancestral inference II: The founders of Tristan da Cunha. *Annals of Human Genetics*, 42:239–253.

Thompson, E.A. (1981). Pedigree analysis of Hodgkin's disease in a Newfoundland genealogy. *Annals of Human Genetics*, 45:279–292.

Thompson, E.A. (1983). Gene extinction and allelic origins in complex genealogies. *Philosophical Transactions of the Royal Society of London (Series B)*, 219:241–251.

Thompson, E.A. (1994). Monte Carlo estimation of multilocus autozygosity probabilities. In *Proceedings of the 1994 Interface conference* (J. Sall and A. Lehman, Eds.). Fairfax Station, VA, pp. 498–506.

Thompson, E.A. (2000). *Statistical Inferences from Genetic Data on Pedigrees*, volume 6 of *NSF–CBMS Regional Conference Series in Probability and Statistics.* Institute of Mathematical Statistics, Beachwood, OH.

Thompson, E.A. (2003). Information from data on pedigree structures. In *Science of Modeling: Proceedings of AIC 2003*, pp. 307–316. Research Memorandum of the Institute of Statistical Mathematics, Tokyo, Japan.

Thompson, E.A. (2004). The importance of Why? *The American Statistician*, 58:198.

Thompson, E.A. (2011). The structure of genetic linkage data: From LIPED to 1M SNPs. *Human Heredity*, 71:88–98.

Thompson, E.A. and Heath, S.C. (1999). Estimation of conditional multilocus gene identity among relatives. In *Statistics in Molecular Biology and Genetics: Selected Proceedings of a 1997 Joint AMS-IMS-SIAM Summer Conference on Statistics in Molecular Biology* (F. Seillier-Moiseiwitsch, Ed.). IMS Lecture Note–Monograph Series Vol. 33, pp. 95–113. Institute of Mathematical Statistics, Hayward, CA.

Thompson, E.A., Kravitz, K., Hill, J. and Skolnick, M.H. (1978). Linkage and the power of a pedigree structure. In *Genetic Epidemiology* (N.E. Morton, Ed.). Academic Press, New York, pp. 247–253.

Thornton, T. and McPeek, M.S. (2007). Case-control association testing with related individuals: A more powerful quasi-likelihood score test. *American Journal of Human Genetics*, 81:321–337.

Thornton, T. and McPeek, M.S. (2010). ROADTRIPS: Case-control association testing with partially or completely unknown population and pedigree structure. *American Journal of Human Genetics*, 86:172–184.

Tibshirani, R.J., Saunders, M., Rosset, S., Zhu, J., and Knight, K. (2005). Sparsity and smoothness via the fused lasso. *Journal of the Royal Statistical Society, Series B*, 67:91–108.

Tong, L. and Thompson, E.A. (2008). Multilocus lod scores in large pedigrees: Combination of exact and approximate calculations. *Human Heredity*, 65:142–153.

Wellcome Trust Case Control Consortium (2007). Genome-wide association study of 14,000 cases of seven common diseases and 3,000 shared controls. *Nature*, 447:661–678.

Wright, S. (1921). Correlation and causation. *Journal of Agricultural Research*, 20:557–585.

Wright, S. (1922). Coefficients of inbreeding and relationship. *American Naturalist*, 56:330–338.

40

Targeted learning: From MLE to TMLE

Mark van der Laan
Division of Biostatistics, School of Public Health
University of California, Berkeley, CA

In this chapter I describe some of the essential elements of my past scientific journey from the study of nonparametric maximum likelihood estimation (NPMLE) to the field targeted learning and the resulting new general tool targeted minimum loss based estimation (TMLE). In addition, I discuss our current and future research program involving the further development of targeted learning to deal with dependent data. This journey involved mastering difficult statistical concepts and ideas, and combining them into an evolving roadmap for targeted learning from data under realistic model assumptions. I hope to convey the message that this is a highly inspiring evolving unifying and interdisciplinary project that needs input for many future generations to come, and one that promises to deal with the current and future challenges of statistical inference with respect to a well-defined typically complex targeted estimand based on extremely highly dimensional data structures per unit, complex dependencies between the units, and very large sample sizes.

40.1 Introduction

Statistical practice has been dominated by the application of statistical methods relying on parametric model assumptions such as linear, logistic, and Cox proportional hazards regression methodology. Most of these methods use maximum likelihood estimation, but others rely on estimating equations such as generalized estimating equations (GEE). These maximum likelihood estimators are known to be asymptotically Normally distributed, and asymptotically efficient under weak regularity conditions, beyond the key condition that the true data generating distribution satisfies the restrictions assumed by these parametric models.

When I started my PhD in 1990, my advisor Richard Gill inspired me to work on maximum likelihood estimation and estimating equation methods for nonparametric or semi-parametric statistical models, with a focus on models for censored data (van der Laan, 1996). Specifically, I worked on the construction of a semi-parametric efficient estimator of the bivariate survival function based on bivariate right-censored failure time data, generalizing the Kaplan–Meier estimator of a univariate survival function. At that time the book by Bickel et al. (1997) on semi-parametric models was about to appear and earlier versions were circulating. There was an enormous interest among the theoreticians, and I had the fortune to learn from various inspiring intellectual leaders such as Richard Gill, Aad van der Vaart, Sara van de Geer, Peter Bickel, Jon Wellner, Richard Dudley, David Pollard, James Robins, and many more.

In order to deal with the challenges of these semi-parametric models I had to learn about efficiency theory for semi-parametric models, relying on a so called least favorable parametric submodel for which estimation of the desired finite dimensional estimand is as hard as it is in the actual infinite-dimensional semi-parametric model. I also had to compute projections in Hilbert spaces to calculate efficient influence curves and corresponding least favorable submodels. Richard Gill taught me how to represent an estimator as a functional applied to the empirical distribution of the data, and how to establish functional differentiability of these estimator-functionals. I was taught about the functional delta-method which translates a) the convergence in distribution of the plugged in empirical process, and b) the functional differentiability of the estimator into the convergence in distribution of the standardized estimator to a Gaussian process; see, e.g., Gill et al. (1995). I learned how to compute the influence curve of a given estimator and that it is the object that identifies the asymptotic Gaussian process of the standardized estimators. In addition, Aad van der Vaart taught me about weak convergence of empirical processes indexed by a class of functions, and Donsker classes defined by entropy integral conditions (van der Vaart and Wellner, 1996), while Richard Gill taught me about models for the intensity of counting processes and continuous time martingales (Andersen et al., 1993). Right after my PhD thesis Jamie Robins taught me over the years a variety of clever methods for calculating efficient influence curves in complex statistical models for complex longitudinal data structures, general estimating equation methodology for semi-parametric models, and I learned about causal inference for multiple time-point interventions (Robins and Rotnitzky, 1992; van der Laan and Robins, 2003).

At that time, I did not know about a large statistical community that would have a hard time accepting the formulation of the statistical estimation problem in terms of a true statistical semi-parametric model, and an estimand/target parameter as a functional from this statistical model to the parameter space, as the way forward, but instead used quotes such as "All models are wrong, but some are useful" to justify the application of wrong parametric models for analyzing data. By going this route, this community does not only accept seriously biased methods for analyzing data in which confidence intervals and p-values have no real meaning, but it also obstructs progress by not formulating the true statistical challenge that needs to be addressed to solve the actual estimation problem. In particular, due to this resistance, we still see that most graduate students in biostatistics and statistics programs do not know much about the above mentioned topics, such as efficiency theory, influence curves, and efficient influence curves. I would not have predicted at that time that I would be able to inspire new generations with the very topics I learned in the 1990s.

Throughout my career, my only goal has been to advance my understanding of how to formulate and address the actual estimation problems in a large variety of real world applications, often stumbling on the need for new theoretical and methodological advances. This has been my journey that started with my PhD thesis and is a product of being part of such a rich community of scientists and young dynamic researchers that care about truth and stand for progress. I try and hope to inspire next generations to walk such journeys, each person in their own individual manner fully utilizing their individual talents and skills, since it is a path which gives much joy and growth, and thereby satisfaction.

In the following sections, I will try to describe my highlights of this scientific journey, resulting in a formulation of the field targeted learning (van der Laan and Rose, 2012), and an evolving roadmap for targeted learning (Pearl, 2009; Petersen and van der Laan, 2012; van der Laan and Rose, 2012) dealing with past, current and future challenges that require the input for many generations to come. To do this in a reasonably effective way we start out with providing some succinct definitions of key statistical concepts such as statistical model, model, target quantity, statistical target parameter, and asymptotic linearity of estimators. Subsequently, we will delve into the construction of finite sample robust, asymptotically efficient substitution estimators in realistic semi-parametric models for experiments that generate complex high dimensional data structures that are representative of the current flood of information generated by our society. These estimators of specified estimands utilize the state of the art in machine learning and data adaptive estimation, while preserving statistical inference. We refer to the field that is concerned with construction of such targeted estimators and corresponding statistical inference as targeted learning.

40.2 The statistical estimation problem

40.2.1 Statistical model

The statistical model encodes known restrictions on the probability distribution of the data, and thus represents a set of statistical assumptions. Let's denote the observed random variable on a unit with O and let P_0 be the probability distribution of O. In addition, let's assume that the observed data is a realization of n independent and identically distributed copies $O_1, \ldots, O_n$ of $O \sim P_0$. Formally, a statistical model is the collection of possible probability distributions of O, and we denote this set with $\mathcal{M}$.

Contrary to most current practice, a statistical model should contain the true P_0, so that the resulting estimation problem is a correct formulation, and not a biased approximation of the true estimation problem. The famous quote that all statistical models are wrong represents a false statement, since it is not hard to formulate truthful statistical models that only incorporate true knowledge, such as the nonparametric statistical model that makes no assumptions at all. Of course, we already made the key statistical assumption that the n random variables $O_1, \ldots, O_n$ are independent and identically distributed, and, that assumption itself might need to be weakened to a statistical model for $(O_1, \ldots, O_n) \sim P_0^n$ that contains the true distribution P_0^n. For a historical and philosophical perspective on "models, inference, and truth," we refer to Starmans (2012).

40.2.2 The model encoding both statistical and non-testable assumptions

A statistical model could be represented as $\mathcal{M} = \{P_\theta : \theta \in \Theta\}$ for some mapping $\theta \mapsto P_\theta$ defined on an infinite-dimensional parameter space Θ. We refer to this mapping $\theta \mapsto P_\theta$ as a model, and it implies the statistical model for the true data distribution $P_0 = P_{\theta_0}$. There will always exist many models that are compatible with a particular statistical model. It is important to note that the statistical model is the only relevant information for the statistical estimation problem. Examples of models are censored data and causal inference models in which case the observed data structure $O = \Phi(C, X)$ is represented as a many to one mapping Φ from the full-data X and censoring variable C to O, in which case the observed data distribution is indexed by the full-data distribution P_X and censoring mechanism $P_{C|X}$. So in this case Θ represents the set of possible $(P_X, P_{C|X})$, and P_θ is the distribution of $\Phi(C, X)$ implied by the distribution θ of (C, X). Different models for $(P_X, P_{C|X})$ might imply the same statistical model for the data distribution of O. We note that a model encodes assumptions beyond the statistical model, and we refer to these additional assumptions as non-testable assumptions since they put no restrictions on the distribution of the data. Assumptions such as $O = \Phi(C, X)$, and the coarsening or missing at random assumption on the conditional distribution $P_{C|X}$ are examples of non-testable assumptions that do not affect the statistical model.

40.2.3 Target quantity of interest, and its identifiability from observed data distribution

The importance of constructing a model is that it allows us to define interesting target quantities $\Psi^F(\theta)$ for a given mapping $\Psi^F : \Theta \rightarrow \mathbb{R}^d$ that represent the scientific question of interest that we would like to learn from our data $O_1, \ldots, O_n$. Given such a definition of target parameter $\Psi^F : \Theta \rightarrow \mathbb{R}^d$ one likes to establish that there exists a mapping $\Psi : \mathcal{M} \rightarrow \mathbb{R}^d$ so that $\Psi^F(\theta) = \Psi(P_\theta)$. If such a mapping Ψ exists we state that the target quantity $\Psi^F(\theta)$ is identifiable from the observed data distribution. This is often only possible by making additional non-testable restrictions on θ in the sense that one is only able to write $\Psi^F(\theta) = \Psi(P_\theta)$ for $\theta \in \Theta^* \subset \Theta$.

40.2.4 Statistical target parameter/estimand, and the corresponding statistical estimation problem

This identifiability result defines a statistical target parameter $\Psi : \mathcal{M} \rightarrow \mathbb{R}^d$. The goal is to estimate $\psi_0 = \Psi(P_0)$ based on n i.i.d. observations on $O \sim P_0 \in \mathcal{M}$. The estimand ψ_0 can be interpreted as the target quantity $\Psi^F(\theta_0)$ *if both the non-testable and the statistical model assumptions hold.* Nonetheless, due to the statistical model containing the true data distribution P_0, ψ_0 always has a pure statistical interpretation as the feature $\Psi(P_0)$ of the data distribution P_0. A related additional goal is to obtain a confidence interval for ψ_0. A sensitivity analysis can be used to provide statistical inference for the underlying target quantity ψ_0^F under a variety of violations of the assumptions that were needed to state that $\psi_0^F = \psi_0$, as we will discuss shortly below.

40.3 The curse of dimensionality for the MLE

40.3.1 Asymptotically linear estimators and influence curves

An estimator is a Euclidean valued mapping $\hat{\Psi}$ on a statistical model that contains all empirical probability distributions. Therefore, one might represent an estimator as a mapping $\hat{\Psi} : \mathcal{M}_{NP} \rightarrow \mathbb{R}^d$ from the nonparametric statistical model $\mathcal{M}_{NP}$ into the parameter space. In order to allow for statistical inference, one is particularly interested in estimators that behave in first order as an empirical mean of i.i.d. random variables so that it is asymptotically Normally distributed. An estimator $\hat{\Psi}$ is asymptotically linear at data distribution P_0 with influence curve D_0 if

$$\hat{\Psi}(P_n) - \psi_0 = \frac{1}{n} \sum_{i=1}^{n} D_0(O_i) + o_P(1/\sqrt{n}).$$

Such an estimator is asymptotically Normally distributed in the sense that $\sqrt{n}\,(\psi_n - \psi_0)$ converges in distribution to $\mathcal{N}(0, \sigma^2)$, where σ^2 is the variance of $D_0(O)$.

40.3.2 Efficient influence curve

Efficiency theory teaches us that an estimator is efficient if and only if it is asymptotically linear with influence curve D_0^* equal to the canonical gradient of the pathwise derivative of $\Psi : \mathcal{M} \to \mathbb{R}^d$. This canonical gradient is also called the efficient influence curve. Due to this fundamental property, the efficient influence curve is one of the most important objects in statistics. There are general representation theorems and corresponding methods for calculation of efficient influence curves based on Hilbert space projections; see Bickel et al. (1997) and van der Laan and Robins (2003). Indeed, the efficient influence curve forms a crucial ingredient in any methodology for construction of efficient estimators.

40.3.3 Substitution estimators

Efficient estimators fully utilize the local constraints at P_0 in the model, but they provide no guarantee of full utilization of the fact that $P_0 \in \mathcal{M}$ and that $\psi_0 = \Psi(P_0)$ for some $P_0 \in \mathcal{M}$. The latter global information of the estimation problem is captured by making sure that the estimator is a substitution estimator that can be represented as $\hat{\Psi}(P_n) = \Psi(P_n^*)$ for some $P_n^* \in \mathcal{M}$. Alternatively, if $\Psi(P)$ only depends on P through an (infinite-dimensional) parameter $Q(P)$, and we denote the target parameter with $\Psi(Q)$ again, then a substitution estimator can be represented as $\Psi(Q_n)$ for a Q_n in the parameter space $\{Q(P) : P \in \mathcal{M}\}$. Efficient estimators based on estimating equation methodology (van der Laan and Robins, 2003) provide no guarantee of obtaining substitution estimators, and are thereby not as finite sample robust as efficient substitution estimators we discuss here.

40.3.4 MLE and curse of dimensionality

Maximum likelihood estimators are examples of substitution estimators. Unfortunately, due to the curse of dimensionality of infinite-dimensional statistical models, the MLE is often ill defined. For example, consider the MLE for the bivariate failure time distribution based on bivariate right-censored failure time data. In this case, an MLE can be implicitly defined (through the so called self-consistency equation) as an estimator that assigns mass $1/n$ to each observation and redistributes this mass over the coarsening for the bivariate failure time implied by this observation according to the MLE itself. In this case, the coarsenings are singletons for the doubly uncensored observations (i.e., both failure times are observed), half-lines for the singly-censored observations (i.e., one failure time is censored), and quadrants for the doubly censored observations. For continuously distributed failure times, all the half-lines contain zero uncensored bivariate failure times, and as a consequence the likelihood does essentially provide no information about how these masses $1/n$ should be distributed over the half-lines. That is the MLE is highly non-unique, and thereby also inconsistent.

Similarly, consider the estimation of $\Psi(P_0) = \mathrm{E}_0\mathrm{E}_0(Y|A=1,W)$ based on n i.i.d. observations on $(W,A,Y) \sim P_0$, and suppose the statistical model is the nonparametric model. The MLE of $\mathrm{E}_0(Y|A=1,W=w)$ is the empirical mean of the outcome among the observations with $A_i = 1, W_i = w$, while the MLE of the distribution of W is the empirical distribution of $W_1,\ldots,W_n$. This estimator is ill-defined since most-strata will have no observations.

40.3.5 Regularizing MLE through smoothing

In order to salvage the MLE the literature suggests to regularize the MLE in some manner. This often involves either smoothing or sieve-based MLE where the fine-tuning parameters need to be selected based on some empirical criterion. For example, in our bivariate survival function example, we could put strips around the half-lines of the single censored observations, and compute the MLE as if the half-lines implied by the single censored observations are now these strips. Under this additional level of coarsening, the MLE is now uniquely defined as long as the strips contain at least one uncensored observation. In addition, if one makes sure that the number of observations in the strips converge to infinity as sample size increases, and the width of the strips converges to zero, then the MLE will also be consistent. Unfortunately, there is still a bias/variance trade-off that needs to be resolved in order to arrange that the MLE of the bivariate survival function is asymptotically linear. Specifically, we need to make sure that the width of the strips converges fast enough to zero so that the bias of the MLE with respect to the conditional densities over the half-lines is $o(1/\sqrt{n})$. This would mean that the width of the strips is $o(1/\sqrt{n})$. For an extensive discussion of this estimation problem, and alternative smoothing approach to repair the NPMLE we refer to van der Laan (1996).

Similarly, we could estimate the regression function $\mathrm{E}_0(Y|A=1,W)$ with a histogram regression method. If the dimension of W is k-dimensional, then for the sake of arranging that the bin contains at least one observation, one needs to select a very large width so that the k-dimensional cube with width h contains one observation with high probability. That is, we will need to select h so that $nh^k \to \infty$. This binning causes bias $O(h)$ for the MLE of $\mathrm{EE}(Y|A=1,W)$. As a consequence, we will need that $n^{-1/k}$ converges to zero faster than $n^{-1/2}$, which only holds when $k=1$. In other words, there is no value of the smoothing parameter that results in a regularized MLE that is asymptotically linear.

Even though there is no histogram-regularization possible, there might exist other ways of regularizing the MLE. The statistics and machine learning literature provides many possible approaches to construct estimators of the required objects Q_0, and thereby $\psi_0 = \Psi(Q_0)$. One strategy would be to define a large class of submodels that contains a sequence of submodels that approximates the complete statistical model (a so called sieve), and construct for each submodel an estimator that achieves the minimax rate under the assumption that Q_0 is an element of this submodel. One can now use a data adaptive selector to select among all these submodel-specific candidate estimators. This general strategy, which is often referred to as sieve-based MLE, results in a minimax adaptive estimator of Q_0, i.e., the estimator converges at the minimax rate of the smallest submodel (measured by entropy) that still contains the true Q_0. Such an estimator is called minimax adaptive. We refer to van der Laan and Dudoit (2003) and van der Laan et al. (2006) for such general minimum loss-based estimators relying on cross-validation to select the subspace. This same strategy can be employed with kernel regression estimators that are indexed by the degree of orthogonality of the kernel and a bandwidth, and one can use a data-adaptive selector to select this kernel and bandwidth. In this manner the resulting data adaptive kernel regression estimator will achieve the minimax rate of convergence corresponding with the unknown underlying smoothness of the true regression function.

40.3.6 Cross-validation

Cross-validation is a particularly powerful tool to select among candidate estimators. In this case, one defines a criterion that measures the performance of a given fit of Q_0 on a particular sub-sample: typically, this is defined as an empirical mean of a loss function that maps the fit Q and observation O_i into a real number and is such that the minimizer of the expectation of the loss over all Q equals the desired true Q_0. For each candidate estimator, one trains the estimator on a training sample and one evaluates the resulting fit on the complement of the training sample, which is called the validation sample. This is carried out for a number of sample splits in training and validation samples, and one selects the estimator that has the best average performance across the sample splits. Statistical theory teaches us that this procedure is asymptotically optimal in great generality in the sense that it performs asymptotically as well as an oracle selector that selects the estimator based on the criterion applied to an infinite validation sample; see, e.g., Györfi et al. (2002), van der Laan and Dudoit (2003), van der Laan et al. (2006), and van der Vaart et al. (2006). The key conditions are that the loss-function needs to be uniformly bounded, and the size of the validation sample needs to converge to infinity.

40.4 Super learning

These oracle results for the cross-validation selector teach us that it is possible to construct an ensemble estimator that asymptotically outperforms any user supplied library of candidate estimators. We called this estimator the super-learner due to its theoretical properties: it is defined as a combination of all the candidate estimators where the weights defining the combination (e.g., convex combination) are selected based on cross-validation; see, e.g., van der Laan et al. (2007) and Polley et al. (2012). By using the super-learner as a way to regularize the MLE, we obtain an estimator with a potentially much better rate of convergence to the true Q_0 than a simple regularization procedure such as the one based on binning discussed above. The bias of this super-learner will converge to zero at the same rate as the rate of convergence of the super-learner. The bias of the plug-in estimator of ψ_0 based on this super-learner will also converge at this rate. Unfortunately, if none of our candidate estimators in the library achieve the rate $1/\sqrt{n}$ (e.g., a MLE according to a correctly specified parametric model), then this bias will be larger than $1/\sqrt{n}$, so that this plug-in estimator will not converge at the desired $\sqrt{n}$ rate. To conclude, although super-learner has superior performance in estimation of Q_0, it still results in an overly biased estimator of the target $\Psi(Q_0)$.

40.4.1 Under-smoothing fails as general method

Our binning discussion above argues that for typical definitions of adaptive estimators indexed by a fine-tuning parameter (e.g., bandwidth, number of basis functions), there is no value of the fine-tuning parameter that would result in a bias for ψ_0 of the order $1/\sqrt{n}$. This is due to the fact that the fine-tuning parameter needs to exceed a certain value in order to define an estimator in the parameter space of Q_0. So even when we would have selected the estimator in our library of candidate estimators that minimizes the MSE with respect to ψ_0 (instead of the one minimizing the cross-validated risk), then we would still have selected an estimator that is overly biased for ψ_0.

The problem is that our candidate estimators rely on a fine tuning parameter that controls *overall* bias of the estimator. Instead we need candidate estimators that have an excellent overall fit of Q_0 but also rely on a tuning parameter that only controls the bias of the resulting plug-in estimator for ψ_0, and we need a way to fit this tuning parameter. For that purpose, we need to determine a submodel of fluctuations $\{Q_n(\epsilon) : \epsilon\}$ through a candidate estimator Q_n at $\epsilon = 0$, indexed by an amount of fluctuation ϵ, where fitting ϵ is locally equivalent with fitting ψ_0 in the actual semi-parametric statistical model $\mathcal{M}$. It appears that the least-favorable submodel from efficiency theory can be utilized for this purpose, while ϵ can be fitted with the parametric MLE. This insight resulted in so called targeted maximum likelihood estimators (van der Laan and Rubin, 2006). In this manner, we can map any candidate estimator into a targeted estimator and we can use the super-learner based on the library of candidate targeted estimators. Alternatively, one computes this targeted fit of a super-learner based on a library of non-targeted candidate estimators.

40.5 Targeted learning

The real message is that one needs to make the learning process targeted towards its goal. The goal is to construct a good estimator of $\Psi(Q_0)$, and that is not the same goal as constructing a good estimator of the much more ambitious infinite-dimensional object Q_0. For example, estimators of $\Psi(Q_0)$ will have a variance that behaves as $1/n$, while a consistent estimator of Q_0 at a point will generally only use local data so that its variance converges at a significant slower rate than $1/n$. The bias of an estimator of Q_0 is a function, while the bias of an estimator of ψ_0 is just a finite dimensional vector of real numbers. For parametric maximum likelihood estimators one fits the unknown parameters by solving the score equations. An MLE in a semi-parametric model would aim to solve all (infinite) score equations, but due to the curse of dimensionality such an MLE simply does not exist for finite samples. However, if we know what score equation really matters for fitting ψ_0, then we can make sure that our estimator will solve that ψ_0-specific score equation. The efficient influence curve of the target parameter mapping $\Psi : \mathcal{M} \to \mathbb{R}^d$ represents this score.

40.5.1 Targeted minimum loss based estimation (TMLE)

The above mentioned insights evolved into the following explicit procedure called Targeted Minimum Loss Based Estimation (TMLE); see, e.g., van der Laan and Rubin (2006), van der Laan (2008) and van der Laan and Rose (2012). Firstly, one constructs an initial estimator of Q_0 such as a loss-based super-learner based on a library of candidate estimators of Q_0. One now defines a loss function $L(Q)$ so that $Q_0 = \arg\min_Q P_0 L(Q)$, and a least-favorable submodel $\{Q(\epsilon) : \epsilon\} \subset \mathcal{M}$ so that the generalized score $\frac{d}{d\epsilon} L\{Q(\epsilon)\}\big|_{\epsilon=0}$ equals or spans the efficient influence curve $D^*(Q, g)$. Here we used the notation $P_0 f = \int f(o) dP_0(o)$. This least-favorable submodel might depend on an unknown nuisance parameter $g = g(P)$. One is now ready to target the fit Q_n in such a way that its targeted version solves the efficient score equation $P_n D^*(Q_n^*, g_0) = 0$. That is, one defines $\epsilon_n = \arg\min_\epsilon P_n L\{Q_n(\epsilon)\}$, and the resulting update $Q_n^1 = Q_n(\epsilon_n)$. This updating process can be iterated till convergence at which point $\epsilon_n = 0$ so that the final update Q_n^* solves the score equation at $\epsilon_n = 0$, and thus $P_n D^*(Q_n^*, g_0) = 0$. The efficient influence curve has the general property that

$$P_0 D^*(Q, g) = \Psi(Q_0) - \Psi(Q) + R(Q, Q_0, g, g_0)$$

for a second order term $R(Q, Q_0, g, g_0)$. In fact, in many applications, we have that $R(Q, Q_0, g, g_0)$ equals an integral of $(Q - Q_0)(g - g_0)$ so that it equals zero if either $Q = Q_0$ or $g = g_0$, which is often referred to as double robustness of the efficient influence curve. In that case, $P_n D^*(Q_n^*, g_0) = 0$ implies that $\Psi(Q_n^*)$ is a consistent estimator of ψ_0. In essence, the norm of $P_n D^*(Q_n^*, g_0)$ represents a criterion measuring a distance between $\Psi(Q_n^*)$ and ψ_0, so that minimizing the Euclidean norm of $P_n D^*(Q_n^*, g_n)$ corresponds with fitting ψ_0. Since in many applications, the nuisance parameter g_0 is unknown, one will have to replace g_0 in the updating procedure by an estimator g_n. In that case, we have

$$P_0 D^*(Q_n^*, g_n) = \psi_0 - \Psi(Q_n^*) + R(Q_n^*, Q_0, g_n, g_0),$$

where the remainder is still a second order term but now also involving cross-term differences $(Q_n^* - Q_0)(g_n - g_0)$.

40.5.2 Asymptotic linearity of TMLE

If this second order remainder term $R(Q_n^*, Q_0, g_n, g_0)$ converges to zero in probability at a rate faster than $1/\sqrt{n}$, then it follows that

$$\psi_n^* - \psi_0 = (P_n - P_0) D^*(Q_n^*, g_n) + o_P(1/\sqrt{n}),$$

so that, if $P_0\{D^*(Q_n^*, g_n) - D^*(Q_0, g_0)\}^2 \to 0$ in probability, and the random function $D^*(Q_n^*, g_n)$ of O falls in a P_0-Donsker class, it follows that

$$\psi_n^* - \psi_0 = (P_n - P_0) D^*(Q_0, g_0) + o_P(1/\sqrt{n}).$$

That is, $\sqrt{n}\,(\psi_n^* - \psi_0)$ is asymptotically Normally distributed with mean zero and variance equal to the variance of the efficient influence curve. Thus, if Q_n^*, g_n are consistent at fast enough rates, then ψ_n^* is asymptotically efficient. Statistical inference can now be based on the Normal limit distribution and an estimator of its asymptotic variance, such as $\sigma_n^2 = P_n D^*(Q_n^*, g_n)^2$. This demonstrates that the utilization of the state of the art in adaptive estimation was not a hurdle for statistical inference, but, on the contrary, it is required to establish the desired asymptotic Normality of the TMLE. Establishing asymptotic linearity of TMLE under misspecification of Q_0 (in the context that the efficient influence curve is double robust), while still allowing the utilization of very adaptive estimators of g_0, has to deal with additional challenges resolved by also targeting the fit of g; see, e.g., van der Laan (2012) and van der Laan and Rose (2012).

40.6 Some special topics

40.6.1 Sensitivity analysis

The TMLE methodology provides us with statistical inference for the estimand ψ_0. One typically wants to report findings about the actual target quantity of interest ψ_0^F, but it might not be reasonable to assume that $\psi_0 = \psi_0^F$. One simple way forward we recently proposed is to define the bias $\psi_0 - \psi_0^F$, and for each assumed value δ, one can now estimate ψ_0^F with $\psi_n - \delta$ and report a corresponding confidence interval or p-value for the test that $\mathcal{H}_0 : \psi_0^F = 0$. Subject matter knowledge combined with data analysis and or simulations might now provide a reasonable upper bound for δ and one can then determine if such an upper bound would still provide significant results for the target quantity of interest. This sensitivity analysis can be made more conservative in exchange for an enhanced interpretation of the sensitivity parameter δ by defining δ as a particular upper bound of the causal bias $\psi_0 - \psi_0^F$. Such an upper bound might be easier to interpret and thereby improve the sensitivity analysis. We refer to Diaz and van der Laan (2012) for an introduction of this type of sensitivity analysis, a practical demonstration with a few data examples, and a preceding literature using alternative approaches; see, e.g., Rotnitzky et al. (2001), Robins et al. (1999), and Scharfstein et al. (1999).

40.6.2 Sample size 1 problems

Above we demonstrated that the statistical inference relies on establishing asymptotic linearity and thereby asymptotic Normality of the standardized estimator of ψ_0. The asymptotic linearity was heavily relying on the central limit theorem and uniform probability bounds for sums of independent variables (e.g., Donsker classes). In many applications, the experiment resulting in the observed data cannot be viewed as a series of independent experiments. For example, observing a community of individuals over time might truly be a single experiment since the individuals might be causally connected through a network. In this case, the sample size is one. Nonetheless, one might know for each individual what other individuals it depends on, or one might know that the data at time t only depends on the past through the data collected over the last x months. Such assumptions imply conditional independence restrictions on the likelihood of the data. As another example, in a group sequential clinical trial one might make the randomization probabilities for the next group of subjects a function of the observed data on all the previously recruited individuals. The general field of adaptive designs concerns the construction of a single experiment that involves data adaptive changes in the design in response to previously observed data, and the key challenge of such designs is develop methods that provide honest statistical inference; see, e.g., Rosenblum and van der Laan (2011). These examples demonstrate that targeted learning should also be concerned with data generated by single experiments that have a lot of structure. This requires the development of TMLE to such statistical models, integration of the state of the art in weak convergence theory for dependent data, and advances in computing due to the additional complexity of estimators and statistical inference in such data generating experiments. We refer to a few recent examples of targeted learning in adaptive designs, and to estimate effects of interventions on a single network of individuals; see van der Laan (2008), Chambaz and van der Laan (2010, 2011a,b), van der Laan (2012), and van der Laan et al. (2012).

40.6.3 Big Data

Targeted learning involves super-learning, complex targeted update steps, evaluation of an often complex estimand, and estimation of the asymptotic variance of the estimator. In addition, since the estimation is tailored to each question separately, for example, the assessment of the effect of a variable (such as the effect of a DNA-mutation on a phenotype) across a large collection of variables requires many times repeating these computer intensive estimation procedures. Even for normal size data sets, such data analyses can already be computationally very challenging.

However, nowadays, many applications contain gigantic data sets. For example, one might collect complete genomic profiles on each individual, so that one collects hundreds of thousands or even millions of measurements on one individual, possibly at various time points. In addition, there are various initiatives in building large comprehensive data bases, such as the sentinel project which builds a data base for all American citizens which is used to evaluate safety issues for drugs. Such data sets cover hundreds of millions of individuals. Many companies are involved in analyzing data on the internet, which can result in data sets with billions of records.

40.7 Concluding remarks

The biggest mistake we can make is to give up on sound statistics, and be satisfied with the application of algorithms that can handle these data sets in one way or another, without addressing a well defined statistical estimation problem. As we have seen, the genomic era has resulted in an erosion of sound statistics, and as a counterforce many advocate to only apply very simple statistics such as sample means, and univariate regressions. Neither approach is satisfying, and fortunately, it is not needed to give up on sound and complex statistical estimation procedures targeting interesting questions of interest.

Instead, we need to more fully integrate with the computer science, train our students in software that can handle these immense computational and memory challenges, so that our methods can be implemented and made accessible to the actual users, but simultaneously we need to stick to our identity as statisticians as part of collaborative highly interdisciplinary teams, pushing forward the development of optimal statistical procedures and corresponding statistical inference to answer the questions of interest. The statistician is fulfilling an absolute crucial role in the design of the experiments, the statistical and causal formulation of the question of interest, the estimation procedure and thereby the design of the software, the development of valid statistical tools for statistical inference, and benchmarking these statistical methods with respect to statistical performance (van der Laan and Rose, 2010).

References

Andersen, P., Borgan, O., Gill, R.D., and Keiding, N. (1993). *Statistical Models Based on Counting Processes.* Springer, New York.

Bickel, P.J., Klaassen, C.A.J., Ritov, Y., and Wellner, J.A. (1997). *Efficient and Adaptive Estimation for Semiparametric Models.* Springer, New York.

Chambaz, A. and van der Laan, M. (2010). *Targeting the Optimal Design in Randomized Clinical Trials with Binary Outcomes and no Covariate.* Technical Report 258, Division of Biostatistics, University of California, Berkeley, CA.

Chambaz, A. and van der Laan, M. (2011a). Targeting the optimal design in randomized clinical trials with binary outcomes and no covariate, simulation study. *International Journal of Biostatistics*, 7:1–30.

Chambaz, A. and van der Laan, M. (2011b). Targeting the optimal design in randomized clinical trials with binary outcomes and no covariate, theoretical study. *International Journal of Biostatistics*, 7:1–32.

Diaz, I. and van der Laan, M. (2012). *Sensitivity Analysis for Causal Inference Under Unmeasured Confounding and Measurement Error Problem.* Technical Report 303, Division of Biostatistics, University of California, Berkeley, CA.

Gill, R.D., van der Laan, M., and Wellner, J.A. (1995). Inefficient estimators of the bivariate survival function for three models. *Annales de l'Institut Henri Poincaré: Probabilités et Statistiques*, 31:545–597.

Györfi, L., Kohler, M., Krzyżak, A., and Walk, H. (2002). *A Distribution-free Theory of Nonparametric Regression.* Springer, New York.

Pearl, J. (2009). *Causality: Models, Reasoning, and Inference*, 2nd edition. Cambridge University Press, New York.

Petersen, M. and van der Laan, M. (2012). *A General Roadmap for the Estimation of Causal Effects.* Division of Biostatistics, University of California, Berkeley, CA.

Polley, E., Rose, S., and van der Laan, M. (2012). Super learning. In *Targeted Learning: Causal Inference for Observational and Experimental Data* (M. van der Laan and S. Rose, Eds.). Springer, New York.

Robins, J. and Rotnitzky, A. (1992). Recovery of information and adjustment for dependent censoring using surrogate markers. In *AIDS Epidemiology.* Birkhäuser, Basel.

Robins, J.M., Rotnitzky, A., and Scharfstein, D.O. (1999). Sensitivity analysis for selection bias and unmeasured confounding in missing data and causal inference models. In *Statistical Models in Epidemiology, the Environment and Clinical Trials.* Springer, New York.

Rosenblum, M. and van der Laan, M. (2011). Optimizing randomized trial designs to distinguish which subpopulations benefit from treatment. *Biometrika*, 98:845–860.

Rotnitzky, A., Scharfstein, D.O., Su, T.-L., and Robins, J.M. (2001). Methods for conducting sensitivity analysis of trials with potentially nonignorable competing causes of censoring. *Biometrics*, 57:103–113.

Scharfstein, D.O., Rotnitzky, A., and Robins, J.M. (1999). Adjusting for nonignorable drop-out using semiparametric non-response models (with discussion). *Journal of the American Statistical Association*, 94:1096–1146.

Starmans, R. (2012). Model, inference, and truth. In *Targeted Learning: Causal Inference for Observational and Experimental Data* (M. van der Laan and S. Rose, Eds.). Springer, New York.

van der Laan, M. (1996). *Efficient and Inefficient Estimation in Semiparametric Models.* Centrum voor Wiskunde en Informatica Tract 114, Amsterdam, The Netherlands.

van der Laan, M. (2008). *The Construction and Analysis of Adaptive Group Sequential Designs.* Technical Report 232, Division of Biostatistics, University of California, Berkeley, CA.

van der Laan, M. (2012). *Statistical Inference when using Data Adaptive Estimators of Nuisance Parameters.* Technical Report 302, Division of Biostatistics, University of California, Berkeley, CA.

van der Laan, M., Balzer, L., and Petersen, M. (2012). Adaptive matching in randomized trials and observational studies. *Journal of Statistical Research*, 46:113–156.

van der Laan, M. and Dudoit, S. (2003). *Unified Cross-validation Methodology for Selection Among Estimators and a General Cross-validated Adaptive Epsilon-net Estimator: Finite Sample Oracle Inequalities and Examples.* Technical Report 130, Division of Biostatistics, University of California, Berkeley, CA.

van der Laan, M., Dudoit, S., and van der Vaart, A. (2006). The cross-validated adaptive epsilon-net estimator. *Statistics & Decisions*, 24:373–395.

van der Laan, M., Polley, E., and Hubbard, A.E. (2007). Super learner. *Statistical Applications in Genetics and Molecular Biology*, 6:Article 25.

van der Laan, M. and Robins, J.M. (2003). *Unified Methods for Censored Longitudinal Data and Causality.* Springer, New York.

van der Laan, M. and Rose, S. (2010). Statistics ready for a revolution: Next generation of statisticians must build tools for massive data sets. *Amstat News*, 399:38–39.

van der Laan, M. and Rose, S. (2012). *Targeted Learning: Causal Inference for Observational and Experimental Data.* Springer, New York.

van der Laan, M. and Rubin, D.B. (2006). Targeted maximum likelihood learning. *International Journal of Biostatistics*, 2:Article 11.

van der Vaart, A., Dudoit, S., and van der Laan, M. (2006). Oracle inequalities for multi-fold cross-validation. *Statistics & Decisions*, 24:351–371.

van der Vaart, A. and Wellner, J.A. (1996). *Weak Convergence and Empirical Processes.* Springer, New York.

41

Statistical model building, machine learning, and the ah-ha moment

Grace Wahba
Statistics, Biostatistics and Medical Informatics, and Computer Sciences
University of Wisconsin, Madison, WI

Highly selected "ah-ha" moments from the beginning to the present of my research career are recalled — these are moments when the main idea just popped up instantaneously, sparking sequences of future research activity — almost all of these moments crucially involved discussions/interactions with others. Along with a description of these moments we give unsought advice to young statisticians. We conclude with remarks on issues relating to statistical model building/machine learning in the context of human subjects data.

41.1 Introduction: Manny Parzen and RKHS

Many of the "ah-ha" moments below involve Reproducing Kernel Hilbert Spaces (RKHS) so we begin there. My introduction to RKHS came while attending a class given by Manny Parzen on the lawn in front of the old Sequoia Hall at Stanford around 1963; see Parzen (1962).

For many years RKHS (Aronszajn, 1950; Wahba, 1990) were a little niche corner of research which suddenly became popular when their relation to Support Vector Machines (SVMs) became clear — more on that later. To understand most of the ah-ha moments it may help to know a few facts about RKHS which we now give.

An RKHS is a Hilbert space $\mathcal{H}$ where all of the evaluation functionals are bounded linear functionals. What this means is the following: let the domain of $\mathcal{H}$ be $\mathcal{T}$, and the inner product $< \cdot, \cdot >$. Then, for each $t \in \mathcal{T}$ there exists an element, call it K_t in $\mathcal{H}$, with the property $f(t) =< f, K_t >$ for all f in $\mathcal{H}$. K_t is known as the representer of evaluation at t. Let $K(s,t) =< K_s, K_t >$; this is clearly a positive definite function on $\mathcal{T} \otimes \mathcal{T}$. By the Moore–Aronszajn theorem, every RKHS is associated with a unique positive definite function, as we have just seen. Conversely, given a positive definite function, there exists a unique RKHS (which can be constructed from linear combinations of the $K_t, t \in \mathcal{T}$ and their limits). Given $K(s,t)$ we denote the associated RKHS as $\mathcal{H}_K$. Observe that nothing has been assumed concerning the domain $\mathcal{T}$. A second role of positive definite functions is as the covariance of a zero mean Gaussian stochastic process on $\mathcal{T}$. In a third role that we will come across later — let $O_1, \ldots, O_n$ be a set of n abstract objects. An $n \times n$ positive definite matrix can be used to assign pairwise squared Euclidean distances d_{ij} between O_i and O_j by $d_{ij} = K(i,i) + K(j,j) - 2K(i,j)$. In Sections 41.1.1–41.1.9 we go through some ah-ha moments involving RKHS, positive definite functions and pairwise distances/dissimilarities. Section 41.2 discusses sparse models and the lasso. Section 41.3 has some remarks involving complex interacting attributes, the "Nature-Nurture" debate, Personalized Medicine, Human subjects privacy and scientific literacy, and we end with conclusions in Section 41.4.

I end this section by noting that Manny Parzen was my thesis advisor, and Ingram Olkin was on my committee. My main advice to young statisticians is: choose your advisor and committee carefully, and be as lucky as I was.

41.1.1 George Kimeldorf and the representer theorem

Back around 1970 George Kimeldorf and I both got to spend a lot of time at the Math Research Center at the University of Wisconsin Madison (the one that later got blown up as part of the anti-Vietnam-war movement). At that time it was a hothouse of spline work, headed by Iso Schoenberg, Carl deBoor, Larry Schumaker and others, and we thought that smoothing splines would be of interest to statisticians. The smoothing spline of order m was the solution to: find f in the space of functions with square integral mth derivative to minimize

$$\sum_{i=1}^{n} \{y_i - f(t_i)\}^2 + \lambda \int_0^1 \{f^{(m)}(t)\}^2 \mathrm{d}t, \tag{41.1}$$

where $t_1, \ldots, t_n \in [0,1]$. Professor Schoenberg many years ago had characterized the solution to this problem as a piecewise polynomial of degree $2m-1$ satisfying some boundary and continuity conditions.

Our ah-ha moment came when we observed that the space of functions with square integrable mth derivative on $[0,1]$ was an RKHS with seminorm $\|Pf\|$ defined by

$$\|Pf\|^2 = \int_0^1 \{f^{(m)}(t)\}^2 \mathrm{d}t$$

and with an associated $K(s,t)$ that we could figure out. (A seminorm is exactly like a norm except that it has a non-trivial null space, here the null space of this seminorm is the span of the polynomials of degree $m-1$ or less.) Then by replacing $f(t)$ by $< K_t, f >$ it was not hard to show by a very simple geometric argument that the minimizer of (41.1) was in the span of the $K_{t_1}, \ldots, K_{t_n}$ and a basis for the null space of the seminorm. But furthermore, the very same geometrical argument could be used to solve the more general problem: find $f \in \mathcal{H}_K$, an RKHS, to minimize

$$\sum_{i=1}^{n} C(y_i, L_i f) + \lambda \|Pf\|_K^2, \tag{41.2}$$

where $C(y_i, L_i f)$ is convex in $L_i f$, with L_i a bounded linear functional in $\mathcal{H}_K$ and $\|Pf\|_K^2$ a seminorm in $\mathcal{H}_K$. A bounded linear functional is a linear functional with a representer in $\mathcal{H}_K$, i.e., there exists $\eta_i \in \mathcal{H}_K$ such that $L_i f =< \eta_i, f >$ for all $f \in \mathcal{H}_K$. The minimizer of (41.2) is in the span of the representers η_i and a basis for the null space of the seminorm. That is known as the representer theorem, which turned out to be a key to fitting (mostly continuous) functions in an infinite-dimensional space, given a finite number of pieces of information. There were two things I remember about our excitement over the result: one of us, I'm pretty sure it was George, thought the result was too trivial and not worthwhile to submit, but submit it we did and it was accepted (Kimeldorf and Wahba, 1971) without a single complaint, within three weeks. I have never since then had another paper accepted by a refereed journal within three weeks and without a single complaint. Advice: if you think it is worthwhile, submit it.

41.1.2 Svante Wold and leaving-out-one

Following Kimeldorf and Wahba, it was clear that for practical use, a method was needed to choose the smoothing or tuning parameter λ in (41.1). The natural goal was to minimize the mean square error over the function f, for which its values at the data points would be the proxy. In 1974 Svante Wold visited Madison, and we got to mulling over how to choose λ. It so happened that Mervyn Stone gave a colloquium talk in Madison, and Svante and I were sitting next to each other as Mervyn described using leaving-out-one to decide on the degree of a polynomial to be used in least squares regression. We looked at each other at that very minute and simultaneously said something, I think it was "ah-ha," but possibly "Eureka." In those days computer time was \$600/hour and Svante wrote a computer program to demonstrate that leaving-out-one did a good job. It took the entire Statistics department's computer money for an entire month to get the results in Wahba and Wold (1975). Advice: go to the colloquia, sit next to your research pals.

41.1.3 Peter Craven, Gene Golub and Michael Heath and GCV

After much struggle to prove some optimality properties of leaving-out-one, it became clear that it couldn't be done in general. Considering the data model $y = f + \epsilon$, where $y = (y_1, \ldots, y_n)^\top, f = (f(t_1), \ldots, f(t_n))^\top$ and $\epsilon = (\epsilon_1, \ldots, \epsilon_n)^\top$ is a zero mean i.i.d. Gaussian random vector, the information in the data is unchanged by multiplying left and right hand side by an orthogonal matrix, since $\Gamma\epsilon$ with Γ orthogonal is still white Gaussian noise. But leaving-out-one can give you a different answer. To explain, we define the influence matrix: let f_λ be the minimizer of (41.1) when C is sum of squares. The influence matrix relates the data to the prediction of the data, $f_\lambda = A(\lambda)y$, where $f_\lambda = (f_\lambda(t_1), \ldots, f_\lambda(t_n))$. A heuristic argument fell out of the blue, probably in an attempt to explain some things to students, that rotating the data so that the influence matrix was constant down the diagonal, was the trick. The result was that instead of leaving-out-one, one should minimize the GCV function

$$V(\lambda) = \frac{\sum_{i=1}^{n} \{y_i - f(t_i)\}^2}{[\text{trace}\{I - A(\lambda)\}]^2}$$

(Craven and Wahba, 1979; Golub et al., 1979). I was on sabbatical at Oxford in 1975 and Gene was at ETH visiting Peter Huber, who had a beautiful house in Klosters, the fabled ski resort. Peter invited Gene and me up for the weekend, and Gene just wrote out the algorithm in Golub et al. (1979) on the train from Zürich to Klosters while I snuck glances at the spectacular scenery. Gene was a much loved mentor to lots of people. He was born on February 29, 1932 and died on November 16, 2007. On February 29 and March 1, 2008 his many friends held memorial birthday services at Stanford and 30 other locations around the world. Ker-Chau Li (Li, 1985, 1986, 1987b) and others later proved optimality properties of the GCV and popular codes in R will compute splines and other fits using GCV to estimate λ and other important tuning parameters. Advice: pay attention to important tuning parameters since the results can be very sensitive to them. Advice: appreciate mentors like Gene if you are lucky enough to have such great mentors.

41.1.4 Didier Girard, Mike Hutchinson, randomized trace and the degrees of freedom for signal

Brute force calculation of the trace of the influence matrix $A(\lambda)$ can be daunting to compute directly for large n. Let f_λ^y be the minimizer of (41.1) with the data vector y and let $f_\lambda^{y+\delta}$ be the minimizer of (41.1) given the perturbed data $y + \delta$. Note that

$$\delta^\top (f_\lambda^y - f_\lambda^{y+\delta}) = \delta^\top A(\lambda)(y + \delta) - A(\lambda)(y) = \sum_{i,j=1}^{n} \delta_i \delta_j a_{ij},$$

where δ_i and a_{ij} are the components of δ and $A(\lambda)$ respectively. If the perturbations are i.i.d. with variance 1, then this sum is an estimate of trace $A(\lambda)$. This simple idea was proposed in Girard (1989) and Hutchinson (1989), with further theory in Girard (1991). It was a big ah-ha when I saw these papers because further applications were immediate. In Wahba (1983), p. 139, I defined the trace of $A(\lambda)$ as the "equivalent degrees of freedom for signal," by analogy with linear least squares regression with $p < n$ where the influence matrix is a rank p projection operator. The degrees of freedom for signal is an important concept in linear and nonlinear nonparametric regression, and it was a mistake to hide it inconspicuously in Wahba (1983). Later Brad Efron (2004) gave an alternative definition of degrees of freedom for signal. The definition in Wahba (1983) depends only on the data; Efron's is essentially an expected value. Note that in (41.1),

$$\text{trace}\{A(\lambda)\} = \sum_{i=1}^{n} \frac{\partial \hat{y}_i}{\partial y_i},$$

where $\hat{y}_i$ is the predicted value of y_i. This definition can reasonably be applied to a problem with a nonlinear forward operator (that is, that maps data onto the predicted data) when the derivatives exist, and the randomized trace method is reasonable for estimating the degrees of freedom for signal, although care should be taken concerning the size of δ. Even when the derivatives don't exist the randomized trace can be a reasonable way of getting at the degrees of freedom for signal; see, e.g., Wahba et al. (1995).

41.1.5 Yuedong Wang, Chong Gu and smoothing spline ANOVA

Sometime in the late 80s or early 90s I heard Graham Wilkinson expound on ANOVA (Analysis of Variance), where data was given on a regular d-dimensional grid, viz.

$$y_{ijk}, \quad t_{ijk}, \quad i = 1, \ldots, I, \quad j = 1, \ldots, J, \quad k = 1, \ldots, K,$$

for $d = 3$ and so forth. That is, the domain is the Cartesian product of several one-dimensional grids. Graham was expounding on how fitting a model from observations on such a domain could be described as set of orthogonal projections based on averaging operators, resulting in main effects, two factor interactions, etc. "Ah-ha" I thought, we should be able to do exactly same thing and more where the domain is the Cartesian product $\mathcal{T} = \mathcal{T}_1 \otimes \cdots \otimes \mathcal{T}_d$ of d arbitrary domains. We want to fit functions on $\mathcal{T}$, with main effects (functions of one variable), two factor interactions (functions of two variables), and possibly more terms given scattered observations, and we just need to define averaging operators for each $\mathcal{T}_\alpha$.

Brainstorming with Yuedong Wang and Chong Gu fleshed out the results. Let $\mathcal{H}^\alpha, \alpha = 1, \ldots, d$ be d RKHSs with domains $\mathcal{T}_\alpha$, each $\mathcal{H}^\alpha$ containing the constant functions. $\mathcal{H} = \mathcal{H}^1 \otimes \cdots \otimes \mathcal{H}^d$ is an RKHS with domain $\mathcal{T}$. For each $\alpha = 1, \ldots, d$, construct a probability measure $\mathrm{d}\mu_\alpha$ on $\mathcal{T}_\alpha$, with the property that the symbol $(\mathcal{E}_\alpha f)(t)$, the averaging operator, defined by

$$(\mathcal{E}_\alpha f)(t) = \int_{\mathcal{T}(\alpha)} f(t_1, \ldots, t_d) \mathrm{d}\mu_\alpha(t_\alpha),$$

is well defined and finite for every $f \in \mathcal{H}$ and $t \in \mathcal{T}$. Consider the decomposition of the identity operator:

$$I = \prod_\alpha (\mathcal{E}_\alpha + (I - \mathcal{E}_\alpha)) = \prod_\alpha \mathcal{E}_\alpha + \sum_\alpha (I - \mathcal{E}_\alpha) \prod_{\beta \neq \alpha} \mathcal{E}_\beta$$
$$+ \sum_{\alpha < \beta} (I - \mathcal{E}_\alpha)(I - \mathcal{E}_\beta) \prod_{\gamma \neq \alpha, \beta} \mathcal{E}_\gamma + \cdots + \prod_\alpha (I - \mathcal{E}_\alpha). \quad (41.3)$$

This decomposition of the identity then always generates a unique (ANOVA-like) decomposition of $f \in \mathcal{H}$ of the form

$$f(t) = \mu + \sum_\alpha f_\alpha(t_\alpha) + \sum_{\alpha < \beta} f_{\alpha\beta}(t_\alpha, t_\beta) + \sum_{\alpha < \beta < \gamma} f_{\alpha\beta\gamma}(t_\alpha, t_\beta, t_\gamma) + \cdots \quad (41.4)$$

where the expansion is unique and (usually) truncated in some manner in practice. Here

$$\mu = \left(\prod_\alpha \mathcal{E}_\alpha \right) f, \quad f_\alpha = \left((I - \mathcal{E}_\alpha) \prod_{\beta \neq \alpha} \mathcal{E}_\beta \right) f,$$

$$f_{\alpha\beta} = \left((I - \mathcal{E}_\alpha)(I - \mathcal{E}_\beta) \prod_{\gamma \neq \alpha, \beta} \mathcal{E}_\gamma \right) f,$$

etc., are the mean, main effects, two factor interactions, etc. The result is usually called an SS ANOVA model, although the components are not limited to splines. For details on how to fit the terms see Gu (2002), Gu and Wahba (1993), Wahba (1990), Wang (2011) and the `assist` and `gss` codes in R. Note that nothing has been said about $\mathcal{T}$ and very little regarding $\mathcal{H}^\alpha$, other than that the constant functions are in each of the constituent spaces and averaging operators can be defined.

41.1.6 Vladimir Vapnik, the mystery caller and the SVM

The AMS–IMS–SIAM Joint Summer Research Conference on Adaptive Selection of Models and Statistical Procedures was held on the campus of Mount Holyoke College in South Hadley, Massachusetts on Sunday, June 23 through Thursday, June 27, 1996. On one of those fine days a session met on a grassy lawn of Mount Holyoke College, when Vladimir Vapnik and I were both invited speakers. I talked first, and noted how the solution to the optimization problem (41.2) led to a function involving the span of the representers. Vladimir spoke next, describing the support vector machine (SVM), a well known and highly successful method for classification, describing something he called the "kernel trick." He exhibited an SVM that was fitted in the span of representers in an RKHS. We will explain the SVM in a moment, but the original SVM, as proposed by Vapnik and coworkers (Vapnik, 1995) was derived from an argument nothing like what I am about to give. Somewhere during Vladimir's talk, an unknown voice towards the back of the audience called out "That looks like Grace Wahba's stuff." It looked obvious that the SVM as proposed by Vapnik with the "kernel trick," could be obtained as the the solution to the optimization problem of (41.2) with $C(y_i, L_i f)$ replaced by the so called hinge function, $(1 - y_i f(t_i))_+$, where $(\tau)_+ = \tau$ if $\tau > 0$ and 0 otherwise. Each data point is coded as ± 1 according as it came from the "plus" class or the "minus" class. For technical reasons the null space of the penalty function consists at most of the constant functions. Thus it follows that the solution is in the span of the representers K_{t_i} from the chosen RKHS plus possibly a constant function. Yi Lin and coworkers (Lin et al., 2002a,b) showed that the SVM was estimating the sign of the log odds ratio, just what is needed for two class classification. The SVM may be compared to the case where one desires to estimate the probability that an object is in the plus class. If one begins with the penalized log likelihood of the Bernoulli distribution and codes the data as ± 1 instead of the usual coding as 0 or 1, then we have the same optimization problem with $C(y_i, f(t_i)) = \ln\{1 + e^{-y_i f(t_i)}\}$ instead of $(1 - y_i f(t_i))_+$ with solution in the same finite dimensional space, but it is estimating the log odds-ratio, as opposed to the sign of the log odds ratio. It was actually a big deal that the SVM could be directly compared with penalized likelihood with Bernoulli data, and it provided a pathway for statisticians and computer scientists to breach a major divide between them on the subject of classification, and to understand each others' work.

For many years before the Hadley meeting, Olvi Mangasarian and I would talk about what we were doing in classification, neither of us having any understanding of what the other was doing. Olvi complained that the statisticians dismissed his work, but it turned out that what he was doing was related to the SVM and hence perfectly legitimate not to mention interesting, from a classical statistical point of view. Statisticians and computer scientists have been on the same page on classification ever since.

It is curious to note that several patents have been awarded for the SVM. One of the early ones, issued on July 15, 1997 is "5649068 Pattern recognition system using support vectors." I'm guessing that the unknown volunteer was David Donoho. Advice: keep your eyes open to synergies between apparently disparate fields.

41.1.7 Yoonkyung Lee, Yi Lin and the multi-category SVM

For classification, when one has $k > 2$ classes it is always possible to apply an SVM to compare membership in one class versus the rest of the k classes, running through the algorithm k times. In the early 2000s there were many papers on one-vs-rest, and designs for subsets vs. other subsets, but it is possible to generate examples where essentially no observations will be identified as being in certain classes. Since one-vs-rest could fail in certain circumstances it was something of an open question how to do multi-category SVMs in one optimization problem that did not have this problem. Yi Lin, Yoonkyung Lee and I were sitting around shooting the breeze and one of us said "how about a sum-to-zero constraint?" and the other two said "ah-ha," or at least that's the way I remember it. The idea is to code the labels as k-vectors, with a 1 in the rth position and $-1/(k-1)$ in the $k-1$ other positions for a training sample in class r. Thus, each observation vector satisfies the sum-to-zero constraint. The idea was to fit a vector of functions satisfying the same sum-to-zero constraint. The multi-category SVM fit estimates $f(t) = (f_1(t), \ldots, f_k(t))$, $t \in \mathcal{T}$ subject to the sum-to-zero constraint everywhere and the classification for a subject with attribute vector t is just the index of the largest component of the estimate of $f(t)$. See Lee and Lee (2003) and Lee et al. (2004a,b). Advice: shooting the breeze is good.

41.1.8 Fan Lu, Steve Wright, Sunduz Keles, Hector Corrada Bravo, and dissimilarity information

We return to the alternative role of positive definite functions as a way to encode pairwise distance observations. Suppose we are examining n objects $O_1, \ldots, O_n$ and are given some noisy or crude observations on their pairwise distances/dissimilarities, which may not satisfy the triangle inequality. The goal is to embed these objects in a Euclidean space in such a way as to respect the pairwise dissimilarities as much as possible. Positive definite matrices encode pairwise squared distances d_{ij} between O_i and O_j as

$$d_{ij}(K) = K(i,i) + K(j,j) - 2K(i,j), \tag{41.5}$$

and, given a non-negative definite matrix of rank $d \leq n$, can be used to embed the n objects in a Euclidean space of dimension d, centered at 0 and unique up to rotations. We seek a K which respects the dissimilarity information d_{ij}^{obs} while constraining the complexity of K by

$$\min_{K \in S_n} \sum |d_{ij}^{\text{obs}} - d_{ij}(K)| + \lambda \,\text{trace}(K), \tag{41.6}$$

where S_n is the convex cone of symmetric positive definite matrices. I looked at this problem for an inordinate amount of time seeking an analytic solution but after a conversation with Vishy (S.V.N. Vishwanathan) at a meeting in Rotterdam in August of 2003 I realized it wasn't going to happen. The ah-ha moment came about when I showed the problem to Steve Wright, who right off said it could be solved numerically using recently developed convex cone software. The result so far is Corrada Bravo et al. (2009) and Lu et al. (2005). In Lu et al. (2005) the objects are protein sequences and the pairwise distances are BLAST scores. The fitted kernel K had three eigenvalues that contained about 95% of the trace, so we reduced K to a rank 3 matrix by truncating the smaller eigenvalues. Clusters of four different kinds of proteins were readily separated visually in three-d plots; see Lu et al. (2005) for the details. In Corrada Bravo et al. (2009) the objects are persons in pedigrees in a demographic study and the distances are based on Malecot's kinship coefficient, which defines a pedigree dissimilarity measure. The resulting kernel became part of an SS ANOVA model with other attributes of persons, and the model estimates a risk related to an eye disease. Advice: find computer scientist friends.

41.1.9 Gábor Székely, Maria Rizzo, Jing Kong and distance correlation

The last ah-ha experience that we report is similar to that involving the randomized trace estimate of Section 41.1.4, i.e., the ah-ha moment came about upon realizing that a particular recent result was very relevant to what we were doing. In this case Jing Kong brought to my attention the important paper of Gábor Székely and Maria Rizzo (Székely and Rizzo, 2009). Briefly, this paper considers the joint distribution of two random vectors, X and Y, say, and provides a test, called distance correlation that it factors so that the two random vectors are independent. Starting with n observations from the joint distribution, let $\{A_{ij}\}$ be the collection of double-centered pairwise distances among the $\binom{n}{2}$ X components, and similarly for $\{B_{ij}\}$. The statistic, called distance correlation, is the analogue of the usual sample correlation between the A's and B's. The special property of the test is that it is justified for X and Y in Euclidean p and q space for arbitrary p and q with no further distributional assumptions. In a demographic study involving pedigrees (Kong et al., 2012), we observed that pairwise distance in death age between close relatives was less than that of unrelated age cohorts. A mortality risk score for four lifestyle factors and another score for a group of diseases was developed via SS ANOVA modeling, and significant distance correlation was found between death ages, lifestyle factors and family relationships, raising more questions than it answers regarding the "Nature-Nurture" debate (relative role of genetics and other attributes).

We take this opportunity to make a few important remarks about pairwise distances/dissimilarities, primarily how one measures them can be important, and getting the "right" dissimilarity can be 90% of the problem. We remark that family relationships in Kong et al. (2012) were based on a monotone function of Malecot's kinship coefficient that was different from the monotone function in Corrada Bravo et al. (2009). Here it was chosen to fit in with the different way the distances were used. In (41.6), the pairwise dissimilarities can be noisy, scattered, incomplete and could include subjective distances like "very close, close.. " etc. not even satisfying the triangle inequality. So there is substantial flexibility in choosing the dissimilarity measure with respect to the particular scientific context of the problem. In Kong et al. (2012) the pairwise distances need to be a complete set, and be Euclidean (with some specific metric exceptions). There is still substantial choice in choosing the definition of distance, since any linear transformation of a Euclidean coordinate system defines a Euclidean distance measure. Advice: think about how you measure distance or dissimilarity in any problem involving pairwise relationships, it can be important.

41.2 Regularization methods, RKHS and sparse models

The optimization problems in RKHS are a rich subclass of what can be called regularization methods, which solve an optimization problem which trades fit to the data versus complexity or constraints on the solution. My first encounter with the term "regularization" was Tikhonov (1963) in the context of finding numerical solutions to integral equations. There the L_i of (41.2) were noisy integrals of an unknown function one wishes to reconstruct, but the observations only contained a limited amount of information regarding the unknown function. The basic and possibly revolutionary idea at the time was to find a solution which involves fit to the data while constraining the solution by what amounted to an RKHS seminorm, $(\int \{f''(t)\}^2 \mathrm{d}t)$ standing in for the missing information by an assumption that the solution was "smooth" (O'Sullivan, 1986; Wahba, 1977). Where once RKHS were a niche subject, they are now a major component of the statistical model building/machine learning literature.

However, RKHS do not generally provide sparse models, that is, models where a large number of coefficients are being estimated but only a small but unknown number are believed to be non-zero. Many problems in the "Big Data" paradigm are believed to have, or want to have sparse solutions, for example, genetic data vectors that may have many thousands of components and a modest number of subjects, as in a case-control study. The most popular method for ensuring sparsity is probably the lasso (Chen et al., 1998; Tibshirani, 1996). Here a very large dictionary of basis functions $(B_j(t), j = 1, 2, \ldots)$ is given and the unknown function is estimated as $f(t) = \sum_j \beta_j B_j(t)$ with the penalty functional $\lambda \sum_j |\beta_j|$ replacing an RKHS square norm. This will induce many zeroes in the β_j, depending, among other things on the size of λ. Since then, researchers have commented that there is a "zoo" of proposed variants of sparsity-inducing penalties, many involving assumptions on structures in the data; one popular example is Yuan and Lin (2006). Other recent models involve mixtures of RKHS and sparsity-inducing penalty functionals. One of our contributions to this "zoo" deals with the situation where the data vectors amount to very large "bar codes," and it is desired to find patterns in the bar codes relevant to some outcome. An innovative algorithm which deals with a humongous number of interacting patterns assuming that only a small number of coefficients are non-zero is given in Shi et al. (2012), Shi et al. (2008) and Wright (2012).

As is easy to see here and in the statistical literature, the statistical modeler has overwhelming choices in modeling tools, with many public codes available in the software repository R and elsewhere. In practice these choices must be made with a serious understanding of the science and the issues motivating the data collection. Good collaborations with subject matter researchers can lead to the opportunity to participate in real contributions to the science. Advice: learn absolutely as much as you can about the subject matter of the data that you contemplate analyzing. When you use "black boxes" be sure you know what is inside them.

41.3 Remarks on the nature-nurture debate, personalized medicine and scientific literacy

We and many other researchers have been developing methods for combining scattered, noisy, incomplete, highly heterogenous information from multiple sources with interacting variables to predict, classify, and determine patterns of attributes relevant to a response, or more generally multiple correlated responses.

Demographic studies, clinical trials, and ad hoc observational studies based on electronic medical records, which have familial (Corrada Bravo et al., 2009; Kong et al., 2012), clinical, genetic, lifestyle, treatment and other attributes can be a rich source of information regarding the Nature-Nurture debate, as well informing Personalized Medicine, two popular areas reflecting much activity. As large medical systems put their records in electronic form interesting problems arise as to how to deal with such unstructured data, to relate subject attributes to outcomes of interest. No doubt a gold mine of information is there, particularly with respect to how the various attributes interact. The statistical modeling/machine learning community continues to create and improve tools to deal with this data flood, eager to develop better and more efficient modeling methods, and regularization and dissimilarity methods will no doubt continue to play an important role in numerous areas of scientific endeavor. With regard to human subjects studies, a limitation is the problem of patient confidentiality — the more attribute information available to explore for its relevance, the trickier the privacy issues, to the extent that de-identified data can actually be identified. It is important, however, that statisticians be involved from the very start in the design of human subjects studies.

With health related research, the US citizenry has some appreciation of scientific results that can lead to better health outcomes. On the other hand any scientist who reads the newspapers or follows present day US politics is painfully aware that a non-trivial portion of voters and the officials they elect have little or no understanding of the scientific method. Statisticians need to participate in the promotion of increased scientific literacy in our educational establishment at all levels.

41.4 Conclusion

In response to the invitation from COPSS to contribute to their 50th Anniversary Celebration, I have taken a tour of some exciting moments in my career, involving RKHS and regularization methods, pairwise dissimilarities and distances, and lasso models, dispensing un-asked for advice to new researchers along the way. I have made a few remarks concerning the richness of models based on RKHS, as well as models involving sparsity-inducing penalties with some remarks involving the Nature-Nurture Debate and Personalized Medicine. I end this contribution with thanks to my many coauthors — identified here or not — and to my terrific present and former students. Advice: Treasure your collaborators! Have great students!

Acknowledgements

This work was supported in part by the National Science Foundation Grant DMS 0906818 and the National Institutes of Health Grant R01 EY09946. The author thanks Didier Girard for pointing out a mistake and also noting that his 1987 report entitled "Un algorithme simple et rapide pour la validation croisée généralisée sur des problèmes de grande taille" (Rapport RR669–M, Informatique et mathématiques appliquées de Grenoble, France) predates the reference Girard (1989) given here.

References

Aronszajn, N. (1950). Theory of reproducing kernels. *Transactions of the American Mathematical Society*, 68:337–404.

Chen, S., Donoho, D., and Saunders, M. (1998). Atomic decomposition by basis pursuit. *SIAM Journal of Scientific Computing*, 20:33–61.

Corrado Bravo, H., Leeb, K.E., Klein, B.E., Klein, R., Iyengar, S.K., and Wahba, G. (2009). Examining the relative influence of familial, genetic and environmental covariate information in flexible risk models. *Pro-*

ceedings of the National Academy of Sciences, 106:8128–8133. Open Source at www.pnas.org/content/106/20/8128.full.pdf+html, PMCID: PMC 2677979.

Craven, P. and Wahba, G. (1979). Smoothing noisy data with spline functions: Estimating the correct degree of smoothing by the method of generalized cross-validation. *Numerische Mathematik*, 31:377–403.

Efron, B. (2004). The estimation of prediction error: Covariance penalties and cross-validation. *Journal of the American Statistical Association*, 99:619–632.

Girard, D. (1989). A fast "Monte Carlo cross-validation" procedure for large least squares problems with noisy data. *Numerische Mathematik*, 56:1–23.

Girard, D. (1991). Asymptotic optimality of the fast randomized versions of GCV and C_L in ridge regression and regularization. *The Annals of Statistics*, 19:1950–1963.

Golub, G., Heath, M., and Wahba, G. (1979). Generalized cross validation as a method for choosing a good ridge parameter. *Technometrics*, 21:215–224.

Gu, C. (2002). *Smoothing Spline ANOVA Models.* Springer, New York.

Gu, C. and Wahba, G. (1993). Smoothing spline ANOVA with component-wise Bayesian "confidence intervals." *Journal of Computational and Graphical Statistics*, 2:97–117.

Hutchinson, M. (1989). A stochastic estimator for the trace of the influence matrix for Laplacian smoothing splines. *Communications in Statistics — Simulations*, 18:1059–1076.

Kimeldorf, G. and Wahba, G. (1971). Some results on Tchebycheffian spline functions. *Journal of Mathematical Analysis and its Applications*, 33:82–95.

Kong, J., Klein, B.E., Klein, R., Lee, K.E., and Wahba, G. (2012). Using distance correlation and smoothing spline ANOVA to assess associations of familial relationships, lifestyle factors, diseases and mortality. *PNAS*, pp. 20353–20357.

Lee, Y. and Lee, C.-K. (2003). Classification of multiple cancer types by multi-category support vector machines using gene expression data. *Bioinformatics*, 19:1132–1139.

Lee, Y., Lin, Y., and Wahba, G. (2004). Multicategory support vector machines, theory, and application to the classification of microarray data and satellite radiance data. *Journal of the American Statistical Association*, 99:67–81.

Lee, Y., Wahba, G., and Ackerman, S. (2004). Classification of satellite radiance data by multicategory support vector machines. *Journal of Atmospheric and Oceanic Technology*, 21:159–169.

Li, K.C. (1985). From Stein's unbiased risk estimates to the method of generalized cross-validation. *The Annals of Statistics*, 13:1352–1377.

Li, K.C. (1986). Asymptotic optimality of C_L and generalized cross validation in ridge regression with application to spline smoothing. *The Annals of Statistics*, 14:1101–1112.

Li, K.C. (1987). Asymptotic optimality for C_p, C_L, cross-validation and generalized cross validation: Discrete index set. *The Annals of Statistics*, 15:958–975.

Lin, Y., Lee, Y., and Wahba, G. (2002). Support vector machines for classification in nonstandard situations. *Machine Learning*, 46:191–202.

Lin, Y., Wahba, G., Zhang, H., and Lee, Y. (2002). Statistical properties and adaptive tuning of support vector machines. *Machine Learning*, 48:115–136.

Lu, F., Keleş, S., Wright, S.J., and Wahba, G. (2005). A framework for kernel regularization with application to protein clustering. *Proceedings of the National Academy of Sciences*, 102:12332–12337. Open Source at `www.pnas.org/content/102/35/12332`, PMCID: PMC118947.

O'Sullivan, F. (1986). A statistical perspective on ill-posed inverse problems. *Statistical Science*, 1:502–527.

Parzen, E. (1962). An approach to time series analysis. *The Annals of Mathematical Statistics*, 32:951–989.

Shi, W., Wahba, G., Irizarry, R.A., Corrada Bravo, H., and Wright, S.J. (2012). The partitioned LASSO-patternsearch algorithm with application to gene expression data. *BMC Bioinformatics*, 13-98.

Shi, W., Wahba, G., Wright, S.J., Lee, K., Klein, R., and Klein, B. (2008). LASSO-patternsearch algorithm with application to ophthalmology and genomic data. *Statistics and its Interface*, 1:137–153.

Székely, G. and Rizzo, M. (2009). Brownian distance covariance. *The Annals of Applied Statistics*, 3:1236–1265.

Tibshirani, R.J. (1996). Regression shrinkage and selection via the lasso. *Journal of the Royal Statistical Society, Series B*, 58:267–288.

Tikhonov, A. (1963). Solution of incorrectly formulated problems and the regularization method. *Soviet Mathematics Doklady*, 4:1035–1038.

Vapnik, V. (1995). *The Nature of Statistical Learning Theory.* Springer, Berlin.

Wahba, G. (1977). Practical approximate solutions to linear operator equations when the data are noisy. *SIAM Journal of Numerical Analysis*, 14:651–667.

Wahba, G. (1983). Bayesian "confidence intervals" for the cross-validated smoothing spline. *Journal of the Royal Statistical Society, Series B*, 45:133–150.

Wahba, G. (1990). *Spline Models for Observational Data.* SIAM. CBMS–NSF Regional Conference Series in Applied Mathematics, vol. 59.

Wahba, G., Johnson, D.R., Gao, F., and Gong, J. (1995). Adaptive tuning of numerical weather prediction models: Randomized GCV in three and four dimensional data assimilation. *Monthly Weather Review*, 123:3358–3369.

Wahba, G. and Wold, S. (1975). A completely automatic French curve. *Communications in Statistics*, 4:1–17.

Wang, Y. (2011). *Smoothing Splines: Methods and Applications.* Chapman & Hall, London.

Wright, S.J. (2012). Accelerated block-coordinate relaxation for regularized optimization. *SIAM Journal of Optimization*, 22:159–186. Preprint and software available at `http://pages.cs.wisc.edu/~swright/LPS/`.

Yuan, M. and Lin, Y. (2006). Model selection and estimation in regression with grouped variables. *Journal of the Royal Statistical Society, Series B*, 68:49–67.

42

In praise of sparsity and convexity

Robert J. Tibshirani
Department of Statistics
Stanford University, Stanford, CA

To celebrate the 50th anniversary of the COPSS, I present examples of exciting developments of sparsity and convexity, in statistical research and practice.

42.1 Introduction

When asked to reflect on an anniversary of their field, scientists in most fields would sing the praises of their subject. As a statistician, I will do the same. However, here the praise is justified! Statistics is a thriving discipline, more and more an essential part of science, business and societal activities. Class enrollments are up — it seems that everyone wants to be a statistician — and there are jobs everywhere. The field of machine learning, discussed in this volume by my friend Larry Wasserman, has exploded and brought along with it the computational side of statistical research. Hal Varian, Chief Economist at `Google`, said "I keep saying that the sexy job in the next 10 years will be statisticians. And I'm not kidding." Nate Silver, creator of the *New York Times* political forecasting blog "538" was constantly in the news and on talk shows in the runup to the 2012 US election. Using careful statistical modelling, he forecasted the election with near 100% accuracy (in contrast to many others). Although his training is in economics, he (proudly?) calls himself a statistician. When meeting people at a party, the label "Statistician" used to kill one's chances of making a new friend. But no longer!

In the midst of all this excitement about the growing importance of statistics, there are fascinating developments within the field itself. Here I will discuss one that has been the focus my research and that of many other statisticians.

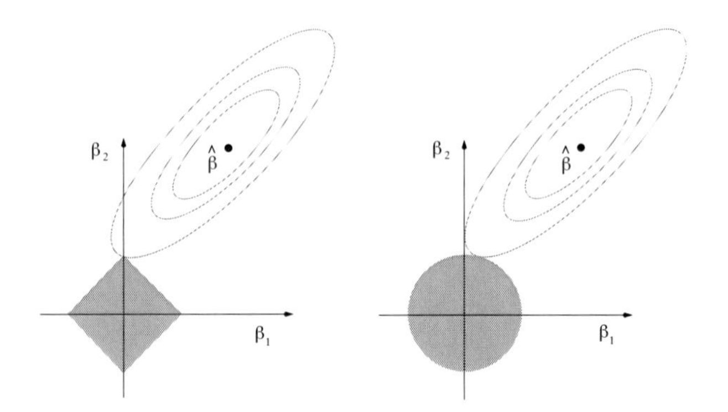

FIGURE 42.1
Estimation picture for the lasso (left) and ridge regression (right). Shown are contours of the error and constraint functions. The solid areas are the constraint regions $|\beta_1| + |\beta_2| \leq t$ and $\beta_1^2 + \beta_2^2 \leq t^2$, respectively, while the ellipses are the contours of the least squares error function. The sharp corners of the constraint region for the lasso yield sparse solutions. In high dimensions, sparsity arises from corners and edges of the constraint region.

42.2 Sparsity, convexity and ℓ_1 penalties

One of the earliest proposals for using ℓ_1 or absolute-value penalties, was the lasso method for penalized regression. Given a linear regression with predictors x_{ij} and response values y_i for $i \in \{1, \ldots, N\}$ and $j \in \{1, \ldots, p\}$, the lasso solves the ℓ_1-penalized regression

$$\text{minimize}_\beta \left\{ \frac{1}{2} \sum_{i=1}^{N} \left(y_i - \sum_{j=1}^{p} x_{ij}\beta_j \right)^2 + \lambda \sum_{j=1}^{p} |\beta_j| \right\}.$$

This is equivalent to minimizing the sum of squares with constraint $|\beta_1| + \cdots + |\beta_p| \leq s$. It is similar to ridge regression, which has constraint $\beta_1^2 + \cdots + \beta_p^2 \leq s$. Because of the form of the ℓ_1 penalty, the lasso does variable selection and shrinkage; while ridge regression, in contrast, only shrinks. If we consider a more general penalty of the form $(\beta_1^q + \cdots + \beta_p^q)^{1/q}$, then the lasso uses $q = 1$ and ridge regression has $q = 2$. Subset selection emerges as $q \to 0$, and the lasso corresponds to the smallest value of q (i.e., closest to subset selection) that yields a convex problem. Figure 42.1 gives a geometric view of the lasso and ridge regression.

The lasso and ℓ_1 penalization have been the focus of a great deal of work recently. Table 42.1, adapted from Tibshirani (2011), gives a sample of this work.

TABLE 42.1
A sampling of generalizations of the lasso.

Method	Authors
Adaptive lasso	Zou (2006)
Compressive sensing	Donoho (2004), Candès (2006)
Dantzig selector	Candès and Tao (2007)
Elastic net	Zou and Hastie (2005)
Fused lasso	Tibshirani et al. (2005)
Generalized lasso	Tibshirani and Taylor (2011)
Graphical lasso	Yuan and Lin (2007b), Friedman et al. (2010)
Grouped lasso	Yuan and Lin (2007a)
Hierarchical interaction models	Bien et al. (2013)
Matrix completion	Candès and Tao (2009), Mazumder et al. (2010)
Multivariate methods	Joliffe et al. (2003), Witten et al. (2009)
Near-isotonic regression	Tibshirani et al. (2011)

The original motivation for the lasso was interpretability: it is an alternative to subset regression for obtaining a sparse model. Since that time, two unforeseen advantages of convex ℓ_1-penalized approaches have emerged: *Computational* and *statistical* efficiency. On the computational side, convexity of the problem and sparsity of the final solution can be used to great advantage. When most parameter estimates are zero in the solution, those parameters can be handled with minimal cost in the search for the solution. Powerful and scalable techniques for convex optimization can be unleashed on the problem, allowing the solution of very large problems. One particularly promising approach is coordinate descent (Fu, 1998; Friedman et al., 2007, 2010), a simple one-at-a-time method that is well-suited to the separable lasso penalty. This method is simple and flexible, and can also be applied to a wide variety of other ℓ_1-penalized generalized linear models, including Cox's proportional hazards model for survival data. Coordinate descent is implemented in the popular `glmnet` package in the R statistical language, written by Jerome Friedman, Trevor Hastie, and myself, with help in the Cox feature from Noah Simon.

On the statistical side, there has also been a great deal of deep and interesting work on the mathematical aspects of the lasso, examining its ability to produce a model with minimal prediction error, and also to recover the true underlying (sparse) model. Important contributors here include Bühlmann, Candès, Donoho, Greenshtein, Johnstone, Meinshausen, Ritov, Wainwright, Yu, and many others. In describing some of this work, Hastie et al. (2001) coined the informal "Bet on Sparsity" principle. The ℓ_1 methods assume that the truth is sparse, in some basis. If the assumption holds true, then the parameters can be efficiently estimated using ℓ_1 penalties. If the assumption does not hold — so that the truth is dense — then no method will be able to recover the underlying model without a large amount of data per parameter. This is typically not the case when $p \gg N$, a commonly occurring scenario.

42.3 An example

I am currently involved in a cancer diagnosis project with researchers at Stanford University. They have collected samples of tissue from 10 patients undergoing surgery for stomach cancer. The aim is to build a classifier that can distinguish three kinds of tissue: normal epithelial, stromal and cancer. Such a classifier could be used to assist surgeons in determining, in real time, whether they had successfully removed all of the tumor. It could also yield insights into the cancer process itself. The data are in the form of images, as sketched in Figure 42.2. A pathologist has labelled each region (and hence the pixels inside a region) as epithelial, stromal or cancer. At each pixel in the image, the intensity of metabolites is measured by a kind of mass spectrometry, with the peaks in the spectrum representing different metabolites. The spectrum has been finely sampled at about 11,000 sites. Thus the task is to build a classifier to classify each pixel into one of the three classes, based on the 11,000 features. There are about 8000 pixels in all.

For this problem, I have applied an ℓ_1-regularized multinomial model. For each class $k \in \{1, 2, 3\}$, the model has a vector $(\beta_{1k}, \ldots, \beta_{pk})$ of parameters representing the weight given to each feature in that class. I used the `glmnet` package for fitting the model: it computes the entire path of solutions for all values of the regularization parameter λ, using cross-validation to estimate the best value of λ (I left one patient out at a time). The entire computation required just a few minutes on a standard Linux server.

The results so far are encouraging. The classifier shows 93–97% accuracy in the three classes, using only around 100 features. These features could yield insights about the metabolites that are important in stomach cancer. There is much more work to be done — collecting more data, and refining and testing the model. But this shows the potential of ℓ_1-penalized models in an important and challenging scientific problem.

42.4 The covariance test

So far, most applications of the lasso and ℓ_1 penalties seem to focus on large problems, where traditional methods like all-subsets-regression can't deal with the problem computationally. In this last section, I want to report on some

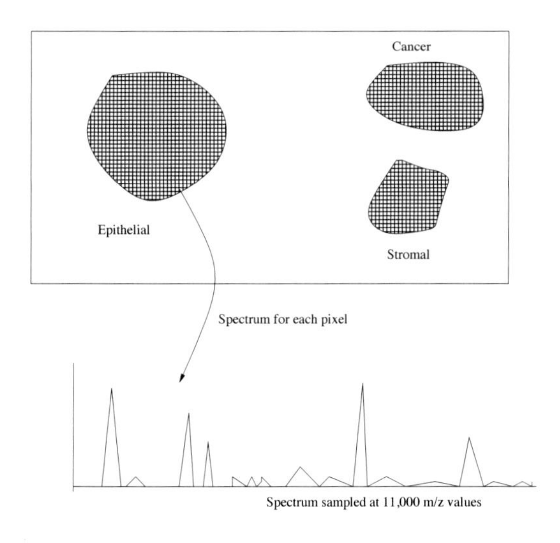

FIGURE 42.2
Schematic of the cancer diagnosis problem. Each pixel in each of the three regions labelled by the pathologist is analyzed by mass spectrometry. This gives a feature vector of 11,000 intensities (bottom panel), from which we try to predict the class of that pixel.

very recent work that suggest that ℓ_1 penalties may have a more fundamental role in classical mainstream statistical inference.

To begin, consider standard forward stepwise regression. This procedure enters predictors one a time, choosing the predictor that most decreases the residual sum of squares at each stage. Defining RSS to be the residual sum of squares for the model containing j predictors and denoting by RSS_{null} the residual sum of squares for the model omitting the predictor $k(j)$, we can form the usual statistic

$$R_j = (\text{RSS}_{\text{null}} - \text{RSS})/\sigma^2$$

(with σ assumed known for now), and compare it to a $\chi^2_{(1)}$ distribution.

Although this test is commonly used, we all know that it is wrong. Figure 42.3 shows an example. There are 100 observations and 10 predictors in a standard Gaussian linear model, in which all coefficients are actually zero. The left panel shows a quantile-quantile plot of 500 realizations of the statistic R_1 versus the quantiles of the $\chi^2_{(1)}$ distribution. The test is far too liberal and the reason is clear: the $\chi^2_{(1)}$ distribution is valid for comparing two fixed nested linear models. But here we are adaptively choosing the best predictor, and comparing its model fit to the null model.

In fact it is difficult to correct the chi-squared test to account for adaptive selection: half-sample splitting methods can be used (Meinshausen et al., 2009;

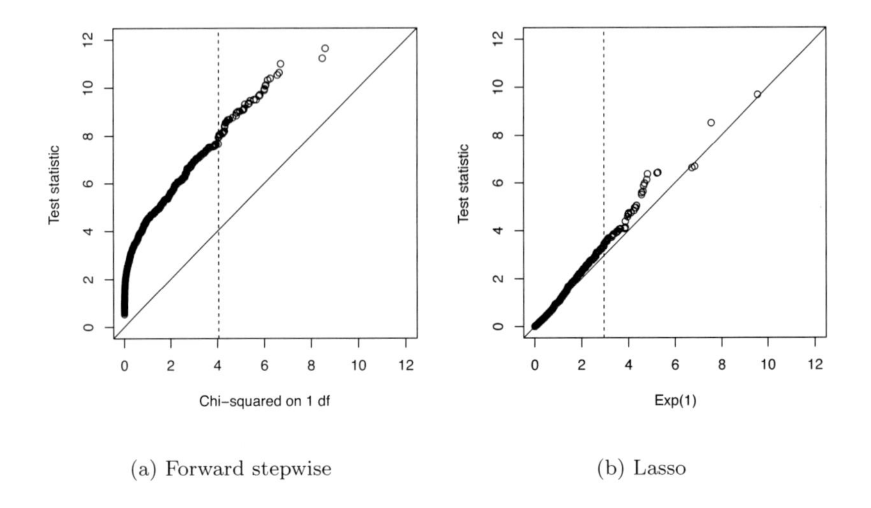

(a) Forward stepwise (b) Lasso

FIGURE 42.3
A simple example with $n = 100$ observations and $p = 10$ orthogonal predictors. All true regression coefficients are zero, $\beta^* = 0$. On the left is a quantile-quantile plot, constructed over 1000 simulations, of the standard chi-squared statistic R_1, measuring the drop in residual sum of squares for the first predictor to enter in forward stepwise regression, versus the χ_1^2 distribution. The dashed vertical line marks the 95% quantile of the $\chi_{(1)}^2$ distribution. The right panel shows a quantile-quantile plot of the covariance test statistic T_1 for the first predictor to enter in the lasso path, versus its asymptotic distribution $\mathcal{E}(1)$. The covariance test explicitly accounts for the adaptive nature of lasso modeling, whereas the usual chi-squared test is not appropriate for adaptively selected models, e.g., those produced by forward stepwise regression.

Wasserman and Roeder, 2009), but these may suffer from lower power due to the decrease in sample size.

But the lasso can help us! Specifically, we need the LAR (least angle regression) method for constructing the lasso path of solutions, as the regularization parameter λ is varied. I won't give the details of this construction here, but we just need to know that there are a special set of decreasing knots $\lambda_1 > \cdots > \lambda_k$ at which the active set of solutions (the non-zero parameter estimates) change. When $\lambda > \lambda_1$, the solutions are all zero. At the point $\lambda = \lambda_1$, the variable most correlated with y enters the model. At each successive value λ_j, a variable enters or leaves the model, until we reach λ_k where we obtain the full least squares solution (or one such solution, if $p > N$).

We consider a test statistic analogous to R_j for the lasso. Let $\mathbf{y}$ be the vector of outcome values and $\mathbf{X}$ be the design matrix. Assume for simplicity that the error variance σ^2 is known. Suppose that we have run LAR for $j-1$ steps, yielding the active set of predictors $\mathcal{A}$ at $\lambda = \lambda_j$. Now we take one more step, entering a new predictor $k(j)$, and producing estimates $\hat{\beta}(\lambda_j)$ at λ_{j+1}. We wish to test if the $k(j)$th component $\beta_{k(j)}$ is zero. We refit the lasso, keeping $\lambda = \lambda_{j+1}$ but using just the variables in $\mathcal{A}$. This yields estimates $\hat{\beta}_{\mathcal{A}}(\lambda_{j+1})$. Our proposed covariance test statistic is defined by

$$T_j = \frac{1}{\sigma^2}\{\langle \mathbf{y}, \mathbf{X}\hat{\beta}(\lambda_{j+1})\rangle - \langle \mathbf{y}, \mathbf{X}_{\mathcal{A}}\hat{\beta}_{\mathcal{A}}(\lambda_{j+1})\rangle\}. \tag{42.1}$$

Roughly speaking, this statistic measures how much of the covariance between the outcome and the fitted model can be attributed to the $k(j)$th predictor, which has just entered the model.

Now something remarkable happens. Under the null hypothesis that all signal variables are in the model: as $p \to \infty$, T_j converges to an exponential random variable with unit mean, $\mathcal{E}(1)$. The right panel of Figure 42.3 shows the same example, using the covariance statistic. This test works for testing the first variable to enter (as in the example), or for testing noise variables after all of the signal variables have entered. And it works under quite general conditions on the design matrix. This result properly accounts for the adaptive selection: the shrinkage in the ℓ_1 fitting counteracts the inflation due to selection, in just the right way to make the degrees of freedom (mean) of the null distribution exactly equal to 1 asymptotically. This idea can be applied to a wide variety of models, and yields honest p-values that should be useful to statistical practitioners.

In a sense, the covariance test and its exponential distribution generalize the RSS test and its chi-squared distribution, to the adaptive regression setting.

This work is very new, and is summarized in Lockhart et al. (2014). The proofs of the results are difficult, and use extreme-value theory and Gaussian processes. They suggest that the LAR knots λ_k may be fundamental in understanding the effects of adaptivity in regression. On the practical side, regression software can now output honest p-values as predictors enter a model, that properly account for the adaptive nature of the process. And all of this may be a result of the convexity of the ℓ_1-penalized objective.

42.5 Conclusion

In this chapter I hope that I have conveyed my excitement for some recent developments in statistics, both in its theory and practice. I predict that convexity and sparsity will play an increasing important role in the development of statistical methodology.

References

Bien, J., Taylor, J., and Tibshirani, R.J. (2013). A lasso for hierarchical interactions. *The Annals of Statistics*, 41:1111–1141.

Candès, E.J. (2006). Compressive sampling. In *Proceedings of the International Congress of Mathematicians*, Madrid, Spain. `www.acm.caltech.edu/~emmanuel/papers/CompressiveSampling.pdf`

Candès, E.J. and Tao, T. (2007). The Dantzig selector: Statistical estimation when p is much larger than n. *The Annals of Statistics*, 35: 2313–2351.

Candès, E.J. and Tao, T. (2009). The power of convex relaxation: Near-optimal matrix completion. `http://www.citebase.org/abstract?id=oai:arXiv.org:0903.1476`

Donoho, D.L. (2004). *Compressed Sensing.* Technical Report, Statistics Department, Stanford University, Stanford, CA. `www-stat.stanford.edu/~donoho/Reports/2004/CompressedSensing091604.pdf`

Friedman, J.H., Hastie, T.J., and Tibshirani, R.J. (2007). Pathwise coordinate optimization. *The Annals of Applied Statistics*, 1:302–332.

Friedman, J.H., Hastie, T.J., and Tibshirani, R.J. (2010). Regularization paths for generalized linear models via coordinate descent. *Journal of Statistical Software*, 33:1–22.

Fu, W. (1998). Penalized regressions: The bridge versus the lasso. *Journal of Computational and Graphical Statistics*, 7:397–416.

Hastie, T.J., Tibshirani, R.J., and Friedman, J.H. (2001). *The Elements of Statistical Learning: Data Mining, Inference and Prediction.* Springer, New York.

Joliffe, I.T., Trendafilov, N.T., and Uddin, M. (2003). A modified principal component technique based on the lasso. *Journal of Computational and Graphical Statistics*, 12:531–547.

Lockhart, R.A., Taylor, J., Tibshirani, R.J., and Tibshirani, R.J. (2014). A significance test for the lasso (with discussion). *The Annals of Statistics*, in press.

Mazumder, R., Hastie, T.J., and Tibshirani, R.J. (2010). Spectral regularization algorithms for learning large incomplete matrices. *Journal of Machine Learning Research*, 11:2287–2322.

Meinshausen, N., Meier, L., and Bühlmann, P. (2009). *P*-values for high-dimensional regression. *Journal of the American Statistical Association*, 104:1671–1681.

Tibshirani, R.J. (2011). Regression shrinkage and selection via the lasso: A retrospective. *Journal of the Royal Statistical Society, Series B*, 73:273–282.

Tibshirani, R.J., Hoefling, H., and Tibshirani, R.J. (2011). Nearly-isotonic regression. *Technometrics*, 53:54–61.

Tibshirani, R.J., Saunders, M., Rosset, S., Zhu, J., and Knight, K. (2005). Sparsity and smoothness via the fused lasso. *Journal of the Royal Statistical Society, Series B*, 67:91–108.

Tibshirani, R.J. and Taylor, J. (2011). The solution path of the generalized lasso. *The Annals of Statistics*, 39:1335–1371.

Wasserman, L.A. and Roeder, K. (2009). High-dimensional variable selection. *Journal of the American Statistical Association*, 37:2178–2201.

Witten, D.M., Tibshirani, R.J., and Hastie, T.J. (2009). A penalized matrix decomposition, with applications to sparse principal components and canonical correlation analysis. *Biometrika*, 10:515–534.

Yuan, M. and Lin, Y. (2007a). Model selection and estimation in regression with grouped variables. *Journal of the Royal Statistical Society, Series B*, 68:49–67.

Yuan, M. and Lin, Y. (2007b). Model selection and estimation in the Gaussian graphical model. *Biometrika*, 94:19–35.

Zou, H. (2006). The adaptive lasso and its oracle properties. *Journal of the American Statistical Association*, 101:1418–1429.

Zou, H. and Hastie, T.J. (2005). Regularization and variable selection via the elastic net. *Journal of the Royal Statistical Society, Series B*, 67:301–320.

43

Features of Big Data and sparsest solution in high confidence set

Jianqing Fan
Department of Operations Research and Financial Engineering
Princeton University, Princeton, NJ

This chapter summarizes some of the unique features of Big Data analysis. These features are shared neither by low-dimensional data nor by small samples. Big Data pose new computational challenges and hold great promises for understanding population heterogeneity as in personalized medicine or services. High dimensionality introduces spurious correlations, incidental endogeneity, noise accumulation, and measurement error. These unique features are very distinguished and statistical procedures should be designed with these issues in mind. To illustrate, a method called a sparsest solution in high-confidence set is introduced which is generally applicable to high-dimensional statistical inference. This method, whose properties are briefly examined, is natural as the information about parameters contained in the data is summarized by high-confident sets and the sparsest solution is a way to deal with the noise accumulation issue.

43.1 Introduction

The first decade of this century has seen the explosion of data collection in this age of information and technology. The technological revolution has made information acquisition easy and cheap through automated data collection processes. Massive data and high dimensionality characterize many contemporary statistical problems from biomedical sciences to engineering and social sciences. For example, in disease classification using microarray or proteomics data, tens of thousands of expressions of molecules or proteins are potential predictors; in genome-wide association studies, hundreds of thousands of SNPs are potential covariates; in machine learning, tens of thousands of features are extracted from documents, images and other objects; in spatial-temporal problems encountered in economics and earth sciences, time series from hundreds or thousands of regions are collected. When interactions are considered, the dimensionality grows even more quickly. Other examples of massive data include high-resolution images, high-frequency financial data, fMRI data, e-commerce data, marketing data, warehouse data, functional and longitudinal data, among others. For an overview, see Hastie et al. (2009) and Bühlmann and van de Geer (2011).

Salient features of Big Data include both large samples and high dimensionality. Furthermore, Big Data are often collected over different platforms or locations. This generates issues with heterogeneity, measurement errors, and experimental variations. The impacts of dimensionality include computational cost, algorithmic stability, spurious correlations, incidental endogeneity, noise accumulations, among others. The aim of this chapter is to introduce and explain some of these concepts and to offer a sparsest solution in high-confident set as a viable solution to high-dimensional statistical inference.

In response to these challenges, many new statistical tools have been developed. These include boosting algorithms (Freund and Schapire, 1997; Bickel et al., 2006), regularization methods (Tibshirani, 1996; Chen et al., 1998; Fan and Li, 2001; Candès and Tao, 2007; Fan and Lv, 2011; Negahban et al., 2012), and screening methods (Fan and Lv, 2008; Hall et al., 2009; Li et al., 2012). According to Bickel (2008), the main goals of high-dimensional inference are to construct as effective a method as possible to predict future observations, to gain insight into the relationship between features and response for scientific purposes, and hopefully, to improve prediction.

As we enter into the Big Data era, an additional goal, thanks to large sample size, is to understand heterogeneity. Big Data allow one to apprehend the statistical properties of small heterogeneous groups, termed "outliers" when sample size is moderate. It also allows us to extract important but weak signals in the presence of large individual variations.

43.2 Heterogeneity

Big Data enhance our ability to find commonalities in a population, even in the presence of large individual variations. An example of this is whether drinking a cup of wine reduces health risks of certain diseases. Population structures can be buried in the presence of large statistical noise in the data. Nevertheless, large sample sizes enable statisticians to mine such hidden structures. What also makes Big Data exciting is that it holds great promises for understanding population heterogeneity and making important discoveries, say about molecular mechanisms involved in diseases that are rare or affecting small populations. An example of this kind is to answer the question why chemotherapy is helpful for certain populations, while harmful or ineffective for some other populations.

Big Data are often aggregated from different sites and different platforms. Experimental variations need to be accounted for before their full analysis. Big Data can be thought of as a mixture of data arising from many heterogeneous populations. Let k be the number of heterogeneous groups, $\mathbf{X}$ be a collection of high-dimensional covariates, and y be a response. It is reasonable to regard Big Data as random realizations from a mixture of densities, viz.

$$p_1 f_1(y; \boldsymbol{\theta}_1(\mathbf{x})) + \cdots + p_k f_k(y; \boldsymbol{\theta}_k(\mathbf{x})),$$

in which $f_j(y; \boldsymbol{\theta}_j(\mathbf{x}))$ is the conditional density of Y given $\mathbf{X} = \mathbf{x}$ in population $j \in \{1, \ldots, k\}$, and the function $\boldsymbol{\theta}_j(\mathbf{x})$ characterizes the dependence of the distribution on the covariates. Gaussian mixture models are a typical example; see, e.g., Khalili and Chen (2007) or Städler et al. (2010).

When the sample size is moderate, data from small groups with small p_j rarely occur. Should such data be sampled, they are usually regarded as statistical outliers or buried in the larger groups. There are insufficient amounts of data to infer about $\theta_j(\mathbf{x})$. Thanks to Big Data, when n is so large that np_j is also large, there are sufficient amounts of data to infer about commonality $\boldsymbol{\theta}_j(\mathbf{x})$ in this rare subpopulation. In this fashion, Big Data enable us to discover molecular mechanisms or genetic associations in small subpopulations, opening the door to personalized treatments. This holds true also in consumer services where different subgroups demand different specialized services.

The above discussion further suggests that Big Data are paramountly important in understanding population heterogeneity, a goal that would be illusory when the sample size is only moderately large. Big Data provide a way in which heterogeneous subpopulations can be distinguished and personalized treatments can be derived. It is also an important tool for the discovery of weak population structures hidden in large individual variations.

43.3 Computation

Large-scale computation plays a vital role in the analysis of Big Data. High-dimensional optimization is not only expensive but also unstable in computation, in addition to slowness in convergence. Algorithms that involve iterative inversions of large matrices are infeasible due to instability and computational costs. Scalable and stable implementations of high-dimensional statistical procedures must be sought. This relies heavily on statistical intuition, large-scale screening and small-scale optimization. An example is given in Fan et al. (2009).

Large numbers of observations, which can be in the order of tens of thousands or even millions as in genomics, neuro-informatics, marketing, and online learning studies, also give rise to intensive computation. When the sample size is large, the computation of summary statistics such as correlations among all variables is expensive. Yet statistical methods often involve repeated evaluations of such functions. Parallel computing and other updating techniques are required. Therefore, scalability of techniques to both dimensionality and the number of cases should be borne in mind when developing statistical procedures.

43.4 Spurious correlation

Spurious correlation is a feature of high dimensionality. It refers to variables that are not correlated theoretically but whose sample correlation is high. To illustrate the concept, consider a random sample of size $n = 50$ of p independent standard $\mathcal{N}(0, 1)$ random variables. Thus the population correlation between any two random variables is zero and their corresponding sample correlation should be small. This is indeed the case when the dimension is small in comparison with the sample size. When p is large, however, spurious correlations start to appear. To illustrate this point, let us compute

$$\hat{r} = \max_{j \geq 2} \widehat{\text{corr}}(Z_1, Z_j)$$

where $\widehat{\text{corr}}(Z_1, Z_j)$ is the sample correlation between variables Z_1 and Z_j. Similarly, we can compute

$$\hat{R} = \max_{|\mathcal{S}|=5} \widehat{\text{corr}}(Z_1, \mathbf{Z}_{\mathcal{S}}), \tag{43.1}$$

which is the maximum multiple correlation between Z_1 and $\mathbf{Z}_{\mathcal{S}}$ with $1 \notin \mathcal{S}$, namely, the correlation between Z_1 and its best linear predictor using $\mathbf{Z}_{\mathcal{S}}$. In the implementation, we use the forward selection algorithm as an approximation to compute $\hat{R}$, which is no larger than $\hat{R}$ but avoids computing all $\binom{p}{5}$ multiple R^2 in (43.1). This experiment is repeated 200 times.

The empirical distributions of $\hat{r}$ and $\hat{R}$ are shown in Figure 43.1. The spurious correlation $\hat{r}$ is centered around .45 for $p = 1000$ and .55 for $p =$ 10,000. The corresponding values are .85 and .91 when the multiple correlation $\hat{R}$ is used. Theoretical results on the order of the spurious correlation $\hat{r}$ are given in Cai and Jiang (2012) and Fan et al. (2012), but the order of $\hat{R}$ remains unknown.

The impact of spurious correlation includes false scientific discoveries and false statistical inferences. In terms of scientific discoveries, Z_1 and $\mathbf{Z}_{\hat{\mathcal{S}}}$ are practically indistinguishable when $n = 50$, given that their correlation is around .9 for a set $\hat{S}$ with $|\hat{S}| = 5$. If Z_1 represents the expression of a gene that is responsible for a disease, we can discover five genes $\hat{S}$ that have a similar predictive power even though they are unrelated to the disease. Similarly,

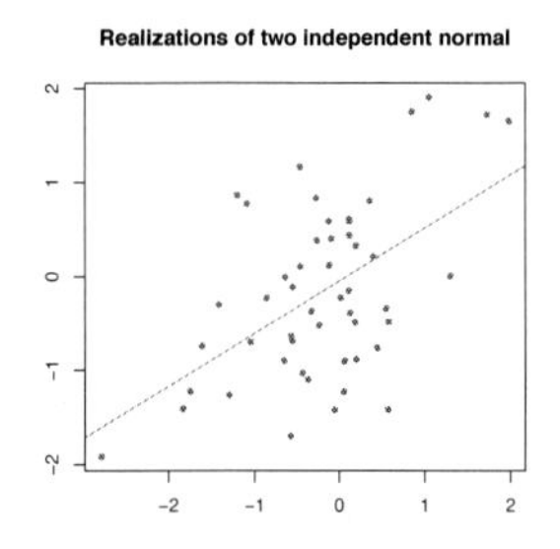

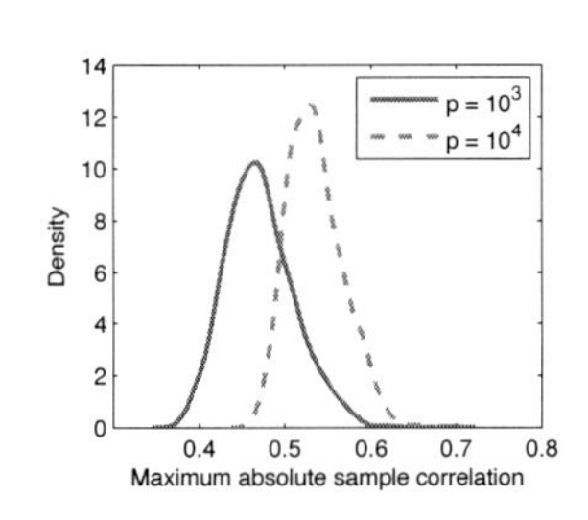

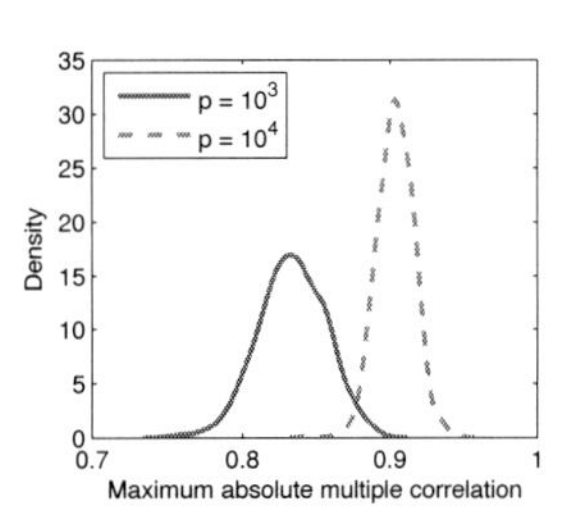

FIGURE 43.1
Illustration of spurious correlation. Left panel: a typical realization of Z_1 with its most spuriously correlated variable ($p = 1000$); middle and right panels: distributions of $\hat{r}$ and $\hat{R}$ for $p = 1000$ and $p = 10{,}000$. The sample size is $n = 50$.

if the genes in $\hat{S}$ are truly responsible for a disease, we may end up wrongly pronouncing Z_1 as the gene that is responsible for the disease.

We now examine the impact of spurious correlation on statistical inference. Consider a linear model

$$Y = \mathbf{X}^\top \beta + \varepsilon, \sigma^2 = \mathrm{var}(\varepsilon).$$

The residual variance based on a selected set $\hat{\mathcal{S}}$ of variables is

$$\hat{\sigma}^2 = \frac{1}{n - |\hat{\mathcal{S}}|} \mathbf{Y}^\top (\mathbf{I}_n - \mathbf{P}_{\hat{\mathcal{S}}}) \mathbf{Y}, \qquad \mathbf{P}_{\hat{\mathcal{S}}} = \mathbf{X}_{\hat{\mathcal{S}}} (\mathbf{X}_{\hat{\mathcal{S}}}^\top \mathbf{X}_{\hat{\mathcal{S}}})^{-1} \mathbf{X}_{\hat{\mathcal{S}}}^\top.$$

When the variables are not data selected and the model is unbiased, the degree of freedom adjustment makes the residual variance unbiased. However, the situation is completely different when the variables are data selected. For example, when $\beta = 0$, one has $\mathbf{Y} = \epsilon$ and all selected variables are spurious. If the number of selected variables $|\hat{\mathcal{S}}|$ is much smaller than n, then

$$\hat{\sigma}^2 = \frac{1}{n - |\hat{\mathcal{S}}|} (1 - \gamma_n^2) \|\epsilon\|^2 \approx (1 - \gamma_n^2) \sigma^2,$$

where $\gamma_n^2 = \epsilon^\top \mathbf{P}_{\hat{\mathcal{S}}} \epsilon / \|\epsilon\|^2$. Therefore, σ^2 is underestimated by a factor of γ_n^2.

Suppose that we select only one spurious variable. This variable must then be mostly correlated with $\mathbf{Y}$ or, equivalently, ϵ. Because the spurious correlation is high, the bias is large. The two left panels of Figure 43.2 depict the distributions of γ_n along with the associated estimates of $\hat{\sigma}^2$ for different choices of p. Clearly, the bias increases with the dimension, p.

When multiple spurious variables are selected, the biases of residual variance estimation become more pronounced, since the spurious correlation gets larger as demonstrated in Figure 43.1. To illustrate this, consider the linear

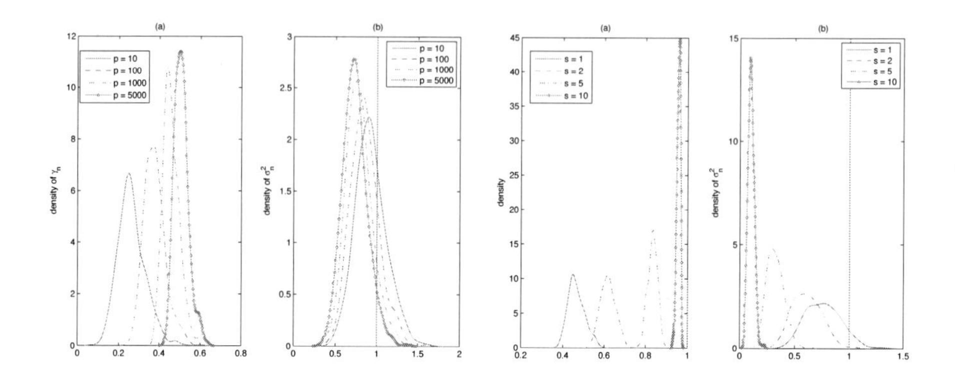

FIGURE 43.2
Distributions of spurious correlations. Left panel: Distributions of γ_n for the null model when $|\hat{\mathcal{S}}| = 1$ and their associated estimates of $\sigma^2 = 1$ for various choices of p. Right panel: Distributions of γ_n for the model $Y = 2X_1 + .3X_2 + \varepsilon$ and their associated estimates of $\sigma^2 = 1$ for various choices of $|\hat{\mathcal{S}}|$ but fixed $p = 1000$. The sample size $n = 50$. Adapted from Fan et al. (2012).

model $\mathbf{Y} = 2\mathbf{X}_1 + .3\mathbf{X}_2 + \epsilon$ and use the stepwise selection method to recruit variables. Again, the spurious variables are selected mainly due to their spurious correlation with ϵ, the unobserved but realized vector of random noises. As shown in the two right panels of Figure 43.2, the spurious correlation is very large and $\hat{\sigma}^2$ gets notably more biased when $|\hat{\mathcal{S}}|$ gets larger.

Underestimation of residual variance leads the statistical inference astray. Variables are declared statistically significant that are not in reality, and this leads to faulty scientific conclusions.

43.5 Incidental endogeneity

High dimensionality also gives rise to incidental endogeneity. Scientists collect covariates that are potentially related to the response. As there are many covariates, some of those variables can be incidentally correlated with the residual noise. This can cause model selection inconsistency and incorrect selection of genes or SNPs for understanding molecular mechanism or genetic associations.

Let us illustrate this problem using the simple linear model. The idealized model for variable selection is that there is a small subset $\mathcal{S}_0$ of variables that explains a large portion of the variation in the response Y, viz.

$$Y = \mathbf{X}^\top \beta_0 + \varepsilon, \quad \mathrm{E}(\varepsilon \mathbf{X}) = 0, \tag{43.2}$$

in which the true parameter vector β_0 has support $\mathcal{S}_0$. The goal of variable selection is to find the set $\mathcal{S}_0$ and estimate the regression coefficients β_0.

To be more concrete, let us assume that the data generating process is $Y = X_1 + X_2 + \varepsilon$, so that $\mathcal{S}_0 = \{1, 2\}$. As we do not know which variables are related to Y in the joint model, we collect as many covariates as possible that we deem to be potentially related to Y, in the hope of including all members in $\mathcal{S}_0$. Some of those X_j are incidentally correlated with $Y - X_1 - X_2$ or ε. This makes model (43.2) invalid. The rise of incidental endogeneity is due to high dimensionality, making the specifications $\mathrm{E}(\varepsilon \mathbf{X}) = 0$ invalid for some collected covariates, unintentionally. The more covariates are collected, the more unlikely this assumption is.

Does incidental endogeneity arise in practice? Can the exogeneity assumption $\mathrm{E}(\varepsilon \mathbf{X}) = 0$ be validated? After data collection, variable selection techniques such as the lasso (Tibshirani, 1996; Chen et al., 1998) and folded concave penalized least squares (Fan and Li, 2001; Zou and Li, 2008) are frequently used before drawing conclusions. The model is rarely validated. Indeed, the residuals were computed based only on a small set of the selected variables. Unlike with ordinary least squares, the exogeneity assumption in (43.2) cannot be validated empirically because most variables are not used to compute the residuals. We now illustrate this fact with an example.

Consider the gene expressions of 90 western Europeans from the international "HapMap" project (Thorisson et al., 2005); these data are available on `ftp://ftp.sanger.ac.uk/pub/genevar/`. The normalized gene expression data were generated with an Illumina Sentrix Human-6 Expression Bead Chip (Stranger et al., 2007). We took the gene expressions of *CHRNA6*, cholinergic receptor, nicotinic, alpha 6, as the response variable, and the remaining expression profiles of 47,292 as covariates. The left panel of Figure 43.3 presents the correlation between the response variable and its associated covariates.

Lasso is then employed to find the genes that are associated with the response. It selects 23 genes. The residuals $\hat{\varepsilon}$ are computed, which are based on those genes. The right panel of Figure 43.3 displays the distribution of the sample correlations between the covariates and the residuals. Clearly, many of them are far from zero, which is an indication that the exogeneity assumption in (43.2) cannot be validated. That is, incidental endogeneity is likely present. What is the consequence of this endogeneity? Fan and Liao (2014) show that this causes model selection inconsistency.

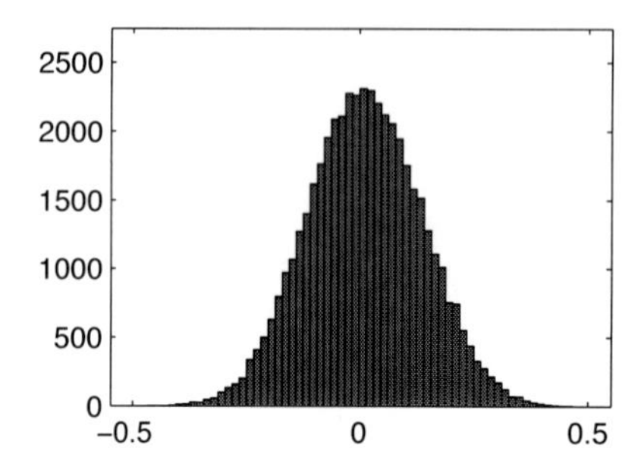

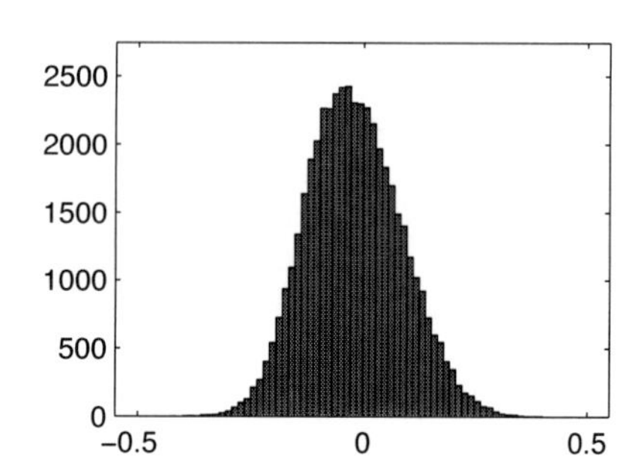

FIGURE 43.3
Distributions of sample correlations. Left panel: Distributions of the sample correlation $\widehat{\text{corr}}(X_j, Y)$ ($j = 1, \ldots,$ 47,292). Right panel: Distribution of the sample correlation $\widehat{\text{corr}}(X_j, \hat{\varepsilon})$, in which $\hat{\varepsilon}$ represents the residuals after the lasso fit.

How do we deal with endogeneity? Ideally, we hope to be able to select consistently $\mathcal{S}_0$ under only the assumption that

$$Y = \mathbf{X}_{\mathcal{S}_0}^\top \beta_{\mathcal{S}_0,0} + \varepsilon, \quad \mathrm{E}(\varepsilon \mathbf{X}_{\mathcal{S}_0}) = 0,$$

but this assumption is too weak to recover the set $\mathcal{S}_0$. A stronger assumption is

$$Y = \mathbf{X}_{\mathcal{S}_0}^\top \beta_{\mathcal{S}_0,0} + \varepsilon, \quad \mathrm{E}(\varepsilon | \mathbf{X}_{\mathcal{S}_0}) = 0. \tag{43.3}$$

Fan and Liao (2014) use over identification conditions such as

$$\mathrm{E}(\varepsilon \mathbf{X}_{\mathcal{S}_0}) = 0 \quad \text{and} \quad \mathrm{E}(\varepsilon \mathbf{X}_{\mathcal{S}_0}^2) = 0 \tag{43.4}$$

to distinguish endogenous and exogenous variables, which are weaker than the condition in (43.3). They introduce the Focused Generalized Method of Moments (FGMM) which uses the over identification conditions to select consistently the set of variables $\mathcal{S}_0$. The readers can refer to their paper for technical details. The left panel of Figure 43.4 shows the distribution of the correlations between the covariates and the residuals after the FGMM fit. Many of the correlations are still non-zero, but this is fine, as we assume only (43.4) and merely need to validate this assumption empirically. For this data set, FGMM

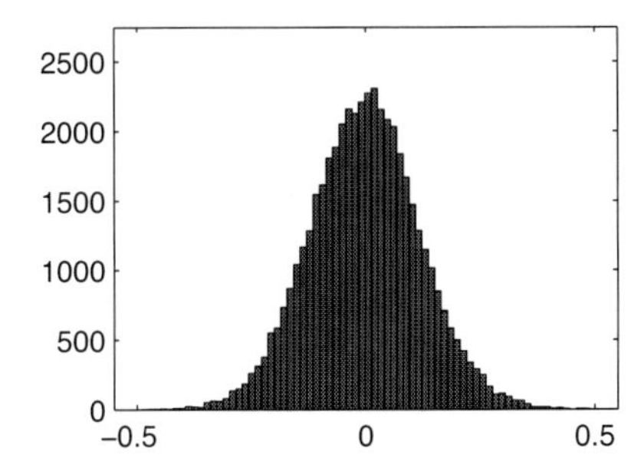

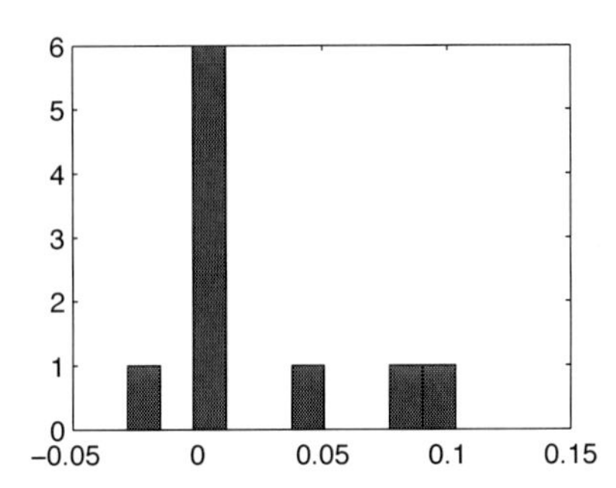

FIGURE 43.4
Left panel: Distribution of the sample correlation $\widehat{\text{corr}}(X_j, \hat{\varepsilon})$, in which $\hat{\varepsilon}$ represents the residuals after the FGMM fit. Right panel: Distribution of the sample correlation $\widehat{\text{corr}}(X_j, \hat{\varepsilon})$ for only selected 5 genes by FGMM.

selects five genes. Therefore, we need only validate 10 empirical correlations specified by conditions (43.4). The empirical correlations between the residuals after the FGMM fit and the five selected covariates are zero, and their correlations with squared covariates are small. The results are displayed in the right panel of Figure 43.4. Therefore, our model assumptions and model diagnostics are consistent.

43.6 Noise accumulation

When a method depends on the estimation of many parameters, the estimation errors can accumulate. For high-dimensional statistics, noise accumulation is more severe and can even dominate the underlying signals. Consider, for example, a linear classification which assigns the class label $\mathbf{1}(\mathbf{x}^\top \beta > 0)$ for each new data point $\mathbf{x}$. This rule can have high discrimination power when β is known. However, when an estimator $\hat{\beta}$ is used instead, the classification rule can be as bad as a random guess due to the accumulation of errors in estimating the high-dimensional vector $\hat{\beta}$.

As an illustration, we simulate n data points respectively from the population $\mathcal{N}(\boldsymbol{\mu}_0, \mathbf{I}_p)$ and $\mathcal{N}(\boldsymbol{\mu}_1, \mathbf{I}_p)$, in which $p = 4500$, $\boldsymbol{\mu}_0 = 0$, and $\boldsymbol{\mu}_1$ is a realization of a mixture of point mass 0 with probability .98 and the standard double exponential distribution with probability .02. Therefore, most components have no discriminative power, yet some components are very powerful in classification. Indeed, among 2% or 90 realizations from the double exponential distributions, several components are very large, and many components are small.

The distance-based classifier, which classifies $\mathbf{x}$ to class 1 when

$$\|\mathbf{x} - \boldsymbol{\mu}_1\|^2 \leq \|\mathbf{x} - \boldsymbol{\mu}_0\|^2 \quad \text{or} \quad \beta^\top(\mathbf{x} - \boldsymbol{\mu}) \geq 0,$$

where $\beta = \boldsymbol{\mu}_1 - \boldsymbol{\mu}_0$ and $\boldsymbol{\mu} = (\boldsymbol{\mu}_0 + \boldsymbol{\mu}_1)/2$. Letting Φ denote the cumulative distribution function of a standard Normal random variable, we find that the misclassification rate is $\Phi(-\|\boldsymbol{\mu}_1 - \boldsymbol{\mu}_0\|/2)$, which is effectively zero because by the Law of Large Numbers,

$$\|\boldsymbol{\mu}_1 - \boldsymbol{\mu}_0\| \approx \sqrt{4500 \times .02 \times 1} \approx 9.48.$$

However, when β is estimated by the sample mean, the resulting classification rule behaves like a random guess due to the accumulation of noise.

To help the intuition, we drew $n = 100$ data points from each class and selected the best m features from the p-dimensional space, according to the absolute values of the components of $\boldsymbol{\mu}_1$; this is an infeasible procedure, but can be well estimated when m is small (Fan and Fan, 2008). We then projected the m-dimensional data on their first two principal components. Figure 43.5 presents their projections for various values of m. Clearly, when $m = 2$, these two projections have high discriminative power. They still do when $m = 100$, as there are noise accumulations and also signal accumulations too. There are about 90 non-vanishing signals, though some are very small; the expected values of those are approximately 9.48 as noted above. When $m = 500$ or 4500, these two projections have no discriminative power at all due to noise accumulation. See also Hall et al. (2005) for a geometric representation of high dimension and low sample size data for further intuition.

43.7 Sparsest solution in high confidence set

To attenuate the noise accumulation issue, we frequently impose the sparsity on the underlying parameter β_0. At the same time, the information on β_0 contained in the data is through statistical modeling. The latter is summarized by confidence sets of β_0 in $\mathbb{R}^p$. Combining these two pieces of information, a general solution to high-dimensional statistics is naturally the sparsest solution in high-confidence set.

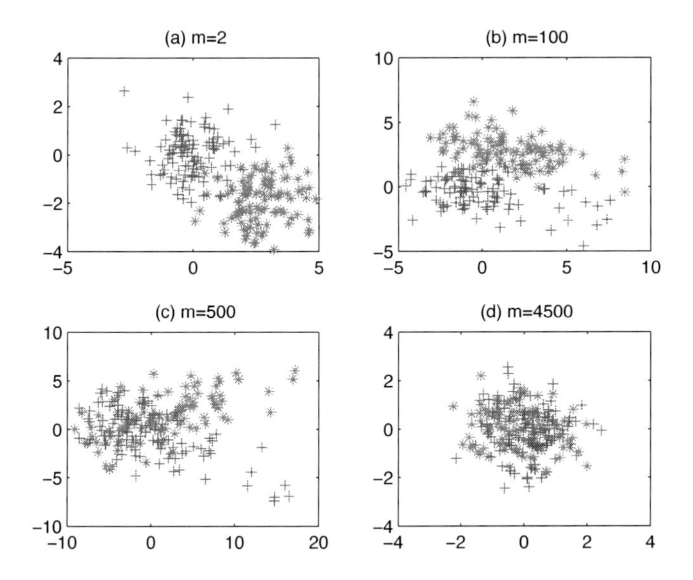

FIGURE 43.5
Scatter plot of projections of observed data ($n = 100$ from each class) onto the first two principal components of the m-dimensional selected feature space.

43.7.1 A general setup

We now elaborate the idea. Assume that the Big Data are collected in the form $(\mathbf{X}_1, Y_1), \ldots, (\mathbf{X}_n, Y_n)$, which can be regarded as a random sample from the population $(\mathbf{X}, Y)$. We wish to find an estimate of the sparse vector $\beta_0 \in \mathbb{R}^p$ such that it minimizes $L(\beta) = \mathrm{E}\{L(\mathbf{X}^\top \beta, Y)\}$, in which the loss function is assumed convex in the first argument so that $L(\beta)$ is convex. The setup encompasses the generalized linear models (McCullagh and Nelder, 1989) with $L(\theta, y) = b(\theta) - \theta y$ under the canonical link where $b(\theta)$ is a model-dependent convex function, robust regression with $L(\theta, y) = |y - \theta|$, the hinge loss $L(\theta, y) = (1 - \theta y)_+$ in the support vector machine (Vapnik, 1999) and exponential loss $L(\theta, y) = \exp(-\theta y)$ in AdaBoost (Freund and Schapire, 1997; Breiman, 1998) in classification in which y takes values ± 1, among others. Let

$$L_n(\beta) = \frac{1}{n} \sum_{i=1}^{n} L(\mathbf{X}_i^\top \beta, Y_i)$$

be the empirical loss and $L_n'(\beta)$ be its gradient. Given that $L'(\beta_0) = 0$, a natural confidence set is of form

$$\mathcal{C}_n = \{\beta \in \mathbb{R}^p : \|L_n'(\beta)\|_\infty \le \gamma_n\}$$

for some given γ_n that is related to the confidence level. Here $L'_n(\beta) = 0$ can be regarded as the estimation equations. Sometimes, it is handy to construct the confidence sets directly from the estimation equations.

In principle, any norm can be used in constructing confidence set. However, we take the L_∞-norm as it is the conjugate norm to the L_1-norm in Hölder's inequality. It also makes the set $\mathcal{C}_n$ convex, because $|L'_n(\beta)|$ is nondecreasing in each argument. The tuning parameter γ_n is chosen so that the set $\mathcal{C}_n$ has confidence level $1 - \delta_n$, viz.

$$\Pr(\beta_0 \in \mathcal{C}_n) = \Pr\{\|L'_n(\beta_0)\|_\infty \le \gamma_n\} \ge 1 - \delta_n. \tag{43.5}$$

The confidence region $\mathcal{C}_n$ is called a high confidence set because $\delta_n \to 0$ and can even be zero. Note that the confidence set is the interface between the data and parameters; it should be applicable to all statistical problems, including those with measurement errors.

The set $\mathcal{C}_n$ is the summary of the data information about β_0. If in addition we assume that β_0 is sparse, then a natural solution is the intersection of these two pieces of information, namely, finding the sparsest solution in the high-confidence region, viz.

$$\min_{\beta \in \mathcal{C}_n} \|\beta\|_1 = \min_{\|L'_n(\beta)\|_\infty \le \gamma_n} \|\beta\|_1. \tag{43.6}$$

This is a convex optimization problem. Here, the sparsity is measured by the L_1-norm, but it can also be measured by other norms such as the weighted L_1-norm (Zou and Li, 2008). The idea is related to that in Negahban et al. (2012), where a nice framework for analysis of high-dimensional M-estimators with decomposable regularizers is established for restricted convex losses.

43.7.2 Examples

The Danzig selector (Candès and Tao, 2007) is a specific case of problem (43.6) in which the loss is quadratic $L(x, y) = (x - y)^2$ and $\delta_n = 0$. This provides an alternative view to the Danzig selector. If $L(x, y) = \rho(|x - y|)$ for a convex function ρ, then the confidence set implied by the data is

$$\mathcal{C}_n = \{\beta \in \mathbb{R}^p : \|\rho'(|\mathbf{Y} - \mathbf{X}\beta|)\mathbf{X}^\top \operatorname{svn}(\mathbf{Y} - \mathbf{X}\beta)\|_\infty \le \gamma_n\}$$

and the sparsest solution in the high confidence set is now given by

$$\min \|\beta\|_1, \quad \text{subject to } \|\rho'(|\mathbf{Y} - \mathbf{X}\beta|)\mathbf{X}^\top \operatorname{svn}(\mathbf{Y} - \mathbf{X}\beta)\|_\infty \le \gamma_n.$$

In particular, when $\rho(\theta) = \theta$ and $\rho(\theta) = \theta^2/2$, they correspond to the L_1-loss and L_2-loss (the Danzig selector).

Similarly, in construction of sparse precision $\mathbf{\Theta} = \mathbf{\Sigma}^{-1}$ for the Gaussian graphic model, if $L(\Theta, \mathbf{S}_n) = \|\Theta\mathbf{S}_n - \mathbf{I}_p\|_F^2$ where $\mathbf{S}_n$ is the sample covariance matrix and $\|\cdot\|_F$ is the Frobenius norm, then the high confidence set provided by the data is

$$\mathcal{C}_n = \{\mathbf{\Theta} : \|\mathbf{S}_n \cdot (\mathbf{\Theta}\mathbf{S}_n - \mathbf{I}_p)\|_\infty \le \gamma_n\},$$

where $\cdot$ denotes the componentwise product (a factor 2 of off-diagonal elements is ignored). If we construct the high-confidence set based directly on the estimation equations $L_n'(\mathbf{\Theta}) = \mathbf{\Theta}\mathbf{S}_n - \mathbf{I}_p$, then the sparse high-confidence set becomes

$$\min_{\|\mathbf{\Theta}\mathbf{S}_n - \mathbf{I}_p\|_\infty \le \gamma_n} \|\text{vec}(\mathbf{\Theta})\|_1.$$

If the matrix L_1-norm is used in (43.6) to measure the sparsity, then the resulting estimator is the CLIME estimator of Cai et al. (2011), viz.

$$\min_{\|\mathbf{\Theta}\mathbf{S}_n - \mathbf{I}_p\|_\infty \le \gamma_n} \|\mathbf{\Theta}\|_1.$$

If we use the Gaussian log-likelihood, viz.

$$L_n(\mathbf{\Theta}) = -\ln(|\mathbf{\Theta}|) + \text{tr}(\mathbf{\Theta}\mathbf{S}_n),$$

then $L_n'(\mathbf{\Theta}) = -\mathbf{\Theta}^{-1} + \mathbf{S}_n$ and $\mathcal{C}_n = \{\|\mathbf{\Theta}^{-1} - \mathbf{S}_n\|_\infty \le \gamma_n\}$. The sparsest solution is then given by

$$\min_{\|\mathbf{\Theta}^{-1} - \mathbf{S}_n\|_\infty \le \gamma_n} \|\mathbf{\Theta}\|_1.$$

If the relative norm $\|\mathbf{A}\|_\infty = \|\mathbf{\Theta}^{1/2}\mathbf{A}\mathbf{\Theta}^{1/2}\|_\infty$ is used, the solution can be more symmetrically written as

$$\min_{\|\mathbf{\Theta}^{1/2}\mathbf{S}_n\mathbf{\Theta}^{1/2} - \mathbf{I}_p\|_\infty \le \gamma_n} \|\mathbf{\Theta}\|_1.$$

In the construction of the sparse linear discriminant analysis from two Normal distributions $\mathcal{N}(\boldsymbol{\mu}_0, \mathbf{\Sigma})$ and $\mathcal{N}(\boldsymbol{\mu}_1, \mathbf{\Sigma})$, the Fisher classifier is linear and of the form $\mathbf{1}\{\beta^\top(\mathbf{X} - \boldsymbol{\mu}) > 0\}$, where $\boldsymbol{\mu} = (\boldsymbol{\mu}_0 + \boldsymbol{\mu}_1)/2$, $\boldsymbol{\delta} = \boldsymbol{\mu}_1 - \boldsymbol{\mu}_0$, and $\beta = \mathbf{\Sigma}^{-1}\boldsymbol{\delta}$. The parameters $\boldsymbol{\mu}$ and $\boldsymbol{\delta}$ can easily be estimated from the sample. The question is how to estimate β, which is assumed to be sparse. One direct way to construct confidence interval is to base directly the estimation equations $L_n'(\beta) = \mathbf{S}_n\beta - \hat{\boldsymbol{\delta}}$, where $\mathbf{S}_n$ is the pooled sample covariance and $\hat{\boldsymbol{\delta}}$ is the difference of the two sample means. The high-confidence set is then

$$\mathcal{C}_n = \{\beta : \|\mathbf{S}_n\beta - \hat{\boldsymbol{\delta}}\|_\infty \le \gamma_n\}. \tag{43.7}$$

Again, this is a set implied by data with high confidence. The sparsest solution is the linear programming discriminant rule by Cai et al. (2011).

The above method of constructing confidence is neither unique nor the smallest. Observe that (through personal communication with Dr Emre Barut)

$$\|\mathbf{S}_n\beta - \hat{\boldsymbol{\delta}}\|_\infty = \|(\mathbf{S}_n - \mathbf{\Sigma})\beta + \boldsymbol{\delta} - \hat{\boldsymbol{\delta}}\|_\infty \le \|(\mathbf{S}_n - \mathbf{\Sigma})\|_\infty\|\beta\|_1 + \|\boldsymbol{\delta} - \hat{\boldsymbol{\delta}}\|_\infty.$$

Therefore, a high confidence set can be taken as

$$\mathcal{C}_n = \{\|\mathbf{S}_n\beta - \hat{\boldsymbol{\delta}}\|_\infty \le \gamma_{n,1}\|\beta\|_1 + \gamma_{n,2}\}, \tag{43.8}$$

where $\gamma_{n,1}$ and $\gamma_{n,2}$ are the high confident upper bound of $\|(\mathbf{S}_n - \mathbf{\Sigma})\|_\infty$ and $\|\boldsymbol{\delta} - \hat{\boldsymbol{\delta}}\|_\infty$. The set (43.8) is smaller than the set (43.7), since a further bound $\|\beta\|_1$ in (43.8) by a constant $\gamma_{n,3}$ yields (43.7).

43.7.3 Properties

Let $\hat{\beta}$ be a solution to (43.6) and $\hat{\mathbf{\Delta}} = \hat{\boldsymbol{\beta}} - \beta_0$. As in the Danzig selection, the feasibility of β_0 implied by (43.5) entails that

$$\|\beta_0\|_1 \geq \|\hat{\boldsymbol{\beta}}\|_1 = \|\beta_0 + \hat{\mathbf{\Delta}}\|_1. \tag{43.9}$$

Letting $\mathcal{S}_0 = \mathrm{supp}(\beta_0)$, we have

$$\|\beta_0 + \hat{\mathbf{\Delta}}\|_1 \;=\; \|(\beta_0 + \hat{\mathbf{\Delta}})_{\mathcal{S}_0}\|_1 + \|\hat{\mathbf{\Delta}}_{\mathcal{S}_0^c}\|_1 \geq \|\beta_0\|_1 - \|\hat{\mathbf{\Delta}}_{\mathcal{S}_0}\|_1 + \|\hat{\mathbf{\Delta}}_{\mathcal{S}_0^c}\|_1.$$

This together with (43.9) yields

$$\|\hat{\mathbf{\Delta}}_{\mathcal{S}_0}\|_1 \geq \|\hat{\mathbf{\Delta}}_{\mathcal{S}_0^c}\|_1, \tag{43.10}$$

i.e., $\hat{\mathbf{\Delta}}$ is sparse or "restricted." In particular, with $s = |\mathcal{S}_0|$,

$$\|\hat{\mathbf{\Delta}}\|_2 \geq \|\hat{\mathbf{\Delta}}_{\mathcal{S}_0}\|_2 \geq \|\hat{\mathbf{\Delta}}_{\mathcal{S}_0}\|_1/\sqrt{s} \geq \|\hat{\mathbf{\Delta}}\|_1/(2\sqrt{s}), \tag{43.11}$$

where the last inequality uses (43.10). At the same time, since $\hat{\boldsymbol{\beta}}$ and β_0 are in the feasible set (43.5), we have

$$\|L_n'(\hat{\boldsymbol{\beta}}) - L_n'(\beta_0)\|_\infty \leq 2\gamma_n$$

with probability at least $1 - \delta_n$. By Hölder's inequality, we conclude that

$$|[L_n'(\hat{\boldsymbol{\beta}}) - L_n'(\beta_0)]^\top \hat{\mathbf{\Delta}}| \leq 2\gamma_n \|\hat{\mathbf{\Delta}}\|_1 \leq 4\sqrt{s}\gamma_n \|\hat{\mathbf{\Delta}}\|_2 \tag{43.12}$$

with probability at least $1 - \delta_n$, where the last inequality utilizes (43.11). By using the Taylor's expansion, we can prove the existence of a point β^* on the line segment between β_0 and $\hat{\boldsymbol{\beta}}$ such that $L_n'(\hat{\boldsymbol{\beta}}) - L_n'(\beta_0) = L_n''(\beta^*)\hat{\mathbf{\Delta}}$. Therefore,

$$|\hat{\mathbf{\Delta}}^\top L_n''(\beta^*)\hat{\mathbf{\Delta}}| \leq 4\sqrt{s}\gamma_n \|\hat{\mathbf{\Delta}}\|_2.$$

Since $\mathcal{C}_n$ is a convex set, $\beta^* \in \mathcal{C}_n$. If we generalize the restricted eigenvalue condition to the generalized restricted eigenvalue condition, viz.

$$\inf_{\|\mathbf{\Delta}_{\mathcal{S}_0}\|_1 \geq \|\mathbf{\Delta}_{\mathcal{S}_0^c}\|_1} \; \inf_{\beta \in \mathcal{C}_n} |\mathbf{\Delta}^\top L_n''(\beta)\mathbf{\Delta}|/\|\mathbf{\Delta}\|_2^2 \geq a, \tag{43.13}$$

then we have

$$\|\hat{\mathbf{\Delta}}\|_2 \leq 4a^{-1}\sqrt{s}\gamma_n. \tag{43.14}$$

The inequality (43.14) is a statement on the L_2-convergence of $\hat{\boldsymbol{\beta}}$, with probability at least $1 - \delta_n$. Note that each component of

$$L_n'(\hat{\boldsymbol{\beta}}) - L_n'(\beta_0) = L_n'(\beta_0 + \hat{\mathbf{\Delta}}) - L_n'(\beta_0)$$

in (43.12) has the same sign as the corresponding component of $\hat{\mathbf{\Delta}}$. Condition (43.13) can also be replaced by the requirement

$$\inf_{\|\mathbf{\Delta}_{\mathcal{S}_0}\|_1 \geq \|\mathbf{\Delta}_{\mathcal{S}_0^c}\|_1} |[L_n'(\beta_0 + \Delta) - L_n'(\beta_0)]^\top \Delta| \geq a\|\Delta\|^2.$$

This facilitates the case where L_n'' does not exist and is a specific case of Negahban et al. (2012).

43.8 Conclusion

Big Data arise from many frontiers of scientific research and technological developments. They hold great promise for the discovery of heterogeneity and the search for personalized treatments. They also allow us to find weak patterns in presence of large individual variations.

Salient features of Big Data include experimental variations, computational cost, noise accumulation, spurious correlations, incidental endogeneity, and measurement errors. These issues should be seriously considered in Big Data analysis and in the development of statistical procedures.

As an example, we offered here the sparsest solution in high-confidence sets as a generic solution to high-dimensional statistical inference and we derived a useful mean-square error bound. This method combines naturally two pieces of useful information: the data and the sparsity assumption.

Acknowledgement

This project was supported by the National Institute of General Medical Sciences of the National Institutes of Health through Grants R01–GM072611 and R01–GMR01GM100474. Partial funding in support of this work was also provided by National Science Foundation grant DMS–1206464. The author would like to thank Ahmet Emre Barut, Yuan Liao, and Martin Wainwright for help and discussion related to the preparation of this chapter. The author is also grateful to Christian Genest for many helpful suggestions.

References

Bickel, P.J. (2008). Discussion on the paper "Sure independence screening for ultrahigh dimensional feature space" by Fan and Lv. *Journal of the Royal Statistical Society, Series B*, 70:883–884.

Bickel, P.J., Ritov, Y., and Zakai, A. (2006). Some theory for generalized boosting algorithms. *The Journal of Machine Learning Research*, 7:705–732.

Breiman, L. (1998). Arcing classifier. *The Annals of Statistics*, 26:801–849.

Bühlmann, P. and van de Geer, S. (2011). *Statistics for High-Dimensional Data: Methods, Theory and Applications.* Springer, Berlin.

Cai, T. and Jiang, T. (2012). Phase transition in limiting distributions of coherence of high-dimensional random matrices. *Journal of Multivariate Analysis*, 107:24–39.

Cai, T., Liu, W., and Luo, X. (2011). A constrained ℓ_1 minimization approach to sparse precision matrix estimation. *Journal of the American Statistical Association*, 106:594–607.

Candès, E.J. and Tao, T. (2007). The Dantzig selector: Statistical estimation when p is much larger than n. *The Annals of Statistics*, 35:2313–2351.

Chen, S.S., Donoho, D.L., and Saunders, M.A. (1998). Atomic decomposition by basis pursuit. *SIAM Journal on Scientific Computing*, 20:33–61.

Fan, J. and Fan, Y. (2008). High-dimensional classification using features annealed independence rules. *The Annals of Statistics*, 36:2605.

Fan, J., Guo, S., and Hao, N. (2012). Variance estimation using refitted cross-validation in ultrahigh dimensional regression. *Journal of the Royal Statistical Society, Series B*, 74:37–65.

Fan, J. and Li, R. (2001). Variable selection via non-concave penalized likelihood and its oracle properties. *Journal of the American Statistical Association*, 96:1348–1360.

Fan, J. and Liao, Y. (2014). Endogeneity in ultrahigh dimension. *Journal of the American Statistical Association*, to appear.

Fan, J. and Lv, J. (2008). Sure independence screening for ultrahigh dimensional feature space (with discussion). *Journal of the Royal Statistical Society, Series B*, 70:849–911.

Fan, J. and Lv, J. (2011). Nonconcave penalized likelihood with *np*-dimensionality. *IEEE Transactions on Information Theory*, 57:5467–5484.

Fan, J., Samworth, R., and Wu, Y. (2009). Ultrahigh dimensional feature selection: Beyond the linear model. *The Journal of Machine Learning Research*, 10:2013–2038.

Freund, Y. and Schapire, R.E. (1997). A decision-theoretic generalization of on-line learning and an application to boosting. *Journal of Computer and System Sciences*, 55:119–139.

Hall, P., Marron, J.S., and Neeman, A. (2005). Geometric representation of high dimension, low sample size data. *Journal of the Royal Statistical Society, Series B*, 67:427–444.

Hall, P., Titterington, D.M., and Xue, J.-H. (2009). Tilting methods for assessing the influence of components in a classifier. *Journal of the Royal Statistical Society, Series B*, 71:783–803.

Hastie, T., Tibshirani, R.J., and Friedman, J. (2009). *The Elements of Statistical Learning.* Springer, New York.

Khalili, A. and Chen, J. (2007). Variable selection in finite mixture of regression models. *Journal of the American Statistical Association*, 102:1025–1038.

Li, R., Zhong, W., and Zhu, L. (2012). Feature screening via distance correlation learning. *Journal of the American Statistical Association*, 107:1129–1139.

McCullagh, P. and Nelder, J.A. (1989). *Generalized Linear Models.* Chapman & Hall, London.

Negahban, S.N., Ravikumar, P., Wainwright, M.J., and Yu, B. (2012). A unified framework for high-dimensional analysis of M-estimators with decomposable regularizers. *Statistical Science*, 27:538–557.

Städler, N., Bühlmann, P., and van de Geer, S. (2010). ℓ_1-penalization for mixture regression models (with discussion). *Test*, 19:209–256.

Stranger, B.E., Nica, A.C., Forrest, M.S., Dimas, A., Bird, C.P., Beazley, C., Ingle, C.E., Dunning, M., Flicek, P., Koller, D., Montgomery, S., Tavaré, S., Deloukas, P., and Dermitzakis, E.T. (2007). Population genomics of human gene expression. *Nature Genetics*, 39:1217–1224.

Thorisson, G.A., Smith, A.V., Krishnan, L., and Stein, L.D. (2005). The International HapMap Project Web Site. *Genome Research*, 15:1592–1593.

Tibshirani, R.J. (1996). Regression shrinkage and selection via the lasso. *Journal of the Royal Statistical Society, Series B*, 58:267–288.

Vapnik, V. (1999). *The Nature of Statistical Learning Theory.* Springer, Berlin.

Zou, H. and Li, R. (2008). One-step sparse estimates in nonconcave penalized likelihood models. *The Annals of Statistics*, 36:1509–1533.

44

Rise of the machines

Larry A. Wasserman
Department of Statistics
Carnegie Mellon University, Pittsburgh, PA

On the 50th anniversary of the COPSS, I reflect on the rise of the field of machine learning and what it means for statistics. Machine learning offers a plethora of new research areas, new applications areas and new colleagues to work with. Our students now compete with those in machine learning for jobs. I am optimistic that visionary statistics departments will embrace this emerging field; those that ignore or eschew machine learning do so at their own risk and may find themselves in the rubble of an outdated, antiquated field.

44.1 Introduction

Statistics is the science of learning from data. Machine learning (ML) is the science of learning from data. These fields are identical in intent although they differ in their history, conventions, emphasis and culture.

There is no denying the success and importance of the field of statistics for science and, more generally, for society. I'm proud to be a part of the field. The focus of this essay is on one challenge (and opportunity) to our field: the rise of machine learning.

During my twenty-five year career I have seen machine learning evolve from being a collection of rather primitive (yet clever) set of methods to do classification, to a sophisticated science that is rich in theory and applications.

A quick glance at *The Journal of Machine Learning Research* (`jmlr.csail.mit.edu`) and NIPS (`books.nips.cc`) reveals papers on a variety of topics that will be familiar to statisticians such as conditional likelihood, sequential design, reproducing kernel Hilbert spaces, clustering, bioinformatics, minimax theory, sparse regression, estimating large covariance matrices, model selection, density estimation, graphical models, wavelets, nonparametric regression. These could just as well be papers in our flagship statistics journals.

This sampling of topics should make it clear that researchers in machine learning — who were at one time somewhat unaware of mainstream statistical methods and theory — are now not only aware of, but actively engaged in, cutting edge research on these topics.

On the other hand, there are statistical topics that are active areas of research in machine learning but are virtually ignored in statistics. To avoid becoming irrelevant, we statisticians need to (i) stay current on research areas in ML and (ii) change our outdated model for disseminating knowledge and (iii) revamp our graduate programs.

44.2 The conference culture

ML moves at a much faster pace than statistics. At first, ML researchers developed expert systems that eschewed probability. But very quickly they adopted advanced statistical concepts like empirical process theory and concentration of measure. This transition happened in a matter of a few years. Part of the reason for this fast pace is the conference culture. The main venue for research in ML is refereed conference proceedings rather than journals.

Graduate students produce a stream of research papers and graduate with hefty CV's. One of the reasons for the blistering pace is, again, the conference culture.

The process of writing a typical statistics paper goes like this: you have an idea for a method, you stew over it, you develop it, you prove some results about it, and eventually you write it up and submit it. Then the refereeing process starts. One paper can take years.

In ML, the intellectual currency is conference publications. There are a number of deadlines for the main conferences (NIPS, AISTATS, ICML, COLT). The threat of a deadline forces one to quit ruminating and start writing. Most importantly, all faculty members and students are facing the same deadline so there is a synergy in the field that has mutual benefits. No one minds if you cancel a class right before the NIPS deadline. And then, after the deadline, everyone is facing another deadline: refereeing each other's papers and doing so in a timely manner. If you have an idea and don't submit a paper on it, then you may be out of luck because someone may scoop you.

This pressure is good; it keeps the field moving at a fast pace. If you think this leads to poorly written papers or poorly thought out ideas, I suggest you look at `nips.cc` and read some of the papers. There are some substantial, deep papers. There are also a few bad papers. Just like in our journals. The papers are refereed and the acceptance rate is comparable to our main journals. And if an idea requires more detailed follow-up, then one can always write a longer journal version of the paper.

Absent this stream of constant deadlines, a field moves slowly. This is a problem for statistics not only for its own sake but also because it now competes with ML.

Of course, there are disadvantages to the conference culture. Work is done in a rush, and ideas are often not fleshed out in detail. But I think that the advantages outweigh the disadvantages.

44.3 Neglected research areas

There are many statistical topics that are dominated by ML and mostly ignored by statistics. This is a shame because statistics has much to offer in all these areas. Examples include semi-supervised inference, computational topology, online learning, sequential game theory, hashing, active learning, deep learning, differential privacy, random projections and reproducing kernel Hilbert spaces. Ironically, some of these — like sequential game theory and reproducing kernel Hilbert spaces — started in statistics.

44.4 Case studies

I'm lucky. I am at an institution which has a Machine Learning Department (within the School of Computer Science) and, more importantly, the ML department welcomes involvement by statisticians. So I've been fortunate to work with colleagues in ML, attend their seminars, work with ML students and teach courses in the ML department.

There are a number of topics I've worked on at least partly due to my association with ML. These include, statistical topology, graphical models, semi-supervised inference, conformal prediction, and differential privacy.

Since this paper is supposed to be a personal reflection, let me now briefly discuss two of these ML problems that I have had the good fortune to work on. The point of these examples is to show how statistical thinking can be useful for machine learning.

44.4.1 Case study I: Semi-supervised inference

Suppose we observe data $(X_1, Y_1), \ldots, (X_n, Y_n)$ and we want to predict Y from X. If Y is discrete, this is a classification problem. If Y is real-valued, this is a regression problem. Further, suppose we observe more data

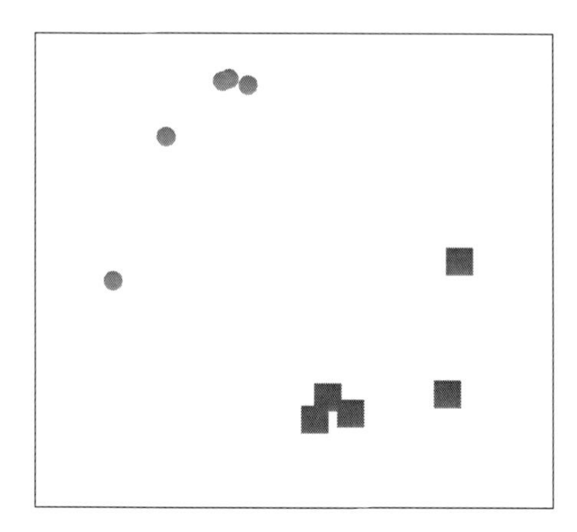

FIGURE 44.1
Labeled data.

$X_{n+1}, \ldots, X_N$ without the corresponding Y values. We thus have labeled data $\mathcal{L} = \{(X_1, Y_1), \ldots, (X_n, Y_n)\}$ and unlabeled data $\mathcal{U} = \{X_{n+1}, \ldots, X_N\}$. How do we use the unlabeled data in addition to the labeled data to improve prediction? This is the problem of *semi-supervised inference.*

Consider Figure 44.1. The covariate is $x = (x_1, x_2) \in \mathbb{R}^2$. The outcome in this case is binary as indicated by the circles and squares. Finding the decision boundary using only the labeled data is difficult. Figure 44.2 shows the labeled data together with some unlabeled data. We clearly see two clusters. If we make the additional assumption that $\Pr(Y = 1|X = x)$ is smooth relative to the clusters, then we can use the unlabeled data to nail down the decision boundary accurately.

There are copious papers with heuristic methods for taking advantage of unlabeled data. To see how useful these methods might be, consider the following example. We download one-million webpages with images of cats and dogs. We randomly select 100 pages and classify them by hand. Semi-supervised methods allow us to use the other 999,900 webpages to construct a good classifier.

But does semi-supervised inference work? Or, to put it another way, under what conditions does it work? In Azizyan et al. (2012), we showed the following (which I state informally here).

Suppose that $X_i \in \mathbb{R}^d$. Let $\mathcal{S}_n$ denote the set of supervised estimators; these estimators use only the labeled data. Let $\mathcal{SS}_N$ denote the set of semi-supervised estimators; these estimators use the labeled data and unlabeled data. Let m be the number of unlabeled data points and suppose that $m \geq n^{2/(2+\xi)}$ for some $0 < \xi < d - 3$. Let $f(x) = \mathrm{E}(Y|X = x)$. There is a large, nonparametric class of distributions $\mathcal{P}_n$ such that the following is true:

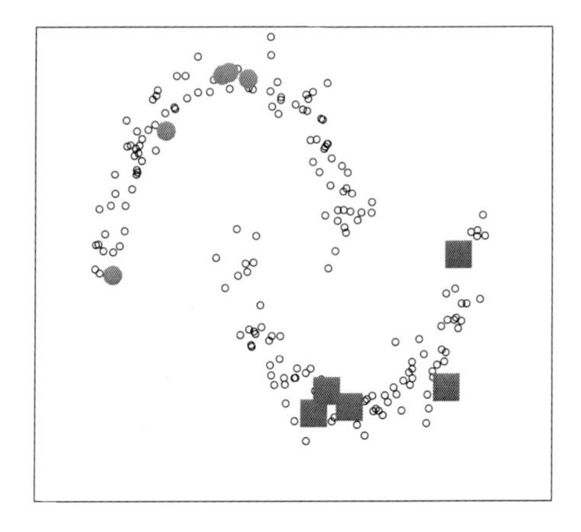

FIGURE 44.2
Labeled and unlabeled data.

1. There is a semi-supervised estimator $\widehat{f}$ such that

$$\sup_{P \in \mathcal{P}_n} R_P(\widehat{f}) \leq \left(\frac{C}{n}\right)^{\frac{2}{2+\xi}}, \tag{44.1}$$

   where $R_P(\widehat{f}) = \mathrm{E}\{\widehat{f}(X) - f(X)\}^2$ is the risk of the estimator $\widehat{f}$ under distribution P.

2. For supervised estimators $\mathcal{S}_n$, we have

$$\inf_{\widehat{f} \in \mathcal{S}_n} \sup_{P \in \mathcal{P}_n} R_P(\widehat{f}) \geq \left(\frac{C}{n}\right)^{\frac{2}{d-1}}. \tag{44.2}$$

3. Combining these two results, we conclude that

$$\frac{\inf_{\widehat{f} \in \mathcal{SS}_N} \sup_{P \in \mathcal{P}_n} R_P(\widehat{f})}{\inf_{\widehat{f} \in \mathcal{S}_n} \sup_{P \in \mathcal{P}_n} R_P(\widehat{f})} \leq \left(\frac{C}{n}\right)^{\frac{2(d-3-\xi)}{(2+\xi)(d-1)}} \longrightarrow 0 \tag{44.3}$$

   and hence, semi-supervised estimation dominates supervised estimation.

The class $\mathcal{P}_n$ consists of distributions such that the marginal for X is highly concentrated near some lower dimensional set and such that the regression function is smooth on this set. We have not proved that the class must be of this form for semi-supervised inference to improve on supervised inference but we suspect that is indeed the case. Our framework includes a parameter α that characterizes the strength of the semi-supervised assumption. We showed that, in fact, one can use the data to adapt to the correct value of α.

44.4.2 Case study II: Statistical topology

Computational topologists and researchers in Machine Learning have developed methods for analyzing the shape of functions and data. Here I'll briefly review some of our work on estimating manifolds (Genovese et al., 2012b,a,c).

Suppose that M is a manifold of dimension d embedded in $\mathbb{R}^D$. Let $X_1, \ldots, X_n$ be a sample from a distribution in P supported on M. We observe

$$Y_i = X_i + \epsilon_i, \quad i \in \{1, \ldots, n\} \tag{44.4}$$

where $\epsilon_1, \ldots, \epsilon_n \sim \Phi$ are noise variables.

Machine learning researchers have derived many methods for estimating the manifold M. But this leaves open an important statistical question: how well do these estimators work? One approach to answering this question is to find the minimax risk under some loss function. Let $\widehat{M}$ be an estimator of M. A natural loss function for this problem is Hausdorff loss:

$$H(M, \widehat{M}) = \inf\Big\{\epsilon : \ M \subset \widehat{M} \oplus \epsilon \ \text{ and } \ \widehat{M} \subset M \oplus \epsilon\Big\}. \tag{44.5}$$

Let $\mathcal{P}$ be a set of distributions. The parameter of interest is $M = \text{support}(P)$ which we assume is a d-dimensional manifold. The minimax risk is

$$R_n = \inf_{\widehat{M}} \sup_{P \in \mathcal{P}} \mathrm{E}_P[H(\widehat{M}, M)]. \tag{44.6}$$

Of course, the risk depends on what conditions we assume on M and on the noise Φ.

Our main findings are as follows. When there is no noise — so the data fall on the manifold — we get $R_n \asymp n^{-2/d}$. When the noise is perpendicular to M, the risk is $R_n \asymp n^{-2/(2+d)}$. When the noise is Gaussian the rate is $R_n \asymp 1/\log n$. The latter is not surprising when one considers the similar problem of estimating a function when there are errors in variables.

The implications for machine learning are that, the best their algorithms can do is highly dependent on the particulars of the type of noise.

How do we actually estimate these manifolds in practice? In Genovese et al. (2012c) we take the following point of view: If the noise is not too large, then the manifold should be close to a d-dimensional hyper-ridge in the density $p(y)$ for Y. Ridge finding is an extension of mode finding, which is a common task in computer vision.

Let p be a density on $\mathbb{R}^D$. Suppose that p has k modes $m_1, \ldots, m_k$. An integral curve, or path of steepest ascent, is a path $\pi : \mathbb{R} \to \mathbb{R}^D$ such that

$$\pi'(t) = \frac{d}{dt}\,\pi(t) = \nabla p\{\pi(t)\}. \tag{44.7}$$

Under weak conditions, the paths π partition the space and are disjoint except at the modes (Irwin, 1980; Chacón, 2012).

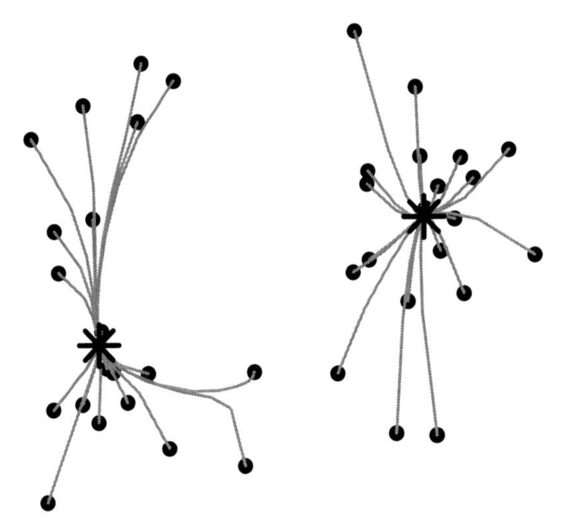

FIGURE 44.3
The mean shift algorithm. The data points move along trajectories during iterations until they reach the two modes marked by the two large asterisks.

The *mean shift algorithm* (Fukunaga and Hostetler, 1974; Comaniciu and Meer, 2002) is a method for finding the modes of a density by following the steepest ascent paths. The algorithm starts with a mesh of points and then moves the points along gradient ascent trajectories towards local maxima. A simple example is shown in Figure 44.3.

Given a function $p : \mathbb{R}^D \to \mathbb{R}$, let $g(x) = \nabla p(x)$ denote the gradient at x and let $H(x)$ denote the Hessian matrix. Let

$$\lambda_1(x) \geq \cdots \geq \lambda_D(x) \tag{44.8}$$

denote the eigenvalues of $H(x)$ and let $\Lambda(x)$ be the diagonal matrix whose diagonal elements are the eigenvalues. Write the spectral decomposition of $H(x)$ as $H(x) = U(x)\Lambda(x)U(x)^\top$. Fix $0 \leq d < D$ and let $V(x)$ be the last $D-d$ columns of $U(x)$ (i.e., the columns corresponding to the $D-d$ smallest eigenvalues). If we write $U(x) = [V_\diamond(x) : V(x)]$ then we can write $H(x) = [V_\diamond(x) : V(x)]\Lambda(x)[V_\diamond(x) : V(x)]^\top$. Let $L(x) = V(x)V(x)^\top$ be the projector on the linear space defined by the columns of $V(x)$. Define the *projected gradient*

$$G(x) = L(x)g(x). \tag{44.9}$$

If the vector field $G(x)$ is Lipschitz then by Theorem 3.39 of Irwin (1980), G defines a global flow as follows. The flow is a family of functions $\phi(x,t)$ such that $\phi(x,0) = x$ and $\phi'(x,0) = G(x)$ and $\phi(s,\phi(t,x)) = \phi(s+t,x)$. The flow lines, or integral curves, partition the space and at each x where $G(x)$ is non-null, there is a unique integral curve passing through x. The intuition is that the flow passing through x is a gradient ascent path moving towards higher values of p. Unlike the paths defined by the gradient g which move towards modes, the paths defined by G move towards ridges.

The paths can be parameterized in many ways. One commonly used parameterization is to use $t \in [-\infty, \infty]$ where large values of t correspond to higher values of p. In this case $t = \infty$ will correspond to a point on the ridge. In this parameterization we can express each integral curve in the flow as follows. A map $\pi : \mathbb{R} \to \mathbb{R}^D$ is an integral curve with respect to the flow of G if

$$\pi'(t) = G\{\pi(t)\} = L\{\pi(t)\}g\{\pi(t)\}. \tag{44.10}$$

Definition. The *ridge* R consists of the destinations of the integral curves: $y \in R$ if $\lim_{t\to\infty} \pi(t) = y$ for some π satisfying (44.10).

As mentioned above, the integral curves partition the space and for each $x \notin R$, there is a unique path π_x passing through x. The ridge points are zeros of the projected gradient: $y \in R$ implies that $G(y) = (0, \ldots, 0)^\top$. Ozertem and Erdogmus (2011) derived an extension of the mean-shift algorithm, called the *subspace constrained mean shift* algorithm that finds ridges which can be applied to the kernel density estimator. Our results can be summarized as follows:

1. Stability. We showed that if two functions are sufficiently close together then their ridges are also close together (in Hausdorff distance).

2. We constructed an estimator $\widehat{R}$ such that

$$H(R, \widehat{R}) = O_P\left(\left(\frac{\log n}{n}\right)^{\frac{2}{D+8}}\right) \tag{44.11}$$

where H is the Hausdorff distance. Further, we showed that $\widehat{R}$ is topologically similar to R. We also construct an estimator $\widehat{R}_h$ for $h > 0$ that satisfies

$$H(R_h, \widehat{R}_h) = O_P\left(\left(\frac{\log n}{n}\right)^{\frac{1}{2}}\right), \tag{44.12}$$

where R_h is a smoothed version of R.

3. Suppose the data are obtained by sampling points on a manifold and adding noise with small variance σ^2. We showed that the resulting density p has a ridge R_σ such that

$$H(M, R_\sigma) = O\left(\sigma^2 \log^3(1/\sigma)\right) \tag{44.13}$$

and R_σ is topologically similar to M. Hence when the noise σ is small, the ridge is close to M. It then follows that

$$H(M, \widehat{R}) = O_P\left(\left(\frac{\log n}{n}\right)^{\frac{2}{D+8}}\right) + O\left(\sigma^2 \log^3(1/\sigma)\right). \tag{44.14}$$

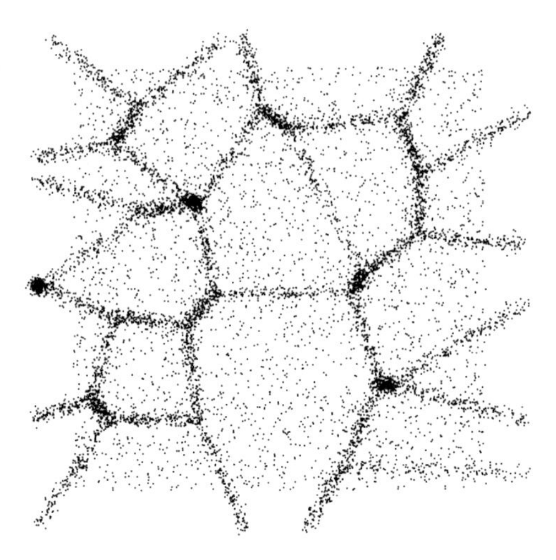

FIGURE 44.4
Simulated cosmic web data.

An example can be found in Figures 44.4 and 44.5. I believe that statistics has much to offer to this area especially in terms of making the assumptions precise and clarifying how accurate the inferences can be.

44.5 Computational thinking

There is another interesting difference that is worth pondering. Consider the problem of estimating a mixture of Gaussians. In statistics we think of this as a solved problem. You use, for example, maximum likelihood which is implemented by the EM algorithm. But the EM algorithm does not solve the problem. There is no guarantee that the EM algorithm will actually find the MLE; it's a shot in the dark. The same comment applies to MCMC methods.

In ML, when you say you've solved the problem, you mean that there is a polynomial time algorithm with provable guarantees. There is, in fact, a rich literature in ML on estimating mixtures that do provide polynomial time algorithms. Furthermore, they come with theorems telling you how many observations you need if you want the estimator to be a certain distance from the truth, with probability at least $1-\delta$. This is typical for what is expected of an estimator in ML. You need to provide a provable polynomial time algorithm and a finite sample (non-asymptotic) guarantee on the estimator.

ML puts heavier emphasis on computational thinking. Consider, for example, the difference between P and NP-hard problems. This is at the heart

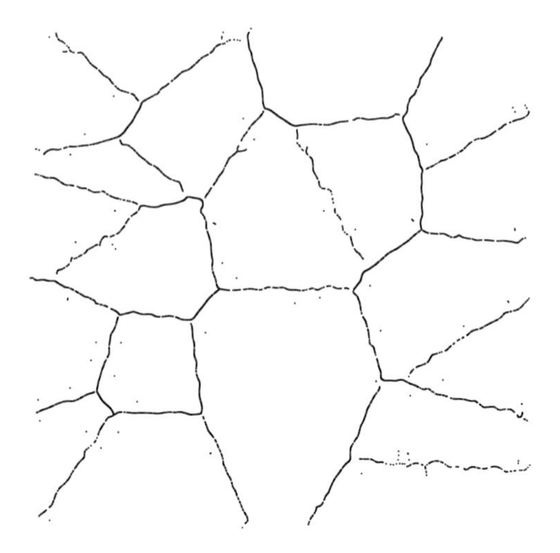

FIGURE 44.5
Ridge finder applied to simulated cosmic web data.

of theoretical computer science and ML. Running an MCMC on an NP-hard problem might be meaningless. Instead, it may be better to approximate the NP-hard problem with a simpler problem. How often do we teach this to our students?

44.6 The evolving meaning of data

For most of us in statistics, data means numbers. But data now includes images, documents, videos, web pages, twitter feeds and so on. Traditional data — numbers from experiments and observational studies — are still of vital importance but they represent a tiny fraction of the data out there. If we take the union of all the data in the world, what fraction is being analyzed by statisticians? I think it is a small number.

This comes back to education. If our students can't analyze giant datasets like millions of twitter feeds or millions of web pages, then other people will analyze those data. We will end up with a small cut of the pie.

44.7 Education and hiring

The goal of a graduate student in statistics is to find an advisor and write a thesis. They graduate with a single data point; their thesis work.

The goal of a graduate student in ML is to find a dozen different research problems to work on and publish many papers. They graduate with a rich data set; many papers on many topics with many different people.

Having been on hiring committees for both statistics and ML, I can say that the difference is striking. It is easier to choose candidates to interview in ML. You have a lot of data on each candidate and you know what you are getting. In statistics, it is a struggle. You have little more than a few papers that bear their advisor's footprint.

The ML conference culture encourages publishing many papers on many topics which is better for both the students and their potential employers. And now, statistics students are competing with ML students, putting statistics students at a significant disadvantage.

There are a number of topics that are routinely covered in ML that we rarely teach in statistics. Examples are: Vapnik–Chervonenkis theory, concentration of measure, random matrices, convex optimization, graphical models, reproducing kernel Hilbert spaces, support vector machines, and sequential game theory. It is time to get rid of antiques like UMVUE, complete statistics and so on, and teach modern ideas.

44.8 If you can't beat them, join them

I don't want to leave the reader with the impression that we are in some sort of competition with ML. Instead, we should feel blessed that a second group of statisticians has appeared. Working with ML and adopting some of their ideas enriches both fields.

ML has much to offer statistics. And statisticians have a lot to offer ML. For example, we put much emphasis on quantifying uncertainty (standard errors, confidence intervals, posterior distributions), an emphasis that is perhaps lacking in ML. And sometimes, statistical thinking casts new light on existing ML methods. A good example is the statistical view of boosting given in Friedman et al. (2000). I hope we will see collaboration and cooperation between the two fields thrive in the years to come.

Acknowledgements

I'd like to thank Aaditya Ramdas, Kathryn Roeder, Rob Tibshirani, Ryan Tibshirani, Isa Verdinelli, a referee and readers of my blog for reading a draft of this essay and providing helpful suggestions.

References

Azizyan, M., Singh, A., and Wasserman, L.A. (2013). Density-sensitive semi-supervised inference. *The Annals of Statistics*, 41:751–771.

Chacón, J.E. (2012). Clusters and water flows: A novel approach to modal clustering through morse theory. *arXiv preprint arXiv:1212.1384*.

Comaniciu, D. and Meer, P. (2002). Mean shift: A robust approach toward feature space analysis. *Pattern Analysis and Machine Intelligence, IEEE Transactions on*, 24:603 –619.

Friedman, J., Hastie, T., and Tibshirani, R.J. (2000). Additive logistic regression: A statistical view of boosting (with discussion). *The Annals of Statistics*, 28:337–407.

Fukunaga, K. and Hostetler, L.D. (1975). The estimation of the gradient of a density function, with applications in pattern recognition. *IEEE Transactions on Information Theory*, 21:32–40.

Genovese, C.R., Perone-Pacifico, M., Verdinelli, I., and Wasserman, L.A. (2012). Manifold estimation and singular deconvolution under hausdorff loss. *The Annals of Statistics*, 40:941–963.

Genovese, C.R., Perone-Pacifico, M., Verdinelli, I., and Wasserman, L.A. (2012). Minimax manifold estimation. *Journal of Machine Learning Research*, 13:1263–1291.

Genovese, C.R., Perone-Pacifico, M., Verdinelli, I., and Wasserman, L.A. (2012). Nonparametric ridge estimation. *arXiv preprint arXiv:1212.5156*.

Irwin, M.C. (1980). *Smooth Dynamical Systems*. Academic Press, New York.

Ozertem, U. and Erdogmus, D. (2011). Locally defined principal curves and surfaces. *Journal of Machine Learning Research*, 12:1249–1286.

45

A trio of inference problems that could win you a Nobel Prize in statistics (if you help fund it)

Xiao-Li Meng
Department of Statistics
Harvard University, Cambridge, MA

Statistical inference is a field full of problems whose solutions require the same intellectual force needed to win a Nobel Prize in other scientific fields. Multi-resolution inference is the oldest of the trio. But emerging applications such as individualized medicine have challenged us to the limit: infer estimands with resolution levels that far exceed those of any feasible estimator. Multi-phase inference is another reality because (big) data are almost never collected, processed, and analyzed in a single phase. The newest of the trio is multi-source inference, which aims to extract information in data coming from very different sources, some of which were never intended for inference purposes. All of these challenges call for an expanded paradigm with greater emphases on qualitative consistency and relative optimality than do our current inference paradigms.

45.1 Nobel Prize? Why not COPSS?

The title of my chapter is designed to grab attention. But why Nobel Prize (NP)? Wouldn't it be more fitting, for a volume celebrating the 50th anniversary of COPSS, to entitle it "A Trio of Inference Problems That Could Win You a COPSS Award (and you don't even have to fund it)?" Indeed, some media and individuals have even claimed that the COPSS Presidents' Award is the NP in Statistics, just as they consider the Fields Medal to be the NP in Mathematics.

No matter how our egos might wish such a claim to be true, let us face the reality. There is no NP in statistics, and worse, the general public does not seem to appreciate statistics as a "rocket science" field. Or as a recent blog (August 14, 2013) in *Simply Statistics* put it: "Statistics/statisticians need better marketing" because (among other reasons)

> "Our top awards don't get the press they do in other fields. The Nobel Prize announcements are an international event. There is always speculation/intense interest in who will win. There is similar interest around the Fields Medal in mathematics. But the top award in statistics, the COPSS award, doesn't get nearly the attention it should. Part of the reason is lack of funding (the Fields is \$15K, the COPSS is \$1K). But part of the reason is that we, as statisticians, don't announce it, share it, speculate about it, tell our friends about it, etc. The prestige of these awards can have a big impact on the visibility of a field."

The fact that there is more public interest in the Fields than in COPSS should make most statisticians pause. No right mind would downplay the centrality of mathematics in scientific and societal advancement throughout human history. Statistics seems to be starting to enjoy a similar reputation as being at the core of such endeavors as we move deeper into the digital age. However, the attention around top mathematical awards such as the Fields Medal has hardly been about their direct or even indirect impact on everyday life, in sharp contrast to our emphasis on the practicality of our profession. Rather, these awards arouse media and public interest by featuring how ingenious the awardees are and how difficult the problems they solved, much like how conquering Everest bestows admiration not because the admirers care or even know much about Everest itself but because it represents the ultimate physical feat. In this sense, the biggest winner of the Fields Medal is mathematics itself: enticing the brightest talent to seek the ultimate intellectual challenges.

And *that* is the point I want to reflect upon. Have we statisticians adequately conveyed to the media and general public the depth and complexity of our beloved subject, in addition to its utility? Have we tried to demonstrate that the field of statistics has problems (e.g., modeling ignorance) that are as intellectually challenging as the Goldbach conjecture or Riemann Hypothesis, and arguably even more so because our problems cannot be formulated by mathematics alone? In our effort to make statistics as simple as possible for general users, have we also emphasized adequately that reading a couple of stat books or taking a couple of stat courses does not qualify one to teach statistics?

In recent years I have written about making statistics as easy to learn as possible. But my emphasis (Meng, 2009b) has been that we must make a tremendous collective effort to change the perception that "Statistics is easy to teach, but hard (and boring) to learn" to a reality of "Statistics is hard to teach, but easy (and fun) to learn." Statistics is hard to teach because it is intellectually a very demanding subject, and to teach it well requires both depth in theory and breadth in application. It is easy and fun to learn because it is directly rooted in everyday life (when it is conveyed as such) and it builds upon many common logics, not because it lacks challenging problems or deep theory.

Therefore, the invocation of NP in the title is meant to remind ourselves that we can also attract the best minds to statistics by demonstrating how intellectually demanding it is. As a local example, my colleague Joe Blitzstein turned our Stat110 from an enrollment of about 80 to over 480 by making it both more real-life rooted and more intellectually demanding. The course has become a Harvard sensation, to the point that when our students' newspaper advises freshmen "how to make 20% effort and receive 80% grade," it explicitly states that Stat110 is an exception and should be taken regardless of the effort required. And of course the NPs in the natural and social sciences are aimed at work with enormous depth, profound impact, and ideally both. The trio of inference problems described below share these features — their solutions require developing some of the deepest theory in inference, and their impacts are immeasurable because of their ubiquity in quantitative scientific inquiries.

The target readership of this chapter can best be described by a Chinese proverb: "Newborn calves are unafraid of tigers," meaning those young talents who are particularly curious and courageous in their intellectual pursuits. I surely hope that future COPSS (if not NP) winners are among them.

45.2 Multi-resolution inference

To borrow an engineering term, a central task of statistical inference is to separate signal from noise in the data. But what is signal and what is noise? Traditionally, we teach this separation by writing down a regression model, typically linear,

$$Y = \sum_{i=0}^{p} \beta_i X_i + \epsilon,$$

with the regression function $\sum_{i=0}^{p} \beta_i X_i$ as signal, and ϵ as noise. Soon we teach that the real meaning of ϵ is anything that is not captured by our designated "signal," and hence the "noise" ϵ could still contain, in real terms, signals of interest or that should be of interest.

This seemingly obvious point reminds us that the concepts of signal and noise are relative — noise for one study can be signal for another, and vice versa. This relativity is particularly clear for those who are familiar with multi-resolution methods in engineering and applied mathematics, such as wavelets (see Daubechies, 1992; Meyer, 1993), where we use wavelet coefficients below or at a primary resolution for estimating signals. The higher frequency ones are treated as noise and used for variance estimation; see Donoho and Johnstone (1994), Donoho et al. (1995) and Nason (2002). Therefore what counts for signal or noise depends entirely on our choice of the primary resolution. The multi-resolution framework described below is indeed inspired by my learning of wavelets and related multi-resolution methods (Bouman et al., 2005, 2007; Lee and Meng, 2005; Hirakawa and Meng, 2006), and motivated by the need to deal with Big Data, where the complexity of emerging questions has forced us to go diving for perceived signals in what would have been discarded as noise merely a decade ago.

But how much of the signal that our inference machine recovers will be robust to the assumptions we make (e.g., via likelihood, prior, estimating equations, etc.) and how much will wash out as noise with the ebb and flow of our assumptions? Such a question arose when I was asked to help analyze a large national survey on health, where the investigator was interested in studying men over 55 years old who had immigrated to the US from a particular country, among other such "subpopulation analyses." You may wonder what is so special about wanting such an analysis. Well, nothing really, except that there was not a single man in the dataset who fit the description! I was therefore brought in to deal with the problem because the investigator had learned that I could perform the magic of multiple imputation. (Imagine how much data collection resource could have been saved if I could multiply impute myself!)

Surely I could (and did) build some hierarchical model to "borrow information," as is typical for small area estimations; see Gelman et al. (2003) and Rao (2005). In the dataset, there were men over 55, men who immigrated from that country, and even men over 55 who immigrated from a neighboring country. That is, although we had no direct data from the subpopulation of interest, we had plenty of indirect data from related populations, however defined. But how confident should I be that whatever my hierarchical machine produces is reproducible by someone who actually has direct data from the target subpopulation?

Of course you may ask why did the investigator want to study a subpopulation with no direct data whatsoever? The answer turned out to be rather simple and logical. Just like we statisticians want to work on topics that are new and/or challenging, (social) scientists want to do the same. They are much less interested in repeating well-established results for large populations than in making headway on subpopulations that are difficult to study. And what could be more difficult than studying a subpopulation with no data? Indeed, political scientists and others routinely face the problem of empty cells in contingency tables; see Gelman and Little (1997) and Lax and Phillips (2009).

If you think this sounds rhetorical or even cynical, consider the rapidly increasing interest in individualized medicine. If I am sick and given a choice of treatments, the central question to me is which treatment has the best chance to cure me, not some randomly selected 'representative' person. There is no logical difference between this desire and the aforementioned investigator's desire to study a subpopulation with no observations. The clinical trials testing these treatments surely did not include a subject replicating my description exactly, but this does not stop me from desiring individualized treatments. The grand challenge therefore is how to infer an estimand with granularity or resolution that (far) exceeds what can be estimated directly from the data, i.e., we run out of enough sample replications (way) before reaching the desired resolution level.

45.2.1 Resolution via filtration and decomposition

To quantify the role of resolution for inference, consider an outcome variable Y living on the same probability space as an information filtration $\{\mathcal{F}_r, r = 0, \ldots, R\}$. For example, $\mathcal{F}_r = \sigma(X_0, \ldots, X_r)$, the σ-field generated by covariates $\{X_0, \ldots, X_r\}$, which perhaps is the most common practical situation. The discussion below is general, as long as $\mathcal{F}_{r-1} \subset \mathcal{F}_r$, where $r \in \{1, \ldots, R\}$ can be viewed as an index of resolution. Intuitively, we can view $\mathcal{F}_r$ as a set of specifications that restrict our target population — the increased specification/information as captured by $\mathcal{F}_r$ allows us to zoom into more specific subpopulations; here we assume $\mathcal{F}_0$ is the trivial zero-information filter, i.e., X_0 represents the constant intercept term, and $\mathcal{F}_R$ is the maximal filter, e.g., with infinite resolution to identify a unique individual, and R can be infinite. Let

$$\mu_r = \mathrm{E}(Y|\mathcal{F}_r) \quad \text{and} \quad \sigma_r^2 = \mathrm{var}(Y|\mathcal{F}_r)$$

be the conditional mean (i.e., regression) and conditional variance (or covariance) of Y given $\mathcal{F}_r$, respectively. When $\mathcal{F}_r$ is generated by $\{X_0, \ldots, X_r\}$, we have the familiar $\mu_r = \mathrm{E}(Y|X_0, \ldots, X_r)$ and $\sigma_r^2 = \mathrm{var}(Y|X_0, \ldots, X_r)$.

Applying the familiar EVE law

$$\mathrm{var}(Y|\mathcal{F}_r) = \mathrm{E}\{\mathrm{var}(Y|\mathcal{F}_s)|\mathcal{F}_r\} + \mathrm{var}\{\mathrm{E}(Y|\mathcal{F}_s)|\mathcal{F}_r\},$$

where $s > r$, we obtain the conditional ANOVA decomposition

$$\sigma_r^2 = \mathrm{E}(\sigma_s^2|\mathcal{F}_r) + \mathrm{E}\{(\mu_s - \mu_r)^2|\mathcal{F}_r\}. \tag{45.1}$$

This key identity reveals that the (conditional) variance at resolution r is the sum of an estimated variance and an estimated (squared) bias. In particular, we use the information in $\mathcal{F}_r$ (and our model assumptions) to estimate the variance at the higher resolution s and to estimate the squared bias incurred from using μ_r to proxy for μ_s. This perspective stresses that σ_r^2 is itself also an estimator, in fact our best guess at the reproducibility of our indirect data inference at resolution r by someone with direct data at resolution s.

This dual role of being simultaneously an estimand (of a lower resolution estimator) and an estimator (of a higher resolution estimand) is the essence of the multi-resolution formulation, unifying the concepts of variance and bias, and of model estimation and model selection. Specifically, when we set up a model with the signal part at a particular resolution r (e.g., $r = p$ for the linear model), we consider μ_r to be an acceptable estimate for any μ_s with $s > r$. That is, even though the difference between μ_s and μ_r reflects systematic variation, we purposely re-classify it as a component of random variation.

In the strictest sense, bias results whenever real information remains in the residual variation (e.g., the ϵ term in the linear model). However, statisticians have chosen to further categorize bias in this strict sense depending on whether it occurs above or below/at the resolution level r. When the information in the residual variation resides in resolutions higher than r then we use the term "variance" for the price of failing to include that information. When the residual information resides in resolutions lower than or at r, then we keep the designation "bias." This categorization, just as the mathematician's O notation, serves many useful purposes, but we should not forget that it is ultimately artificial.

This point is most clear when we apply (45.1) in a telescopic fashion (by first making $s = r + 1$ and then summing over r) and when $R = \infty$:

$$\sigma_r^2 = \mathrm{E}(\sigma_\infty^2 | \mathcal{F}_r) + \sum_{i=r}^{\infty} \mathrm{E}\{(\mu_{i+1} - \mu_i)^2 | \mathcal{F}_r\}. \tag{45.2}$$

The use of $R = \infty$ is a mathematical idealization of the situations where our specifications can go on indefinitely, such as with individualized medicine, where we have height, weight, age, gender, race, education, habit, all sorts of medical test results, family history, genetic compositions, environmental factors, etc. That is, we switch from the hopeless $n = 1$ (i.e., a single individual) case to the hopeful $R = \infty$ scenario. The σ_∞^2 term captures the variation of the population at infinite resolution. Whether σ_∞^2 should be set to zero or not reflects whether we believe the world is fundamentally stochastic or appears to be stochastic because of our human limitation in learning every mechanism responsible for variations, as captured by $\mathcal{F}_\infty$. In that sense σ_∞^2 can be viewed as the intrinsic variance with respect to a given filtration. Everything else in the variance at resolution r are merely biases (e.g., from using μ_i to estimate μ_{i+1}) accumulated at higher resolutions.

45.2.2 Resolution model estimation and selection

When $\sigma_\infty^2 = 0$, the infinite-resolution setup essentially is the same as a potential outcome model (Rubin, 2005), because the resulting population is of size one and hence comparisons on treatment effects must be counterfactual. This is exactly the right causal question for individualized treatments: what would be my (health, test) outcome if I receive one treatment versus another? In order to estimate such an effect, however, we must lower the resolution to a finite and often small degree, making it possible to estimate average treatment effects, by averaging over a population that permits some degrees of replication. We then hope that the attributes (i.e., predictors) left in the "noise" will not contain enough real signals to alter our quantitative results, as compared to if we had enough data to model those attributes as signals, to a degree that would change our qualitative conclusions, such as choosing one treatment versus another.

That is, when we do not have enough (direct) data to estimate μ_R, we first choose a $\mathcal{F}_{\tilde{r}}$, and then estimate μ_R by $\hat{\mu}_{\tilde{r}}$. The "double decoration" notation $\hat{\mu}_{\tilde{r}}$ highlights two kinds of error:

$$\hat{\mu}_{\tilde{r}} - \mu_R = (\hat{\mu}_{\tilde{r}} - \mu_{\tilde{r}}) + (\mu_{\tilde{r}} - \mu_R). \tag{45.3}$$

The first parenthesized term in (45.3) represents the usual model estimation error (for the given $\tilde{r}$), and hence the usual "hat" notation. The second is the bias induced by the resolution discrepancy between our actual estimand and intended estimand, which represents the often forgotten model selection error. As such, we use the more ambiguous "tilde" notation $\tilde{r}$, because its construction cannot be based on data alone, and it is not an estimator of R (e.g., we hope $\tilde{r} \ll R$).

Determining $\tilde{r}$, as a model selection problem, then inherits the usual bias-variance trade-off issue. Therefore, any attempt to find an "automated" way to determine $\tilde{r}$ would be as disappointing as those aimed at automated procedures for optimal bias-variance trade-off (see Meng, 2009a; Blitzstein and Meng, 2010). Consequently, we must make assumptions in order to proceed. Here the hope is that the resolution formulation can provide alternative or even better ways to pose assumptions suitable for quantifying the trade-off in practice and for combating other thorny issues, such as nuisance parameters. In particular, if we consider the filtration $\{\mathcal{F}_r, r = 0, 1, \ldots\}$ as a cumulative "information basis," then the choice of $\tilde{r}$ essentially is in the same spirit as finding a sparse representation in wavelets, for which there is a large literature; see, e.g., Donoho and Elad (2003), Poggio and Girosi (1998), and Yang et al. (2009). Here, though, it is more appropriate to label $\mu_{\tilde{r}}$ as a parsimonious representation of μ_R.

As usual, we can impose assumptions via prior specifications (or penalty for penalized likelihood). For example, we can impose a prior on the model complexity $\tilde{R}_\delta$, the smallest (fixed) r such that $\mathrm{E}\{(\mu_r - \mu_R)^2\} \leq \delta$, where δ represents the acceptable trade-off between granularity and model complexity (e.g., involving more X's) and the associated data and computational cost. Clearly $\tilde{R}_\delta$ always exists but it may be the case that $\tilde{R}_\delta = R$, which means that no lower-resolution approximation is acceptable for the given δ.

Directly posing a prior for $\tilde{R}_\delta$ is similar to using L_0-regularization (Lin et al., 2010). Its usefulness depends on whether we can expect all X_r's to be more or less exchangeable in terms of their predictive power. Otherwise, the resolution framework reminds us to consider putting a prior on the ordering of the X_i's (in terms of predictive power). Conditional on the ordering, we impose priors on the predictive power of incremental complexity, $\Delta_r = \mu_{r+1} - \mu_r$. These priors should reflect our expectation for Δ_r^2 to decay with r, such as imposing $\mathrm{E}(\Delta_r^2) > \mathrm{E}(\Delta_{r+1}^2)$. If monotonicity seems too strong an assumption, we could first break the X_i's into groups, assume exchangeability within each group, and then order the groups according to predictive power. That is to say, finding a complete ordering of the X_i's may require prior knowledge that is too refined. We weaken this knowledge requirement by seeking only an ordering over equivalence classes of the X_i's where each equivalence class represents a set of variables which we are not able to a priori distinguish with respect to predictive power. The telescoping additivity in (45.2) implies that imposing a prior on the magnitude of Δ_r will induce a control over the "total resolution bias" (TRB)

$$\mathrm{E}(\mu_{\tilde{R}_\delta} - \mu_R)^2 = \sum_{r=\tilde{R}_\delta}^{R} \mathrm{E}(\mu_r - \mu_{r+1})^2,$$

which holds because Δ_r and Δ_s are orthogonal (i.e., uncorrelated) when $s \neq r$.

A good illustration of this rationale is provided when $\mathcal{F}_r$ is generated by a series of binary variables $\{X_0, \ldots, X_r\}$ with $r \in \{0, \ldots, R\}$. In such cases, our multi-resolution setup is equivalent to assuming a weighted binary tree model with total depth R; see Knuth (1997) and Garey (1974). Here each node is represented by a realization of $\vec{X}_r = (X_0, \ldots, X_r)$, $\vec{x}_r = (x_0, \ldots, x_r)$, at which the weights of its two (forward) branches are given by $w_{\vec{x}_r}(x) = \mathrm{E}(Y|\vec{X}_r = \vec{x}_r, X_{r+1} = x)$ respectively with $x = 0, 1$. It is then easy to show that

$$\mathrm{E}(\Delta_r^2) \leq \frac{1}{4}\mathrm{E}\{w_{\vec{X}_r}(1) - w_{\vec{X}_r}(0)\}^2 \equiv \frac{1}{4}\mathrm{E}\{D^2(\vec{X}_r)\},$$

where $D^2(\vec{X}_r)$ is a measure of the predictive power of X_{r+1} that is not already contained in $\vec{X}_r$. For the previous linear regression, $D^2(\vec{X}_r) = \beta_{r+1}^2$. Thus putting a prior on $D^2(\vec{X}_r)$ can be viewed as a generalization of putting a prior on the regression coefficient, as routinely done in Bayesian variable selection; see Mitchell and Beauchamp (1988) and George and McCulloch (1997).

It is worthwhile to emphasize that Bayesian methods, or at least the idea of introducing assumptions on Δ_r's, seems inevitable. This is because "pure" data-driven type of methods, such as cross-validation (Arlot and Celisse, 2010), are unlikely to be fruitful here — the basic motivation of a multi-resolution framework is the lack of sufficient replications at high resolutions (unless we impose non-testable exchangeability assumptions to justify synthetic replications, but then we are just being Bayesian). It is equally important to point out that the currently dominant practice of pretending $\mu_{\tilde{R}} = \mu_R$ makes the strongest Bayesian assumption of all: the TRB, and hence any Δ_r $(r \geq \tilde{R})$, is exactly zero. In this sense, using a non-trivial prior for Δ_r makes less extreme assumptions than currently done in practice.

In a nutshell, a central aim of putting a prior on Δ_r to regulate the predictive power of the covariates is to identify practical ways of ordering a set of covariates to form the filtration $\{\mathcal{F}_r, r \geq 0\}$ to achieve rapid decay of $\mathrm{E}(\Delta_r^2)$ as r increases, essentially the same goal as for stepwise regression or principal component analysis. By exploring the multi-resolution formulation we hope to identify viable alternatives to common approaches such as LASSO. In general, for the multi-resolution framework to be fruitful beyond the conceptual level, many fundamental and methodological questions must be answered. The three questions below are merely antipasti to whet your appetite (for NP, or not):

(a) For what classes of models on $\{Y, X_j, j = 0, \ldots, R\}$ and priors on ordering and predictive power, can we determine practically an order $\{X_{(j)}, j \geq 0\}$ such that the resulting $\mathcal{F}_r = \sigma(X_{(j)}, j = 0, \ldots, r)$ will ensure a parsimonious representation of μ_R with quantifiably high probability?

(b) What should be our guiding principles for making a trade-off between sample size n and recorded/measured data resolution R, when we have the choice between having more data of lower quality (large n, small R) or less data of higher quality (small n, large R)?

(c) How do we determine the appropriate resolution level for hypothesis testing, considering that hypotheses testing involving higher resolution estimands typically lead to larger multiplicity? How much multiplicity can we reasonably expect our data to accommodate, and how do we quantify it?

45.3 Multi-phase inference

Most of us learned about statistical modelling in the following way. We have a data set that can be described by a random variable Y, which can be modelled by a probability function or density $\Pr(Y|\theta)$. Here θ is a model parameter, which can be of infinite dimension when we adopt a non-parametric or semi-parametric philosophy. Many of us were also taught to resist the temptation of using a model just because it is convenient, mentally, mathematically, or computationally. Instead, we were taught to learn as much as possible about the data generating process, and think critically about what makes sense substantively, scientifically, and statistically. We were then told to check and re-check the goodness-of-fit, or rather the lack of fit, of the model to our data, and to revise our model whenever our resources (time, energy, and funding) permit.

These pieces of advice are all very sound. Indeed, a hallmark of statistics as a scientific discipline is its emphasis on critical and principled thinking about the entire process from data collection to analysis to interpretation to communication of results. However, when we take our proud way of thinking (or our reputation) most seriously, we will find that we have not practiced what we have preached in a rather fundamental way.

I wish this were merely an attention-grabbing statement like the title of my chapter. But the reality is that when we put down a single model $\Pr(Y|\theta)$, however sophisticated or "assumption-free," we have already simplified too much. The reason is simple. In real life, especially in this age of Big Data, the data arriving at an analyst's desk or disk are almost never the original raw data, however defined. These data have been pre-processed, often in multiple phases, because someone felt that they were too dirty to be useful, or too large to pass on, or too confidential to let the user see everything, or all of the above! Examples range from microarrays to astrophysics; see Blocker and Meng (2013).

"So what?" Some may argue that all this can be captured by our model $\Pr(Y|\theta)$, at least in theory, if we have made enough effort to learn about the entire process. Putting aside the impossibility of learning about everything in practice (Blocker and Meng, 2013), we will see that the single-model formulation is simply not rich enough to capture reality, even if we assume that every pre-processor and analyst have done everything correctly. The trouble here is that pre-processors and analysts have different goals, have access to different data resources, and make different assumptions. They typically do not and cannot communicate with each other, resulting in separate (model) assumptions that no single probabilistic model can coherently encapsulate. We need a multiplicity of models to capture a multiplicity of incompatible assumptions.

45.3.1 Multiple imputation and uncongeniality

I learned about these complications during my study of the multiple imputation (MI) method (Rubin, 1987), where the pre-processor is the imputer. The imputer's goal was to preserve as much as possible in the imputed data the joint distributional properties of the original complete data (assuming, of course, the original complete-data samples were scientifically designed so that their properties are worthy of preservation). For that purpose, the imputer should and will use anything that can help, including confidential information, as well as powerful predictive models that may not capture the correct causal relations.

In addition, because the imputed data typically will be used for many purposes, most of which cannot be anticipated at the time of imputation, the imputation model needs to include as many predictors as possible, and be as saturated as the data and resources permit; see Meng (1994) and Rubin (1996). In contrast, an analysis model, or rather an approach (e.g., given by software), often focuses on specific questions and may involve only a (small) subset of the variables used by the imputer. Consequently, the imputer's model and the user's procedure may be uncongenial to each other, meaning that no model can be compatible with both the imputer's model and the user's procedure. The technical definitions of congeniality are given in Meng (1994) and Xie and Meng (2013), which involve embedding an analyst's procedure (often of frequentist nature) into an imputation model (typically with Bayesian flavor). For the purposes of the following discussion, two models are "congenial" if their implied imputation and analysis procedures are the same. That is, they are operationally, though perhaps not theoretically, equivalent.

Ironically, the original motivation of MI (Rubin, 1987) was a separation of labor, asking those who have more knowledge and resources (e.g., the US Census Bureau) to fix/impute the missing observations, with the hope that subsequent analysts can then apply their favorite complete-data analysis procedures to reach valid inferences. This same separation creates the issue of uncongeniality. The consequences of uncongeniality can be severe, from both theoretical and practical points of view. Perhaps the most striking example is that the very appealing variance combining rule for MI inference derived under congeniality (and another application of the aforementioned EVE law), namely,

$$\mathrm{var}_{\mathrm{Total}} = \mathrm{var}_{\mathrm{Between-imputation}} + \mathrm{var}_{\mathrm{Within-imputation}} \tag{45.4}$$

can lead to seriously invalid results in the presence of uncongeniality, as reported initially by Fay (1992) and Kott (1995).

Specifically, the so-called Rubin's variance combining rule is based on (45.4), where

$$\mathrm{var}_{\mathrm{Between-imputation}} \quad \text{and} \quad \mathrm{var}_{\mathrm{Within-imputation}}$$

are estimated by $(1+m^{-1})B_m$ and $\bar{U}_m$, respectively (Rubin, 1987). Here the $(1+m^{-1})$ factor accounts for the Monte Carlo error due to finite m, B_m is the sampling variance of $\hat{\theta}^{(\ell)} \equiv \hat{\theta}_A(Y_{\mathrm{com}}^{(\ell)})$ and $\bar{U}_m$ is the sample average of $U(Y_{\mathrm{com}}^{(\ell)}), \ell = 1, \ldots, m$, where $\hat{\theta}_A(Y_{\mathrm{com}})$ is the analyst's complete-data estimator for θ, $U(Y_{\mathrm{com}})$ is its associated variance (estimator), and $Y_{\mathrm{mis}}^{(\ell)}$ are i.i.d. draws from an imputation model $P_I(Y_{\mathrm{mis}}|Y_{\mathrm{obs}})$. Here, for notational convenience, we assume the complete data Y_{com} can be decomposed into the missing data Y_{mis} and observed data Y_{obs}. The left-hand side of (45.4) then is meant to be an estimator, denoted by T_m, of the variance of the MI estimator of θ, i.e., $\bar{\theta}_m$, the average of $\{\hat{\theta}^{(\ell)}, \ell = 1, \ldots, m\}$.

To understand the behavior of $\bar{\theta}_m$ and T_m, let us consider a relatively simple case where the missing data are missing at random (Rubin, 1976), and the imputer does not have any additional data. Yet the imputer has adopted a Bayesian model uncongenial to the analyst's complete-data likelihood function, $P_A(Y_{\mathrm{com}}|\theta)$, even though both contain the true data-generating model as a special case. For example, the analyst may have correctly assumed that two subpopulations share the same mean, an assumption that is not in the imputation model; see Meng (1994) and Xie and Meng (2013). Furthermore, we assume the analyst's complete-data procedure is the fully efficient MLE $\hat{\theta}_A(Y_{\mathrm{com}})$, and $U_A(Y_{\mathrm{com}})$, say, is the usual inverse of Fisher information.

Clearly we need to take into account both the sampling variability and imputation uncertainty, and for consistency we need to take both imputation size $m \to \infty$ and data size $n \to \infty$. That is, we need to consider replications generated by the hybrid model (note $P_I(Y_{\mathrm{mis}}|Y_{\mathrm{obs}})$ is free of θ):

$$P_H(Y_{\mathrm{mis}}, Y_{\mathrm{obs}}|\theta) = P_I(Y_{\mathrm{mis}}|Y_{\mathrm{obs}})P_A(Y_{\mathrm{obs}}|\theta), \tag{45.5}$$

where $P_A(Y_{\mathrm{obs}}|\theta)$ is derived from the analyst's complete-data model $P_A(Y_{\mathrm{com}}|\theta)$.

To illustrate the complication caused by uncongeniality, let us assume $m = \infty$ to eliminate the distraction of Monte Carlo error due to finite m. Writing

$$\bar{\theta}_\infty - \theta = \{\bar{\theta}_\infty - \hat{\theta}_A(Y_{\text{com}})\} + \{\hat{\theta}_A(Y_{\text{com}}) - \theta\},$$

we have

$$\begin{aligned} \text{var}_H(\bar{\theta}_\infty) &= \text{var}_H\{\bar{\theta}_\infty - \hat{\theta}_A(Y_{\text{com}})\} + \text{var}_H\{\hat{\theta}_A(Y_{\text{com}})\} \\ &\quad +2\,\text{cov}_H\{\bar{\theta}_\infty - \hat{\theta}_A(Y_{\text{com}}), \hat{\theta}_A(Y_{\text{com}})\}, \end{aligned} \tag{45.6}$$

where all the expectations are with respect to the hybrid model defined in (45.5). Since we assume both the imputer's model and the analyst's model are valid, it is not too hard to see intuitively — and to prove under regularity conditions, as in Xie and Meng (2013) — that the first term and second term on the right-hand side of (45.6) are still estimated consistently by B_m and $\bar{U}_m$, respectively. However, the trouble is that the cross term as given in (45.6) is left out by (45.4), so unless this term is asymptotically negligible, Rubin's variance estimator of $\text{var}_H(\bar{\theta}_\infty)$ via (45.4) cannot be consistent, an observation first made by Kott (1995).

Under congeniality, this term is indeed negligible. This is because, under our current setting, $\bar{\theta}_\infty$ is asymptotically (as $n \to \infty$) the same as the analyst's MLE based on the observed data Y_{obs}; we denote it, with an abuse of notation, by $\hat{\theta}_A(Y_{\text{obs}})$. But $\hat{\theta}_A(Y_{\text{obs}}) - \hat{\theta}_A(Y_{\text{com}})$ and $\hat{\theta}_A(Y_{\text{com}})$ must be asymptotically orthogonal (i.e., uncorrelated) under P_A, which in turn is asymptotically the same as P_H due to congeniality (under the usual regularity conditions that guarantee the equivalence of frequentist and Bayesian asymptotics). Otherwise there must exist a linear combination of $\hat{\theta}_A(Y_{\text{obs}}) - \hat{\theta}_A(Y_{\text{com}})$ and $\hat{\theta}_A(Y_{\text{com}})$ — and hence of $\hat{\theta}_A(Y_{\text{obs}})$ and $\hat{\theta}_A(Y_{\text{com}})$ — that is asymptotically more efficient than $\hat{\theta}_A(Y_{\text{com}})$, contradicting the fact that $\hat{\theta}_A(Y_{\text{com}})$ is the full MLE under $P_A(Y_{\text{com}}|\theta)$.

When uncongeniality arises, it becomes entirely possible that there exists a linear combination of $\bar{\theta}_\infty - \hat{\theta}_A(Y_{\text{com}})$ and $\hat{\theta}_A(Y_{\text{com}})$ that is more efficient than $\theta_A(Y_{\text{com}})$ at least under the actual data generating model. This is because $\bar{\theta}_\infty$ may inherit, through the imputed data, additional (valid) information that is not available to the analyst, and hence is not captured by $P_A(Y_{\text{com}}|\theta)$. Consequently, the cross-term in (45.6) is not asymptotically negligible, making (45.4) an inconsistent variance estimator; see Fay (1992), Meng (1994), and Kott (1995).

The above discussion also hints at an issue that makes the multi-phase inference formulation both fruitful and intricate, because it indicates that consistency can be preserved when the imputer's model does not bring in additional (correct) information. This is a much weaker requirement than congeniality, because it is satisfied, for example, when the analyst's model is nested within (i.e., less saturated than) the imputer's model. Indeed, in Xie and Meng (2013) we established precisely this fact, under regularity conditions. However, when we assume that the imputer model is nested within the analyst's model, we can prove only that (45.4) has a positive bias. But even this weaker result requires an additional assumption — for multivariate θ — that the loss of information is the same for all components of θ. This additional requirement for multivariate θ was both unexpected and troublesome, because in practice there is little reason to expect that the loss of information will be the same for different parameters.

All these complications vividly demonstrate both the need for and challenges of the multi-phase inference framework. By multi-phase, our motivation is not merely that there are multiple parties involved, but more critically that the phases are sequential in nature. Each phase takes the output of its immediate previous phase as the input, but with little knowledge of how other phases operate. This lack of mutual knowledge reality leads to uncongeniality, which makes any single-model framework inadequate for reasons stated before.

45.3.2 Data pre-processing, curation and provenance

Taking this multi-phase perspective but going beyond the MI setting, we (Blocker and Meng, 2013) recently explored the steps needed for building a theoretical foundation for pre-processing in general, with motivating applications from microarrays and astrophysics. We started with a simple but realistic two-phase setup, where for the pre-processor phase, the input is Y and the output is $T(Y)$, which becomes the input of the analysis phase. The pre-process is done under an "observation model" $P_Y(Y|X,\xi)$, where X represents the ideal data we do not have (e.g., true expression level for each gene), because we observe only a noisy version of it, Y (e.g., observed probe-level intensities), and where ξ is the model parameter characterizing how Y is related to X, including how noises were introduced into the observation process (e.g., background contamination). The downstream analyst has a "scientific model" $P_X(X|\theta)$, where θ is the scientific estimand of interest (e.g., capturing the organism's patterns of gene expression). To the analyst, both X and Y are missing, because only $T(Y)$ is made available to the analyst. For example, $T(Y)$ could be background corrected, normalized, or aggregated Y. The analyst's task is then to infer θ based on $T(Y)$ only.

Given such a setup, an obvious question is what $T(Y)$ should the pre-processor produce/keep in order to ensure that the analyst's inference of θ will be as sharp as possible? If we ignore practical constraints, the answer seems to be rather trivial: choose $T(Y)$ to be a (minimal) sufficient statistic for

$$P_Y(y|\theta,\xi) = \int P_Y(y|x;\xi)P_X(x|\theta)\mu(\mathrm{d}x). \tag{45.7}$$

But this does not address the real problem at all. There are thorny issues of dealing with the nuisance (to the analyst) parameter ξ, as well as the issue of computational feasibility and cost. But most critically, because of the separation of the phases, the scientific model $P_X(X|\theta)$ and hence the marginal model $P_Y(Y|\theta, \xi)$ of (45.7) is typically unknown to the pre-processor. At the very best, the pre-processor may have a working model $\tilde{P}_X(X|\eta)$, where η may not live even on the same space as θ. Consequently, the pre-processor may produce $T(Y)$ as a (minimal) sufficient statistic with respect to

$$\tilde{P}_Y(y|\eta, \xi) = \int P_Y(y|x; \xi)\tilde{P}_X(x|\eta)\mu(\mathrm{d}x). \tag{45.8}$$

A natural question then is what are sufficient and necessary conditions on the pre-processor's working model such that a $T(Y)$ (minimally) sufficient for (45.8) will also be (minimally) sufficient for (45.7). Or to use computer science jargon, when is $T(Y)$ a lossless compression (in terms of statistical efficiency)?

Evidently, we do not need the multi-phase framework to obtain trivial and useless answers such as setting $T(Y) = Y$ (which will be sufficient for any model of Y only) or requiring the working model to be the same as the scientific model (which tells us nothing new). The multi-phase framework allows us to formulate and obtain theoretically insightful and practically relevant results that are unavailable in the single-phase framework. For example, in Blocker and Meng (2013), we obtained a non-trivial sufficient condition as well as a necessary condition (but they are not the same) for preserving sufficiency under a more general setting involving multiple (parallel) pre-processors during the pre-process phase. The sufficient condition is in the same spirit as the condition for consistency of Rubin's variance rule under uncongeniality. That is, in essence, sufficiency under (45.8) implies sufficiency under (45.7) when the working model is more saturated than the scientific model. This is rather intuitive from a multi-phase perspective, because the fewer assumptions we make in earlier phases, the more flexibility the later phases inherit, and consequently, the better the chances these procedures preserve information or desirable properties.

There is, however, no free lunch. The more saturated our model is, the less compression it achieves by statistical sufficiency. Therefore, in order to make our results as practically relevant as possible, we must find ways to incorporate computational efficiency into our formulation. However, establishing a general theory for balancing statistical and computational efficiency is an extremely challenging problem. The central difficulty is well known: statistical efficiency is an inherent property of a procedure, but the computational efficiency can vary tremendously across computational architectures and over time.

For necessary conditions, the challenge is of a different kind. Preserving sufficiency is a much weaker requirement than preserving a model, even for minimal sufficiency. For example, $\mathcal{N}(\mu, 1)$ and Poisson(λ) do not share even the same state space. However, the sample mean is a minimal sufficient statistic for both models. Therefore, a pre-processing model could be seriously flawed yet still lead to the best possible pre-processing (this could be viewed as a case of action consistency; see Section 45.5). This type of possibility makes building a multi-phase inference theory both intellectually demanding and intriguing.

In general, "What to keep?" or "Who will share what, with whom, when, and why?" are key questions for the communities in information and computer sciences, particularly in the areas of data curation and data provenance; see Borgman (2010) and Edwards et al. (2011). Data/digital curation, as defined by the US National Academies, is "the active management and enhancement of digital information assets for current and future use," and data provence is "a record that describes the people, institutions, entities, and activities involved in producing, influencing, or delivering a piece of data or a thing" (Moreau et al., 2013). Whereas these fields are clearly critical for preserving data quality and understanding the data collection process for statistical modelling, currently there is little dialogue between these communities and statisticians despite shared interests. For statisticians to make meaningful contributions, we must go beyond the single-phase/single-model paradigm because the fundamental problems these fields address involve, by default, multiple parties, who do not necessarily (or may not even be allowed to) share information, and yet they are expected to deliver scientifically useful data and digital information.

I believe the multi-phase inference framework will provide at least a relevant formulation to enter the conversation with researchers in these areas. Of course, there is a tremendous amount of foundation building to be done, even just to sort out which results in the single-phase framework are directly transferable and which are not. The three questions below again are just an appetizer:

(a) What are practically relevant theoretical criteria for judging the quality of pre-processing, without knowing how many types of analyses ultimately will be performed on the pre-processed data?

(b) What are key considerations and methods for formulating generally uncongeniality for multi-phase inference, for quantifying the degrees of uncongeniality, and for setting up a threshold for a tolerable degree?

(c) How do we quantify trade-offs between efficiencies that are designed for measuring different aspects of the multi-phase process, such as computational efficiency for pre-processing and statistical efficiency for analysis?

45.4 Multi-source inference

As students of statistics, we are all taught that a scientific way of collecting data from a population is to take a probabilistic sample. However, this was not the case a century ago. It took about half a century since its formal introduction in 1895 by Anders Nicolai Kiær (1838–1919), the founder of Statistics Norway, before probabilistic sampling became widely understood and accepted (see Bethlehem, 2009). Most of us now can explain the idea intuitively by analogizing it with common practices such as that only a tiny amount of blood is needed for any medical test (a fact for which we are all grateful). But it was difficult then for many — and even now for some — to believe that much can be learned about a population by studying only, say, a 5% random sample. Even harder was the idea that a 5% random sample is better than a 5% "quota sample," i.e., a sample purposefully chosen to mimic the population. (Very recently a politician dismissed an election pool as "non-scientific" because "it is random.")

Over the century, statisticians, social scientists, and others have amply demonstrated theoretically and empirically that (say) a 5% probabilistic/random sample is better than any 5% non-random samples in many measurable ways, e.g., bias, MSE, confidence coverage, predictive power, etc. However, we have not studied questions such as "Is an 80% non-random sample 'better' than a 5% random sample in measurable terms? 90%? 95%? 99%?"

This question was raised during a fascinating presentation by Dr. Jeremy Wu, then (in 2009) the Director of LED (Local Employment Dynamic), a pioneering program at the US Census Bureau. LED employed synthetic data to create an OnTheMap application that permits users to zoom into any local region in the US for various employee-employer paired information without violating the confidentiality of individuals or business entities. The synthetic data created for LED used more than 20 data sources in the LEHD (Longitudinal Employer-Household Dynamics) system. These sources vary from survey data such as a monthly survey of 60,000 households, which represent only .05% of US households, to administrative records such as unemployment insurance wage records, which cover more than 90% of the US workforce, to census data such as the quarterly census of earnings and wages, which includes about 98% of US jobs (Wu, 2012 and personal communication from Wu).

The administrative records such as those in LEHD are not collected for the purpose of statistical inference, but rather because of legal requirements, business practice, political considerations, etc. They tend to cover a large percentage of the population, and therefore they must contain useful information for inference. At the same time, they suffer from the worst kind of selection biases because they rely on self-reporting, convenient recording, and all sorts of other "sins of data collection" that we tell everyone to avoid.

But statisticians cannot avoid dealing with such complex combined data sets, because they are playing an increasingly vital role for official statistical systems and beyond. For example, the shared vision from a 2012 summit meeting, between the government statistical agencies from Australia, Canada, New Zealand, the United Kingdom, and the US, includes

> "Blending together multiple available data sources (administrative and other records) with traditional surveys and censuses (using paper, internet, telephone, face-to-face interviewing) to create high quality, timely statistics that tell a coherent story of economic, social and en-

vironmental progress must become a major focus of central government statistical agencies." (Groves, February 2, 2012)

Multi-source inference therefore refers to situations where we need to draw inference by using data coming from different sources and some (but not all) of which were not collected for inference purposes. It is thus broader and more challenging than multi-frame inference, where multiple data sets are collected for inference purposes but with different survey frames; see Lohr and Rao (2006). Most of us would agree that the very foundation of statistical inference is built upon having a representative sample; even in notoriously difficult observational studies, we still try hard to create pseudo "representative" samples to reduce the impact of confounding variables. But the availability of a very large subpopulation, however biased, poses new opportunities as well as challenges.

45.4.1 Large absolute size or large relative size?

Let us consider a case where we have an administrative record covering f_a percent of the population, and a simple random sample (SRS) from the same population which only covers f_s percent, where $f_s \ll f_a$. Ideally, we want to combine the maximal amount of information from both of them to reach our inferential conclusions. But combining them effectively will depend critically on the relative information content in them, both in terms of how to weight them (directly or implied) and how to balance the gain in information with the increased analysis cost. Indeed, if the larger administrative dataset is found to be too biased relative to the cost of processing it, we may decide to ignore it. Wu's question therefore is a good starting point because it directly asks how the relative information changes as their relative sizes change: how large should f_a/f_s be before an estimator from the administrative record dominates the corresponding one from the SRS, say in terms of MSE?

As an initial investigation, let us denote our finite population by $\{x_1, \ldots, x_N\}$. For the administrative record, we let $R_i = 1$ whenever x_i is recorded and zero otherwise; and for SRS, we let $I_i = 1$ if x_i is sampled, and zero otherwise, where $i \in \{1, \ldots, N\}$. Here we assume $n_a = \sum_{i=1}^N R_i \gg n_s = \sum_{i=1}^N I_i$, and both are considered fixed in the calculations below. Our key interest here is to compare the MSEs of two estimators of the finite-sample population mean $\bar{X}_N$, namely,

$$\bar{x}_a = \frac{1}{n_a}\sum_{i=1}^N x_i R_i \quad \text{and} \quad \bar{x}_s = \frac{1}{n_s}\sum_{i=1}^N x_i I_i.$$

Recall for finite-population calculations, all x_i's are fixed, and all the randomness comes from the response/recording indicator R_i for $\bar{x}_a$ and the sampling indicator I_i for $\bar{x}_s$. Although the administrative record has no probabilistic mechanism imposed by the data collector, it is a common strategy to model the responding (or recording or reporting) behavior via a probabilistic model.

Here let us assume that a probit regression model is adequate to capture the responding behavior, which depends on only the individual's x value. That is, we can express $R_i = \mathbf{1}(Z_i \le \alpha + \beta x_i)$, where the Z_i's form an i.i.d sample from $\mathcal{N}(0,1)$. We could imagine Z_i being, e.g., the ith individual's latent "refusal tendency," and when it is lower than a threshold that is linear in x_i, the individual responds. The intercept α allows us to model the overall percentage of respondents, with larger α implying more respondents. The slope β models the strength of the self-selecting mechanism. In other words, as long as $\beta \neq 0$, we have a non-ignorable missing-data mechanism (Rubin, 1976).

Given that $\bar{x}_s$ is unbiased, its MSE is the same as its variance (Cochran, 2007), viz.

$$\mathrm{var}(\bar{x}_s) = \frac{1-f_s}{n_s}\, S_N^2(x), \quad \text{where } S_N^2(x) = \frac{1}{N-1}\sum_{i=1}^{N}(x_i - \bar{x}_N)^2. \tag{45.9}$$

The MSE of $\bar{x}_a$ is more complicated, mostly because R_i depends on x_i. But under our assumption that N is very large and $f_a = n_a/N$ stays (far) away from zero, the MSE is completely dominated by the squared bias term of $\bar{x}_a$, which itself is well approximated by, again because N (and hence n_a) is very large,

$$\mathrm{Bias}^2(\bar{x}_a) = \left\{ \frac{\sum_{i=1}^{N}(x_i - \bar{x}_N)p(x_i)}{\sum_{i=1}^{N} p(x_i)} \right\}^2, \tag{45.10}$$

where $p(x_i) = \mathrm{E}(R_i|x_i) = \Phi(\alpha + \beta x_i)$, and Φ is the CDF for $\mathcal{N}(0,1)$.

To get a sense of how this bias depends on f_a, let us assume that the finite population $\{x_1, \ldots, x_N\}$ itself can be viewed as an SRS of size N from a super population $X \sim \mathcal{N}(\mu, \sigma^2)$. By the Law of Large Numbers, the bias term in (45.10) is essentially the same as (again because N is very large)

$$\frac{\mathrm{cov}\{X, p(X)\}}{\mathrm{E}\{p(X)\}} = \frac{\sigma \mathrm{E}\{Z\Phi(\tilde{\alpha} + \tilde{\beta}Z)\}}{\mathrm{E}\{\Phi(\tilde{\alpha} + \tilde{\beta}Z)\}} = \frac{\sigma\tilde{\beta}}{\sqrt{1+\tilde{\beta}^2}} \frac{\phi\left(\dfrac{\tilde{\alpha}}{\sqrt{1+\tilde{\beta}^2}}\right)}{\Phi\left(\dfrac{\tilde{\alpha}}{\sqrt{1+\tilde{\beta}^2}}\right)}, \tag{45.11}$$

where $\tilde{\alpha} = \alpha + \beta\mu, \tilde{\beta} = \sigma\beta$, $Z \sim \mathcal{N}(0,1)$, and ϕ is its density function. Integration by parts and properties of Normals are used for arriving at (45.11).

An insight is provided by (45.11) when we note $\Phi\{\tilde{\alpha}/(1+\tilde{\beta}^2)^{1/2}\}$ is well estimated by f_a because N is large, and hence $\tilde{\alpha}/(1+\tilde{\beta}^2)^{1/2} \approx \Phi^{-1}(f_a) = z_{f_a}$, where z_q is the qth quantile of $\mathcal{N}(0,1)$. Consequently, we have from (45.11),

$$\frac{\mathrm{MSE}(\bar{x}_a)}{\sigma^2} \approx \frac{\mathrm{Bias}^2(\bar{x}_a)}{\sigma^2} = \frac{\tilde{\beta}^2}{1+\tilde{\beta}^2}\frac{\phi^2(z_{f_a})}{f_a^2} = \frac{\tilde{\beta}^2}{1+\tilde{\beta}^2}\frac{e^{-z_{f_a}^2}}{2\pi f_a^2}, \tag{45.12}$$

which will be compared to (45.9) after replacing $S_N^2(X)$ by σ^2. That is,

$$\frac{\text{MSE}(\bar{x}_s)}{\sigma^2} = \frac{1}{n_s} - \frac{1}{N} \approx \frac{1}{n_s}, \tag{45.13}$$

where $1/N$ is ignored for the same reason that $\text{var}(\bar{x}_a) = O(N^{-1})$ is ignored.

It is worthy to point out that the seemingly mismatched units in comparing (45.12), which uses relative size f_a, with (45.13), which uses the absolute size n_s, reflects the different natures of non-sampling and sampling errors. The former can be made arbitrarily small only when the relative size f_a is made arbitrarily large, that is $f_a \to 1$; just making the absolute size n_a large will not do the trick. In contrast, as is well known, we can make (45.13) arbitrarily small by making the absolute size n_s arbitrarily large even if $f_s \to 0$ when $N \to \infty$. Indeed, for most public-use data sets, f_s is practically zero. For example, with respect to the US population, an $f_s = .01\%$ would still render n_s more than 30,000, large enough for controlling sampling errors for many practical purposes. Indeed, (45.13) will be no greater than .000033. In contrast, if we were to use an administrative record of the same size, i.e., if $f_a = .01\%$, then (45.12) will be greater than 3.13, almost 100,000 times (45.13), if $\tilde{\beta} = .5$.

However, if $f_a = 95\%$, $z_{f_a} = 1.645$, (45.12) will be .00236, for the same $\tilde{\beta} = .5$. This implies that as long as n_s does not exceed about 420, the estimator from the biased sample will have a smaller MSE (assuming, of course, $N \gg 420$). The threshold value for n_s will drop to about 105 if we increase $\tilde{\beta}$ to 2, but will increase substantially to about 8,570 if we drop $\tilde{\beta}$ to .1. We must be mindful, however, that these comparisons assume the SRS and more generally the survey data have been collected perfectly, which will not be the case in reality because of both non-responses and response biases; see Liu et al. (2013). Hence in reality it would take a smaller f_a to dominate the probabilistic sample with f_s sampling fraction, precisely because the latter has been contaminated by non-probabilistic selection errors as well. Nevertheless, a key message here is that, as far as statistical inference goes, what makes a "Big Data" set big is typically not its absolute size, but its relative size to its population.

45.4.2 Data defect index

The sensitivity of our comparisons above to $\tilde{\beta}$ is expected because it governs the self-reporting mechanism. In general, whereas closed-form expressions such as (45.12) are hard to come by, the general expression in (45.10) leads to

$$\frac{\text{Bias}^2(\bar{x}_a)}{S_N^2(x)} = \rho_N^2(x,p)\left\{\frac{S_N(p)}{\bar{p}_N}\right\}^2\left(\frac{N-1}{N}\right)^2 < \rho_N^2(x,p)\frac{1-\bar{p}_N}{\bar{p}_N}, \tag{45.14}$$

where $\bar{p}_N$ is the mean of p_i, $\rho_N(x,p)$ is the correlation between x_i and p_i, and the term inside the first set of brackets is the coefficient of variation of p_i, all of which are with respect to the finite population, i.e., the uniform distribution over the index space $\{1,\ldots,N\}$. This explains the notation $\rho_N(x,p)$, in contrast to $\rho(X,p(X))$, which is with respect to X from the super population.

The (middle) re-expression of the bias given in (45.14) in terms of the correlation between sampling variable x and sampling/response probability p is a standard strategy in the survey literature; see Hartley and Ross (1954) and Meng (1993). Although mathematically trivial, it provides a greater statistical insight, i.e., the sample mean from an arbitrary sample is an unbiased estimator for the target population mean if and only if the sampling variable x and the data collection mechanism $p(x)$ are uncorrelated. In this sense we can view $\rho_N(x,p)$ as a "defect index" for estimation (using sample mean) due to the defect in data collection/recording. This result says that we can reduce estimation bias of the sample mean for non-equal probability samples or even non-probability samples as long as we can reduce the magnitude of the correlation between x and $p(x)$. This possibility provides an entryway into dealing with a large but biased sample, and exploiting it may require less knowledge about $p(x)$ than required for other bias reduction techniques such as (inverse probability) weighting, as in the Horvitz-Thompson estimator.

The (right-most) inequality in (45.14) is due to the fact that for any random variable satisfying $U \in [0,1]$, $\text{var}(U) \leq \text{E}(U)\{1 - \text{E}(U)\}$. This bound allows us to control the bias using only the proportion $\bar{p}_N$, which is well estimated by the observed sample fraction f_a. It says that we can also control the bias by letting f_a approach one. In the traditional probabilistic sampling context, this observation would only induce a "duhhh" response, but in the context of multi-source inference it is actually a key reason why an administrative record can be very useful despite being a non-probabilistic sample.

Cautions are much needed however, because (45.14) also indicates that it is not easy at all to use a large f_a to control the bias (and hence MSE). By comparing (45.13) and the bound in (45.14) we will need (as a sufficient condition)

$$f_a > \frac{n_s \rho_N^2(x,p)}{1 + n_s \rho_N^2(x,p)}$$

in order to guarantee $\text{MSE}(\bar{x}_a) < \text{MSE}(\bar{x}_s)$. For example, even if $n_s = 100$, we would need over 96% of the population if $\rho_N = .5$. This reconfirms the power of probabilistic sampling and reminds us of the danger in blindly trusting that "Big Data" must give us better answers. On the other hand, if $\rho_N = .1$, then we will need only 50% of the population to beat a SRS with $n_s = 100$. If $n_s = 100$ seems too small in practice, the same $\rho_N = .1$ also implies that a 96% subpopulation will beat a SRS as large as $n_s = \rho_N^{-2}\{f_a/(1 - f_a)\} = 2400$, which is no longer a practically irrelevant sample size.

Of course all these calculations depend critically on knowing the value of ρ_N, which cannot be estimated from the biased sample itself. However, recall for multi-source inference we will also have at least a (small) probabilistic sample. The availability of both small random sample(s) and large non-random sample(s) opens up many possibilities. The following (non-random) sample of questions touch on this and other issues for multi-source inference:

(a) Given partial knowledge of the recording/response mechanism for a (large) biased sample, what is the optimal way to create an intentionally biased sub-sampling scheme to counter-balance the original bias so the resulting sub-sample is guaranteed to be less biased than the original biased sample in terms of the sample mean, or other estimators, or predictive power?

(b) What should be the key considerations when combining small random samples with large non-random samples, and what are the sensible "corner-cutting" guidelines when facing resource constraints? How can the combined data help to estimate $\rho_N(x,p)$? In what ways can such estimators aid multi-source inference?

(c) What are theoretically sound and practically useful defect indices for prediction, hypothesis testing, model checking, clustering, classification, etc., as counterparts to the defect index for estimation, $\rho_N(x,p)$? What are their roles in determining information bounds for multi-source inference? What are the relevant information measures for multi-source inference?

45.5 The ultimate prize or price

Although we have discussed the trio of inference problems separately, many real-life problems involve all of them. For example, the aforementioned OnTheMap application has many resolution levels (because of arbitrary zoom-in), many sources of data (more than 20 sources), and many phases of pre-process (even God would have trouble keeping track of all the processing that these twenty some survey, census, and administrative data sets have endured!), including the entire process of producing the synthetic data themselves. Personalized medicine is another class of problems where one typically encounters all three types of complications. Besides the obvious resolution issue, typically the data need to go through pre-processing in order to protect the confidentiality of individual patients (beyond just removing the patient's name). Yet individual level information is most useful. To increase the information content, we often supplement clinical trial data with observational data, for example, on side effects when the medications were used for another disease.

To bring the message home, it is a useful exercise to imagine ourselves in a situation where our statistical analysis would actually be used to decide the best treatment for a serious disease for a loved one or even for ourselves. Such a "personalized situation" emphasizes that it is my interest/life at stake, which should encourage us to think more critically and creatively, not just to publish another paper or receive another prize. Rather, it is about getting to the bottom of what we do as statisticians — to transform whatever empirical observations we have into the best possible quantitative evidence for scientific understanding and decision making, and more generally, to advance science, society, and civilization. That is our ultimate prize.

However, when we inappropriately formulate our inference problems for mental, mathematical, or computational convenience, the chances are that someone or, in the worst case, our entire society will pay the ultimate price. We statisticians are quick to seize upon the 2008 world-wide financial crisis as an ultimate example in demonstrating how a lack of understanding and proper accounting for uncertainties and correlations leads to catastrophe. Whereas this is an extreme case, it is unfortunately not an unnecessary worry that if we continue to teach our students to think only in a single-resolution, single-phase, single-source framework, then there is only a single outcome: they will not be at the forefront of quantitative inference. When the world is full of problems with complexities far exceeding what can be captured by our theoretical framework, our reputation for critical thinking about the entirety of the inference process, from data collection to scientific decision, cannot stand.

The "personalized situation" also highlights another aspect that our current teaching does not emphasize enough. If you really had to face the unfortunate I-need-treatment-now scenario, I am sure your mind would not be (merely) on whether the methods you used are unbiased or consistent. Rather, the type of questions you may/should be concerned with are (1) "Would I reach a different conclusion if I use another analysis method?" or (2) "Have I really done the best given my data and resource constraints?" or (3) "Would my conclusion change if I were given all the original data?"

Questions (1) and (2) remind us to put more emphasis on relative optimality. Whereas it is impossible to understand all biases or inconsistencies in messy and complex data, knowledge which is needed to decide on the optimal method, we still can and should compare methods relative to each other, as well as relative to the resources available (e.g., time, energy, funding). Equally important, all three questions highlight the need to study much more qualitative consistency or action consistency than quantitative consistency (e.g., the numerical value of our estimator reaching the exact truth in the limit). Our methods, data sets, and numerical results can all be rather different (e.g., a p-value of .2 versus .8), yet their resulting decisions and actions can still be identical because typically there are only two (yes and no) or at most a handful of choices.

It is this "low resolution" of our action space in real life which provides flexibility for us to accept quantitative inconsistency caused by defects such as resolution discrepancy, uncongeniality or selection bias, yet still reach scientifically useful inference. It permits us to move beyond single-phase, single-source, or single resolution frameworks, but still be able to obtain theoretically elegant and practically relevant results in the same spirit as those NP-worthy findings in many other fields. I therefore very much hope you will join me for this intellectually exciting and practically rewarding research journey, unless, of course, you are completely devoted to fundraising to establish an NP in statistics.

Acknowledgements

The material on multi-resolution inference benefitted greatly from critical comments by Alex Blocker and Keli Liu, both of whom also provided many insightful comments throughout, as did David Jones. The joint work with Alex Blocker and Xianchao Xie (cited in the reference list) shaped the formulation of the multi-phase inference, which was greatly encouraged by Christine Borgman, who also taught me, together with Alyssa Goodman, Paul Groth, and Margaret Hedstrom, data curation and data provenance. Dr. Jeremy Wu inspired and encouraged me to formulate the multi-source inference, and provided extremely helpful information and insights regarding the LED/LEHD program. Keli Liu also provided invaluable editing and proofreading, as did Steven Finch. "Good stuff!" coming from my academic twin brother Andrew Gelman was all the encouragement I needed to squeeze out every possible minute between continental breakfasts and salmon/chicken dinners. I give them 100% thanks, but 0% liability for any naïveté, wishful thinking, and sign of lack of sleep — this has been the most stressful paper I have ever written. I also thank the NSF for partial financial support, and the Editors, especially Xihong Lin and Geert Molenberghs, for help and extraordinary patience.

References

Arlot, S. and Celisse, A. (2010). A survey of cross-validation procedures for model selection. *Statistics Surveys*, 4:40–79.

Bethlehem, J. (2009). *The Rise of Survey Sampling.* CBS Discussion Paper No. 9015.

Blitzstein, J. and Meng, X.-L. (2010). Nano-project qualifying exam process: An intensified dialogue between students and faculty. *The American Statistician*, 64:282–290.

Blocker, A.W. and Meng, X.-L. (2013). The potential and perils of preprocessing: Building new foundations. *Bernoulli*, 19:1176–1211.

Borgman, C.L. (2010). Research data: Who will share what, with whom, when, and why? *China-North America Library Conference, Beijing*, People's Republic of China.

Bouman, P., Dukic, V. and Meng, X.-L. (2005). A Bayesian multiresolution hazard model with application to an AIDS reporting delay study. *Statistica Sinica*, 15:325–357.

Bouman, P., Meng, X.-L., Dignam, J., and Dukić, V. (2007). A multiresolution hazard model for multicenter survival studies: Application to tamoxifen treatment in early stage breast cancer. *Journal of the American Statistical Association*, 102:1145–1157.

Cochran, W.G. (2007). *Sampling Techniques*. Wiley, New York.

Daubechies, I. (1992). *Ten Lectures on Wavelets*. SIAM.

Donoho, D.L. and Elad, M. (2003). Optimally sparse representation in general (nonorthogonal) dictionaries via ℓ_1 minimization. *Proceedings of the National Academy of Sciences*, 100:2197–2202.

Donoho, D.L. and Johnstone, I.M. (1994). Ideal spatial adaptation by wavelet shrinkage. *Biometrika*, 81:425–455.

Donoho, D.L., Johnstone, I.M., Kerkyacharian, G., and Picard, D. (1995). Wavelet shrinkage: Asymptopia? (with discussion). *Journal of the Royal Statistical Society, Series B*, 57:301–369.

Edwards, P.N., Mayernik, M.S., Batcheller, A.L., Bowker, G.C., and Borgman, C.L. (2011). Science friction: Data, metadata, and collaboration. *Social Studies of Science*, 41:667–690.

Fay, R.E. (1992). When are inferences from multiple imputation valid? *Proceedings of the Survey Research Methods Section*, American Statistical Association, Washington, DC, pp. 227–232.

Garey, M. (1974). Optimal binary search trees with restricted maximal depth. *SIAM Journal on Computing*, 3:101–110.

Gelman, A., Carlin, J.B., Stern, H.S., and Rubin, D.B. (2003). *Bayesian Data Analysis*. Chapman & Hall, London.

Gelman, A. and Little, T.C. (1997). Poststratification into many categories using hierarchical logistic regression. *Survey Methodology*, 23:127–35.

George, E.I. and McCulloch, R.E. (1997). Approaches for Bayesian variable selection. *Statistica Sinica*, 7:339–373.

Groves, R.M. (February 2, 2012). National statistical offices: Independent, identical, simultaneous actions thousands of miles apart US Census Bureau Director's Blog, `http://blogs.census.gov/directorsblog/`.

Hartley, H. and Ross, A. (1954). Unbiased ratio estimators. *Nature*, 174:270–271.

Hirakawa, K. and Meng, X.-L. (2006). An empirical Bayes EM-wavelet unification for simultaneous denoising, interpolation, and/or demosaicing. In *Image Processing, 2006 IEEE International Conference on*. IEEE, pp. 1453–1456.

Knuth, D. (1997). *The Art of Computer Programming*, Vol 1. *Fundamental Algorithms*, 3rd edition. Addison-Wesley, Reading, MA.

Kott, P.S. (1995). A paradox of multiple imputation. *Proceedings of the Survey Research Methods Section*, American Statistical Association, Washington, DC, pp. 380–383.

Lax, J.R. and Phillips, J.H. (2009). How should we estimate public opinion in the states? *American Journal of Political Science*, 53:107–121.

Lee, T.C. and Meng, X.-L. (2005). A self-consistent wavelet method for denoising images with missing pixels. In *Proceedings of the 30th IEEE International Conference on Acoustics, Speech, and Signal Processing*, 2:41–44.

Lin, D., Foster, D.P., and Ungar, L.H. (2010). *A Risk Ratio Comparison of ℓ_0 and ℓ_1 Penalized Regressions.* Technical Report, University of Pennsylvania, Philadelphia, PA.

Liu, J., Meng, X.-L., Chen, C.-N. and Alegrita, M. (2013). Statistics can lie but can also correct for lies: Reducing response bias in NLAAS via Bayesian imputation. *Statistics and Its Interface*, 6:387–398.

Lohr, S. and Rao, J.N.K. (2006). Estimation in multiple-frame surveys. *Journal of the American Statistical Association*, 101:1019–1030.

Meng, X.-L. (1993). On the absolute bias ratio of ratio estimators. *Statistics & Probability Letters*, 18:345–348.

Meng, X.-L. (1994). Multiple-imputation inferences with uncongenial sources of input (with discussion). *Statistical Science*, 9:538–558.

Meng, X.-L. (2009a). Automated bias-variance trade-off: Intuitive inadmissibility or inadmissible intuition? In *Frontiers of Statistical Decision Making and Bayesian Analysis* (M.H. Chen, D.K. Dey, P. Mueller, D. Sun, and K. Ye, Eds.). Springer, New York, pp. 95–112.

Meng, X.-L. (2009b). Desired and feared — What do we do now and over the next 50 years? *The American Statistician*, 63:202–210.

Meyer, Y. (1993). Wavelets-algorithms and applications. *Wavelets-Algorithms and Applications*, 1:142.

Mitchell, T.J. and Beauchamp, J.J. (1988). Bayesian variable selection in linear regression. *Journal of the American Statistical Association*, 83:1023–1032.

Moreau, L., Belhajjame, K., B'Far, R., Cheney, J., Coppens, S., Cresswell, S., Gil, Y., Groth, P., Klyne, G., Lebo, T., McCusker, J., Miles, S., Myers, J., Sahoo, S., and Tilmes, C., Eds (2013). *PROV-DM: The PROV Data Model.* Technical Report, World Wide Web Consortium.

Nason, G.P. (2002). Choice of wavelet smoothness, primary resolution and threshold in wavelet shrinkage. *Statistics and Computing*, 12:219–227.

Poggio, T. and Girosi, F. (1998). A sparse representation for function approximation. *Neural Computation*, 10:1445–1454.

Rao, J.N.K. (2005). *Small Area Estimation.* Wiley, New York.

Rubin, D.B. (1976). Inference and missing data. *Biometrika*, 63:581–592.

Rubin, D.B. (1987). *Multiple Imputation for Nonresponse in Surveys.* Wiley, New York.

Rubin, D.B. (1996). Multiple imputation after 18+ years. *Journal of the American Statistical Association*, 91:473–489.

Rubin, D.B. (2005). Causal inference using potential outcomes. *Journal of the American Statistical Association*, 100:322–331.

Wu, J. (2012). 21st century statistical systems. *Blog: NotRandomThought*, August 1, 2012. Available at `http://jeremyswu.blogspot.com/`.

Xie, X. and Meng, X.-L. (2013). Dissecting multiple imputation from a multi-phase inference perspective: What happens when there are three uncongenial models involved? *The Annals of Statistics*, under review.

Yang, J., Peng, Y., Xu, W., and Dai, Q. (2009). Ways to sparse representation: An overview. *Science in China Series F: Information Sciences*, 52:695–703.

Part V

Advice for the next generation

46

Inspiration, aspiration, ambition

C.F. Jeff Wu
School of Industrial and Systems Engineering
Georgia Institute of Technology, Atlanta, GA

46.1 Searching the source of motivation

One can describe the motivation or drive for accomplishments or scholarship at three levels: inspiration, aspiration, and ambition. They represent different (but not necessarily exclusive) mindsets or *modi operandi.* Let me start with the *Merriam–Webster Dictionary* definitions of the three words.

(a) Inspiration is "the action or power of moving the intellect or emotions." In its religious origin, inspiration can be described as "a divine influence or action... to qualify him/her to receive and communicate sacred revelation." It works at the spiritual level even in describing work or career.

(b) Aspiration is "a strong desire to achieve something high or great." It has a more concrete aim than inspiration but still retains an idealistic element.

(c) Ambition is "the desire to achieve a particular end" or "an ardent desire for rank, fame, or power." It has a utilitarian connotation and is the most practical of the three. Ambition can be good when it drives us to excel, but it can also have a negative effect. Aspiration, being between the two, is more difficult to delineate.

Before I go on, I would like to bring your attention to a convocation speech (Wu, 2008) entitled "Idealism or pragmatism" that I gave in 2008 at the University of Waterloo. This speech is reproduced in the Appendix. Why or how is this related to the main theme of this chapter? Idealism and pragmatism are two ideologies we often use to describe how we approach life or work. They represent different mindsets but are not mutually exclusive. Inspiration is clearly idealistic, ambition has a pragmatic purpose, and aspiration can be found in both. The speech can be taken as a companion piece to this chapter.

46.2 Examples of inspiration, aspiration, and ambition

To see how inspiration, aspiration, and ambition work, I will use examples in the statistical world for illustration. Jerzy Neyman is an embodiment of all three. Invention of the Neyman–Pearson theory and confidence intervals is clearly inspirational. Neyman's success in defending the theory from criticism by contemporaries like Sir Ronald A. Fisher was clearly an act of aspiration. His establishment of the Berkeley Statistics Department as a leading institution of learning in statistics required ambition in addition to aspiration.

The personality of the individual often determines at what level(s) he/she operates. Charles Stein is a notable example of inspiration as evidenced by his pioneering work in Stein estimation, Stein–Chen theory, etc. But he did not possess the necessary attribute to push for his theory. It is the sheer originality and potential impact of his theoretical work that helped his contributions make their way to wide acceptance and much acclaim.

Another example of inspiration, which is more technical in nature, is the Cooley–Tukey algorithm for the Fast Fourier Transform (FFT); see Cooley and Tukey (1965). The FFT has seen many applications in engineering, science, and mathematics. Less known to the statistical world is that the core technical idea in Tukey's development of the algorithm came from a totally unrelated field. It employed Yates' algorithm (Yates, 1937) for computing factorial effects in two-level factorial designs.

In Yates' time, computing was very slow and therefore he saw the need to find a fast algorithm (in fact, optimal for the given problem) to ease the burden on mechanical calculators. About thirty years later, Tukey still felt the need to develop a fast algorithm in order to compute the discrete Fourier transform over many frequency values. Even though the stated problems are totally different, their needs for faster algorithm (relative to the technology in their respective times) were similar. By some coincidence, Yates' early work lent a good hand to the later development of the FFT.

As students of the history of science, we can learn from this example. If work has structural elegance and depth, it may find good and unexpected applications years later. One cannot and should not expect an instant gratification from the work. Alas, this may come too late for the ambitious.

Examples of ambition without inspiration abound in the history of science. Even some of the masters in statistics could not stay above it. Here are two examples. In testing statistical independence in $r \times c$ contingency table, Karl Pearson used $rc - 1$ as the degrees of freedom. Fisher showed in 1922 that, when the marginal proportions are estimated, the correct degrees of freedom should be $(r-1)(c-1)$. Pearson did not react kindly. He said in the same year "Such a view is entirely erroneous. [$\cdots$] I trust my critic will pardon me for comparing him with Don Quixote tilting at the windmill" (Pearson, 1922, p. 191). Fisher's retort came much later. In a 1950 volume of his collected works, he wrote of Pearson: "If peevish intolerance of free opinion in others is a sign of senility, it is one which he had developed at an early age" (Fisher, 1950). Even the greatest ever statistician could not be more magnanimous.

46.3 Looking to the future

In 2010, I gave a speech (Wu, 2010) whose main motivation was the discomforting trend I have witnessed in the last 15–20 years in the US and elsewhere. Compared to back when I started my career, there has been an increasing emphasis on the number of papers, journal rankings, citations, and funding. Back then, a new PhD could secure a tenure-track post in a top department with no paper published or accepted as long as the letters were good and the work in the thesis was considered to have good quality. Not anymore. We now see most leading candidates in the applicant pool to have several papers in top journals (and who ranks these journals?) Is this due to inflation or is the new generation really smarter or work harder than mine? Admittedly most of the top departments still judge the candidates by the merits of the work. But the new and unhealthy emphasis has affected the community by and large.

There are some obvious culprits, mostly due to the environment we are in. The funding agencies give preference to large team projects, which require a large number of papers, patents, etc. The widespread use of internet such as Scientific Citation Index (SCI) and Google Scholar has led to instant comparisons and rankings of researchers. Unfortunately this obsession with numerics has led to several widely used rankings of universities in the world. In many countries (US being one lucky exception), university administrators pressure researchers to go for more citations in order to boost their ranking.

In the statistical world, some countries list the "Big Four" (i.e., *The Annals of Statistics*, *Biometrika*, the *Journal of the American Statistical Association* and the *Journal of the Royal Statistical Society, Series B*) as the most desirable journals for promotion and awards. The detrimental impact on the development of long lasting work is obvious but young researchers can't afford to work or think long term. Immediate survival is their primary concern.

What can be done to mitigate this negative effect? I am not optimistic about the environment that spawned this trend. The widespread use of the internet can only exacerbate the trend. I hope that the scientific establishment and policy makers of countries that aspire to join the league of scientific powers will soon realize that sheer numbers of papers or citations alone do not lead to major advances and discoveries. They should modify their reward systems accordingly.

The leading academic departments, being good practitioners, bear a great responsibility in convincing the community not to use superficial numeric measures. At the individual level, good education at an early stage can help. Professors should serve as role models and advise students to go for quality over quantity. This theme should be featured visibly in conferences or sessions for new researchers. My final advice for the aspiring young researchers is to always look inward to your inspiration, aspiration, and ambition to plan and assess your work and career.

Appendix: Idealism or pragmatism (Wu, 2008)

Let me join my colleague Professor Rao in thanking the University, the Chancellor, and the President for bestowing such an honor upon us and in congratulating this year's graduating class for their hard work and achievements. Since I am younger, Jon said that I should stand here longer [laugh].

I would like to share some thoughts on our responsibilities to the society and to ourselves. When I was a bit younger than you are now, I faced two distinct choices: medicine and pure mathematics. This was Taiwan in the 1960s, and most of my relatives urged my parents to nudge me toward medicine, not because they thought I would make a good doctor, but because it would provide a secure career and high income. Thanks to my parents, though, I was able to follow my passion and pursue mathematics. At that time I did not consider the financial consequences because I enjoyed doing math. I am not here to suggest that you follow this romanticism in your career planning — in fact many of you have probably lined up some good jobs already [laugh]. Rather, I want to discuss the role of idealism and pragmatism in our lives.

At the many turning points of our lives, we are often faced with choosing one or the other. Most will heed the call of pragmatism and shun idealism. For example, some of us may find that we disagree with a policy or decision at work. Yet it will be our job to follow or implement this policy. A pragmatist would not go out of his way to show disapproval, even if this policy goes against his conscience. On the other hand, an idealist in this situation is likely to show her disapproval, even if it puts her livelihood at risk.

One of the most shining examples of idealism, of course, is Nelson Mandela, who fought for freedom in South Africa. Apartheid was designed to intimidate minorities into submission. Even something as simple as membership in a legal political organization could lead to consequences such as loss of income and personal freedom. Knowing these risks fully well, Mandela and countless others embarked on that freedom struggle, which lasted for decades.

While I do not expect or suggest that many can follow the most idealistic route, pragmatism and idealism are not incompatible. For example, researchers can channel their efforts to finding new green energy solutions. Even humble statisticians like us can help these environmental researchers design more efficient experiments. Successful business people, which many of you will become, can pay more attention to corporate social responsibility, and not focus exclusively on the bottom line.

Perhaps it is naive, but I truly believe that most often we can strike a balance between what is good for others and what is good for us. If we can keep this spirit and practice it, the world will be a much better and more beautiful place!

Thank you for your attention and congratulations to you once again.

References

Cooley, J.W. and Tukey, J.W. (1965). An algorithm for the machine calculation of complex Fourier series. *Mathematics of Computation*, 19:297–301.

Fisher, R.A. (1950). *Contributions to Mathematical Statistics.* Chapman & Hall, London.

Pearson, K. (1922). On the χ^2 test of goodness of fit. *Biometrika*, 14:186–191.

Wu, C.F.J. (2008). Convocation speech delivered on June 13, 2008 at the graduation ceremony of the Faculty of Mathematics, University of Waterloo, Ontario, Canada.

Wu, C.F.J. (2010). Plenary talk given on October 22, 2010 during the triennial Chinese Conference in Probability and Statistics held at Nankai University, Tianjin, People's Republic of China.

Yates, F. (1937). *The Design and Analysis of Factorial Experiments.* Imperial Bureau of Soil Sciences, Tech. Comm. No. 35.

47

Personal reflections on the COPSS Presidents' Award

Raymond J. Carroll

Department of Statistics

Texas A&M University, College Station, TX

47.1 The facts of the award

I received the COPSS Presidents' Award in 1988, one year after Jeff Wu and one year before Peter Hall. I was the eighth recipient.

I had just moved to Texas A&M in fall 1987, and did not know that Cliff Spiegelman, also still at Texas A&M, had nominated me. I remember very clearly being told about this honor. I was working at home, and out of the blue I received a phone call from the head of the Presidents' Award Committee, about a month before the Joint Statistical Meetings in New Orleans, and he asked, roughly, "are you going to the JSM in New Orleans?" I actually had not planned on it, and he told me I probably should since I had won the Presidents' Award. I remember being very happy, and I know I took the rest of the day off and just floated around.

I am not by nature a very reflective person, preferring instead to look ahead and get on with the next project. However, the invitation to write for the COPSS 50th Anniversary Book Project motivated me to reflect a bit on what I had done prior to 1988, and if there were any morals to the story that I could share.

47.2 Persistence

Persistence I have. My first six submitted papers were rejected, and some in a not very nice way. Having rejected two of my papers, the then-Editor of *The Annals of Statistics* wrote to tell me that he thought I had no possibility of a successful career in academics, and that I would be better off going into industry; at the time, this was a grand insult from an academic. The only effect that had on me was that I worked harder, and I swore that one day that $\#^{\#}\&\#$ Editor would invite me to give a talk at his university, which finally happened in 1990. By then, I had calmed down.

Over the years, editors have become less judgmental, but it is easy to forget how devastating it can be for a new PhD to have his/her thesis paper rejected. I have seen very talented students who left academia after their thesis paper had its initial rejection. As the Editor of the "Theory and Methods" Section of the *Journal of the American Statistical Association* (JASA) and then later *Biometrics*, I was always on the lookout for new PhD's, and would work to get their papers published, even in very abbreviated form.

Unfortunately nowadays, it is routine in some sub-areas of statistics to see applicants for an initial appointment at the Assistant Professor level with approximately six papers in top journals, without a postdoc. I remain skeptical that this is really their work. In any case, this makes it harder to be generous, and it has led to a bit of a coarsening effect in the review of the first paper of many new PhD's.

47.3 Luck: Have a wonderful Associate Editor

We all know that many Associate Editors are merely mailboxes, but by no means all. My experience is that the quality of Associate Editors is a stationary process. I wrote a paper in 1976 or so for *The Annals of Statistics* which had a rather naive proof about expansions of what were then called M-estimators (I guess they still are). The Associate Editor, who knew I was a new Assistant Professor and who I later found out was Willem van Zwet, to whom I remain grateful, wrote a review that said, in effect, "Nice result, it is correct, but too long, here is a two-page proof." The paper appeared (Carroll, 1978), and for years I wondered who my benefactor was. I was at a conference at Purdue about 15 years later, and Bill came up to me and said, and this is a quote "I wrote a better paper than you did." He was right: the published paper is five pages long!

47.4 Find brilliant colleagues

I was and am extremely lucky in my choice of colleagues, and there is both plan and serendipity in this. A partial list of collaborators includes Presidents' Award winners Ross Prentice, Jeff Wu, Peter Hall, Kathryn Roeder, Jianqing Fan, Xihong Lin, and Nilanjan Chatterjee. Anyone who writes papers with them is, by definition, brilliant! However, from 1974 to 1984, the number of statistical coauthors of what I consider good papers totaled exactly one, namely David Ruppert (now at Cornell). Even by 1988, after sabbaticals and the Presidents' Award, for what I considered really good papers, the total number of statistical coauthors who were not my students was only eight. To appreciate how the world has changed, or how I have changed, from 2012 until now my serious statistics papers not with students or postdocs have had 26 statistical coauthors! It is an amazing thing to reflect upon the massive change in the way statistical methodology has evolved.

I received my PhD in 1974, and by 1978 I was tenured. I had written a number of sole-authored papers in major journals, but I was not very satisfied with them, because I seemed, to my mind, to be lurching from one technical paper to the next without much of a plan. It just was not much fun, and I started toying with the idea of going to medical school, which seemed a lot more interesting, as well as more remunerative.

My statistical world changed in the fall of 1977, when David Ruppert became my colleague. It was a funny time that fall because both of our wives were away working on postdocs/dissertations, and our offices were next to one another in a corner. We became friends, but in those days people did not naturally work together. That fall I had a visitor and was trying to understand a topic that at the time was fashionable in robust statistics: what happens to the least squares estimator obtained after deleting a percentage of the data with the largest absolute residuals from an initial least squares fit? This was perceived wisdom as a great new statistical technique. David heard us talking about the problem, came in to participate, and then my visitor had to leave. So, David and I sat there staring at the blackboard, and within two hours we had solved the problem, other than the technical details (there went two months). It was fun, and I realized that I did not much like working alone, but wanted to share the thrill of discovery, and pick other people's brains. The best part was that David and I had the same mentality about methodology, but his technical skill set was almost orthogonal to mine.

The paper (Carroll and Ruppert, 1980) also included some theory about quantile regression. The net effect was that we showed that trimming some large residuals after an initial fit is a terrible idea, and the method quickly died the death that it deserved. A fun paper along these same lines is He and Portnoy (1992). The paper also had my first actual data set, the salinity data in the Pamlico Sound. We published the data in a table, and it has made its way into numerous textbooks, but without a citation! In the nine years we were colleagues, we wrote 29 papers, including two papers on transformation of data (Carroll and Ruppert, 1981, 1984). Overall, David and I are at 45 joint papers and three books. It is trite to give advice like "Get lucky and find a brilliant colleague at the rarified level of David Ruppert," but that's what I did. Lucky I am!

47.5 Serendipity with data

Just before David and I became colleagues, I had my first encounter with data. This will seem funny to new researchers, but this was in the era of no personal computers and IBM punch cards.

It was late 1976, I was the only Assistant Professor in the department, and I was sitting in my office happily minding my own business, when two very senior and forbidding faculty members came to my office and said "Carroll, come here, we want you to meet someone" (yes, in the 1970s, people really talked like that, especially at my institution). In the conference room was a very distinguished marine biologist, Dirk Frankenberg (now deceased), who had come over for a consult with *senior* people, and my colleagues said, in effect, "Talk to this guy" and left. He was too polite to say "Why do I want to talk to a 26-year old who knows nothing?" but I could tell that was what he was thinking.

Basically, Dirk had been asked by the North Carolina Department of Fisheries (NCDF) to build a model to predict the shrimp harvest in the Pamlico Sound for 1977 or 1978, I forget which. The data, much of it on envelopes from fishermen, consisted of approximately $n = 12$ years of monthly harvests, with roughly four time periods per year, and $p = 3$ covariates: water temperature in the crucial estuary, water salinity in that estuary, and the river discharge into the estuary, plus their lagged versions. I unfortunately (fortunately?) had never taken a linear model course, and so was too naive to say the obvious: "You cannot do that, n is too small!" So I did.

In current lingo, it is a "small p, small n," the very antithesis of what is meant to be modern. I suspect 25% of the statistical community today would scoff at thinking about this problem because it was not "small n, large p," but it actually was a problem that needed solving, as opposed to lots of what is going on. I noticed a massive discharge that would now be called a high leverage point, and I simply censored it at a reasonable value. I built a model, and it predicted that 1978 (if memory serves) would be the worst year on record, ever (Hunt et al., 1980), and they should head to the hills. Dirk said "Are you sure?" and me in my naïveté said "yes," and like a gambler, it hit: it was the terrible year. The NCDF then called it the NCDF model! At least in our report we said that the model should be updated yearly (my attempt at full employment and continuation of the research grant), but they then fired us. The model did great for two more years (blind luck), then completely missed the fourth year, wherein they changed the title of the model to reflect where I was employed at the time. You can find Hunt et al. (1980) at `http://www.stat.tamu.edu/~carroll/2012.papers.directory/Shrimp_Report_1980.pdf`.

This is a dull story, except for me, but it also had a moral: the data were clearly heteroscedastic. This led me to my fascination with heteroscedasticity, which later led to my saying that "variances are not nuisance parameters"

(Carroll, 2003). In the transformation world, it led David and me to a paper (Carroll and Ruppert, 1984), and also led to 1/2 of our first book: *Transformation and Weighting in Regression*. Dirk later set us up on another project, managing the Atlantic menhaden fishery, with a brilliant young colleague of his named Rick Deriso, now Chief Scientist, Tuna-Billfish Program, at the Inter-American Tropical Tuna Commission (IATTC) (Reish et al., 1985; Ruppert et al., 1984, 1985).

47.6 Get fascinated: Heteroscedasticity

From the experience with what I call the salinity data, I became fascinated with the concept of heteroscedasticity. I went on a sabbatical to Heidelberg in 1980, and bereft of being able to work directly with David, started thinking hard about modeling variances. I asked what I thought was an obvious question: can one do weighted least squares efficiently without positing a parametric model for the *variances*? David and I had already figured out that the common practice of positing a model for the variances as a function of the mean and then doing normal-theory maximum likelihood was non-model-robust if the variance function was misspecified (Carroll and Ruppert, 1982). I sat in my nice office in Heidelberg day after day, cogitating on the problem. For a month I did nothing else (the bliss of no email), and wrote a paper (Carroll, 1982) that has many references. In modern terms it is not much of a technical "tour de force," and modern semiparametric statisticians have recognized that this is a case where adaptation is obvious, but at the time it was very new and surprising, and indeed a very, very senior referee did not think it was true, but he could not find the flaw. The Editor of *The Annals of Statistics*, to whom I am forever grateful for sticking up for me, insisted that I write out a very detailed proof, which turned out to be over 100 pages by hand, and be prepared to make it available, make a copy and send it along. I still have it! The paper is mostly cited by econometricians, but it was fun.

Later, with my then student Marie Davidian, currently the ASA President, we worked out the theory and practice of parametric variance function estimation (Davidian and Carroll, 1987).

47.7 Find smart subject-matter collaborators

In many fields of experimental science, it is thought that the only way to advance is via solving a so-called "major" problem. In statistics though, "major" problems are not defined *a priori*. If they exist, then many very smart people are working on them, which seems a counter-productive strategy to me. I once spent a fruitless year trying to define something "major," and all I ended up with was feeling stupid and playing golf and going fishing.

I now just float: folks come to me to talk about their problems, and I try to solve theirs and see if there is a statistics paper in it.

What I do like though is personal paradigm shifts when researchers wander into my office with a "simple" problem. This happened to me on a sabbatical in 1981–82 at the National Heart, Lung and Blood Institute in Bethesda, Maryland. I was a visitor and all the regular statisticians had gone off for a retreat, and one day in walks Rob Abbott, one of the world's great cardiovascular epidemiologists (with a PhD in statistics, so he speaks our language). He asked "are you a statistician," I admitted it (I never do at a party), and he wanted to talk with someone about a review he had gotten on a paper about coronary heart disease (CHD) and systolic blood pressure (SPB). If you go to a doctor's office and keep track of your measured SBP, you will be appalled about the variability of it. My SBP has ranged from 150 to 90 in the past three years, as an example. A referee had asked "what is the effect of measurement error in SBP on your estimate of relative risk of CHD?" In the language of current National Football League beer commercials, I said "I love you guy."

I will quote Larry Shepp, who "discovered" a formula that had been discovered many times before, and who said "yes, but when I discovered it, it stayed discovered!" You can find this on the greatest source of statistics information, Wikipedia.

I was convinced at the time (I have since found out this is not exactly true) that there was no literature on nonlinear models with measurement error. So, I dived in and have worked on this now for many years. The resulting paper (Carroll et al., 1984), a very simple paper, has a fair number of citations, and many papers after this one have more. How many times in one's life does a stranger wander in and say "I have a problem," and you jump at it?

Actually, to me, this happens a lot, although not nearly with the same consequences. In the late 1990s, I was at a reception for a toxicological research center at Texas A&M, and feeling mighty out of place, since all the lab scientists knew one another and were doing what they do. I saw a now long-term colleague in Nutrition, Nancy Turner, seeming similarly out of place. I wandered over, asked her what she did, and she introduced me to the world of molecular biology in nutrition. She drew a simple little graph of what statisticians now call "hierarchical functional data," and we have now written many papers together (six in statistics journals), including a series of papers on functional data analysis (Morris et al., 2001; Morris and Carroll, 2006).

47.8 After the Presidents' Award

Since the COPSS Award, my main interests have migrated to problems in epidemiology and statistical methods to solve those problems. The methods include deconvolution, semiparametric regression, measurement error, and functional data analysis, which have touched on problems in nutritional epidemiology, genetic epidemiology, and radiation epidemiology. I have even become a committed Bayesian in a fair amount of my applied work (Carroll, 2013).

I have found problems in nutritional epidemiology particularly fascinating, because we "*know*" from animal studies that nutrition is important in cancer, but finding these links in human longitudinal studies has proven to be surprisingly difficult. I remember an exquisite experiment done by Joanne Lupton (now a member of the US Institute of Medicine) and Nancy Turner where they fed animals a diet rich in corn oil (the American potato chip diet) versus a diet rich in fish oil, exposed them to a carcinogen, and within 12 hours after exposure all the biomarkers (damage, repair, apoptosis, etc.) lit up as different between the two diets, with corn oil always on the losing end. When the microarray became the gold standard, in retrospect a sad and very funny statement, they found that without doing anything to the animals, 10% of the genes were different at a false discovery rate of 5%. Diet matters!

There are non-statisticians such as Ed Dougherty who think the field of statistics lost its way when the microarray came in and thinking about hypotheses/epistemology went out (Dougherty, 2008; Dougherty and Bittner, 2011): "Does anyone really believe that data mining could produce the general theory of relativity?" I recently had a discussion with a very distinguished computer scientist who said, in effect, that it is great that there are many computer scientists who understand (Bayesian) statistics, but would it not be great if they understood what they are doing scientifically? It will be very interesting to see how this plays out. Statistical reasoning, as opposed to computation, while not the total domain of statisticians, seems to me to remain crucial. To quote from Dougherty and Bittner (2011),

> "The lure of contemporary high-throughput technologies is that they can measure tens, or even hundreds, of thousands of variables simultaneously, thereby spurring the hope that complex patterns of interaction can be sifted from the data; however, two limiting problems immediately arise. First, the vast number of variables implies the existence of an exponentially greater number of possible patterns in the data, the majority of which likely have nothing to do with the problem at hand and a host of which arise spuriously on account of variation in the measurements, where even slight variation can be disastrous owing to the number of variables being considered. A second problem is that the mind cannot conceptualize the vast number of variables. Sound experimental design constrains the number of variables to facilitate

> finding meaningful relations among them. Recall Einstein's comment that, for science 'truly creative principle resides in mathematics.' The creativity of which Einstein speaks resides in the human mind. There appears to be an underlying assumption to data mining that the mind is inadequate when it comes to perceiving salient relations among phenomena and that machine-based pattern searching will do a better job. This is not a debate between which can grope faster, the mind or the machine, for surely the latter can grope much faster. The debate is between the efficacy of mind in its creative synthesizing capacity and pattern searching, whether by the mind or the machine."

What success I have had comes from continuing to try to find research problems by working on applications and finding important/interesting applied problems that cannot be solved with existing methodology. I am spending much of my time these days working on developing methods for dietary patterns research, since nutritional epidemiologists have found that dietary patterns are important predictors of cancer. I use every tool I have, and engage many statistical colleagues to help solve the problems.

References

Carroll, R.J. (1978). On almost sure expansion for M-estimates. *The Annals of Statistics*, 6:314–318.

Carroll, R.J. (1982). Adapting for heteroscedasticity in linear models. *The Annals of Statistics*, 10:1124–1233.

Carroll, R.J. (2003). Variances are not always nuisance parameters: The 2002 R.A. Fisher lecture. *Biometrics*, 59:211–220.

Carroll, R.J. (2014). Estimating the distribution of dietary consumption patterns. *Statistical Science*, 29:in press.

Carroll, R.J and Ruppert, D. (1980). Trimmed least squares estimation in the linear model. *Journal of the American Statistical Association*, 75:828–838.

Carroll, R.J. and Ruppert, D. (1981). Prediction and the power transformation family. *Biometrika*, 68:609–616.

Carroll, R.J. and Ruppert, D. (1982). A comparison between maximum likelihood and generalized least squares in a heteroscedastic linear model. *Journal of the American Statistical Association*, 77:878–882.

Carroll, R.J. and Ruppert, D. (1984). Power transformations when fitting theoretical models to data. *Journal of the American Statistical Association*, 79:321–328.

Carroll, R.J., Spiegelman, C.H., Lan, K.K.G., Bailey, K.T., and Abbott, R.D. (1984). On errors-in-variables for binary regression models. *Biometrika*, 71:19–26.

Davidian, M. and Carroll, R.J. (1987). Variance function estimation. *Journal of the American Statistical Association*, 82:1079–1092.

Dougherty, E.R. (2008). On the epistemological crisis in genomics. *Current Genomics*, 9:67–79.

Dougherty, E.R. and Bittner, M.L. (2011). *Epistemology of the Cell: A Systems Perspective on Biological Knowledge.* Wiley–IEEE Press.

He, X. and Portnoy, S. (1992). Reweighted ls estimators converge at the same rate as the initial estimator. *The Annals of Statistics*, 20:2161–2167.

Hunt, J.H., Carroll, R.J., Chinchilli, V., and Frankenberg, D. (1980). Relationship between environmental factors and brown shrimp production in Pamlico Sound, North Carolina. *Report to the Division of Marine Fisheries, North Carolina Department of Natural Resources.*

Morris, J.S. and Carroll, R.J. (2006). Wavelet-based functional mixed models. *Journal of the Royal Statistical Society, Series B*, 68:179–199.

Morris, J.S., Wang, N., Lupton, J.R., Chapkin, R.S., Turner, N.D., Hong, M.Y., and Carroll, R.J. (2001). Parametric and nonparametric methods for understanding the relationship between carcinogen-induced DNA adduct levels in distal and proximal regions of the colon. *Journal of the American Statistical Association*, 96:816–826.

Reish, R.L., Deriso, R.B., Ruppert, D., and Carroll, R.J. (1985). An investigation of the population dynamics of Atlantic menhaden (*Brevoortia tyrannus*). *Canadian Journal of Fisheries and Aquatic Sciences*, 42:147–157.

Ruppert, D., Reish, R.L., Deriso, R.B., and Carroll, R.J. (1984). Monte Carlo optimization by stochastic approximation, with application to harvesting of Atlantic menhaden. *Biometrics*, 40:535–545.

Ruppert, D., Reish, R.L., Deriso, R.B., and Carroll, R.J. (1985). A stochastic model for managing the Atlantic menhaden fishery and assessing managerial risks. *Canadian Journal of Fisheries and Aquatic Sciences*, 42:1371–1379.

48

Publishing without perishing and other career advice

Marie Davidian
Department of Statistics
North Carolina State University, Raleigh, NC

In my 25-plus years as an academic statistician, I have had the good fortune of serving in a variety of roles, including as statistical consultant and collaborator, Editor, Chair of grant review panels, and organizer of and participant in workshops for junior researchers. Drawing on this experience, I share my thoughts and advice on two key career development issues: Balancing research with demands on one's time and on the importance of cultivating and developing one's communication skills.

48.1 Introduction

A career in statistical research is both exciting and challenging. Contributing to the advance of knowledge in our field is extremely rewarding. However, many junior researchers report having difficulty balancing this objective with their many other responsibilities, including collaborative work on funded projects in other disciplines, teaching, and service. And rightly so — our field is unique in the sense that many of us are expected to engage in methodological research in our own discipline *and* to contribute to research in other disciplines through participation in substantive projects. The latter, along with instructional and service responsibilities, can be difficult to navigate for young researchers in the first few years of their careers.

Also unique to our field is the need for outstanding ability to communicate effectively not only with each other but across diverse disciplinary boundaries. To meet this dual challenge, we must be excellent writers and speakers. Publishing our own work, writing successful grant applications in support of our research, and assisting our collaborators with communicating the results of their research and applying for funding is a critical skill that we statisticians must develop.

This volume commemorating the 50th anniversary of the Committee of Presidents of Statistical Societies (COPSS) presents an excellent opportunity for me to share my experience, for the most part learned the hard way, on balancing the competing demands we face and on being an effective communicator. As you'll see in the next two sections, it took me some time in my own career to develop these skills. Despite my slow start, I have subsequently been very fortunate to have served as a journal editor, a chair of NIH grant review panels, and as a consulting and collaborating statistician, through which I have learned a great deal about both of these topics.

With many colleagues over the past decade, including authors of some other chapters of this book, I have served as a senior participant in what is now called the ENAR Workshop for Junior Biostatisticians in Health Research, which has been supported by grants from the National Institutes of Health. (Xihong Lin and I wrote the first grant application, and the grant has subsequently been renewed under the expert direction of Amy Herring.) Although targeted to biostatisticians, this workshop covers skills that are essential to all young researchers. Much of what I have to say here has been shaped by not only my own career but by the insights of my fellow senior participants.

48.2 Achieving balance, and how you never know

Embarking on a career as a statistical researcher can be daunting, probably considerably more so today than it was for me back in 1987. I had just received my PhD in statistics from the University of North Carolina at Chapel Hill and had accepted a position in the Department of Statistics at North Carolina State University, barely 25 miles away. I was excited to have the opportunity to become a faculty member and to teach, consult with other scientists on campus, and carry out statistical methods research.

And, at the same time, I was, frankly, terrified. Sure, I'd done well in graduate school and had managed to garner job offers in several top departments. But could I really *do* this? In particular, could I really do *research*?

I was extremely fortunate to have had a thesis advisor, Ray Carroll, who was what we would call today an outstanding mentor. Ray had not only introduced me to what at the time was a cutting-edge methodological area through my dissertation research, he had also been a great role model. I'll tell you more about Ray in the next section. He seemed confident in my prospects for success in academia and urged me to forge ahead, that I would do just fine.

But I couldn't help being plagued by self-doubt. While I was in graduate school, Ray was always there. He proposed the area in which I did my dissertation research. He was available to help when I got stuck and to discuss the next step. And now I was supposed to do this all on my own? Moreover, I didn't know the first thing about collaborating with scientists in other disciplines.

The first year wasn't easy as I made the transition from student in a very theoretical department to faculty member in a much more applied one, in a position in which I was expected to serve as a consultant to faculty in other departments across campus. It was a bit like a "trial by fire" as I struggled to learn what is truly the art of being a good applied statistician and collaborator, a skill light-years removed from my training in Chapel Hill. Simultaneously, as the papers from my dissertation were accepted, the realization that I needed to move forward with research loomed. Sure, I had some extensions of my dissertation work I was pursuing, but after that, what would I do? I couldn't keep doing variance function estimation forever. The amount of time I spent on my mostly routine but extensive statistical consulting and teaching the two-semester sequence for PhD students in agriculture and life sciences left me little time to ponder new research problems. To top it off, I was asked to serve on the university's Undergraduate Courses and Curriculum Committee, and, not knowing any better, I agreed. I now know that a faculty member as junior as I was should not be asked to serve on a committee that meets every few weeks for several hours and focuses solely on administrative activities completely tangential to research or collaboration.

I will admit to spending many evenings sitting on the balcony of my Raleigh apartment looking out at the parking lot and wondering how I would ever compile a record worthy of promotion and tenure a scant six years later.

But the most amazing thing happened. A student in the Department of Crop Science who was taking my statistics course approached me after class and asked if she could make an appointment to discuss her dissertation research, which involved development of a new, experimental strain of soybean. She had conducted a field experiment over the last three growing seasons in which she had collected longitudinal data on measures of plant growth of both the experimental and a commercial strain and was unsure of how to conduct an analysis that would address the question of whether the two competing soybean varieties had different specific features of their growth patterns. The growth trajectories showed an "S-shaped" pattern that clearly couldn't be described by regression models she knew, and it did not seem to her that analysis of variance methods would address the questions. Could I help? (Of course, at that point I could have no input on the design, which is sadly still often the case to this day, but luckily the experiment had been well-designed and conducted.)

At about this same time, I regularly had been bemoaning my feelings of inadequacy and being overwhelmed to my good friend David Giltinan, who had graduated from Chapel Hill three years ahead of me and taken a job in nonclinical research at the pharmaceutical company Merck. David recognized that some of my dissertation research was relevant to problems he was seeing and introduced me to the subject-matter areas and his collaborators. He had begun working with pharmacokineticists and insisted that I needed to learn about this field and that I could have a lot to contribute. I grudgingly agreed to look at some of the papers he sent to me.

So what does this have to do with soybeans? Everything. As it turned out, despite the disparate application areas, the statistical problem in both the soybean experiment and pharmacokinetics was basically the same. Longitudinal trajectories that exhibited patterns that could be well-described in models nonlinear in parameters arising from solutions to differential equations but where obviously the parameters took on different values across plants or subjects. Questions about the typical behavior of specific features of the trajectories and how variable this is across plants of subjects and changes systematically with the characteristics of the plants or subjects (like strain, weight, or kidney function). And so on.

These two chance events, a consulting client needing help analyzing data from a soybean experiment and a friend insisting I learn about an application area I previously did not even know existed, led me to a fascinating and rewarding area of methodological research. The entire area of nonlinear mixed effects modeling and analysis, the groundwork for which had been laid mostly by pharmacokineticists in their literature, was just being noticed by a few statisticians. Fortuitously, David and I were among that small group. The need for refinement, new methodology, and translation to other subject matter areas (like crop science) was great. I went from fretting over what to work on next to frustration over not having enough time to pursue simultaneously all the interesting challenges to which I thought I could make a contribution.

My determination led me to figure out how to make the time. I'd found a niche where I *knew* I could do useful research, which would have never happened had I not been engaged in subject-matter challenges through my consulting and friendship with David. I was no longer sitting on the balcony; instead, I spent some of those evenings working. I realized that I did not have to accommodate every consulting client's preferred meeting time, and I adopted a firm policy of blocking off one day per week during which I would not book consulting appointments or anything else, no matter what. And when my term on the university committee concluded, I declined when approached about a similar assignment.

To make a long story short, I am proud that David and I were among the many statisticians who developed methods that brought nonlinear mixed effects models into what is now routine use. Our most exciting achievement was when John Kimmel, who was then a Statistics Editor with Chapman & Hall, approached us about writing a book on the topic. Write a book? That had not dawned on either of us (back then, writing a book on one's research was much less common than it is today). For me, was this a good idea, given I would be coming up for tenure in a year? Writing a book is a significant, time-consuming undertaking; would this be a sensible thing to do right now? As scared as I was about tenure, the opportunity to work with David on putting together a comprehensive account of this area, all in one place, and make it accessible to practitioners and researchers who could benefit, was just too compelling. As it turned out, I ended up leaving North Carolina for Harvard (for personal reasons) shortly after we agreed to do the book, but I made it a priority and continued my policy of blocking off a day to work on it, despite being a clinical trials statistician with 13 protocols — I did waffle a few times, but for the most part I simply made it clear I was not available for conference calls or meetings on that day. Remarkably, we stuck to our vow of completing the book in a year (and I did get tenure and eventually moved back to North Carolina State). Our book (Davidian and Giltinan, 1995), although it is now somewhat outdated by all the advances in this area that have followed it, is one of my most satisfying professional accomplishments to this day.

As I said at the outset, starting out in a career in statistical research can be overwhelming. Balancing so many competing demands — collaborative projects, consulting, teaching, methodological research, committee responsibilities — is formidable, particularly for new researchers transitioning from graduate school. It may seem that you will never have enough time for methodological research, and you may even find identifying worthy research problems to be challenging, like I did. My story is not unique, and it taught me many lessons, on which I, along with my fellow senior participants in the ENAR junior researchers workshop, have dispensed advice over the years. Here are just a few of the key points we always make.

Number 1: Set aside time for your own interests, no matter what. It can be an entire day or an afternoon, whatever your position will permit. Put it in your calendar, and block it off. And do not waver. If a collaborator wants to schedule a meeting during that time, politely say that you are already committed. Your research is as important as your other responsibilities, and thus merits dedicated time, just as do meetings, teaching, and committee activities.

Along those same lines, learn that it is okay to say "no" when the alternative is being over-committed. Do not agree to take on new projects or responsibilities unless you are given time and support commensurate with the level of activity. If you are in a setting in which statisticians are asked to be part of a project for a percentage of their time, insist on that percentage being adequate — no project will ever involve just five percent of your effort. If you are being asked to serve on too many departmental or university committees, have an honest talk with your Department Chair to establish a realistic expectation, and then do not exceed it.

Finally, never pre-judge. When I set up the appointment with the crop scientist, I assumed it would be just another routine consulting encounter, for which I'd propose and carry out standard analyses and which would just add to the pile of work I already had. When David insisted I learn about pharmacokinetics, I was skeptical. As statisticians engaged in collaboration, we will always do many routine things, and we eventually develop radar for identifying the projects that will likely be routine. But you never know when that next project is going to reveal a new opportunity. And, as it did for me, alter the course of one's career. Be judicious to the extent that you can, but, unless you have very good reason, never write off anything. And never say never. If you'd told me back in 1987 that I would have published a best-selling *book* a mere eight years later, I would have asked you what you had been smoking!

48.3 Write it, and write it again

I'd always liked writing — in fact, when I was in high school, I toyed with the idea of being an English major. But my love of math trumped that idea, and I went on to major first in mechanical engineering at the University of Virginia, and then, realizing I was pretty bored, switched to applied mathematics. It was in the last semester of my senior year that, by chance, I took a statistics course from a relatively new Assistant Professor named David Harrington. I was hooked, and Dave was such a spectacular instructor that I ended up hanging around for an additional year and getting a Master's degree. Because I was in an Applied Mathematics Department in an Engineering School back in 1980, and because there was no Statistics Department at UVa back then, I ended up taking several courses from Dave (one of the only statisticians on the entire campus), including a few as reading courses. You may know of Dave — he eventually left Virginia for the Dana Farber Cancer Center and the Department of Biostatistics at Harvard School of Public Health — and among his many other accomplishments, he wrote a best-selling book on survival analysis (Fleming and Harrington, 1991) with *his* good friend from graduate school, Tom Fleming.

I mention Dave because he was the first to ever talk to me explicitly about the importance of a statistician being a good writer. I had to write a Master's thesis as part of my degree program, and of course Dave was my advisor. It was mainly a large simulation study, which I programmed and carried out (and which was fun) — the challenge was to write up the background and rationale, the design of the simulations, and the results and their interpretation in a clear and logical fashion. I will always remember Dave's advice as I set out to do this for the first time: "Write it, and write it again." Meaning that one can *always* improve on what one has written to make it more accessible and understandable to the reader. And that one should always strive to do this. It's advice I give to junior researchers and my graduate students to this day.

I learned a lot from Dave about clear and accessible writing through that Master's thesis. And fortunately for me, my PhD advisor, Ray Carroll, picked up where Dave left off. He insisted that I develop the skill of writing up results as I obtained them in a formal and organized way, so that by the time I had to begin preparing my dissertation, I had a large stack of self-contained documents, neatly summarizing each challenge, derivation, and result. Ray always emphasized the importance of clarity and simplicity in writing and speaking (which he demonstrated by editing everything I wrote for my dissertation and the papers arising from it). His motto for writing a good journal article was "Tell 'em what you'll tell 'em, tell 'em, and tell 'em what you told 'em." As you'll see shortly, I've adopted that one as a guiding principle as well.

I learned a lot from both Dave and Ray that laid the groundwork for my own strong interest in effective scientific writing. I am certain that, had I not had the benefit of their guidance, I would not have developed my own skills to the point that I eventually had the opportunity to serve as a Journal Editor. In my three years as Coordinating Editor of *Biometrics* in 2000–02 and my current role (since 2006) as Executive Editor, I have read and reviewed probably well over 1000 papers and have seen the entire spectrum, from those that were a joy to read to those that left me infuriated. And ditto for my time spent on NIH study sections (grant review panels). During my many years on what is currently the NIH Biostatistical Methods and Research Design study section, including three years as its Chair, I read grant applications that were so clear and compelling that I almost wanted to write my own personal check to fund them, but others that left me questioning the audacity of the investigators for expecting the taxpayers to support a project that they could not even convincingly and clearly describe.

What is it that makes one article or grant application so effective and another one so dreadful? Of course, the methodological developments being presented must have a sound basis. But even if they are downright brilliant and path-breaking, if they are not communicated in a way that the intended audience can unambiguously understand, they are not going to be appreciated. Given what one has to say is worthy, then, it is the quality of the *writing* that plays the primary role in whether or not a paper gets published or a grant gets funded. I'll concentrate on writing here, but most of what I say can be adapted equally well to oral presentation.

So how does one become a good writer? Admittedly, some people are just naturally gifted communicators, but most of us must practice and perfect our writing skills. And they can be perfected! Here is a synopsis of the points I and my colleagues stress to junior researchers when discussing effective writing of journal articles and grant applications.

First and foremost, before you even begin, identify and understand your target audience. If you are writing a journal article, you have two types of target readers. The Editor, Associate Editor, and referees at the journal, some of whom will be experts in the area and all of whom must be convinced of your work's relevance and novelty; and, ultimately, readers of the journal, who may span the range from experts like you to others with a general background who are hoping to learn something new. If you are writing a grant application, it is likely that many on the review panel will have only passing familiarity with your area while a few will be experts. Your presentation must be accessible to all of them, providing the novices with the background they need to understand your work while communicating the key advances to experts who already have that background. And you need to do this while respecting a nonnegotiable restriction on the length of your article or research proposal. That's a pretty tall order.

To address it, take some time and think carefully about the main message you want to convey and what you can reasonably hope to communicate effectively in the space allotted. You must acknowledge that you cannot pack in everything that you'd like or give the full background. So step into the shoes of your different readers. What background is essential for a novice to appreciate the premise of your work? Some of this you may be able to review briefly and explicitly, but most likely you will need to refer these readers to references where that background is presented. In that case, what references would be most appropriate? What results would an expert be willing to accept without seeing all the technical details (that might be more than a novice would need to see anyway)? What aspects of your work would be the most exciting to expert readers and should be highlighted, and which could be just mentioned in passing? Careful assessment of this will help you to establish what you must include to reach everyone and what you can omit or downplay but still communicate the main message. For journal articles, the option of supplementary material allows you the luxury of presenting much more, but always keep in mind that not all readers will consult it, so the main article must always contain the most critical material.

Once you have an idea of your audience and what they should take away, the key is to *tell the story*. For a journal article, a good principle to follow is the one Ray espouses; for a grant application, the format is more regimented, but the same ideas apply. First, "tell 'em what you'll tell 'em!" The introductory section to an article or proposal is often the hardest to write but the most important. This is where you motivate and excite all readers and give them a reason to want to keep reading! The opening sentence should focus immediately on the context of the work; for example, the renowned paper on generalized estimating equations by Liang and Zeger (1986) starts with "Longitudinal data sets are comprised...," which leaves no doubt in a reader's mind about the scope of the work. After setting the stage like this, build up the background. Why is the problem important? What are the major challenges? What is known? What are the limitations of current methods? What gaps in understanding need to be filled? For novice readers, note critical concepts and results that must be understood to appreciate your work, and provide key references where these readers may obtain this understanding. It is often very helpful to cite a substantive application that exemplifies the challenge (some journals even require this); this may well be an example that you will return to later to illustrate your approach.

The next step is to "tell 'em." You've made the case for why your audience should be interested in your story, now, tell it! Here, organization and logical flow are critical. Organize your presentation into sections, each having a clear focus and purpose that naturally leads to what follows. Completeness is critical; at any point along the way, the reader should have all the information s/he needs to have followed up to that point. Motivate and describe the steps leading to your main results, and relegate any derivations or side issues that could distract from the main flow of ideas to supplementary material for a journal article (or don't include them at all in a grant application). Relate complex concepts to concrete examples or simple special cases to assist novice readers grasp the main ideas. This is especially effective in grant applications, where reviewers are likely not to be experts.

The following principles seem obvious, but you would be surprised how often authors violate them! Do not refer to ideas or concepts until after you have introduced them. State your assumptions up front and before or when you need them for the first time. Do not use acronyms, terms, symbols, or notation until after they have been defined; for that matter, be sure to define every acronym, term, and symbol you use. And only define notation you really need. The less clutter and information a reader has to remember, the better.

Be as clear, concise, and helpful as you can. With limited space, every sentence and equation counts and must be understandable and unambiguous. Avoid "flowery" words if simpler ones are available, and if you catch yourself writing long sentences, strive to break them into several. Paraphrase and interpret mathematical results in plain English to give a sense of what results mean and imply. Use a formal, scientific style of writing (different from that used in this chapter). In particular, do not use contractions such as "it's" and "don't," and use only complete sentences; although these constructions may be used in a "popular" piece of writing like this one, they are *not* appropriate in scientific writing. Grammar and punctuation should be formal and correct (ask a colleague for help if English is not your native language), and be sure to spell check. Consult the articles in your target journal for examples of stylistic and grammatical conventions.

When reporting empirical studies, be sure that everything a reader would need to reproduce a simulation scenario him- or herself is presented. Do not display mind-numbing tables of numbers with little explanation; instead, choose to present limited results that illustrate the most important points and provide detailed interpretation, emphasizing how the results support the premise of your story. In fact, consider if it is feasible to present some results graphically, which can often be more efficient and effective than a tabular format.

In summary, do not leave your reader guessing! One useful practice to adopt is to step into your audience's shoes often. Read what you have written, and ask yourself: "Would I be able to understand what comes next given what I have presented so far?" Be honest, and you'll identify ways you could do a better job at conveying your message.

You may not have this luxury in a grant application, but in a journal article, you do. Once you've told the story, "tell 'em what you told 'em." Usually, this would be done in a final Discussion or Conclusions section. Restate what you set out to accomplish and review what was done to address it. Highlight the key findings, and discuss their significance and impact. It is just as important to note the limitations of what you have done and to identify what remains to be done. This summary does not have to be very long, but it should leave the reader with a clear understanding of why s/he bothered to read your work.

How does one go about getting started? The prospect of sitting down to write a journal article or research proposal can be daunting. Keep in mind that authors who simply sit down and start writing are rare. Most good writers, either literally, or figuratively in their minds, formulate an outline establishing the basic organization. Some will do this by beginning a LaTeX document with tentative section headings that correspond to the main ideas and results, and then filling in details. Also know that most good writers do not do this in order. They may write up the results first and leave the introduction until later, after they have a sense of what follows that material. If you find yourself grasping for the right word and/or agonizing over a detail, do not allow yourself to get stuck; make a note to come back to it later. As your work takes shape, you'll realize that you may want to move some material to another place, and the words and details you struggled with begin to gel.

Finally, I'll return to Dave's motto: "Write it, and write it again." No writer, no matter how talented or skilled, produces a perfect first draft. Once you have your first draft, review it carefully and critically, and be ruthless! Evaluate every sentence, and make sure that it is really necessary and, if it is, that it says exactly what you mean. Be on the lookout for repetition and redundancy — have you repeated something that you said earlier that doesn't bear repeating? This is a waste of precious space that could be put to better use. Be your own worst critic! Ask yourself: Have I cited the relevant literature and background sufficiently? Are there gaps in my logic or storytelling that would impede understanding? Are there parts that could be confusing or unclear? Is the overall message obvious, and have I made a convincing case? The bottom line is that you can *always* improve on what you have written. At some point, of course, you must let go and declare what you have to be the finished product, but a few rounds of putting your work aside and reading it again in a day or two can be very effective toward refining it to the point where any further improvements would be minimal.

48.4 Parting thoughts

A career in statistical research can be incredibly rewarding. The challenges are many, but the skills required can be mastered. I've touched on just two key elements — balancing competing responsibilities and effective writing — that have played a major role in my own career. I hope that my experience is helpful to the next generation of statistical scientists, to whom I leave one final piece of advice. Have fun! In spite of the challenges and occasional frustrations, enjoy what you do, and, if you don't, look for a change. We in our field are lucky that, at least currently, our skills are in high demand. Regardless of what type of position you find is right for you, becoming skilled at finding balance and being a good communicator will always serve you well.

References

Davidian, M. and Giltinan, D.M. (1995). *Nonlinear Models for Repeated Measurement Data.* Chapman & Hall, London.

Fleming, T.R. and Harrington, D.P. (1991). *Counting Processes and Survival Analysis.* Wiley, New York.

Liang, K.Y. and Zeger, S.L. (1986). Longitudinal data analysis using generalized linear models. *Biometrika*, 73:13–22.

49

Converting rejections into positive stimuli

Donald B. Rubin
Department of Statistics
Harvard University, Cambridge, MA

"It's not that I'm so smart, it's just that I stay with problems longer."
– Albert Einstein

At first glance this Einstein quotation may seem to have little to do with my title, but those readers who know something of Einstein's early life will recall that these years were not full of recognized scientific successes, but he kept working on his problems. And that is certainly related to why I chose the quote, but there is more to it. I have been fortunate to have had many journal publications, but less than one percent were accepted at first submission — far more were immediately rejected, followed closely by those that were rejected accompanied with the suggestion that it would not be wise to resubmit. However, I cannot think of an instance where this nasty treatment of my magnificent (self-assessed) work (sometimes joint) did not lead to a markedly improved publication, somewhere. In fact, I think that the drafts that have been repeatedly rejected by many different journals possibly represent my best contributions! Certainly the repeated rejections, combined with my trying to address various comments, led to better exposition and sometimes better problem formulation as well.

So here, in an attempt to inspire younger researchers to stay the course, I'll relay some of my stories on the topic, of course using some of my own publications as examples. I'll give only a short summary of each example, hopefully just enough for the reader to get the basic idea of the work (or possibly even read it, or as my wonderful PhD advisor, Bill Cochran, used to say, "I'd prefer if you read it and understood it, but if not, please read it; failing that, just cite it!"). For potential interest, I'll insert the approximate number of `Google Scholar` cites as of August 1, 2013. These counts may be of interest because the relationship between the number of citations and my memory of the paper's ease of acceptance appears to me to be zero (excluding the EM outlier). So young writers, if you think you have a good idea that reviewers do not appreciate, you're not alone, and quite possibly on to a very good idea, especially if the reviewers come across as real experts in their reports, but appear to have off-target comments.

49.1 My first attempt

> "A non iterative algorithm for least squares estimation of missing values in any analysis of variance design." *Journal of the Royal Statistical Society, Series C*, vol. 21 (1972), pp. 136–141. [Number of citations: 58]

This was my first sole-authored submission, and of course, I thought it was very clever, combining simple matrix manipulations with simple computations to generalize an old "Rothamsted" (to use Cochran's word) method to fill in missing data in an experimental design with their least squares estimates — a standard objective in those days (see the target article or Little and Rubin (2002, Chapter 2), for the reason for this objective). When I submitted this, I was still a PhD student, and when I received the report and saw "tentative reject," I was not a happy camper. Cochran calmed me down, and gave me some advice that he learned as a wee Scottish lad on the links: Keep your eye on the ball! Meaning, the objective when writing is to communicate with your readers, and the reviewers are making useful suggestions for improved communication. He went on to say:

> "The Editor is not your enemy — at this point in time, he has no idea who you even are! The Editor sent your draft to people who are more experienced than you, and they are reading it without pay to help you and the journal."

I was calm and the paper was accepted, a revision or two later. I was only fully calm, however, until the next "tentative reject" letter a few months later.

49.2 I'm learning

> "Matching to remove bias in observational studies." *Biometrics*, vol. 29 (1973), pp. 159–183. Printer's correction note in vol. 30 (1974), p. 728. [Number of citations: 392]

> "The use of matched sampling and regression adjustment to remove bias in observational studies." *Biometrics*, vol. 29 (1973), pp. 184–203. [Number of citations: 321]

This pair of submissions was based on my PhD thesis written under Bill's direction — back-to-back submissions, meaning both were submitted at the same time, with the somewhat "aggressive" suggestion to publish them back-to-back if they were acceptable. Both were on matched sampling, which at the time was really an unstudied topic in formal statistics. The only publication that was close was the wonderful classic Cochran (1968). Once again, a tentative rejection, but this time with all sorts of misunderstandings, criticisms and suggestions, that would take voluminous amounts of time to implement, and because at the time I was faculty in the department, I was a busy boy! I again told Bill how furious I was about these reviews, and Bill once again calmed me down and told me to remember what he had said earlier, and moreover, I should realize that these reviewers had spent even more time trying to help me, and that's why their comments were so long. Of course, Bill was correct, and both papers were greatly improved by my addressing the comments — not necessarily accepting the suggestions but addressing them. This lesson is important — if a reviewer complains about something and makes a suggestion as to how things should be changed, you as the author, needn't accept the reviewer's suggestion, but you should fix that thing to avoid the criticism. I was beginning to learn how to communicate, which is the entire point of writing journal articles or books.

49.3 My first JASA submission

> "Characterizing the estimation of parameters in incomplete data problems." *Journal of the American Statistical Association*, vol. 69 (1974), pp. 467–474. [Number of citations: 177]

This article concerns factoring likelihoods with missing data, which presented generalizations and extensions of prior work done by Anderson (1957) and Lord (1955) concerning the estimation of parameters with special patterns of missing data. Here, the editorial situation was interesting because, when I submitted the draft in 1970, the *JASA* Theory and Methods Editor was Brad Efron, whom I had met a couple of years earlier when he visited Harvard, and the Associate Editor was Paul Holland, my good friend and colleague at Harvard. So, I thought, finally, I will get a fast and snappy acceptance, maybe even right away!

No way! Paul must have (I thought) selected the most confused mathematical statisticians in the world — these reviewers didn't grasp any of the insights in my wonderous submission! And they complained about all sorts of irrelevant things. There is no doubt that if it hadn't been for Paul and Brad, it would have taken years more to get it into *JASA*, or would have followed the path of Rubin (1976) described below, or far worse. They were both helpful in explaining that the reviewers were not idiots, and actually they had some decent suggestions, properly interpreted — moreover, they actually liked the paper — which was very difficult to discern from the reports written for my eyes. Another set of lessons were apparent. First, read between the lines of a report: Editors do not want to over commit for fear the author won't pay attention to the suggestions. Second, reinterpret editorial and reviewer's suggestions in ways that you believe improve the submission. Third, thank them in your reply for suggestions that improved the paper — they did spend time writing reports, so acknowledge it. Fourth, it does help to have friends in positions of power!

49.4 Get it published!

"Estimating causal effects of treatments in randomized and nonrandomized studies." *Journal of Educational Psychology*, vol. 66 (1974), pp. 688–701. [Number of citations: 3084]

This paper is the one that started my publishing trail to use the potential outcomes notation to define formally causal effects in all situations, not just in randomized experiments as in Neyman (1923). Actually Neyman said he never made that generalization because he never thought of it, and anyway, doing so would be too speculative; see Rubin (2010) for the story on this. Everyone dealing with non-randomized studies for causal effects was using the observed value notation with one outcome (the observed value of the outcome) and one indicator variable for treatments until this paper. So in fact, Rubin (1974a) was the initiating reason for the phrase "Rubin Causal Model" — RCM, coined in Holland (1986).

I wrote this in some form when I was still at Harvard, more as notes for an introductory statistics course for psychologists. Someone suggested that I spruce it up a bit and submit it for publication. I did, but then couldn't get it published anywhere! Every place that I submitted the piece, rejected it. Sometimes the reason was that "every baby statistics student knows this" (I agreed that they should, but then show me where it is written!); sometimes the reason was "it's completely wrong"! And, in fact, I just received (July 2013) an email stating that "the Rubin definition of 'causality' is not appealing to many eminent statisticians." Sometimes the comments were even insulting, especially so because I was submitting statistical work from my position at Educational Testing Service (ETS) rather than a respected university statistics department. I asked around ETS and someone suggested the place, the *Journal of Educational Statistics*, where it ended up — I think that the acceptance was because of some high level intervention from someone who did like the paper but, more importantly, wanted to get me off his back — I honestly do not remember whom to thank.

There are several lessons here. First, it demonstrates that if a publication is good and good people find out about it (again, it helps to know good people), it will get read and cited. So if you are having this kind of problem with something that you are convinced is decent, get it published somewhere, and start citing it yourself in your own publications that are less contentious, and nag your friends to do so! Second, if you are repeatedly told by some reviewers that everyone knows what you are saying, but without specific references, and other reviewers are saying what you are writing is completely wrong but without decent reasons, you are probably on to something. This view is reinforced by the next example. And it reiterates the point that it does help to connect with influential and wise people.

49.5 Find reviewers who understand

> "Inference and missing data." *Biometrika*, vol. 63 (1976), pp. 581–592 (with discussion and reply). [Number of citations: 4185]

This article is extremely well known because it established the basic terminology for missing data situations, which is now so standard that this paper often isn't cited for originating the ideas, although often the definitions are summarized somewhat incorrectly. As Molenberghs (2007) wrote: "... it is fair to say that the advent of missing data methodology as a genuine field within statistics, with its proper terminology, taxonomy, notation and body of results, was initiated by Rubin's (1976) landmark paper." But was this a bear to get published! It was rejected, I think twice, from both sides of *JASA*; also from *JRSS B* and I believe *JRSS A*. I then decided to make it more "mathy," and I put in all this measure theory "window dressing" (a.s., a.e., both with respect to different measures because I was doing Bayesian, repeated sampling and likelihood inference). Then it got rejected twice from *The Annals of Statistics*, where I thought I had a chance because I knew the Editor — knowing important people doesn't always help. But when I told him my woes after the second and final rejection from *The Annals*, and I asked his advice on where I should send it next, he suggested "Yiddish Weekly" — what a great guy!

But I did not give up even though all the comments I received were very negative; but to me, these comments were also very confused and very wrong. So I tried *Biometrika* — home run! David Cox liked it very much, and he gave it to his PhD student, Rod Little, to read and to contribute a formal comment. All those prior rejections created, not only a wonderful publication, but lead to two wonderful friendships. The only real comment David had as the Editor was to eliminate all that measure theory noise, not because it was wrong but rather because it just added clutter to important ideas. Two important messages: First, persevere if you think that you have something important to say, especially if the current reviewers seem not up to speed. Second, try to find a sympathetic audience, and do not give up.

49.6 Sometimes it's easy, even with errors

> "Maximum likelihood from incomplete data via the EM algorithm." *Journal of the Royal Statistical Society, Series B*, vol. 39 (1977), pp. 1–38 (joint work with A.P. Dempster and N. Laird, published with discussion and reply). [Number of citations: 34,453]

Those early years at ETS were tough with respect to getting articles accepted, and I think it *is* tougher submitting from less academically prestigious places. But publishing things became a bit easier as I matured. For example, the EM paper was accepted right away, with even invited discussion. It was to be a read paper in London in 1976, the trip where I met Rod Little and David Cox in person — the latter mentioned that he really wasn't fond of the title of the already accepted Rubin (1976) because something that's missing can't be "given" — the Latin meaning of data. And this rapid acceptance for the EM paper was despite having one of its proofs wrong — misapplication of the triangle inequality! Wu (1983) corrected this error, which was not critical to the fundamental ideas in the paper about the generality of the missing data perspective. In statistics, ideas trump mathematics — see Little's (2013) Fisher lecture for more support for this position. In this case, a rapid acceptance allowed an error to be published and corrected by someone else. If this can be avoided it should be, even if it means withdrawing an accepted paper; three examples of this follow.

49.7 It sometimes pays to withdraw the paper!

It sometimes pays to withdraw a paper. It can be good, it can be important, and even crucial at times, as the following examples show.

49.7.1 It's good to withdraw to complete an idea

> "Parameter expansion to accelerate EM: The PX-EM algorithm." *Biometrika*, vol. 85 (1998), pp. 755–770 (joint work with C.H. Liu and Y.N. Wu). [Number of citations: 243]

This submission was done jointly with two exceptionally talented former PhD students of mine, Chuanhai Liu and Ying Nian Wu. It was a technically very sound article, which introduced the PX-EM algorithm, an extension of EM. If correctly implemented, it always converged in fewer steps than EM — nice. But after the submission was accepted by an old friend, Mike Titterington at *Biometrika*, there was an intuitive connection that I knew had to be there, but that we had not included formally; this was the connection between PX-EM and ANCOVA, which generally creates more efficient estimated treatment effects by estimating a parameter whose value is known to be zero (e.g., the difference in the expected means of covariates in the treatment and control groups in a completely randomized experiment is zero, but ANCOVA estimates it by the difference in sample means). That's what PX-EM does — it introduces a parameter whose value is known, but estimates that known value at each iteration, and uses the difference between the estimate and the known value to obtain a larger increase in the actual likelihood. But we hadn't done the formal math; so we withdrew the accepted paper to work on that.

Both Chuanhai and Yingnian were fine with that decision. My memory is that we basically destroyed part of a Christmas holiday getting the idea down correctly. We were now ready to resubmit, and it was not surprising that it was re-accepted overnight, I think. Another lesson: Try to make each publication as clean as possible — you and your coauthors will have to live with the published result forever, or until someone cleans it up!

49.7.2 It's important to withdraw to avoid having a marginal application

> "Principal stratification for causal inference with extended partial compliance: Application to Efron–Feldman data." *Journal of the American Statistical Association*, vol. 103 (2008), pp. 101–111 (joint work with H. Jin). [Number of citations: 65]

This paper re-analyzed a data set from an article (Efron and Feldman, 1991) on noncompliance, but I think that Hui Jin and I approached it more appropriately using principal stratification (Frangakis and Rubin, 2002). I had a decade to ponder the issues, the benefit of two great economics coauthors in the interim (Angrist et al., 1996), a wonderful PhD student (Constantine Frangakis) to help formulate a general framework, and a great PhD student to work on the example. The submission was accepted fairly quickly, but as it was about to go to the Copy Editors, I was having my doubts about the last section, which I really liked in principle, but the actual application didn't make complete scientific sense, based on my experience consulting on various pharmaceutical projects. So I wanted to withdraw and to ask my coauthor, who had done all the extensive computing very skillfully, to do all sorts of new computing. Her initial reaction was something like: Had I lost my mind? Withdraw a paper already accepted in *JASA*?! But wasn't the objective of writing and rewriting to get the paper accepted? But after listening to the reasons, she went along with my temporary insanity, and she executed the final analyses that made scientific sense with great skill and care. Of course the paper was re-accepted. And it won the Mitchell prize at the Joint Statistical Meetings in 2009 for the best Bayesian paper.

The message here is partly a repeat of the above one regarding publishing the best version that you can, but it is more relevant to junior authors anxious for publications. I surely know how difficult it can be, certainly in the early years, to build a CV and get promoted; but that's short term. Eventually real quality will triumph, and don't publish anything that you think may haunt you in the future, even if it's accepted in a top journal. As Pixar's Jay Shuster put it: "Pain is temporary, 'suck' is forever." By the way, Hui Jin now has an excellent position at the International Monetary Fund.

49.7.3 It's really important to withdraw to fix it up

> "Multiple imputation by ordered monotone blocks with application to the Anthrax Vaccine Research Program," *Journal of Computational and Graphical Statistics*; 2013; in press (joint work with F. Li, M. Baccini, F. Mealli, E.R. Zell, and C.E. Frangakis)

This publication hasn't yet appeared, at least at the time of my writing this, but it emphasizes the same point, with a slightly different twist because of the multiplicity of coauthors of varying seniority. This paper grew out of a massive joint effort by many people, each doing different things on a major project. I played the role of the MI-guru and organizer, and the others were absorbed with various computing, writing, and data analytic roles. Writing a document with five major actors was complicated and relatively disorganized — the latter issue, my fault. But then all of a sudden, the paper was written, submitted, and remarkably the first revision was accepted! I now had to read the entire thing, which had been "written" by a committee of six, only two of which were native English speakers! Although some of the writing was good, there were parts that were confusing and other parts that appeared to be contradictory. Moreover, on closer examination, there were parts where it appeared that mistakes had been made, mistakes that would take vast amounts of time to correct fully, but that only affected a small and orthogonal part of the paper. These problems were really only evident to someone who had an overview of the entire project (e.g., me), not reviewers of the submission.

I emailed my coauthors (some of whom were across the Atlantic) that I wanted to withdraw and rewrite. Initially, there seemed to be some shock (but wasn't the purpose of writing to get things published?), but they agreed — the more senior authors essentially immediately, and more junior ones after a bit of contemplation. The Editor who was handling this paper (Richard Levine) made the whole process as painless as possible. The revision took months to complete; and it was re-accepted over a weekend. And I'm proud of the result. Same message, in some sense, but wise Editors want to publish good things just as much as authors want to publish in top flight journals.

49.8 Conclusion

I have been incredibly fortunate to have access to sage advice from wonderful mentors, obviously including advice about how to react to rejected submissions. It may not always be true, and I do know of some gross examples, but in the vast majority of cases, Editors and reviewers are giving up their time to try to help authors, and, I believe, are often especially generous and helpful to younger or inexperienced authors. Do not read into rejection letters personal attacks, which are extremely rare. The reviewers may not be right, but only in rare situations, which I believe occur with submissions from more senior authors, who are "doing battle" with the current Editors, is there any personal animus. As Cochran pointed out to me about 1970, they probably don't even know anything about you, especially if you're young. So my advice is: Quality trumps quantity, and stick with good ideas even when you have to do battle with the Editors and reviewers — they are not perfect judges but they are, almost uniformly, on your side.

Whatever wisdom is offered by this "fireside chat" on dealing with rejections of journal submissions, owes a huge debt to the advice of my mentors and very respected folks along my path. So with the permission of the Editors of this volume, I will follow with a description of my incredible good fortune to meet such folks. As one of the wisest folks in our field (his name is hidden among the authors of the additional references) once said to me: If you ask successful people for their advice on how to be successful, their answers are, "Be more like me." I agree, but with the addition: "And meet wonderful people." This statement creates a natural transition to the second part of my contribution to this 50th anniversary volume, on the importance of listening to wise mentors and sage colleagues. I actually wrote the second part before the first part, but on rereading it, I feared that it suffered from two problems; one, it sounded too self-congratulatory and almost elitist. The Editors disagreed and thought it actually could be a helpful chapter for some younger readers, perhaps because it does illustrate how good fortune plays such an important role, and I certainly have had that with respect to the wonderful influences I've had in my life. The advice: Take advantage of such good fortune!

References

Angrist, J.D., Imbens, G.W., and Rubin, D.B. (1996). Identification of causal effects using instrumental variables. *Journal of the American Statistical Association*, 94:444–472.

Anderson, T.W. (1957). Maximum likelihood estimates for a multivariate normal distribution when some observations are missing. *Journal of the American Statistical Association*, 52:200–203.

Cochran, W.G. (1968). The effectiveness of adjustment by subclassification in removing bias in observational studies. *Biometrics*, 24:295–313.

Dempster, A.P., Laird, N., and Rubin, D.B. (1977). Maximum likelihood from incomplete data via the EM Algorithm. *Journal of the Royal Statistical Society, Series B*, 39:1–38.

Efron, B. and Feldman, D. (1991). Compliance as an explanatory variable in clinical trials. *Journal of the American Statistical Association*, 86:9–17.

Frangakis, C.E. and Rubin, D.B. (2002). Principal stratification in causal inference. *Biometrics*, 58:21–29.

Holland, P.M. (1986). Statistics and causal inference. *Journal of the American Statistical Association*, 81:945–960.

Jin, H. and Rubin, D.B. (2008). Principal stratification for causal inference with extended partial compliance: Application to Efron–Feldman data. *Journal of the American Statistical Association*, 103:101–111.

Li, F., Baccini, M., Mealli, F., Zell, E.R., Frangakis, C.E., and Rubin, D.B. (2013). Multiple imputation by ordered monotone blocks with application to the Anthrax Vaccine Research Program. *Journal of Computational and Graphical Statistics*, in press.

Little, R.J.A. (2013). In praise of simplicity not mathematistry! Ten simple powerful ideas for the statistical scientist. *Journal of the American Statistical Association*, 108:359–369.

Little, R.J.A. and Rubin, D.B. (2002). *Statistical Analysis with Missing Data*, 2nd edition. Wiley, New York.

Liu, C.H., Rubin, D.B., and Wu, Y.N. (1998). Parameter expansion to accelerate EM: The PX-EM Algorithm. *Biometrika*, 85:755–770.

Lord, F.M. (1955). Estimation of parameters from incomplete data. *Journal of the American Statistical Association*, 50:870–876.

Molenberghs, G. (2007). What to do with missing data? *Journal of the Royal Statistical Society, Series A*, 170:861–863.

Neyman, J. (1923). On the application of probability theory to agricultural experiments: Essay on principles. Section 9. *Roczniki Nauk Rolniczych*, 10:1–51. [English translation of the original Polish article available in *Statistical Science*, 5:465–472.]

Rubin, D.B. (1972). A non iterative algorithm for least squares estimation of missing values in any analysis of variance design. *Journal of the Royal Statistical Society, Series C*, 21:136–141.

Rubin, D.B. (1973a). Matching to remove bias in observational studies. *Biometrics*, 29:159–183. [Printer's correction note 30, p. 728.]

Rubin, D.B. (1973b). The use of matched sampling and regression adjustment to remove bias in observational studies. *Biometrics*, 29:184–203.

Rubin, D.B. (1974a). Characterizing the estimation of parameters in incomplete data problems. *The Journal of the American Statistical Association*, 69:467–474.

Rubin, D.B. (1974b). Estimating causal effects of treatments in randomized and nonrandomized studies. *Journal of Educational Psychology*, 66:688–701.

Rubin, D.B. (1976). Inference and missing data. *Biometrika*, 63:581–592.

Rubin, D.B. (2010). Reflections stimulated by the comments of Shadish (2009) and West & Thoemmes (2009). *Psychological Methods*, 15:38–46.

Wu, C.F.J. (1983). On the convergence properties of the EM algorithm. *The Annals of Statistics*, 11:95–103.

50

The importance of mentors

Donald B. Rubin
Department of Statistics
Harvard University, Cambridge, MA

On this 50th anniversary of the COPSS, I feel incredibly fortunate to have stumbled into the career path that I followed, which appears, even in hindsight, like an overgrown trail in the woods that somehow led to a spectacular location. In some sense, there is an odd coincidence in that it is also roughly the 50th anniversary of my recognizing the field of statistics as a valuable one. When thinking about what I could say here that might interest or help others, and is not available in other places, I eventually decided to write about my path and the importance, to me at least, of having wonderful "mentors" with different backgrounds, which allowed me to appreciate many different modes of productive thinking. Probably the characteristic of the field of statistics that makes it so appealing to me is the wonderful breadth of intellectual topics that it touches. Many of my statistical mentors had a deep appreciation for this, and for that I will always be grateful, but also I have always felt very fortunate to have had admirable mentors from other disciplines as well.

50.1 My early years

I grew up in Evanston, Illinois, home of Northwestern University. As a kid, I was heavily influenced intellectually by a collection of people from various directions. My father was one of four brothers, all of whom were lawyers, and we used to have stimulating arguments about all sorts of topics; arguing was not a hostile activity but rather an intellectually engaging one. Probably the most argumentative was Uncle Sy, from DC, who had framed personal letters of thanks for service from, eventually, all the presidents starting with Harry Truman going through Gerald Ford, as well as some contenders, such as Adlai Stevenson, and various Supreme Court Justices, and even Eleanor Roosevelt with whom I shook hands back then. It was a daunting experience, not only because of her reputation, but also because, according to my memory, she was twice as tall as I was! The relevance of this to the field of statistics is that it created in me a deep respect for the principles of our legal system, to which I find statistics deeply relevant, for example, concerning issues as diverse as the death penalty, affirmative action, tobacco litigation, ground water pollution, wire fraud, etc.

But the uncle who was most influential on my eventual interest in statistics was my mother's brother, a dentist (then a bachelor), who loved to gamble small amounts, either in the bleachers at Wrigley Field, betting on the outcome of the next pitch while we watched the Cubs lose, or at Arlington Race track, where I was taught as a wee lad how to read the Racing Form and estimate the "true" odds from the various displayed betting pools while losing two dollar bets. Wednesday and Saturday afternoons, during the warm months then, were times to learn statistics — even if at various bookie joints that were sometimes raided. As I recall, I was a decent student of his, but still lost small amounts — this taught me to never gamble with machines. Later, as a PhD student at Harvard, I learned never to "gamble" with "Pros." From those days I am reminded of the W.C. Fields line who, when playing poker for money on a public train car, was chastised by an older woman: "Young man, don't you know that gambling is a sin?" He replied, "The way I play it, it's not gambling." The Harvard Pro's were not gambling when playing against me.

There were two other important influences on my statistical interests from the late 1950s and early 1960s. First there was an old friend of my father's from their government days together, a Professor Emeritus of Economics at the University of California Berkeley, George Mehren, with whom I had many entertaining and educational (to me) arguments, which generated a respect for economics, which continues to grow to this day. And second, my wonderful teacher of physics at Evanston Township High School — Robert Anspaugh — who tried to teach me to think like a scientist, and how to use mathematics in pursuit of science. So by the time I left high school for college, I appreciated some probabilistic thinking from gambling, some scientific thinking from physics, and I had deep respect for disciplines other than formal mathematics. These, in hindsight, are exposures that were crucial to the kind of formal statistics to which I gravitated as I matured.

50.2 The years at Princeton University

When I entered Princeton in 1961, like many kids at the time, I had a pile of advanced placements, which lined me up for a BA in three years, but unknown to me before I entered, I was also lined up for a crazy plan to get a PhD in physics in five years, in a program being proposed by John Wheeler, a well-known professor of Physics there (and Richard Feynman's PhD advisor years earlier). Wheeler was a fabulous teacher, truly inspirational. Within the first week, he presented various "paradoxes" generated by special relativity, introduced the basics of quantum mechanics, gave homework problems designed to stimulate intuitive but precise thinking (e.g., estimate — by careful reasoning — how far a wild goose can fly), pointed out errors in our current text (e.g., coherent light cannot be created — it can — lasers were invented about a year after this edition was published), and many other features of scientific thinking that are critically important but often nearly absent from some people's statistical thinking, either because they do not have the requisite mathematical background (and sometimes appear to think that algorithmic thinking is a substitute) or because they are still enamored with thoughtless mathematical manipulations, or perhaps some other reason.

In any case, my physics lessons from Anspaugh and Wheeler were crucial to my thinking, especially the naturalness of two closely related messages: Time marches on, and despite more recent "back to the future" movies, we cannot go back in time, which leads directly to the second message — in any scientific problem, there will always exist missing data, the "observer effect" (intuitively related to the Heisenberg uncertainty principle, but different). That is, you cannot precisely measure both position and momentum at the same point in time, say t, because the physical act of measuring one of them at t affects the other's value after t; this is just like the fact that you cannot go back in time to give the other treatment in a causal inference problem, and the choice of notation and problem formulation should reflect these facts. All of statistics should be formulated as missing data problems (my view since about 1970, although not everyone's).

But like many kids of that age, I was torn by competing demands about how to grow up, as well as larger social issues of that time, such as our involvement in Vietnam. And Wheeler took a leave of absence, I think to visit Texas Austin in my second year, so I switched fields. My exact reasoning from that time is a bit fuzzy, and although I continued to take some more advanced physics courses, I switched from Physics to Psychology towards the end of my second year, where my mathematical and scientific background seemed both rare and appreciated, whereas in math and physics, at least in my cohort, both skill sets were good, especially so in physics, but not rare. This decision was an immature one (not sure what a mature one would have been), but a fine decision because it introduced me to some new ways of thinking as well as to new fabulous academic mentors.

First, there was a wonderful Psychologist, Silvan Tomkins, author of the three volume "Affect, Imagery, Consciousness," who introduced me to Sigmund Freud's work, and other philosopher/psychologists on whose work his own book built. I was amazed that interpreting dreams of strangers actually worked much of the time; if I asked the right questions about their dreams, I could quite often tell things about strangers such as recondite fears or aspirations! There may really exist a "collective unconscious" to use Jung's phrase. In any case, I developed a new respect for psychology, including for their neat experiments to assess creativity and such "soft" concepts — there was real scientific understanding among many psychologists, and so much yet to learn about the mind! So armed with that view that there is much to do in that direction, in my final year, I actually applied to PhD programs in psychology, and was accepted at Stanford, the University of Michigan, and Harvard.

Stanford was the strongest technically, with a very quiet but wonderful professor who subsequently moved to Harvard, Bill Estes. Michigan had a very strong mathematical psychology program, and when I visited there in spring of 1965, I was hosted primarily by a very promising graduating PhD student, Amos Tversky, who subsequently wrote extremely influential and Nobel Prize (in economics) winning work with Danny Kahneman. Amos' work, even in 1965, was obviously great stuff, but I decided on Harvard, for the wrong reason (girlfriend on the East coast), but meeting Bill and Amos, and hearing the directions of their work, confirmed the idea that being in psychology was going to work out well — until I got to Harvard.

50.3 Harvard University — the early years

My start at Harvard in the Department of Social Relations, which was the home of psychology back then, was disappointing, to say the least. First, all sorts of verbal agreements, established on my visit only months before with a senior faculty member, were totally forgotten! I was told that my undergraduate education was scientifically deficient because it lacked "methods and statistics" courses, and I would have to take them at Harvard or withdraw. Because of all the math and physics that I'd had at Princeton, I was insulted! And because I had independent funding from an NSF graduate fellowship, I found, what was essentially, a Computer Science (CS) program, which seemed happy to have me, probably because I knew `Fortran`, and had used it extensively at Princeton; but I also found some real math courses and ones in CS on "mathy" topics, such as computational complexity, more interesting than the CS ones, although it was clear that computers, as they were evolving, were going to change the way much of science was done.

But what to do with my academic career? The military draft was still in place, and neither Vietnam nor Canada seemed appealing. And I had picked up a Master's degree from CS in the spring of 1966.

A summer job in Princeton in 1966 lead to an interesting suggestion. I was doing some programming for John Tukey and some consulting for a Princeton Sociology Professor, Robert Althauser, basically writing programs to do matched sampling; Althauser seemed impressed by my ability to program and to do mathematics, and we discussed my future plans — he mentioned Fred Mosteller and the decade old Statistics Department at Harvard; he suggested that I look into it. I did, and by fall of 1968, I was trying my third PhD program at Harvard, still using the NSF funding that was again renewed. A great and final field change!

But my years in CS were good ones too. And the background in CS was extremely useful in Statistics — for doing my own work and for helping other PhD students. But an aside about a lesson I learned then: When changing jobs, never admit you know anything about computing or you will never have any time to yourself! After a couple of years of denying any knowledge about computers, no one will ask, and furthermore, by then, you will be totally ignorant about anything new and practical in the world of computing, in any case — at least I was.

50.4 My years in statistics as a PhD student

These were great years, with superb mentoring by senior folks: Fred Mosteller, who taught me about the value of careful, precise writing and about responsibilities to the profession; Art Dempster, who continued the lessons about scientific thinking I learned earlier, by focusing his statistics on principles rather than ad hoc procedures; and of course, Bill Cochran, a wonderfully wise and kind person with a fabulous dry sense of humor, who really taught me what the field of statistics, at least to him, concerned. Also important was meeting life-long friends, such as Paul Holland, as a junior faculty member. Also, there were other faculty with whom I became life-long friends, in particular, Bob Rosenthal, a professor in psychology — we met in a Cochran seminar on experimental design. Bob has great statistical insights, especially in design, but did not have the mathematical background to do any "heavy-lifting" in this direction, but this connection helped to preserve the long-term interests in psychology. Bob was a mentor in many ways, but one of the most important was how to be a good professor for your students — they deserve access to your time and mind and its accumulated wisdom.

Another psychology faculty member, whom I met in the summer of 1965 and greatly influenced me, was Julian Jaynes from Princeton, who became relatively famous for his book "The Origin of Consciousness in the Breakdown of the Bicameral Mind" — a spectacularly interesting person, with whom I became very close during my post-graduate years when I was at ETS (the Educational Testing Services) in Princeton. A bit more, shortly, on his influence on my thinking about the importance of bridging ideas across disciples.

After finishing my graduate work in 1970, I stayed around Harvard Statistics for one more year as a faculty member co-teaching with Bob Rosenthal the "Statistics for Psychologists" course that, ironically, the Social Relations Department wanted me to take five years earlier, thereby driving me out of their program! I decided after that year that being a junior faculty member, even in a great department, was not for me. So I ended up accepting a fine position at ETS in Princeton, New Jersey, where I also taught part-time at Princeton's young Statistics Department, which renewed my friendship with Tukey; between the two positions, my annual salary was more than twice what I could be offered at Harvard to stay there as junior faculty.

50.5 The decade at ETS

The time at ETS really encouraged many of my earlier applied and theoretical connections — it was like an academic position with teaching responsibilities replaced by consulting on ETS's social science problems, including psychological and educational testing ones; and I had the academic connection at Princeton, where for several years I taught one course a year. My ETS boss, Al Beaton, had a Harvard Doctorate in Education, and had worked with Dempster on computational issues, such as the "sweep operator." Al was a very nice guy with deep understanding of practical computing issues. These were great times for me, with tremendous freedom to pursue what I regarded as important work. Also in those early years I had the freedom to remain in close contact with Cochran, Dempster, Holland, and Rosenthal, which was very important to me and fully encouraged by Beaton. I also had a Guggenheim fellowship in 1978, during which I spent a semester teaching causal inference back at Harvard. A few years before I had visited the University of California Berkeley for a semester, where I was given an office next to Jerzy Neyman, who was then retired but very active — a great European gentleman, who clearly knew the difference between mathematical statistics for publishing and real statistics for science — there is no doubt that I learned from him, not a mentor as such, but as a patient and kind scholar interested in helping younger people, even one from ETS.

Here's where Julian Jaynes re-enters the picture in a major way. We became very close friends, having dinner and drinks together several times a week at a basement restaurant/bar in Princeton called the Annex. We would have long discussions about psychology and scientific evidence, e.g., what makes for consciousness. His knowledge of history and of psychology was voluminous, and he, in combination with Rosenthal and the issues at ETS, certainly cemented my fascination with social science generally. A different style mentor, with a truly eye-opening view of the world.

50.6 Interim time in DC at EPA, at the University of Wisconsin, and the University of Chicago

Sometime around 1978 I was asked to be the Coordinating and Applications Editor of *JASA*. Stephen Stigler was then Theory and Methods Editor. I had previously served as an Associate Editor for Morrie DeGroot, and was becoming relatively more well known, for things like the EM algorithm and different contributions that appeared in various statistical and social science journals, so more options were arising. I spent two weeks in December 1978 at the Environmental Protection Agency in the Senior Executive Service (very long story), but like most things that happened to me, it was very fortunate. I was in charge of a couple of statistical projects with connections to future mentors, one with a connection to Herman Chernoff (then at MIT), and one with a connection to George Box; George and I really hit it off, primarily because of his insistence on statistics having connections to real problems, but also because of his wonderful sense of humor, which was witty and ribald, and his love of good spirits.

Previously I had met David Cox, via my 1976 *Biometrika* paper "Inference and missing data" discussed by Cox's then PhD student and subsequently my great coauthor, Rod Little. I found that the British style of statistics fit fabulously with my own interests, and the senior British trio, Box, Cochran and Cox, were models for the kind of statistician I wanted to be. I also participated with Box in several Gordon Conferences on Statistics and Chemistry in the late 1970s and early 1980s, where George could unleash his "casual" side. Of some importance to my applied side, at one of these I met, and became good friends with, Lewis Sheiner, UCSF Pharmacology Professor. Lewis was a very wise doctor with remarkably good statistical understanding, who did a lot of consulting for FDA and for pharmaceutical companies, which opened up another connection to an applied discipline for me, in which I am still active, with folks at FDA and individuals in the pharmaceutical world.

In any case, the EPA position led to an invitation to visit Box at the Math Research Center at the University of Wisconsin, which I gladly accepted. Another great year with long-term friends and good memories. But via Steve Stigler and other University of Chicago connections, a full professor position was offered, jointly in Statistics and in the Department of Education. I was there for only two years, but another wonderful place to be with more superb mentors; David Wallace and Paul Meier, in particular, were especially helpful to me in my first full-time full professor position. I also had a connection to the National Opinion Research Corporation, which was important. It not only was the home of the first grant to support multiple imputation, but because they did real survey work, they were actually interested in my weird ideas about surveys! And because they also did work in economics, this initiated a bridge to that wonderful field that is still growing for me. Great times.

50.7 The three decades at Harvard

I'm just completing my 30th year at the Harvard Department of Statistics, and these years have been fabulous ones, too. The first of those years renewed and reinforced my collaborations with Bob Rosenthal, through our co-teaching a "statistics for psychologists" course and our Thursday "brown-bag consulting" lunch. Other psychologists there have been influential as well, such as Jerry Kagan, a wonderfully thoughtful guy with a fabulous sense of humor, who was a great mentor regarding personality theory, as was Phil Holzman with his focus on schizophrenia. We would all meet at Bill and Kay Estes's spectacular Christmas parties at their "William James" house, with notable guests such as Julia Child, who lived down the block and reminded me of Eleanor Roosevelt. These personal connections to deep-thinking psychologists clearly affect the way I approach problems.

These early years as Professor at Harvard also saw a real attempt to create something of a bridge to economics in Cambridge, initially through some 1990s efforts with Bob Solow and then Josh Angrist, both at MIT, and of course my close colleague Guido Imbens now at Stanford, and then again with Guido more recently in the context of our causal book and our co-taught course. Also, economist Eric Maskin, who recently returned to Harvard after a stint in Princeton, convinced me to teach a "baby causal" course in the "core" for undergraduates who had no background in anything technical — it was good for me and my teaching fellows, and hopefully some of those who took the course. Another economist who influenced me was the Dean who "hired me" — Henry Rosovsky — one of the wisest and most down-to-earth men I have ever met; we shared many common interests, such as classic cars and good lunches. A wonderful mentor about academic life! Every academic should read his book: "The University: An Owner's Manual."

And of course there were the senior folks in Statistics: Art Dempster with his principled approach to statistics, was always a pleasure to observe; Fred Mosteller and his push for collaborations and clear writing; and Herman Chernoff (whom I hired; he used to refer to me as his "boss" — hmm, despite my being over 20 years his junior). Herman attended and still, at 90 years, attends most of our seminars and offers penetrating comments — a fabulous colleague with a fabulous mind and subtle and clever sense of humor. And old friend Carl Morris — always a great colleague.

50.8 Conclusions

I have intentionally focused on mentors of mine who were (or are) older, despite the undeniable fact that I have learned tremendous amounts from my colleagues and students. I also apologize for any mentors whom I have accidentally omitted — I'm sure that there are some. But to all, thanks so much for the guidance and advice that have led me to being a statistician with a variety of interests. My career would have told a very different story if I had not had all the wonderful guidance that I have received. I would have probably ended up in some swamp off that tangled path in the woods. I think that my realizing that fact has greatly contributed to my own desire to help guide my own students and younger colleagues. I hope that I continue to be blessed with mentors, students and colleagues like the ones I've had in the past, until we all celebrate the 100th anniversary of the COPSS!

51

Never ask for or give advice, make mistakes, accept mediocrity, enthuse

Terry Speed
Division of Bioinformatics
Walter and Eliza Hall Institute of Medical Research
and
Department of Statistics
University of California, Berkeley, CA

Yes, that's my advice to statisticians. Especially you, dear reader under 40, for you are one of the people most likely to ask an older statistician for advice. But also you, dear reader over 40, for you are one of the people to whom younger statisticians are most likely to turn for advice.

Why 40? In the 1960s, which I lived through, the mantra was *Never trust anyone over 30.* Times change, and now 40 is (approximately) the cut-off for the COPSS Presidents' Award, so I think it's a reasonable dividing line for separating advisors and advisees. Of course people can and do give and take advice at any age, but I think we regard advice from peers very differently from advice from... advisors. That's what I'm advising against. Please don't get me wrong: I'm not being ageist here, at least not consciously. I'm being a splitter.

Where am I going with all this? There is a sentence that used to be heard a lot on TV shows, both seriously and in jest: "Don't try this at home." It was usually said after showing a stupid or dangerous act, and was a way of disclaiming liability, as they knew it wouldn't work well for most viewers. I often feel that people who give advice should act similarly, ending their advice with "Don't take my advice!"

51.1 Never ask for or give advice

What's wrong with advice? For a start, people giving advice lie. That they do so with the best intentions doesn't alter this fact. This point has been summarized nicely by Radhika Nagpal (2013). I say trust the people who tell you "I have no idea what I'd do in a comparable situation. Perhaps toss a coin." Of course people don't say that, they tell you what they'd like to do or wish they had done in some comparable situation. You can hope for better.

What do statisticians do when we have to choose between treatments A and B, where there is genuine uncertainty within the expert community about the preferred treatment? Do we look for a statistician over 40 and ask them which treatment we should choose? We don't, we recommend running a randomized experiment, ideally a double-blind one, and we hope to achieve a high adherence to the assigned treatment from our subjects. So, if you really don't know what to do, forget advice, just toss a coin, and do exactly what it tells you. But you are an experiment with $n = 1$, you protest. Precisely. What do you prefer with $n = 1$: an observational study or a randomized trial? (It's a pity the experiment can't be singly, much less doubly blinded.)

You may wonder whether a randomized trial is justified in your circumstances. That's a very important point. Is it true that there is genuine uncertainty within the expert community (i.e., you) about the preferred course of action? If not, then choosing at random between your two options is not only unethical, it's stupid. And who decides whether or not there is genuine uncertainty in your mind: you or the people to whom you might turn for advice? This brings me to the most valuable role potential advisors can play for potential advisees, the one I offer when people ask me for advice. I reply "I don't give advice, but I'm very happy to listen and talk. Let's begin." This role cannot be replaced by words in a book like this, or on a website.

51.2 Make mistakes

What if it turns out that you made a wrong decision? I'll pass over the important question of how you learned that it was the wrong decision, of how you tell that the other decision would have been better. That would take me into the world of counterfactuals and causal inference, and I've reserved my next lifetime for a close study of that topic. But let's suppose you really did make a mistake: is that so bad?

There is a modest literature on the virtues of making mistakes, and I like to refer people to it as often as possible. Why? Because I find that too many people in our business — especially young people — seem to be unduly risk averse. It's fine not wanting to lose your money in a casino (though winning has a certain appeal), but always choosing the safe course throughout a career seems sad to me. I think there's a lot to be gained from a modest amount of risk-taking, especially when that means doing something *you* would like to do, and not what your advisor or department chair or favorite COPSS award winner thinks you should do. However, to call it literature might be overly generous. Perhaps a better description is body of platitudes, slogans and epigrams by pundits and leaders from business, sport, and the arts. One thing is certain: professors do not figure prominently in this "literature." *Nothing ventured, nothing gained*, catches the recurrent theme. The playwright George Bernard Shaw wrote: *A life spent making mistakes is not only more honorable, but more useful than a life spent doing nothing*, and many others have echoed his words. Another Irish playwright, Oscar Wilde was more forthright: *Most people die of a sort of creeping common sense, and discover when it is too late that the only things one never regrets are one's mistakes.*

Then there is the view of mistakes as essential for learning. That is nowhere better illustrated than in the process of learning to be a surgeon. Everyone should read the chapter in the book by Atul Gawande (Gawande, 2002) entitled *When Doctors Make Mistakes.* Again, my "mistake literature" is clear on this. Oscar Wilde once more: *Experience is the name everyone gives to their mistakes.* As statisticians, we rarely get to bury our mistakes, so let's all make a few more!

51.3 Accept mediocrity

What's so good about mediocrity? Well, it applies to most of us. Remember the bell curve? Where is it highest? Also, when we condition upon something, we regress towards "mediocrity," the term chosen by Galton (Galton, 1886). Let's learn to love it. When I was younger I read biographies (Gauss, Abel, Kovaleskaya, von Neumann, Turing, Ramanujan, ...) and autobiographies (Wiener, Hardy, Russell, ...) of famous mathematicians. I found them all inspiring, interesting and informative, but light-years from me, for they were all great mathematicians, whereas I was very mediocre one.

At the time I thought I might one day write *Memoirs of a Mediocre Mathematician*, to encourage others like myself, people near the mode of the curve. However, I didn't stay a mathematician long enough for this project to get off the ground. Mediocre indeed. Later I considered writing of *Stories from a Second-Rate Statistician*, but rejected that as too immodest. Perhaps *Tales from a Third-Rate Theorist*, or *Confessions of a C-Grade Calculator*, or *Diary of a D-Grade Data Analyst*, or *News from an Nth-rate Number Cruncher*?

You can see my goal: to have biographical material which can both inspire, interest and inform, but at the same time, encourage, not discourage young statisticians. To tell my readers: I do not live on another planet, I'm like you, both feet firmly planted on Planet Earth. Maybe my goal is mistaken (see above), but I do remember enjoying reading *The Diary of a Nobody* many years ago. You see, I do not believe we can all be whatever we want to be, that all that matters is that we want to be something or be someone, and that if we want it enough, we can achieve it. Without wishing to discourage younger readers who have yet to notice, our ability to make ourselves faster, higher and stronger, not to mention smarter and deeper, is rather limited. Yes, it would be great to win a Gold Medal at the Olympics, or a Nobel Prize or a Fields Medal, or even to win the COPSS Presidents' Award, but my view is that for all but a vanishingly small number of us, such goals are unachievable, no matter how much time and effort we put in. This is not to say that the people who do achieve these goals can do so without expending considerable time and effort, for there is plenty of literature (no quotes now) suggesting that they must. My point is that time and effort are usually not sufficient to bring about dramatic changes in us, that we are what we are.

Also when I was young, I read a statement along the following lines: *Worldly acclaim is the hallmark of mediocrity.* I don't remember where I saw it, and can't find it now, but I liked it, and it has stuck in my head. I used to think of it every time I saw someone else get a prize or receive some other kind of acclaim. I would think "Don't feel too bad, Terry. Galois, Van Gogh, Mozart, Harrison, Mendel and many others all had to wait until after they died to be acclaimed geniuses; your time will come." Of course I always knew that I wasn't in the same class as these geniuses, and, as if to prove that, I came in due course to win some awards. But it still sticks in my mind: that true recognition is what comes after we die, and we shouldn't be too concerned with what comes in our lifetime. I think we'd all be better off accepting what we are, and trying to be a better one of those, than trying to achieve the unachievable. If that means accepting mediocrity, so be it, but then let's aim to be the best – fill in your name – on the planet. Let's be ourselves first and foremost. I think being happy with what we are, while working to make realistic improvements, is a great start to achieving more than we might initially think we can achieve. Unfortunately I can't leave this theme without pointing out that our profession is multidimensional, not one-dimensional, so it is likely that the concept of "best" doesn't even make sense here. We don't have competitions and rankings like chess or tennis players; we try to bring all our skill and experience to bear on any given statistical problem, in the hope that we can find a good answer. But we never have to say we're certain. On the other hand, there may well be a dimension along which you can be the best.

51.4 Enthuse

Why enthuse? Enjoyment of our job is one of the things that distinguish people like us — teachers, researchers, scholars — from the majority of our fellow human beings. We can find our work engaging, challenging, stimulating, rewarding, and fulfilling. It can provide opportunities for expressing our creative sides, for harnessing our competitive urges, for exhibiting our altruistic spirits, and for finding enjoyment in working with others and in solitary pursuits. If we accept all this, then surely we have very good reason to be enthusiastic about what we do. Why not show it?

Not so long ago, I would feel slightly deflated when people complimented me on my enthusiasm for what I do. I would think "Why are they telling me they like my enthusiasm? Why aren't they telling me how much they admire that incredibly original, deep and useful research I was expounding? Is my work no good, is all they see a crazy person waving his arms around wildly?" I've since got over worrying about that, and these days feel very happy if I am able to convey my enthusiasm to others, especially if I can make them smile or laugh at the same time. It's not hard to get a laugh with a weak joke, but I prefer to do it using a mix of slightly comic enthusiasm, and irony. My experience now is that a strong show of enthusiasm sets the stage for laughter, which I think is great. Perhaps I'm drifting from a would-be scholar to a would-be entertainer, but we all have it so good, I think we can afford to share our joy with others. Nowadays I'd rather be remembered as a person who made others laugh in his lectures, than one who impressed everyone with his scholarship.

My summary paraphrases the song popularized by Frank Sinatra (Sinatra, 1969). Read and enjoy all the contributions in this book, but "Do it your way."

Acknowledgments

Many thanks are to Xihong Lin, Jane-Ling Wang, and Bin Yu for their supportive and helpful comments on earlier drafts of this small essay.

References

Galton, F. (1886). Regression towards mediocrity in hereditary stature. *The Journal of the Anthropological Institute of Great Britain and Ireland*, 15:246–263.

Gawande, A. (2002). *Complications: A Surgeon's Notes on an Imperfect Science.* Picador, New York.

Nagpal, R. (2013). The awesomest 7–year postdoc or: How I learned to stop worrying and learned to love the tenure-track faculty life. *Scientific American Guest Blog*, July 21, 2013.

Sinatra, F. (1969). "My Way" `http://www.youtube.com/watch?v=1t8kAbUg4t4`

52

Thirteen rules

Bradley Efron
Department of Statistics
Stanford University, Stanford, CA

52.1 Introduction

When I was five or six my father paraded me around the neighborhood as a mental marvel able to multiply three-digit numbers. I think he enjoyed it more than I did (my savant powers are seriously limited) but it did give my little boat a first push into the big river of statistics.

So, after all these years, am I grateful for the push? Oh yes (thanks Dad!). Statistics is a uniquely fascinating intellectual discipline, poised uneasily as it is at the triple point of mathematics, philosophy, and science. The field has been growing slowly but steadily in influence for a hundred years, with an increased upward slope during the past few decades. "Buy stat futures" would be my advice to ambitious deans and provosts.

At this point I was supposed to come across with some serious advice about the statistical life and how to live it. But a look at some of the other volume entries made it clear that the advice quota was being well met. (I particularly enjoyed Hall, Rubin, and Reid's pieces.) Instead, let me offer some hard-earned rules garnered from listening to thousands of scholarly presentations.

52.2 Thirteen rules for giving a really bad talk

1. Don't plan too carefully, "improv" is the name of the game with technical talks.

2. Begin by thanking an enormous number of people, including blurry little pictures if possible. It comes across as humility.

3. Waste a lot of time at first on some small point, like the correct spelling of "Chebychev." Who ever heard of running out of time? (See Rule 13.)

4. An elaborate outline of the talk to come, phrased in terms the audience hasn't heard yet, really sets the stage, and saves saying "I'm going to present the beginning, the middle, and the end."

5. Don't give away your simple motivating example early on. That's like stepping on your own punchline.

6. A good way to start is with the most general, abstract statement possible.

7. The best notation is the most complete notation — don't skimp on those subscripts!

8. Blank space on the screen is wasted space. There should be an icon for everything — if you say the word "apple," an apple should tumble in from the right, etc. And don't forget to read every word on the screen out loud.

9. Humans are incredibly good at reading tables, so the more rows and columns the better. Statements like "you probably can't make out these numbers but they are pretty much what I said" are audience confidence builders.

10. Don't speak too clearly. It isn't necessary for those in the front row.

11. Go back and forth rapidly between your slides. That's what God made computers for.

12. Try to get across everything you've learned in the past year in the few minutes allotted. These are college grads, right?

13. Oh my, you are running out of time. Don't skip anything, show every slide even if it's just for a millisecond. Saying "This is really interesting stuff, I wish I had time for it" will make people grateful for getting "Chebychev" right.